Infosys Science Foundation Series

Infosys Science Foundation Series in Mathematical
Sciences

The *Infosys Science Foundation Series*

in Mathematical Sciences, a Scopus-indexed book series, is a sub-series of the *Infosys Science Foundation Series*. This sub-series focuses on high-quality content in the domain of mathematical sciences and various disciplines of mathematics, statistics, bio-mathematics, financial mathematics, applied mathematics, operations research, applied statistics and computer science. All content published in the sub-series are written, edited, or vetted by the laureates or jury members of the Infosys Prize. With this series, Springer and the Infosys Science Foundation hope to provide readers with monographs, handbooks, professional books and textbooks of the highest academic quality on current topics in relevant disciplines. Literature in this sub-series will appeal to a wide audience of researchers, students, educators, and professionals across mathematics, applied mathematics, statistics and computer science disciplines.

More information about this subseries at http://www.springer.com/series/13817

Ramji Lal

Algebra 3

Homological Algebra and Its Applications

 Springer

Ramji Lal
University of Allahabad
Prayagraj, Uttar Pradesh, India

ISSN 2363-6149   ISSN 2363-6157 (electronic)
Infosys Science Foundation Series
ISSN 2364-4036   ISSN 2364-4044 (electronic)
Infosys Science Foundation Series in Mathematical Sciences
ISBN 978-981-33-6328-1   ISBN 978-981-33-6326-7 (eBook)
https://doi.org/10.1007/978-981-33-6326-7

This Springer imprint is published by the registered company Springer Nature Singapore Pte Ltd.
The registered company address is: 152 Beach Road, #21-01/04 Gateway East, Singapore 189721,
Singapore

*Dedicated to the memory of
my father like brother*

(Late) Sri Gopal Lal

Preface

Algebra has played a central and a decisive role in formulating and solving the problems in all branches of mathematics, science, and engineering. My earlier plan was to write a series of three volumes on algebra covering a wide spectrum to cater the need of students and researchers at various levels. The two initial volumes have already appeared. However, looking at the size and the contents to be covered, we decided to split the third volume into two volumes, Algebra 3 and Algebra 4. Algebra 3 concentrates on the homological algebra together with its important applications in mathematics, whereas Algebra 4 is about Lie algebras, Chevalley groups, and their representation theory.

Homological algebra has played and is playing a pivotal role in understanding and classifying (up to certain equivalences) the mathematical structures such as topological, geometrical, arithmetical, and the algebraic structures by associating computable algebraic invariants to these structures. Indeed, it has also shown its deep intrinsic presence in dealing with the problems in physics, in particular, in string theory and quantum theory. The present volume, Algebra 3, the third volume in the series, is devoted to introduce the homological methods and to have some of its important applications in geometry, topology, algebraic geometry, algebra, and representation theory. It contains category theory, abelian categories, and homology theory in abelian categories, the n-fold extension functors $EXT^n(-,-)$, the torsion functors $TOR_n(-,-)$, the theory of derived functors, simplicial and singular homology theories with their applications, co-homology of groups, sheaf theory, sheaf co-homology, some amount of algebraic geometry, E'tale sheaf theory and co-homology, and the ℓ-adic co-homology with a demonstration showing its application in the representation theory. The book can act as a text for graduate and advance graduate students specializing in mathematics.

There is no essential prerequisite to understand this book except some basics in algebra (as in Algebra 1 and Algebra 2) together with some amount of calculus and topology. An attempt to follow the logical ordering has been made throughout the book.

My teacher (Late) Professor B. L. Sharma, my colleague at the University of Allahabad; my friends: Prof. Satyadeo, Prof. S. S. Khare, Prof. H. K. Mukherji, and Dr. H. S. Tripathi; my students: Prof. R. P. Shukla, Prof. Shivdatt, Dr. Brajesh Kumar Sharma, Mr. Swapnil Srivastava, Dr. Akhilesh Yadav, Dr. Vivek Jain, and Dr. Vipul Kakkar; and above all the mathematics students of Allahabad University had always been the motivating force for me to write a book. Without their continuous insistence, it would have not come in the present form. I wish to express my warmest thanks to all of them.

Harish-Chandra Research Institute, Allahabad, has always been a great source for me to learn more and more mathematics. I wish to express my deep sense of appreciation and thanks to HRI for providing me all infrastructural facilities to write these volumes.

Last but not least, I wish to express my thanks to my wife Veena Srivastava who had always been helpful in this endeavor.

In spite of all care, some mistakes and misprints might have crept and escaped my attention. I shall be grateful to any such attention. Criticisms and suggestions for the improvement of the book will be appreciated and gratefully acknowledged.

Prayagraj, Uttar Pradesh, India   Ramji Lal

Contents

About the Author

Ramji Lal is a former professor and head of the Department of Mathematics at the University of Allahabad, India. He is also a former adjunct professor of the Harish-Chandra Research Institute (HRI), Allahabad, India. He started his research career at the Tata Institute of Fundamental Research (TIFR), Mumbai, and served the University of Allahabad in different capacities for over 43 years: as a professor, head of the department and the coordinator of the DSA program. At HRI, he initiated a Postgraduate (PG) program in mathematics and coordinated the Nurture Program of National Board for Higher Mathematics (NBHM) from 1996 to 2000. After his retirement from the University of Allahabad, he had been the advisor cum adjunct professor at the Indian Institute of Information Technology (IIIT), Allahabad, for over three years. His research interests include group theory, algebraic K-theory and representation theory.

Notations

Σ	a Category
$Obj\Sigma$	class of objects in Σ
$Mor_\Sigma(A, B)$	set of morphisms from A to B
$A \xrightarrow{f} B$	a morphism f from A to B
I_A	identity morphism on A
SET	category of sets
GP	category of Groups
AB	category of abelian Groups
$Ring$	category of rings with identities
$Mod - R$	category of right R − modules
$R - Mod$	category of left R − modules
TOP	category of topological spaces
$TOP_\star$	category of pointed topological spaces
ω	the category of finite ordinals
Ω^F	functor category
E_v	evaluation functor
$SpecR$	spectrum of the ring R
$\mathbb{Z}_{(p)}$	ring of p − adic integers
$\mathbb{Q}_{(p)}$	field of p − adic integers
$H_n(X)$	nth homology of the chain complex X
$C(f)$	mapping cone of a chain transformation X
(Σ, S)	a simplicial complex
$\Omega(\Sigma, S)$	ordered chain complex associated to (Σ, S)
$\wedge(\Sigma, S)$	oriented chain complex associated to (Σ, S)
$H_q(\Sigma, S)$	q_{th} homology of the simplicial complex (Σ, S)
Δ^q	standard q simplex
$S_q(X)$	free abelian group generated by singular q − simplexes
$H^n(G, A)$	n_{th} cohomology of the group G with coefficient in a G − module A

$H^n_\Sigma(-,-)$ $\qquad\qquad\qquad$ n_{th} cohomology bi $-$ functor from an abelian category Σ to AB

$EXT^n_\Sigma(-,-)$ $\qquad\qquad$ $n -$ fold extension functor

$Tor_n(-,-)$ $\qquad\qquad$ torsion functors

$L_n F$ $\qquad\qquad\qquad\quad$ n_{th} left derived functor of F

$R^n F$ $\qquad\qquad\qquad\quad$ n_{th} right derived functor of F

(E^r, d^r) $\qquad\qquad\quad$ a spectral sequence

E^∞ $\qquad\qquad\qquad\quad$ infinity term of the spectral sequence

$E_C = \{D, E; \alpha, \beta, \gamma\}$ $\qquad$ an exact couple

$|(\Sigma, S)|$ $\qquad\qquad\quad$ geometric realization of (Σ, S)

St_v $\qquad\qquad\qquad\quad$ star of the vertex v

(Σ_{bd}, S_{bd}) $\qquad\qquad$ barycentric subdivision of (Σ, S)

$\chi(A)$ $\qquad\qquad\qquad\quad$ Euler characteristic of a finitely generated graded abelian group A

$\chi(X)$ $\qquad\qquad\qquad\quad$ Euler Poincare characteristic of a polyhedra X

$L(f, X)$ $\qquad\qquad\quad$ Lefschetz number of a continuous map on X

$I(G)$ $\qquad\qquad\qquad\quad$ augmentation ideal

$B(G)$ $\qquad\qquad\qquad\quad$ standard Bar resolution of G

H^p_{dR} $\qquad\qquad\qquad\quad$ p_{th} de $-$ Rham cohomology functor

ϑ_X $\qquad\qquad\qquad\quad$ a ringed space on X

ϑ^R $\qquad\qquad\qquad\quad$ structure sheaf of specR

F_x $\qquad\qquad\qquad\quad$ stalk of the presheaf / sheaf F at a

Pr_X $\qquad\qquad\qquad\quad$ category of presheaves on X

Sh_X $\qquad\qquad\qquad\quad$ category of sheaves on X

$\vartheta_X - Mod$ $\qquad\qquad$ category of $\vartheta_X -$ modules

SF $\qquad\qquad\qquad\quad$ the sheafification of F

$Pic(\vartheta_X)$ $\qquad\qquad\quad$ Picard group of the ringed space ϑ_X

Sp_X $\qquad\qquad\qquad\quad$ category of sheaf spaces

Γ $\qquad\qquad\qquad\quad$ section functor

$H^p(X, F)$ $\qquad\qquad\quad$ p_{th} sheaf cohomology of X with coefficient in sheaf F

$\check{H}^p(X, F)$ $\qquad\qquad\quad$ Cech cohomology of X with values in a presheaf F

A^K_n $\qquad\qquad\qquad\quad$ affine $n -$ set over the field K

$\vartheta(Y)$ $\qquad\qquad\qquad\quad$ ring of regular functions on Y

$K(Y)$ $\qquad\qquad\qquad\quad$ function field of Y

$\vartheta_{Y,P}$ $\qquad\qquad\qquad\quad$ local ring of Y at P

$D_{\bar{v}}$ $\qquad\qquad\qquad\quad$ derivation at the point $\bar{v}$

$T_{\bar{a}}(Y) = Der(\Gamma(Y), K_{\bar{a}})$ $\quad$ tangent space of Y at $\bar{a}$

$ASCH$ $\qquad\qquad\qquad$ category of affine schemes

SCH $\qquad\qquad\qquad\quad$ category of schemes

LRS $\qquad\qquad\qquad\quad$ category of locally ringed spaces

RS $\qquad\qquad\qquad\quad$ category of ringed spaces

$k_X(x)$ $\qquad\qquad\qquad\quad$ residue field of X at x

$Z(X, t)$ $\qquad\qquad\qquad$ zeta function of X

$\acute{e}tPr_X$ $\qquad\qquad\qquad$ category of etale presheaves on X

$\acute{e}tSh_X$	category of etale sheaves on X
$H^r\ (X_{\acute{e}t}, F)$	r_{th} étale cohomology of X with coefficient in étale sheaf F
$F_{\hat{\sigma}}$	stalk at the geometric point $\hat{\sigma}$
$H_c^r(X_{\acute{e}t}, F)$	r_{th} étale cohomology of X with compact support
$H_c^m(X, \overline{\mathbb{Q}_{(l)}})$	$l-$ adic cohomology with compact support
$R(G)$	radical of an algebraic group
B	Borel subgroup
T	the maximal torus
F	the Frobenius map
G^F	the finite group of Lie type
$R_{T,\Theta}$	Deligne − Lusztig virtual character

Chapter 1
Homological Algebra 1

This is the first chapter which develops the basic language of homological algebra. It introduces categories, abelian categories, and the homology theory in an abelian category. Further, it also introduces the bi-functor $EXT(-, -)$.

1.1 Categories and Functors

The category theory gives us a unified, general, and abstract setting for mathematical structures such as groups, rings, modules, vector spaces, topological spaces, etc. Quite often, in mathematics, the concrete results are expressed in the language of category theory. The Gödel–Bernays axiomatic system for sets is the most suitable axiomatic system for the category theory. As described in Chap. 2 of Algebra 1, class is a primitive term in this axiomatic system instead of sets. Indeed, sets are simply the members of classes. The classes which are not sets are termed as proper classes.

Definition 1.1.1 A **category** Σ consists of the following:

1. A class $Obj\,\Sigma$ called the class of objects of Σ.
2. For each pair A, B in $Obj\,\Sigma$, we have a set $Mor_\Sigma(A, B)$ called the set of morphisms from the object A to the object B. Further,

$$Mor_\Sigma(A, B) \bigcap Mor_\Sigma(A', B') \neq \emptyset \ only \ if \ A = A' \ and \ B = B'.$$

3. For each triple A, B, C in $Obj\,\Sigma$, we have a map $\cdot$ from $Mor_\Sigma(B, C) \times Mor_\Sigma(A, B)$ to $Mor_\Sigma(A, C)$ called the law of composition. We denote the image $\cdot(g, f)$ of the pair (g, f) under the map $\cdot$ by gf. Further, the law of composition is associative in the sense that if $f \in Mor_\Sigma(A, B)$, $g \in Mor_\Sigma(B, C)$ and $h \in Mor_\Sigma(C, D)$, then $(hg)f = h(gf)$.

© Springer Nature Singapore Pte Ltd. 2021
R. Lal, *Algebra 3*, Infosys Science Foundation Series,
https://doi.org/10.1007/978-981-33-6326-7_1

4. For each $A \in Obj\Sigma$, there is an element I_A in $Mor_\Sigma(A, A)$ such that $f I_A = f$ for all morphisms f from A, and $I_A g = g$ for all morphisms g to A.

If $f \in Mor_\Sigma(X, Y)$, then X is called the **domain** of f and Y is called the **co-domain** of f. We also represent the morphism f by $X \xrightarrow{f} Y$.

Clearly, for each object A of Σ, I_A is the unique morphism, and it is called the identity morphism on A. The category Σ is called a **small category** if $Obj\Sigma$ is a set.

Example 1.1.2 We have the category SET whose objects are sets, the morphisms from a set A to a set B are precisely the maps from A to B, and the composition law is the composition of maps. This is termed as the category of sets. GP denotes the category whose objects are groups, the morphisms from a group H to a group K are precisely homomorphisms from H to K, and the composition law is the composition of maps. AB denotes the category of abelian groups whose objects are abelian groups, morphisms are homomorphisms, and the composition law is the composition of maps. $RING$ will denote the category whose objects are rings, the morphisms are ring homomorphisms, and the composition law is the composition of maps. The category of topological spaces is denoted by TOP. Thus, the objects of TOP are topological spaces, and the morphisms are continuous maps, and again the composition law is the composition of maps.

Example 1.1.3 Let R be a ring with identity. Then $Mod\text{-}R$ ($R\text{-}Mod$) denotes the category whose objects are right (left) R-modules, the morphisms are module homomorphisms, and the composition law is the composition of maps.

Example 1.1.4 A group G can also be treated as a category having a single object G. The elements of the group can be taken as morphisms from G to G, and the composition law is the binary operation of G.

Example 1.1.5 Let f and g be homotopic maps from a topological space X to a topological space Y. Let h and k be homotopic maps from a topological space Y to a topological space Z. Then hof and kog are homotopic. Thus, we have a category $[TOP]$ whose objects are topological spaces, morphisms are homotopy classes $[f]$ of maps, and the composition law is given by $[g][f] = [gof]$. This category is termed as the *homotopy category*.

Example 1.1.6 We have the category WST of well-ordered sets whose objects are well-ordered sets, morphisms are order-preserving maps, and again the composition law is the composition of maps. In particular, we have the category ω of all finite ordinals. Clearly, ω is a small category.

Example 1.1.7 Let X be a set, and $\leq$ be a reflexive and transitive relation on X. Then $(X, \leq)$ can be treated as a category whose objects are elements of X. $Mor_X(x, y)$ is $\emptyset$ if $x \nleq y$, and $Mor_X(x, y)$ is the singleton $\{i_x^y\}$ if $x \leq y$. If $x \leq y$ and $y \leq z$, then $i_y^z i_x^y = i_x^z$. In particular, every poset (partially ordered set) can be treated as a category. A partially ordered set $(D, \leq)$ is called a **directed set** if for each pair

$a, b \in D$, there is an element $c \in D$ such that $a \leq c$ and also $b \leq c$. Thus, a directed set can also be treated as a category.

Definition 1.1.8 Let Σ be a category. A morphism f from A to B is said to be a **monomorphism (epimorphism)** if it can be left (right) cancelled in the sense that $fg = fh \ (gf = hf)$ implies that $g = h$. A morphism f from A to B is said to be an **isomorphism** if there is a morphism g from B to A such that $gf = I_A$ and $fg = I_B$. Clearly, such a morphism g is unique, and it is called the inverse of f. The inverse of f, if exists, is denoted by f^{-1}.

Remark 1.1.9 In the category SET of sets, a morphism f from a set A to a set B is a monomorphism (epimorphism) if and only if it is an injective (surjective) map. Also in the category GP, a morphism f from a group H to a group K is a monomorphism if and only if f is injective (prove it). Clearly, an isomorphism is a monomorphism as well as an epimorphism. However, a morphism which is a monomorphism as well as an epimorphism need not be an isomorphism. For example, consider the category $Haus$ of Hausdorff topological spaces. The inclusion map from $\mathbb{Q}$ to $\mathbb{R}$ is a monomorphism as well as an epimorphism (it is an epimorphism because $\mathbb{Q}$ is dense in $\mathbb{R}$) in this category. However, it is not an isomorphism.

Let Σ be a category. An object I of Σ is called an *initial object* if $Mor_\Sigma(I, A)$ is a singleton set for all object A in Σ. An object J of Σ is called an *final object* if $Mor_\Sigma(A, J)$ is a singleton set for all object A in Σ. An object 0 of Σ is called a *zero object* if it is initial as well as a final object. The category SET has no initial object. Every object in the category of singleton sets is a zero object. A poset considered as a category has a initial object if and only if it has the least element. It has a final object if and only if it has the largest element. All singleton sets are final objects in SET. In the category GP of groups, all trivial groups are zero objects. If I and $\tilde{I}$ are two initial objects in a category Σ, then there is a unique morphism f from I to $\tilde{I}$, and a unique morphism g from $\tilde{I}$ to I. But, then gf and I_I are both morphisms from I to I. Since I is an initial object $gf = I_I$. Similarly, $fg = I_{\tilde{I}}$. Thus, I and $\tilde{I}$ are isomorphic objects. It follows that an initial object, if exists, is unique up to isomorphisms. Similarly, final and zero objects are also unique up to isomorphisms provided they exist.

Proposition 1.1.10 *Let Σ be a category with 0 and $0'$ as zero objects. Let X and Y be objects in Σ. Let $i_{X,0}$ denote the unique morphism from X to 0, and $i_{X,0'}$ denote the unique morphism from X to $0'$. Let $i_{0,Y}$ denote the unique morphism from 0 to Y, and $i_{0',Y}$ denote the unique morphism from $0'$ to Y. Then $i_{0,Y}i_{X,0} = i_{0',Y}i_{X,0'}$.*

Proof Since $i_{0,0'}i_{X,0}$ and $i_{X,0'}$ are morphisms from X to $0'$, and $0'$ is a zero object, it follows that

$$i_{0,0'}i_{X,0} = i_{X,0'}.$$

Similarly,

$$i_{0',Y}i_{0,0'} = i_{0,Y}.$$

Hence $i_{0,Y} i_{X,0} = i_{0',Y} i_{0,0'} i_{X,0} = i_{0',Y} i_{X,0'}$. ♯

Definition 1.1.11 The unique morphism $i_{0,Y} i_{X,0}$ from X to Y is called the **zero morphism**, and it is denoted by $0_{X,Y}$.

Corollary 1.1.12 *If f is a morphism from Y to Z, and g is a morphism from Z to X, then*

$$f 0_{X,Y} = 0_{X,Z} \text{ and } {}_{X,Y}g = 0_{Z,Y}.$$

Proof Since 0 is a zero object, $f i_{0,Y} = i_{0,Z}$. Hence

$$f 0_{X,Y} = f i_{0,Y} i_{X,0} = i_{0,Z} i_{X,0} = 0_{X,Z}.$$

Similarly, $0_{X,Y} g = 0_{Z,Y}$. ♯

Definition 1.1.13 Let Σ be a category with a zero object. Let $X \xrightarrow{f} Y$ be a morphism. A morphism $K \xrightarrow{i} X$ is called a **kernel** of $X \xrightarrow{f} Y$ if

$$K \xrightarrow{i} X \xrightarrow{f} Y = K \xrightarrow{0_{K,Y}} Y,$$

and whenever

$$K' \xrightarrow{i'} X \xrightarrow{f} Y = K' \xrightarrow{0_{K',Y}} Y,$$

there exists a unique morphism j from K' to K such that $ij = i'$.

Dually, a morphism $Y \xrightarrow{\nu} L$ is called a **co-kernel** of $X \xrightarrow{f} Y$ if

$$X \xrightarrow{f} Y \xrightarrow{\nu} L = X \xrightarrow{0_{X,L}} L,$$

and whenever

$$X \xrightarrow{f} Y \xrightarrow{\nu'} L' = X \xrightarrow{0_{X,L'}} L,$$

there exists a unique morphism μ from L to L' such that $\mu\nu = \nu'$.

Proposition 1.1.14 *If $K \xrightarrow{i} X$ and $K' \xrightarrow{i'} X$ are both kernels of the morphism $X \xrightarrow{f} Y$, then there is a unique isomorphism j from K to K' such that $i' oj = i$. If $Y \xrightarrow{\nu} L$ and $Y \xrightarrow{\nu'} L'$ are both co-kernels of the morphism $X \xrightarrow{f} Y$, then there is a unique isomorphism μ from L to L' such that $\mu o\nu = \nu'$.*

Proof Since $K' \xrightarrow{i'} X$ is a kernel, there is a unique morphism j from K to K' such that $i' oj = i$. Next, since $K \xrightarrow{i} X$ is also a kernel, there is a unique morphism j' from K' to K such that $i oj' = i'$. But, then $i' oj oj' = i'$. Again, since $K' \xrightarrow{i'} X$ is a kernel, and $i' oj oj' = i' = i' oI_{K'}$, it follows that $j oj' = I_{K'}$. Similarly, $j' oj = I_K$. This proves that j is an isomorphism. Similarly, the rest of the assertion follows. ♯

Proposition 1.1.15 *A kernel of a morphism is always a monomorphism. A co-kernel is always an epimorphism.*

Proof Let $K \xrightarrow{i} X$ be a kernel of $X \xrightarrow{f} Y$. Let η and μ be morphisms from Z to K such that $i o \eta = i o \mu$. Then $f o i o \eta = 0_{Z,Y} = f o i o \mu$. Since $K \xrightarrow{i} X$ is a kernel of $X \xrightarrow{f} Y$, and $i o \eta = i o \mu, \eta = \mu$. This shows that i can be left cancelled. Hence, it is a monomorphism. The second part follows similarly. ♯

Remark 1.1.16 A monomorphism need not be a kernel of a morphism. Indeed, a morphism α from H to G in the category GP of groups is a kernel of a morphism if and only if α is an injective homomorphism such that $\alpha(H)$ is a normal subgroup of G (prove it). Also an epimorphism need not be a co-kernel of a morphism. For example, the inclusion map $\mathbb{Q}$ to $\mathbb{R}$ in the category TG of Hausdorff topological groups is an epimorphism but it is not a co-kernel. Note that it is also not a kernel, though it is a monomorphism.

Example 1.1.17 In the category GP of groups, kernels and co-kernels exist. A monomorphism in GP need not be a kernel; however, an epimorphism is always a co-kernel. Every morphism f in GP is uniquely expressible as $f = gh$, where g is monomorphism and h is an epimorphism. Indeed, h is a co-kernel of a kernel of f. However, g need not be a kernel of a co-kernel of f. In the categories AB, Mod-R, and R-Mod, every monomorphism is a kernel, every epimorphism is a co-kernel, and every morphism f is expressible as $f = gh$, where g is a co-kernel of a kernel of f and g is a kernel of a co-kernel of f. Also a morphism in any of these categories is isomorphism if and only if it is a monomorphism as well as an epimorphism.

Let Σ be a category, and A be an object of Σ. Then $Mor_{\Sigma}(A, A)$ is a monoid with respect to the composition of morphisms. The members of $Mor_{\Sigma}(A, A)$ are called the endomorphisms of A. The monoid $Mor_{\Sigma}(A, A)$ is denoted by $End(A)$. An isomorphism from A to A is called an automorphism of A. The set of all automorphisms of A is denoted by $Aut(A)$, and it is a group under composition of morphisms.

Let Σ be a category. We say that a category Γ is a subcategory of Σ if (i) $Obj\Gamma \subseteq Obj\Sigma$; (ii) for each pair $A, B \in Obj\Gamma$, $Mor_{\Gamma}(A, B) \subseteq Mor_{\Sigma}(A, B)$; and (iii) the law of composition of morphisms in Γ is the restriction of the law of composition of morphisms in Σ to Γ. The subcategory Γ is said to be a full subcategory if $Mor_{\Gamma}(A, B) = Mor_{\Sigma}(A, B)$ for all $A, B \in Obj\Gamma$. AB is a full subcategory of GP.

Functors

Definition 1.1.18 Let Σ and Γ be categories. A **functor** F from Σ to Γ is an association which associates to each member $A \in Obj\Sigma$, a member $F(A)$ of $Obj\Gamma$, and to each morphism $f \in Mor_{\Sigma}(A, B)$, a morphism $F(f) \in Mor_{\Gamma}(F(A), F(B))$ such that the following two conditions hold:

(i) $F(gf) = F(g)F(f)$ whenever the composition gf is defined.
(ii) $F(I_A) = I_{F(A)}$ for all $A \in Obj\Sigma$.

Let Σ be a category. Consider the category Σ^o whose objects are same as those of Σ, $Mor_{\Sigma^o}(A, B) = Mor_{\Sigma}(B, A)$, and the composition fg in Σ^o is same as gf in Σ. The category Σ^o is called the opposite category of Σ. A functor from Σ^o to a category Γ is called a **contra-variant functor** from Σ to Γ.

If Σ is a category, then the identity map $I_{Obj\Sigma}$ from $Obj\Sigma$ to itself defines a functor called the identity functor. Composition of functors are functors.

Example 1.1.19 Let H be a group. Denote its abelianizer $H/[H, H]$ by $Ab(H)$. Let f be a homomorphism from H to a group K. Then f induces a homomorphism $Ab(f)$ from $Ab(H)$ to $Ab(K)$ defined by $Ab(f)(h[H, H]) = f(h)[K, K]$. This defines a functor Ab from the category GP of groups to the category AB. This functor is called the abelianizer functor.

Example 1.1.20 Let H be a group. Denote the commutator $[H, H]$ of H by $Comm(H)$. If f is a homomorphism from H to K, then it induces a homomorphism $Comm(f)$ from $Comm(H)$ to $Comm(K)$ defined by
$Comm(f)([a, b]) = [f(a), f(b)]$. This defines a functor $Comm$ from the category GP to itself. This functor is called the commutator functor.

Example 1.1.21 We have a functor Ω from the category GP of groups to the category SET of sets which simply forgets the group structure and retains the set part of the group. More explicitly, $\Omega((G, o)) = G$. Such a functor is called a forgetful functor. There is another such functor from the category $RING$ of rings to the category AB of abelian groups which forgets the ring structure, but retains the additive group part of the ring. There is still another forgetful functor from the category TOP to the category SET which forgets the topological structure, and retains the set part of the space.

Example 1.1.22 Let f be a map from a set X to a set Y. Then f induces a unique homomorphism $F(f)$ from the free group $F(X)$ to the free group $F(Y)$ whose restriction to X is f. This gives us the functor F from the category SET to the category GP. This functor is called the free group functor.

Example 1.1.23 Let Σ be a category, and A be an object of Σ. For each $B \in Obj\Sigma$, we put $Mor_{\Sigma}(A, -)(B) = Mor_{\Sigma}(A, B)$, and for each morphism f from B to C, we have a map $Mor_{\Sigma}(A, -)(f)$ from $Mor_{\Sigma}(A, B)$ to $Mor_{\Sigma}(A, C)$ defined by $Mor_{\Sigma}(A, -)(f)(g) = fg$. It is easily verified that $Mor_{\Sigma}(A, -)$ defined above is a functor from Σ to the category SET of sets. Similarly, we have a contra-variant functor $Mor_{\Sigma}(-, A)$ from the category Σ to the category SET of sets.

Example 1.1.24 There is a very useful and important functor, viz, the fundamental group functor π_1 from the category $TOP^\star$ of pointed topological spaces to the category GP of groups. It has tremendous application in geometry and topology.

Definition 1.1.25 Let Σ be a category. A functor X from a directed set $(D, \leq)$ considered as a category to the category Σ is called a **directed system** in Σ. We denote this directed system by (X, D). An inverse system in Σ is a contra-variant functor from the directed category $(D, \leq)$ to Σ.

Let (X, D) be a directed system in a category Σ. For $\alpha \in D$, denote $X(\alpha)$ by X_α. For $\alpha \leq \beta$, denote $X(i_\alpha^\beta)$ by f_α^β. Then (i) for $\alpha \leq \beta \leq \gamma$, $f_\beta^\gamma f_\alpha^\beta = f_\alpha^\gamma$ and (ii) $f_\alpha^\alpha = I_{X_\alpha}$. Thus, a directed system in Σ consists of the following: (i) a directed set $(D, \leq)$, (ii) a family $\{X_\alpha \mid \alpha \in D\}$ of objects in Σ, and (iii) a family $\{f_\alpha^\beta \in Mor_\Sigma(X_\alpha, X_\beta) \mid \alpha \leq \beta\}$ of morphisms such that $f_\beta^\gamma f_\alpha^\beta = f_\alpha^\gamma$ and $f_\alpha^\alpha = I_{X_\alpha}$. Also, an inverse system in Σ consists of the following: (i) a directed set $(D, \leq)$, (ii) a family $\{X_\alpha \mid \alpha \in D\}$, and (iii) a family $\{f_\alpha^\beta \in Mor_\Sigma(X_\beta, X_\alpha) \mid \alpha \leq \beta\}$ of morphisms with $f_\alpha^\alpha = I_{X_\alpha}$ such that $f_\alpha^\gamma = f_\alpha^\beta f_\beta^\gamma$.

Example 1.1.26 Let $\{X_\alpha \mid \alpha \in \Lambda\}$ be a family of sets such that for any two members X_α and X_β of the family, there is a member X_γ of the family such that $X_\alpha \bigcup X_\beta \subseteq X_\gamma$. Then the family is a directed set under the inclusion relation. Let i_α^β denote the inclusion map from X_α to X_β provided that $X_\alpha \subseteq X_\beta$. This gives us a directed system in the category SET. Similarly, a family $\{G_\alpha \mid \alpha \in \Lambda\}$ of groups such that for any two members G_α and G_β of the family, there is a member G_γ of the family such that G_α and G_β are subgroups of G_γ defines a directed system in the category GP.

Example 1.1.27 The set $\mathbb{N}$ of natural numbers is a directed set with usual ordering. Let p be a prime. For each $n \in \mathbb{N}$, consider the cyclic group $\mathbb{Z}/p^n\mathbb{Z}$. For each $n \leq m$, we have a homomorphism ν_m^n from $\mathbb{Z}/p^m\mathbb{Z}$ to $\mathbb{Z}/p^n\mathbb{Z}$ given by $\nu_m^n(a + p^m\mathbb{Z}) = a + p^n\mathbb{Z}$. This defines an inverse system in the category AB of abelian groups.

Definition 1.1.28 Let (X, D) be a directed system in a category Σ. An object U in Σ together with a family $\{g_\alpha \in Mor_\Sigma(X_\alpha, U) \mid \alpha \in D\}$ is called a **direct limit** of the directed system if the following hold:
(i) For $\alpha \leq \beta$, $g_\beta f_\alpha^\beta = g_\alpha$.
(ii) If V is an object in Σ together with $\{h_\alpha \in Mor_\Sigma(X_\alpha, V) \mid \alpha \in D\}$ with $h_\beta f_\alpha^\beta = h_\alpha$ for each $\alpha \leq \beta$, then there is a unique morphism μ from U to V such that $h_\alpha = \mu g_\alpha$ for each $\alpha \in D$.

Dually, let (X, D) be an inverse system in a category Σ, i.e., X is a contra-variant functor from the directed category D to Σ. An object U in Σ together with a family $\{g_\alpha \in Mor_\Sigma(U, X_\alpha) \mid \alpha \in D\}$ is called an **inverse limit** of the inverse system if the following hold:
(i) For $\alpha \leq \beta$, $f_\alpha^\beta g_\beta = g_\alpha$.
(ii) If V is an object in Σ together with $\{h_\alpha \in Mor_\Sigma(V, X_\alpha) \mid \alpha \in D$ with $f_\alpha^\beta h_\beta = h_\alpha$ for each $\alpha \leq \beta$, then there is a unique morphism μ from V to U such that $h_\alpha = g_\alpha \mu$ for each $\alpha \in D$.

Evidently, a direct limit of a directed system is unique up to isomorphism. More explicitly, if U together with a family $\{g_\alpha \in Mor_\Sigma(X_\alpha, U) \mid \alpha \in D\}$ and V in Σ

together with a family $\{h_\alpha \in Mor_\Sigma(X_\alpha, V) \mid \alpha \in D\}$ are both direct limits of the directed system, then there is a unique isomorphism μ from U to V such that $h_\alpha = \mu g_\alpha$ for each $\alpha \in D$. The unique direct limit of the directed system (X, D) is denoted by $Lim_\rightarrow(X, D)$. Similarly, inverse limit of an inverse system, if exists, is unique, and it is denoted by $Lim_\leftarrow(X, D)$.

Theorem 1.1.29 *In the categories* SET, GP, AB, $Mod\text{-}R$, *and* TOP, *the limits and the inverse limits exist.*

Proof We prove their existence in SET and GP. The proofs of their existence in the rest of the categories are similar, and they are left as exercises. Let (X, D) be a directed system in SET. Let $\coprod_{\alpha \in D} X_\alpha = \bigcup_{\alpha \in D} X_\alpha \times \{\alpha\}$ denote the disjoint sum of the family $\{X_\alpha \mid \alpha \in D\}$ of sets, and j_α the natural inclusion map from X_α to $\coprod_{\alpha \in D} X_\alpha$ for each $\alpha \in D$. Let R be the equivalence relation on $\coprod_{\alpha \in D} X_\alpha$ generated by the set of ordered pairs of the types $(j_\alpha(a), j_\beta(f_\alpha^\beta(a)))$, $a \in X_\alpha, \alpha \leq \beta$. Let U denote the quotient set $(\coprod_{\alpha \in D} X_\alpha)/R$, and $g_\alpha = \nu o j_\alpha$, where ν is the quotient map. We show that U together with the family $\{g_\alpha \mid \alpha \in D\}$ of maps is a direct limit of (X, D). Clearly, $g_\beta o f_\alpha^\beta = g_\alpha$ for all $\alpha \leq \beta$. Let V be a set together with the family $\{h_\alpha : X_\alpha \rightarrow V \mid \alpha \in D\}$ of maps such that $h_\beta o f_\alpha^\beta = h_\alpha$ for all $\alpha \leq \beta$. We have a unique map h from $\coprod_{\alpha \in D} X_\alpha$ to V given by $h(x_\alpha, \alpha) = h_\alpha(x_\alpha)$, where $x_\alpha \in X_\alpha$ and $\alpha \in D$. Since $h_\beta o f_\alpha^\beta = h_\alpha$ for $\alpha \leq \beta$, the generators of the equivalence relation R belong to the kernel of f. From the fundamental theorem of maps (see Algebra 1), we have a unique map $\overline{h}$ from U to V such that $h_\alpha o \overline{h} = g_\alpha$. This shows that U together with the family $\{g_\alpha \mid \alpha \in D\}$ of maps is a direct limit of (X, D).

Next, we prove the existence of direct limits in GP. Let (X, D) be a directed system in GP. Let G denote the free product $*_{\alpha \in D} \prod X_\alpha$ of the family $\{X_\alpha \mid \alpha \in D\}$ of groups, and j_α the natural inclusion homomorphism from X_α to $*_{\alpha \in D} \prod X_\alpha$ for each $\alpha \in D$. Let H be the normal subgroup of $*_{\alpha \in D} \prod X_\alpha$ generated by the set of elements of the types $j_\alpha(a)^{-1} j_\beta(f_\alpha^\beta(a))$, $a \in X_\alpha, \alpha \leq \beta$. Let U denote the quotient group $*_{\alpha \in D} \prod X_\alpha/H$, and $g_\alpha = \nu o j_\alpha$, where ν is the quotient map. We show that U together with the homomorphisms g_α, $\alpha \in D$ is a direct limit of (X, D).

Clearly, $g_\beta o f_\alpha^\beta = g_\alpha$. Let V be a group together with homomorphisms h_α from X_α to V, $\alpha \in D$ such that $h_\beta o f_\alpha^\beta = h_\alpha$ for $\alpha \leq \beta$. From the universal property of the free product, we have a unique homomorphism h from $*_{\alpha \in D} \prod X_\alpha$ to V given by $h(x_\alpha) = h_\alpha(x_\alpha)$, where $x_\alpha \in X_\alpha$ and $\alpha \in D$. Since $h_\beta o f_\alpha^\beta = h_\alpha$ for $\alpha \leq \beta$, the generators of the normal subgroup H belong to the kernel of f. From the fundamental theorem of homomorphism, we have a unique homomorphism $\overline{h}$ from U to V such that $h_\alpha o \overline{h} = g_\alpha$. This shows that U together with the family $\{g_\alpha \mid \alpha \in D\}$ is a direct limit of (X, D).

Now, we show the existence of inverse limits in SET and GP. Let (X, D) be an inverse system in SET (GP). Then X is a contra-variant functor from the directed category D to SET (GP). Thus, we have a family $\{X_\alpha \mid \alpha \in D\}$ of sets (groups) together with the family $\{f_\alpha^\beta : X_\beta \rightarrow X_\alpha \mid \alpha \leq \beta\}$ of maps (homomorphisms) such that $f_\alpha^\alpha = I_{X_\alpha}$ and $f_\alpha^\beta f_\beta^\gamma = f_\alpha^\gamma$. Consider the Cartesian (direct) product $\prod_{\alpha \in D} X_\alpha$ of the family $\{X_\alpha \mid \alpha \in D\}$ of sets (groups). Let p_α denote the α_{th} projection. Consider the subset (subgroup) $U = \{x \in \prod_{\alpha \in D} X_\alpha \mid f_\alpha^\beta(p_\beta(x)) = p_\alpha(x) \mid \alpha \leq \beta\}$

of $\prod_{\alpha \in D} X_\alpha$. We show that U together with the family $\{g_\alpha = p_\alpha|_U \mid \alpha \in D\}$ is the inverse limit of the inverse system.

Let V be a set (group), and $\{h_\alpha : V \to X_\alpha \mid \alpha \in D\}$ be a family of maps (homomorphisms) such that $f_\alpha^\beta o h_\beta = h_\alpha$ for all $\alpha \leq \beta$. We have the unique map (homomorphism) h from V to $\prod_{\alpha \in D} X_\alpha$ given by $p_\alpha(h(v)) = h_\alpha(v)$. Clearly, $h(V) \subseteq U$. This shows that U together with the family $\{g_\alpha = p_\alpha|_U \mid \alpha \in D\}$ of maps (homomorphisms) is an inverse limit of the inverse system. ♯

Thus, if (X, D) is a directed system of sets, then $Lim_\to(X, D)$ is the set $\{\bar{a}_\alpha \mid a_\alpha \in X_\alpha, \alpha \in D\}$ of equivalence classes, where $\bar{a}_\alpha = \bar{b}_\beta$ if and only if there is a $\gamma \in D$ with $\alpha \leq \gamma$, $\beta \leq \gamma$ such that $f_\alpha^\gamma(a_\alpha) = f_\beta^\gamma(b_\beta)$. If (X, D) is a directed system of groups, then the group operation $\cdot$ in $Lim_\to(X, D)$ is given by $\bar{a}_\alpha \cdot \bar{b}_\beta = \bar{c}_\gamma$, where $c_\gamma = f_\alpha^\gamma(a_\alpha) f_\beta^\gamma(b_\beta)$, $\alpha \leq \gamma$, $\beta \leq \gamma$.

Let F be a functor from a category Σ to a category Γ. The functor F is said to be faithful if for each pair $A, B \in Obj\Sigma$, the induced map $f \mapsto F(f)$ from $Mor_\Sigma(A, B)$ to $Mor_\Gamma(F(A), F(B))$ is injective. The functor F is said to be a full functor if these induced maps are surjective. The forgetful functor from GP to SET is faithful but it is not full. The abelianizer functor Ab is not faithful. A functor F is said to be an isomorphism from the category Σ to the category Γ if there is a functor G from Γ to Σ such that $GoF = I_\Sigma$ and $FoG = I_\Gamma$.

Let Σ and Γ be categories. Then $\Sigma \times \Gamma$ represents the category whose objects are pairs $(A, B) \in Obj\Sigma \times Obj\Gamma$, and a morphism from (A, B) to (C, D) is a pair (f, g), where f is a morphism from A to C in Σ, and g is a morphism from B to D in Γ. The composition law is coordinate-wise. This category is called the product category.

A functor from $\Sigma^o \times \Sigma$ to a category Γ is called a **bi-functor** from Σ to Γ. It can be easily observed that the association Mor_Σ which associates to each object (A, B) of $Obj(\Sigma^o \times \Sigma)$, the set $Mor_\Sigma(A, B)$ defines a bi-functor from Σ to SET. Indeed, a morphism from (A, B) to (C, D) in $\Sigma^o \times \Sigma$ is a pair (f, g), where f is a morphism in Σ from C to A and g is a morphism in Σ from B to D, and then $Mor_\Sigma(f, g)$ is the map from $Mor_\Sigma(A, B)$ to $Mor_\Sigma(C, D)$ given by $Mor_\Sigma(f, g)(h) = ghf$. In particular, Hom defines a bi-functor from the category of groups to the category SET of sets.

Natural Transformations

Definition 1.1.30 Let F and G be functors from a category Σ to a category Γ. A natural transformation η from F to G is a family $\{\eta_A \in Mor_\Gamma(F(A), G(A)) \mid A \in Obj\Sigma\}$ of morphisms in Γ such that the diagram

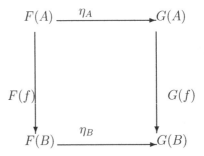

is commutative for all morphisms f in Σ.

Example 1.1.31 Let ν_G denote the quotient homomorphism from G to $G/[G, G]$. Then the family $\{\nu_G \mid G \in ObjGP\}$ defines a natural transformation ν from the identity functor I_{GP} to the abelianizer functor Ab. Here Ab is treated as a functor from GP to GP.

Example 1.1.32 For each set X, we have the inclusion map i_X from X to $\Omega(F(X))$, where F is the free group functor and Ω is the forgetful functor. Evidently, the family $\{i_X \mid X \in ObjSET\}$ of maps defines a natural transformation from the identity functor I_{SET} to the functor ΩoF on the category SET of sets.

Let F and G be two functors from a category Σ to a category Γ. A natural transformation η from F to G is called a *natural equivalence* if η_A is an isomorphism from $F(A)$ to $G(A)$ for all $A \in Obj\Sigma$. This is equivalent to say that there is a natural transformation ρ from G to F such that $\rho_A o \eta_A = I_{F(A)}$ and $\eta_A o \rho_A = I_{G(A)}$ for all objects A of Σ. A functor F from Σ to Γ is called an equivalence from Σ to Γ if there is a functor G from Γ to Σ such that FoG and GoF are naturally equivalent to the corresponding identity functors. Notice that there is a difference between isomorphism and equivalence between categories. An equivalence need not be an isomorphism. However, equivalent categories have same intrinsic properties.

Let F be a functor from a category Σ to a category Γ. Let f be a morphism from C to A in Σ, and g be a morphism from B to D in Γ. This defines a map $Mor_\Gamma(F(f), g)$ from the set $Mor_\Gamma(F(A), B)$ to $Mor_\Gamma(F(C), D)$ given by $Mor_\Gamma(F(f), g)(h) = ghF(f)$. In turn, we get a functor $Mor_\Gamma(F(-), -)$ from the product category $\Sigma^o \times \Gamma$ to the category SET of sets. Similarly, given a functor G from Γ to Σ, we have another functor $Mor_\Sigma(-, G(-))$ from the product category $\Sigma^o \times \Gamma$ to the category SET of sets. We say that F is *left adjoint* to G, or G is *right adjoint* to F if there is a natural isomorphism η from the functor $Mor_\Gamma(F(-), -)$ to the functor $Mor_\Sigma(-, G(-))$. More explicitly, for each object A in Σ and each object B in Γ, we have a bijective map $\eta_{A,B}$ from $Mor_\Gamma(F(A), B)$ to the set $Mor_\Sigma(A, G(B))$ such that $Mor_\Sigma(f, G(g))\eta_{A,B} = \eta_{C,D}Mor_\Gamma(F(f), g)$ for all morphisms (f, g) in $\Sigma^o \times \Gamma$ (look at the corresponding commutative diagram).

Example 1.1.33 Consider the category SET of sets, and the category GP of groups. We have the free group functor F from the category SET to the category GP.

More explicitly, for each set X, we have the free group $F(X)$ on the set X. We also have the forgetful functor Ω from GP to SET. From the universal property of free group, every group homomorphism f from $F(X)$ to G is determined, and it is uniquely determined by its restriction to X. This gives us a bijective map $\eta_{X,G}$ from $Hom(F(X), G)$ to $Map(X, \Omega(G))$. It is easy to observe (using the universal property of a free group) that η, thus obtained, is a natural equivalence. Hence, the free group functor F is left adjoint to the forgetful functor Ω.

Example 1.1.34 We have the forgetful functor Ω from the category AB of abelian groups to the category GP of groups. We also have the abelianizer functor Ab from GP to AB. It can be easily verified that Ab is left adjoint to Ω.

Example 1.1.35 Let (X, x_0) be a pointed topological space. Consider the quotient space of $X \times I$ in which the subset $X \times \{0\} \bigcup X \times \{1\} \bigcup \{x_0\} \times I$ of $X \times I$ is identified to a single point. This quotient space is called the **Reduced suspension** of (X, x_0) and it is denoted by $S(X, x_0)$. The equivalence class determined by (x, t) is denoted by $[x, t]$. We have the special point $[x, 0] = [y, 1] = [x_0, t]$ of $S(X, x_0)$. Further, we have the reduced suspension functor Σ from the category $TOP^\star$ of pointed topological spaces to itself which is given by putting $\Sigma(X, x_0) = (S(X, x_0), [x, 0])$ and $\Sigma(f)([x, t]) = [f(x), t]$, where f is a morphism from (X, x_0) to (Y, y_0). Similarly, we have the function space $\Omega(X, x_0) = (X, x_0)^{(I, \dot{I})}$ with the compact open topology, where $(X, x_0)^{(I, \dot{I})}$ denotes the set of all continuous maps from $(I, \dot{I})$ to (X, x_0). The space $\Omega(X, x_0)$ is called the loop space of X based at x_0 and its members are called loops at x_0. We have the constant loop σ_0 in $\Omega(X, x_0)$. Ω also defines a functor from $TOP^\star$ to itself which is called the **Loop Functor**. For each pair $((X, x_0), (Y, y_0))$ of pointed topological spaces, define a map $\eta_{(X, x_0), (Y, y_0)}$ from $Mor_{Top^\star}(\Sigma(X, x_0), (Y, y_0))$ to $Mor_{Top^\star}((X, x_0), \Omega(Y, y_0))$ by putting $\eta_{(X, x_0), (Y, y_0)}(f)(x) = f([x, t])$. It can be easily seen that $\eta = \{\eta_{(X, x_0), (Y, y_0)} \mid (X, x_0), (Y, y_0) \in ObjTop^\star\}$ is a natural isomorphism between the functors $Mor_{Top^\star}(\Sigma(-), -)$ to $Mor_{Top^\star}(-, \Omega(-))$. This shows that Σ is left adjoint to Ω.

Yoneda Lemma and Yoneda Embedding

Let Σ and Ω be small categories. Since Σ and Ω are small categories, the class of all natural transformations from Σ to Ω forms a set. Thus, we have the category Ω^Σ whose objects are functors from Σ to Ω, and $Mor_{\Omega^\Sigma}(F, G)$ is the set of all natural transformations from F to G. This category is called a **functor category**.

We have a functor E_v from $\Sigma \times \Omega^\Sigma$ to Ω which is given by $E_v(A, F) = F(A)$. This functor is called the **evaluation functor**.

Lemma 1.1.36 *(Yoneda Lemma). For each object (A, F) in $\Sigma \times SET^\Sigma$, we have a bijective map $\eta_{A,F}$ from $Mor_{SET^\Sigma}(Mor_\Sigma(A, -), F)$ to the set $F(A)$ defined by $\eta_{A,F}(\Theta) = \Theta_A(I_A)$.*

Proof All that we need to show is that the map $\eta_{A,F}$ defined in the statement of the lemma is a bijective map for all objects (A, F) in $\Sigma \times SET^\Sigma$. Suppose that $\eta_{A,F}(\Theta) = \eta_{A,F}(\Phi)$. Then $\Theta_A(I_A) = \Phi_A(I_A)$. Since Θ and Φ are natural transformations from $Mor_\Sigma(A, -)$ to F, for each morphism $f \in Mor_\Sigma(A, B)$, the diagram

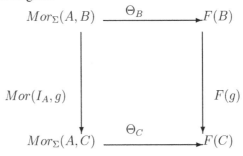

is commutative. Hence

$$\Theta_B(f) = \Theta_B(Mor(I_A, f)(I_A)) = F(f)\Theta_A(I_A) = F(f)\Phi_A(I_A) = \Phi_B(Mor(I_A, f)(I_A)) = \Phi_B(f).$$

This shows that $\Theta = \Phi$. It follows that $\eta_{A,F}$ is injective. To prove that $\eta_{A,F}$ is surjective, let $x \in F(A)$. For each $B \in Obj\Sigma$, define the map Θ_B from $Mor_\Sigma(A, B)$ to $F(B)$ by $\Theta_B(f) = F(f)(x)$. Let g be a morphism from B to C. Let $f \in Mor_\Sigma(A, B)$. Then

$$\Theta_C(Mor(I_A, g)(f)) = \Theta_C(gof) = F(gof)(x) = F(g)F(f)(x) = F(g)\Theta_B(f).$$

This shows that the diagram

$$
\begin{array}{ccc}
Mor_\Sigma(A, B) & \xrightarrow{\;\;\Theta_B\;\;} & F(B) \\
\Big\downarrow{\scriptstyle Mor(I_A, g)} & & \Big\downarrow{\scriptstyle F(g)} \\
Mor_\Sigma(A, C) & \xrightarrow{\;\;\Theta_C\;\;} & F(C)
\end{array}
$$

is commutative for all $g \in Mor_\Sigma(B, C)$. Thus, $\Theta = \{\Theta_B \mid B \in Obj\Sigma\}$ is a natural transformation from $Mor_\Sigma(A, -)$ to F. Evidently, $\Theta_A(I_A) = F(I_A)(x) = x$. This shows that $\eta_{A,F}$ is a bijective map. ♯

Corollary 1.1.37 *Let Σ be a small category. Let A and B be objects in Σ. Then we have a bijective map $\eta_{A,B}$ from $Mor_{SET^\Sigma}(Mor_\Sigma(A, -), Mor_\Sigma(B, -))$ to $Mor_\Sigma(B, A)$ defined by $\eta_{A,B}(\Theta) = \Theta_A(I_A)$. Further, if $\Theta_A(I_A) = f \in Mor_\Sigma (B, A)$, then the map Θ_C from $Mor_\Sigma(A, C)$ to $Mor_\Sigma(B, C)$ is given by $\Theta_C(g) = gof$.*

Proof If we take $F = Mor_\Sigma(B, -)$, then the first part of the result follows from the Yoneda lemma. Again from the proof of the Yoneda lemma, it follows that $\Theta_C(g) = Mor(B, -)(g)(f) = g \circ f$. ♯

Products and Co-products in a Category

Definition 1.1.38 Let A and B be objects in a category Σ. A product of A and B in Σ is a triple (P, f, g), where P is an object of the category Σ, f is morphisms from P to A, and g is a morphism from P to B such that given any such triple (P', f', g'), there is a unique morphism ϕ from P' to P such that $f\phi = f'$ and $g\phi = g'$.

It is easily observed from the definition that if (P, f, g) and (P', f', g') are two products of A and B, then there is an isomorphism ϕ from P' to P such that $f\phi = f'$ and $g\phi = g'$. Thus, the product, if exists, then it is unique up to natural isomorphism. The product of A and B is usually denoted by $A \times B$.

In the category SET of sets, the Cartesian product $A \times B$ with the corresponding projection maps is the product in the category SET. Similarly, the direct product $H \times K$ of the groups H and K together with the corresponding projection maps is the product of H and K in the category GP.

Dually, we have the following.

Definition 1.1.39 Let A and B be objects in a category Σ. A co-product of A and B in Σ is a triple (U, f, g), where U is an object of the category Σ, f is morphisms from A to U, and g is a morphism from B to U such that given any such triple (U', f', g'), there is a unique morphism ϕ from U to U' such that $\phi f = f'$ and $\phi g = g'$.

It is easily observed from the definition that if (U, f, g) and (U', f', g') are two co-products of A and B, then there is an isomorphism ϕ from U to U' with $\phi f = f'$ and $\phi g = g'$. Thus, the co-product, if exists, then it is unique up to natural isomorphism. The co-product of A and B is usually denoted by $A \coprod B$ or by $A \oplus B$.

In the category SET of sets, the disjoint union $(A \times \{0\}) \bigcup (B \times \{1\})$ of A and B with the natural inclusion maps is the co-product of A and B in the category SET. Similarly, the free product $H \star K$ of the groups H and K together with the natural inclusion maps is the co-product of H and K in the category GP (see Exercise 10.4.11 of Algebra 1).

Pullback and Pushout Diagrams

Definition 1.1.40 Let Σ be a category. Let $f \in Mor_\Sigma(A, C)$, and $g \in Mor_\Sigma(B, C)$. A commutative diagram

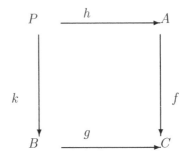

is said to be a pullback diagram of the pair (f, g) of morphisms if given any commutative diagram

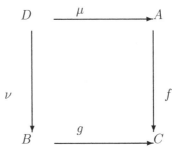

there exists a unique morphism ϕ from D to P such that $h\phi = \mu$ and $k\phi = \nu$.

Dually, a *pushout diagram* is obtained by reversing the arrows in the definition of pullback diagram. The reader is advised to formulate the definition of pushout diagram.

In general, pullback and pushout need not exist in a category. However, they exist in the category SET of sets, and also in the category GP of groups: Let f be a morphism from A to C, and g be a morphism from B to C in the category SET/GP. Consider the product $A \times B$ in the category SET/GP. Let $P = \{(a, b) \in A \times B \mid f(a) = g(b)\}$. Let h denote first projection from P to A, and k denote the second projection from P to B. This gives us a pullback diagram in SET/GP. Similarly, we describe pushout diagrams in the category SET, and also in the category GP. Let f be a map (group homomorphism) from C to A and g be a map (group homomorphism) from C to B. Consider the disjoint union (direct sum) $X = (A \times \{0\}) \bigcup (B \times \{1\})$ $(A \oplus B)$ of A and B. Let R be the equivalence relation (normal subgroup) of X generated by $\{((f(c), 0), (g(c), 1)) \mid c \in C \ (\{(f(c), -g(c)) \mid c \in C\})\}$. Let D denote the quotient X/R and ν the quotient map. Let $\hat{i}$ denote the map from A to D given by $\hat{i}(a) = [a, 0]$ and $\hat{j}$ denote the map from B to D given by $\hat{j}(b) = [0, b]$. Then

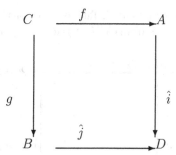

is the required pushout diagram.

Free Objects in a Category

A **concrete category** is a pair (Σ, Λ), where Σ is a category and Λ is a faithful functor from Σ to the category SET of sets.

The pairs (GP, Ω), (AB, Ω), $(Mod - R, \Omega)$, and (TOP, Ω) are all examples of concrete categories, where in each pair Ω denotes the forgetful functor from the respective category to the category SET of sets.

Let (Σ, Λ) be a concrete category, and X be a set. A pair (U, i), where U is an object in Σ and i a map from X to $\Lambda(U)$ is called a **free object** on X if for any pair (V, j), where V is an object in Σ and j is a map from X to $\Lambda(V)$, there is a *unique* morphism ϕ from U to V such that $\Lambda(\phi)oi = j$.

Proposition 1.1.41 *If (U, i) and (V, j) are free objects on X, then there is a unique isomorphism ϕ from U to V such that $\Lambda(\phi)oi = j$.*

Proof Since (U, i) is a free object on X, there is a unique morphism ϕ from U to V such that $\Lambda(\phi)oi = j$. Again, since (V, j) is a free object on X, there is a unique morphism ψ from V to U such that $\Lambda(\psi)oj = i$. In turn, $\Lambda(\psi\phi)oi = i$. Also $\Lambda(I_U)oi = i$. Since (U, i) is a free object on X, $\psi\phi = I_U$. Similarly, $\phi\psi = I_V$. This shows that ϕ is an isomorphism. ♯

A concrete category may not have a free object on any set. For example, the category of fields with forgetful functor has no free object on any set (why?). However, in most of the important concrete categories, there are free objects on each set.

Proposition 1.1.42 *Let (Σ, Λ) be a concrete category. Then free object exists on each set if and only if Λ has a left adjoint F.*

Proof Suppose that F is a functor from the category of sets to Σ which is left adjoint to Λ. Let η be a natural isomorphism from $Mor_\Sigma(F(-), -)$ to $Mor_{SET}(-, \Lambda(-))$. Take $i = \eta_{X, F(X)}(I_{F(X)})$. Then i is a map from X to $\Lambda(F(X))$. We show that the pair $(F(X), i)$ is a free object on X. Let V be an object of Σ, and j be a map from X to $\Lambda(V)$. We have to show the existence of a unique morphism ρ from $F(X)$ to V such that $\Lambda(\rho)oi = j$. The fact that η is a natural isomorphism gives us morphism $(\eta_{X,V})^{-1}(j)$ from $F(X)$ to V. Take $\rho = (\eta_{X,V})^{-1}(j)$. In turn, we have a morphism (I_X, ρ) in $SET^o \times \Sigma$ from $(X, F(X))$ to (X, V), and so also the following commutative diagram:

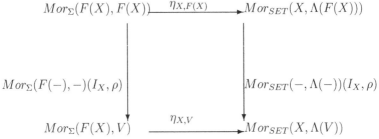

Evidently,

$$j = \eta_{X,V}(Mor_\Sigma(F(-),-)(I_X,\rho)(I_{F(X)}) = Mor_{SET}(-,\Lambda(-))(I_X,\rho)(\eta_{X,F(X)}(I_{F(X)}) = \Lambda(\rho)oi.$$

Clearly, ρ is unique with this property.

Conversely, suppose that free object exist on each set. Let F be an association from $Obj(SET)$ to $Obj(\Sigma)$ together with maps i_X from X to $\Lambda(F(X))$ such that $(F(X), i_X)$ is a free object on X. Let Y be a set and f a map from X to Y. Then $i_Y of$ is a map from X to $\Lambda(F(Y))$. Since $(F(X), i_X)$ is a free object on X, there is a unique morphism $F(f) \in Mor_\Sigma(F(X), F(Y))$ such that $i_Y of = F(f)oi_X$. It is easily observed that F is a functor from SET to Σ which is left adjoint to Λ. ♯

The left adjoint F to the functor Λ in a concrete category (Σ, Λ), if exists, is also called a **free functor**.

Example 1.1.43 In the concrete category (GP, Ω), for each set X, there is a free object $(F(X), i_X)$ on X, where $F(X)$ is free group on X. For the construction, see Chap. 10 of Algebra 1. In the concrete category $(Mod - R, \Omega)$ also, for each set X, there is a free object $(F(X), i_X)$ on X and it is called the free R-module on X. For the construction, see Chap. 7 of Algebra 2.

Definition 1.1.44 Let (Σ, Λ) be a concrete category. A pair (U, x_0), where U is an object of Σ and $x_0 \in \Lambda(U)$, is called a **universal free object** in (Σ, Λ) if for any pair (V, x), where V is an object of Σ and $x \in \Lambda(V)$, there is a unique morphism ϕ from U to V such that $\Lambda(\phi)(x_0) = x$.

Observe that, universal free object, if exists, is unique up to isomorphism. In (GP, Ω), $(\mathbb{Z}, 1)$ (and also $(\mathbb{Z}, -1)$) is universal free object. In $(Mod - R, \Omega)$, $(R, 1)$ is universal free object. Indeed, in most of the important concrete categories, all objects can be obtained from a universal free object by certain universal constructions. For example, every group is quotient of co-product of certain copies of the additive group $\mathbb{Z}$ of integers.

Definition 1.1.45 An object P in a category Σ is called a **projective object** if given any epimorphism β from B to C and a morphism f from P to C, there is a lifting morphism $\overline{f}$ of f from P to B in the sense that $\beta o \overline{f} = f$. An object I in the category Σ is called an **injective object** if given any monomorphism α from A to B and a morphism f from A to I, f can be extended to a morphism $\overline{f}$ from B to I in the sense that $\overline{f} o \alpha = f$.

The axiom of choice ensures that every object in SET is projective as well as injective object.

Proposition 1.1.46 *Let Σ and Γ be categories. Let F be a functor from Σ to Γ which is left adjoint to a functor G from Γ to Σ. Suppose that G takes epimorphisms to epimorphisms. Then F takes projective objects to projective objects.*

Proof Let η be a natural isomorphism from the functor $Mor_\Gamma(F(-), -)$ to $Mor_\Sigma(-, G(-))$. Let P be a projective object in Σ. We have to show that $F(P)$ is a projective object in Γ. Let β be an epimorphism in Γ from B to C and f a morphism from $F(P)$ to C. Under the hypothesis, $G(\beta)$ is an epimorphism from $G(B)$ to $G(C)$ and $\eta_{P,B}(f)$ is a morphism from P to $G(C)$. Since P is projective in Σ, there is a morphism g from P to $G(B)$ such that $G(\beta) o g = \eta_{P,B}(f)$. But then $\beta o (\eta_{P,B}^{-1}(g)) = f$. This shows that P is projective. ♯

Corollary 1.1.47 *Let (Σ, Λ) be a concrete category such that Λ takes an epimorphisms to surjective maps. Then every free object is projective.* ♯

In particular, free objects in the concrete categories (GP, Ω), $(Mod - R, \Omega)$ are projective objects.

Exercises

1.1.1 Show that the composition of any two monomorphism (epimorphism) is a monomorphism (epimorphism).

1.1.2 Describe monomorphisms and epimorphisms in the category of Hausdorff topological groups.

1.1.3 Let Σ be a category. Fix an object A in Σ. Let $\Sigma_A(\Sigma^A)$ denote the class of all monomorphisms (epimorphisms) with co-domain (domain) A. Let $\alpha : K \to A$ ($\alpha : A \to K$) and $\beta : L \to A$ ($\beta : A \to L$) be members of Σ_A (Σ^A). We say that $\alpha \leq \beta$ if there is a morphism γ from K to L such that $\beta\gamma = \alpha$. Show that the relation $\leq$ is reflexive and transitive on Σ_A (Σ^A). We say that α is equivalent to β if $\alpha \leq \beta$ and also $\beta \leq \alpha$. We write $\alpha \approx \beta$ if α is equivalent to β. Show that $\approx$ is an equivalence relation on Σ_A (Σ^A). An equivalence class $[\alpha]$ is called a **subobject(quotient object** of A).

1.1.4 Show that any two left adjoints to a functor are naturally isomorphic.

1.1.5 Show that the left adjoint to the forgetful functor from the category TOP of topological spaces to the category SET of sets exists. Describe it. Describe a free topological space.

1.1.6 Let $TYCH$ denote the category of Tychonoff spaces, and $COMP$ denote the category of compact Hausdorff space. Let $\beta(X)$ denote the Stone–Cech compactification of a Tychonoff space X. Show that β defines a functor from the category $TYCH$ to $COMP$, and it is a left adjoint to the forgetful functor from $COMP$ to $TYCH$.

1.1.7 Let F denote the forgetful functor from the category $RING$ of rings with identities to the category AB of abelian groups. Does there exist a left adjoint to F? If so describe it.

1.1.8 A magma (L, o) with identity is called a *right loop* if the equation $Xoa = b$ has a unique solution for all $a, b \in L$. Thus, every group is a right loop. We have a category RL of right loops whose objects are right loops and morphisms are the operation-preserving maps. Clearly GP is a full subcategory of RL. Show that the forgetful functor F from GP to RL has a left adjoint and describe it.

1.1.9 There are three forgetful functors from the category HG of Hausdorff topological groups. (i) The forgetful functor from HG to SET, (ii) the forgetful functor from HG to GP, and (iii) the forgetful functor from HG to $Haus$. Determine if they have adjoints, and if so, interpret them.

1.1.10 Show that the forgetful functor from the category of fields to the category of commutative integral domains has an adjoint. Describe it.

1.1.11 Show that the forgetful functor from the category of complete metric spaces to the category of metric space has an adjoint. Describe it. This adjoint functor is called the completion.

1.1.12 Show that the forgetful functor from the category of R-modules to the category of abelian groups has an adjoint. Describe it.

1.1.13 Let Σ be a category. Consider the category $\Sigma \times SET^{\Sigma}$. Define an association Ev from the class of objects of $\Sigma \times SET^{\Sigma}$ to the class of sets by $Ev(A, F) = F(A)$. Further for a morphism (f, η) from (A, F) to (B, G), define the map $Ev(f, \eta)$ from $Ev(A, F) = F(A)$ to $Ev(G, B) = G(B)$ by $Ev(f, \eta) = G(f)o\eta_A = \eta_B oF(f)$. Show that Ev is a functor. The functor Ev is called an **evaluation functor**.

1.1.14 Define an association Λ from the class of objects of $\Sigma \times SET^{\Sigma}$ to the class of objects in SET by $\Lambda(A, F) = Mor_{SET^{\Sigma}}(Mor_{\Sigma}(A, -), F)$. Again, for each morphism (f, η) from (A, F) to (B, G), define a map $\Lambda(f, \eta)$ from $\Lambda(A, F)$ to $\Lambda(B, G)$ by

$$\Lambda(f, \eta)(\Theta) = = \eta o\Theta oMor(f, -) = \{\eta_C o\Theta_C oMor(f, I_C) \mid C \in Obj\Sigma\},$$

where Θ is a natural transformation from $Mor_\Sigma(A, -)$ to F. Show that Λ is a functor. Is there a natural isomorphism from Λ to Ev? Support.

1.1.15 Let R be a commutative ring with identity. Let $Spec(R)$ denote the set of all prime ideals of R. Let A be an ideal of R. Let $V(A)$ denote the set of all prime ideals of R containing A. Consider

$$T_{Spec(R)} = \{Spec(R) - V(A) \mid A \; is \; an \; ideal \; of \; R\}.$$

Show that $T_{Spec(R)}$ is a topology on $Spec(R)$. This topology is called the Zariski topology. Let f be a ring homomorphism (all ring homomorphisms are assumed preserve identities) from R to S. Then f defines a map $Spec(f)$ from $Spec(S)$ to $Spec(R)$ by $Spec(f)(p) = f^{-1}(p)$. Show that $Spec(f)$ is a continuous map. Further, show that $Spec$ defines a contra-variant functor from the category of commutative rings to the category of topological spaces. Does it have any adjoint? Support.

1.1.16 Use the axiom of choice to show that the category $VECT$ of vector spaces over a field is equivalent to the category SET of sets. Are they isomorphic?

1.1.17 Treat a group G as a category. Describe products, co-products, pullback diagrams, and pushout diagrams if they exist?

1.1.18 Let $f, g \in Mor_\Sigma(A, B)$. A morphism k from K to A is called an *equalizer* (also called difference kernel) of the pair (f, g) if the following hold: (i) $fk = gk$, (ii) given any morphism k' from K' to A satisfying $fk' = gk'$, there exists a unique morphism μ from K' to K such that $k\mu = k'$. Show that equalizer of a pair of morphisms is a monomorphism. Dually, introduce the concept of co-equalizer (co-kernel), and prove that it is an epimorphism.

1.1.19 Show that equalizers and co-equalizers exist in the categories SET, GP, and $R - Mod$. Describe them.

1.1.20 Let Σ be a category with a zero object. Suppose that equalizers and co-equalizers exist in the category Σ. Show that the kernels and the co-kernels also exist.

In each of the following three exercises, Σ is a category with zero object in which kernel and co-kernel exist.

1.1.21 Let $f : A \rightarrow B$ be a morphism. Let $p : B \rightarrow P$ and $q : B \rightarrow Q$ be co-kernels of f. Let $k : K \rightarrow B$ be a kernel of q. Show that k is also a kernel of p.

1.1.22 Let f be a morphism in Σ. Show that $co\text{-}ker(ker(co\text{-}kerf)) \approx co\text{-}kerf$, and $ker(co\text{-}ker(kerf)) \approx kerf$.

1.1.23 Let $f : A \rightarrow B$ be a morphism. Show that $f = gh$, where $g : C \rightarrow B$ is a kernel of a co-kernel of f, and h is a morphism from A to C. Suppose also that $f = g'h'$, where $g' : D \rightarrow B$ is a kernel of a co-kernel of f, and h' is a morphism from A to D. Show that there is a unique morphism $k : C \to D$ such that $g'k = g$ and $kh = h'$ (observe that k is an isomorphism). In particular, the factorization of f is unique up to certain equivalence. Show by means of an example that h in the above factorization need not be an epimorphism. Further, assume that Σ has equalizers and every monomorphism is a kernel. Show that g in the above factorization is an epimorphism.

1.1.24 Let Σ be a category. Let $\{X_\alpha \mid \alpha \in \Lambda\}$ be a family of objects in Σ. An object P together with a family of morphisms $\{p_\alpha \in Mor_\Sigma(P, X_\alpha) \mid \alpha \in \Lambda\}$ is called a product of the family if given any object Q and a family $f_\alpha \in Mor_\Sigma(Q, X_\alpha) \mid \alpha \in \Lambda\}$, there exists a morphism $\mu \in Mor_\Sigma(Q, P)$ such that $f_\alpha o \mu = p_\alpha$ for all $\alpha \in \Lambda$. Show that the product of an arbitrary family of objects in the categories SET, GP, and R-Mod exist. Dually, formulate the definition of a co-product of an arbitrary family of objects, and show that they also exist in the categories SET, GP, and R-Mod.

1.1.25 Show that the co-product of a family of free objects in a concrete category is a free object on some set. Show also that co-products of projective objects are projective.

1.1.26 Let (Σ, Ω) be a concrete category such that Ω preserves epimorphisms. Let (U, i) be a free object on some set. Let β be an epimorphism in Σ from B to C, and h a morphism from U to C. Show that h can be lifted to a morphism $\overline{h}$ from U to B in the sense that $\beta \overline{h} = h$.

1.1.27 Let (Σ, Ω) be a concrete category. (U, x_0) be a universal free object in (Σ, Ω). Suppose that co-product of an arbitrary family of objects exist in Σ. Show that every free object in (Σ, Ω) is co-product of certain copies of U.

1.1.28 Complete the proof of the remaining cases in Theorem 1.1.29.

1.1.29 Show that every object A of the category SET is the direct limit of the directed system of finite subsets of A. Characterize an object G in GP which is the direct limit of the directed system of finite subgroups of G.

1.1.30 Show that inverse limit exist in the category $TOPG$ of Hausdorff topological groups.

1.1.31 A topological group G is called a **pro-finite group** if it is inverse limit of an inverse system of finite Hausdorff topological groups. Show that a pro-finite group is compact Hausdorff totally disconnected topological group.

1.1.32 Let G be a group, and

$$\Lambda = \{H \mid H \text{ is a normal subgroup of } G \text{ of finite index}\}.$$

Show that Λ is a directed set under the containment $\supseteq$ relation. If $H, K \in \Lambda$ and $H \supseteq K$, then we have a natural homomorphism f_H^K from G/K to G/H. Show that this defines an inverse system of finite groups. Describe the inverse limit $G^\star$ of this inverse system. Thus, $G^\star$ is a pro-finite group, called the **pro-finite completion** of G. Show that the family $\{\nu_H : G \to G/H \mid H \in \Lambda\}$ induces a homomorphism ν from G to $G^\star$. Show that the image $\nu(G)$ is dense in $G^\star$. When can ν be an embedding?

1.1.33 Describe the pro-finite completion $\mathbb{Z}^\star$ of the additive group $\mathbb{Z}$.

1.1.34 Fix a prime p. For each $n \in \mathbb{N}$, we have a finite groups $\mathbb{Z}_{p^n} = \mathbb{Z}/p^n\mathbb{Z}$. For $n \leq m$, we have a homomorphism f_n^m from $\mathbb{Z}_{p^m}$ to $\mathbb{Z}_{p^n}$. This gives an inverse system of finite topological groups in the category of topological groups. The pro-finite group $\mathbb{Z}_{(p)}$, thus obtained, is called the additive group of *p-adic* integers. Describe the group and show that it has an integral domain structure, called the **ring of** *p-adic* **integers**. The field $\mathbb{Q}_{(p)}$ of quotients of $\mathbb{Z}_{(p)}$ is called the **field of** *p-adic* **numbers** .

There is another way to describe $\mathbb{Q}_{(p)}$ as follows: Define a map d_p from $\mathbb{Q} \times \mathbb{Q}$ to $\mathbb{R}^+ \bigcup\{0\}$ by putting $d_p(r, r) = 0$ and putting $d_p(r, s) = (\frac{1}{p})^{\nu_p(r,s)}$, where $r - s = p^{\nu_p(r,s)}\frac{k}{l}$ with k and l co-prime to p, if $r \neq s$. Show that $(\mathbb{Q}, d_p)$ is a metric field in the sense that d_p is a metric (called the *p-adic* **metric**) on $\mathbb{Q}$ and the field operations are continuous. Show that the completion of this metric space has the structure of a metric field which is isomorphic to $\mathbb{Q}_{(p)}$.

1.1.35 Let L be a Galois extension (not necessarily finite) of a field K, and $G(L/K)$ the Galois group (see Chap. 8, Algebra 2). Let

$$\Lambda = \{F \mid F \subseteq L \text{ and } F \text{ is finite Galois extension of } K\}.$$

Show that Λ is directed set under the inclusion relation. For F, $F' \in \Lambda$ with $F \subseteq F'$, we have the restriction homomorphism $j_F^{F'}$ from $G(F'/K)$ to $G(F/K)$. Thus, $\{G(F/K) \mid F \in \Lambda\}$ together with $\{j_F^{F'} \mid F \subseteq F'\}$ is an inverse system of finite discrete groups. Show that the restriction maps from $G(L/K)$ to $G(F/K)$ for all $F \in \Lambda$ induces an isomorphism from $G(L/K)$ to $\overset{Lim}{\to}_{F \in \Lambda} (G(F/K))$. Thus, $G(L/K)$ is a pro-finite group. The topology, thus obtained, is called the **Krull topology**.

1.1.36 Let L be a Galois extension (not necessarily finite) of a field K, and $G(L/K)$ the Galois group. Let $S(G(L/K))$ denote the set closed subgroups of $G(L/K)$, and $SF(L/K)$ the set of intermediary fields. Let $F \in SF(L/K)$. Show that $G(L/F) \in SG(L/K)$. Define a map χ from $SF(L/K)$ to $S(G(L/K))$ by $\chi(F) = G(L/F)$, and a map η from $S(G(L/K))$ to $SF(L/K)$ by $\eta(H) = F(H)$, where $F(H)$ denote the fixed field of H. Show that χ and η are inclusion-reversing maps which are inverses of each other. The result is termed as the fundamental theorem of Galois theory for infinite extensions (see Chap. 8, Algebra 2).

1.1.37 Let G be a first countable (not necessarily Hausdorff) topological group. Call a sequence $\{x_n \mid n \in \mathbb{N}\}$ to be a Cauchy sequence in G if for all neighborhood U of the identity e, there is a natural number n_U such that $n, m \geq n_U$ implies that $x_n^{-1}x_m \in U$. Show that every convergent sequence is a Cauchy sequence. Call G to be complete topological group if every Cauchy sequence is convergent. Assume, further, that G is abelian. Let $\hat{C}(G)$ denote the set of all Cauchy sequences in G. Let $x = \{x_n \mid n \in \mathbb{N}\}$ and $y = \{y_n \mid n \in \mathbb{N}\}$ be Cauchy sequences in G. Show that $x + y = \{x_n + y_n \mid n \in \mathbb{N}\}$ is also a Cauchy sequence, and $\hat{C}(G)$ is an abelian group with respect to $+$. Let $\hat{N}(G)$ denote the subgroup of $\hat{C}(G)$ consisting of those sequences which converge to 0, and $\hat{G}$ denote the quotient group $\hat{C}(G)/\hat{N}(G)$. The group $\hat{G}$ is called the completion of the topological group G. Show that completion defines functor from the category of first countable topological groups to the category of complete topological groups.

1.1.38 Let

$$G = G_0 \supseteq G_1 \supseteq G_2 \supseteq \cdots \supseteq G_n \supseteq G_{n+1} \supseteq \cdots$$

be a chain of subgroups of an abelian group G. Show that G is a topological group with $\{G_n \mid n \geq 0\}$ as fundamental system of neighborhood of the identity e of G. Show that each G_n is a clopen subgroup of G. Let $\hat{G}$ denote the inverse limit of the inverse system $\{(G/G_n, \nu_n) \mid n \in \mathbb{N}\}$, where ν_n is the obvious homomorphism from G/G_n to G/G_{n-1}. Show that $\hat{G}$ is isomorphic to $\hat{\hat{G}}$.

1.1.39 Let R be a ring and A is an ideal of R. We have a chain

$$R = R \supseteq A \supseteq A^2 \supseteq \cdots \supseteq A^n \supseteq A^{n+1} \supseteq \cdots$$

of ideals of R. Show that there is a unique topological ring with $\{A^n \mid n \geq 0\}$ as a fundamental system of neighborhood of o. The topology on R thus obtained is called the $A - adic$ **topology** on R. The completion $\hat{R}$ of this topological ring is called the **A-adic completion**. Let R and R' be rings with ideals A and A', respectively. Let f be a homomorphism from R to R' such that $f(A) \subseteq A'$. Show that f induces a homomorphism from the A-adic completion $\hat{R}$ to $\hat{R}'$.

1.2 Abelian Categories

The category AB of abelian groups and the category Mod-R of modules over a ring R are special type of categories in which morphism sets are equipped with abelian group structures such that the composition laws are bi-additive. There are many such categories. The purpose of this section is to introduce an abstract theory for such categories.

Definition 1.2.1 A category Σ with a zero object together with abelian group structures on $Mor_\Sigma(A, B)$ for all $A, B \in Obj(\Sigma)$ is called an **additive category** if the composition law from $Mor_\Sigma(B, C) \times Mor_\sigma(A, B)$ to $Mor_\Sigma(A, C)$ is bi-additive.

The categories AB and R-Mod are additive categories whereas GP is not an additive category.

Proposition 1.2.2 *Let Σ be an additive category. Let P together with morphisms $p_1 : P \to A$ and $p_2 : P \to B$ represent a product of A and B. Then there exist unique morphisms $i_1 : A \to P$ and $i_2 : B \to P$ such that*

(i) $p_1 i_1 = I_A$, $p_2 i_2 = I_B$ and
(ii) $i_1 p_1 + i_2 p_2 = I_P$.

Also P together with morphisms $i_1 : A \to P$ and $i_2 : B \to P$ represent a co-product of A and B. Conversely, let P be an object together with morphisms p_1, p_2, i_1, and i_2 satisfying the conditions (i) and (ii) above. Then the triple (P, p_1, p_2) represents the product, and the triple (P, i_1, i_2) represents the co-product of A and B.

Proof Suppose that the triple (P, p_1, p_2) represent a product of A and B. We have morphism I_A from A to A, and since Σ has zero object, we have the zero homomorphism $0_{A,B}$ from A to B. From the definition of product, we have a unique homomorphism i_1 from A to P such that $p_1 i_1 = I_A$ and $p_2 i_1 = 0_{A,B}$. Similarly, we have a morphism i_2 from B to P such that $p_2 i_2 = I_B$ and $p_1 i_2 = 0_{B,A}$.

Next,

$$p_1(i_1 p_1 + i_2 p_2) = p_1 i_1 p_1 + p_1 i_2 p_2 = p_1,$$

and also

$$p_2(i_1 p_1 + i_2 p_2) = p_2 i_1 p_1 + p_2 i_2 p_2 = p_2.$$

Also $p_1 I_P = p_1$, and $p_2 I_P = p_2$. From the universal property of a product, $i_1 p_1 + i_2 p_2 = I_P$.

Now, we show that the triple (P, i_1, i_2) represents a co-product of A and B. Let $j_1 : A \to Q$ and $j_2 : B \to Q$ be morphisms. Then $j_1 p_1 + j_2 p_2$ is a morphism from P to Q such that $(j_1 p_1 + j_2 p_2)i_1 = j_1$ and $(j_1 p_1 + j_2 p_2)i_2 = j_2$. Let ϕ is a morphism from P to Q such that $\phi i_1 = j_1$ and $\phi i_2 = j_2$. Then $\phi = \phi I_P = \phi(i_1 p_1 + i_2 p_2) = j_1 p_1 + j_2 p_2$. Thus, $j_1 p_1 + j_2 p_2$ is a unique morphism from P to Q such that $(j_1 p_1 + j_2 p_2)i_1 = j_1$ and $(j_1 p_1 + j_2 p_2)i_2 = j_2$. This shows that the triple (P, i_1, i_2) represents a co-product of A and B.

Conversely, let P be an object together with morphisms p_1, p_2, i_1, and i_2 satisfying the conditions (i) and (ii) of the proposition. Then as above, triple (P, i_1, i_2) represents a co-product of A and B, and triple (P, p_1, p_2) represents a product of A and B. ♯

Corollary 1.2.3 *In an additive category, an object P is a product of A and B if and only if it is a co-product of A and B.* ♯

An additive category Σ is said to be an exact category if any pair of objects in Σ has a product and so also a co-product.

Definition 1.2.4 An exact category Σ is termed as an **abelian category** if the following hold:

(i) Every morphism has a kernel as well as a co-kernel.
(ii) Every monomorphism is a kernel and every epimorphism is co-kernel.

Evidently, opposite category Σ^0 of an abelian category Σ is again an abelian category. Thus, every theorem in an abelian category has a dual theorem in which arrows are reversed.

The category AB and the category R-Mod are examples of abelian categories.

Proposition 1.2.5 *Let Σ be an abelian category. Let $\sigma : A \to B$ be a monomorphism, and $\tau : B \to C$ be a co-kernel of σ. Then σ is a kernel of τ. Dually, if $\sigma : A \to B$ is an epimorphism, and $\tau : K \to A$ is a kernel of σ, then σ is a cokernel of τ.*

Proof Let $\sigma : A \to B$ be a monomorphism, and $\tau : B \to C$ be a co-kernel of σ. Since Σ is an abelian category, $\sigma : A \to B$ is a kernel of a morphism $\mu : B \to D$. Then $\mu\sigma = 0_{A,D}$. Since τ is a co-kernel of σ, there is a unique morphism ν from C to D such that $\nu\tau = \mu$. Let $\rho : L \to B$ be a morphism such that $\tau\rho = 0_{L,C}$. Then $\mu\rho = \nu\tau\rho = 0_{L,D}$. Since σ is a kernel of μ, there is a unique morphism η from L to A such that $\sigma\eta = \rho$. This shows that σ is kernel of τ. Similarly, the rest of the assertion can be proved. ♯

Corollary 1.2.6 *Let A be an object in an abelian category Σ. Let $S(A)$ denote the class of subobjects (see Exercise 1.1.3) of A, and $Q(A)$ denote the class quotient objects of A. Then ker induces a map $\overline{ker}$ from $Q(A)$ to $S(A)$ defined by $\overline{ker}([\sigma]) = [ker\sigma]$, and coker induces a map $\overline{coker}$ from $S(A)$ to $Q(A)$ defined by $\overline{coker}([\sigma]) = [coker\sigma]$ which are inverses of each other. ♯*

The following result is immediate (see Exercise 1.1.23).

Proposition 1.2.7 *Let Σ be an abelian category. Every morphism f is factorizable as $f = gh$, where g is a monomorphism (and so a kernel) and h an epimorphism (and so a co-kernel). Further, the factorization is unique in the sense that if $f = g'h'$, where g' is an monomorphism and h' a epimorphism, then $[g] = [g']$ and $[h] = [h']$.* ♯

Definition 1.2.8 The subobject $[g]$ described in the above proposition is called the **image** of f and the quotient object $[h]$ is called the **co-image** of f.

Proposition 1.2.9 *A morphism in an abelian category is an isomorphism if and only if it is a monomorphism and also an epimorphism.*

Proof Let Σ be an abelian category and σ a morphism in Σ from A to B. Suppose that σ is a monomorphism and also an epimorphism. Let $\tau : K \to A$ be a kernel of σ. Then $\sigma\tau = 0_{K,B} = \sigma 0_{K,A}$. Since σ is a monomorphism, $\tau = 0_{K,A}$. Again, since τ is a kernel $K = 0$ is a zero object. Since every epimorphism in an abelian category is a co-kernel, σ is a co-kernel. Suppose that σ is a co-kernel of $\rho : C \to A$. Then $\sigma\rho = 0_{C,B}$. Since $0_{0,A}$ is a kernel of σ, $\rho = 0_{0,A}0_{C,0} = 0_{C,A}$. Hence, σ is co-kernel of $0_{C,A}$. Since $I_A 0_{C,A} = 0_{C,A}$, there is a unique morphism $\eta : B \to A$ such that $\eta\sigma = I_A$. Similarly, there is a unique morphism $\xi : B \to A$ such that $\sigma\xi = I_B$. Consequently, $\eta = \eta(\sigma\xi) = (\eta\sigma)\xi = \xi$. This shows that σ is an isomorphism. The converse is already seen to be true in any category. ♯

Proposition 1.2.10 *Equalizers and co-equalizers exist in an abelian category.*

Proof Let α and β be morphisms from A to B in an abelian category Σ. Then $\alpha - \beta$ is also a morphism from A to B. Let $\gamma : K \to A$ be a kernel of $\alpha - \beta$. We show that γ is an equalizer of the pair (α, β). Since $(\alpha - \beta)\gamma = 0$, $\alpha\gamma = \beta\gamma$. Next, γ' be a morphism from K' to A such that $\alpha\gamma' = \beta\gamma'$. Then $(\alpha - \beta)\gamma' = 0$. Since γ is kernel of $\alpha - \beta$, there is a unique morphism η from K' to A such that $\gamma\eta = \gamma'$. This shows that γ is an equalizer. Similarly, we can prove the existence of co-equalizers. ♯

Proposition 1.2.11 *Let Σ be an abelian category. Let $\alpha : A \to C$ and $\beta : B \to C$ be morphisms. Then a pullback diagram of the pair (α, β) exists. Dually, if $\alpha : C \to A$ and $\beta : C \to B$ be morphisms, then a pushout diagram of the pair (α, β) exists.*

Proof Let $\alpha : A \to C$ and $\beta : B \to C$ be morphisms. Consider the morphisms αp_1 and βp_2 from P to C, where the triple (P, p_1, p_2) is a product of A and B. Let $\eta : K \to P$ be an equalizer of αp_1 and βp_2. Then $\alpha p_1 \eta = \beta p_2 \eta$. Let D be an object together with morphisms $q_1 : D \to A$ and $q_2 : D \to B$ such that $\alpha q_1 = \beta q_2$. From the definition of a product, there exists a unique morphism ϕ from D to P such that $p_1\phi = q_1$ and $p_2\phi = q_2$. Thus, $\alpha p_1 \phi = \beta p_2 \phi$. Since η is an equalizer of αp_1 and βp_2, there is a unique morphism ψ from D to K such that $p_1 \eta \psi = q_1$ and $p_2 \eta \psi = q_2$. This shows that

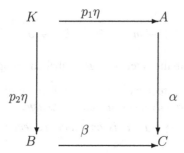

is a pullback diagram of the pair (α, β). Similarly, we can show the existence of a pushout diagram. ♯

An abelian category Σ together with (i) an association which associates to each pair A and B of objects in Σ, a triple (P, f, g) which represents a product of A and B; (ii) an other association which associates to each subobject $[\alpha]$ of an object A (see Exercise 1.1.3), a unique monomorphism σ in the class $[\alpha]$, and an other association which associates to each quotient object $[\beta]$ of an object A, a unique epimorphism τ in the class $[\beta]$ is called a **selective** abelian category.

Thus, in a selective abelian category, every subobject of an object A is determined uniquely by an object B, and a unique monomorphism i from B to A. This morphism will be termed as inclusion morphism. Similarly, a quotient object of A is uniquely determined by an object B and an epimorphism η from A to B.

Using the axiom of choice, we can treat each small abelian category as selective abelian category. The category of R-modules is a selective abelian category. Indeed any subobject of an R-module A is uniquely determined by a submodule B of A, and any quotient object is uniquely determined by a quotient module. To play more with objects, often, we shall assume our abelian category to be a selective abelian category.

Definition 1.2.12 A chain

$$X = \cdots \overset{d_{n-2}}{\to} X_{n-1} \overset{d_{n-1}}{\to} X_n \overset{d_n}{\to} X_{n+1} \overset{d_{n+1}}{\cdots}, n \in \mathbb{Z}$$

of morphisms in an abelian category Σ is called an **exact sequence** at the nth stage if $image d_{n-1} \approx ker d_n$ (equivalently, $[image d_{n-1}] = [ker d_n]$). More explicitly, it is exact at the nth stage if and only if

$$X_{n-1} \overset{d_{n-1}}{\to} X_n = X_{n-1} \overset{\mu_{n-1}}{\to} K_n \overset{\nu_{n-1}}{\to} X_n,$$

where μ_{n-1} is an epimorphism and ν_{n-1} is a kernel of d_n. It is said to be an **exact sequence** if it is exact at each stage. A finite exact sequence of the type

$$0 \to A \overset{\alpha}{\to} B \overset{\beta}{\to} C \to 0$$

is called a **short exact sequence**.

Thus,

$$0 \to m\mathbb{Z} \overset{i}{\to} \mathbb{Z} \overset{\nu}{\to} \mathbb{Z}_m \to 0$$

is a short exact sequence in the category AB of abelian groups.

Proposition 1.2.13 *(i) A sequence $0 \to A \overset{\alpha}{\to} B$ in an abelian category Σ is an exact sequence if and only if α is a monomorphism.*

(ii) A sequence $B \overset{\beta}{\to} C \to 0$ in Σ is an exact sequence if and only if β is an epimorphism.

(iii) A sequence $0 \to A \overset{\alpha}{\to} B \to 0$ in Σ is an exact sequence if and only if α is an isomorphism.

(iv) A sequence $0 \to A \overset{\alpha}{\to} B \overset{\beta}{\to} C \to 0$ in Σ is an exact sequence if and only if α is a monomorphism, and β is a co-kernel of α.

Proof (i) Suppose that $0 \to A \overset{\alpha}{\to} B$ is exact. Since $0 \to A$ is a monomorphism and $0 \to 0$ is an epimorphism, it follows that $0 \to A$ is its image. From the exactness, it follows that $0 \to A$ is a kernel of α. Let β and γ be morphisms from K to A such that $\alpha\beta = \alpha\gamma$. Then $\alpha(\beta - \gamma)$ is the zero morphism from K to B. Again, since $0 \to A$ is a kernel of α, $0_{0,A}0_{K,0} = \beta - \gamma$. This means $\beta - \gamma = 0_{K,A}$. In turn, $\beta = \gamma$. This shows that α is a monomorphism. Conversely, suppose that α is a monomorphism. We show that $0_{0,A}$ is a kernel of α. Let $\beta : K \to A$ be a morphism such that $\alpha\beta = 0_{K,B}$. Also $\alpha 0_{0,A}0_{K,0} = 0_{K,A} = \alpha\beta$. Since α is a monomorphism, $0_{0,A}0_{K,0} = \beta$. This shows that $0_{0,A}$ is a kernel of α.

(ii) This statement is the dual statement of (i).

(iii) Follows from (i), (ii), and Proposition 1.2.9.

(iv) Follows from (i), and the definition of co-kernel. ♯

Definition 1.2.14 A functor F from an abelian category Σ to an abelian category Ω is called an **additive functor** if the induced map $F_{A,B}$ from $Mor_\Sigma(A, B)$ to $Mor_\Omega(F(A), F(B))$ is a group homomorphism for all $A, B \in Obj\Sigma$.

Thus, the forgetful functor from a category $Mod\text{-} R$ of R-modules to the category $Mod\text{-}\mathbb{Z}$ of $\mathbb{Z}$-modules is a faithful additive functor. This is not a full functor.

Example 1.2.15 Let Σ be an abelian category, and D be an object of Σ. Let α be a morphism from A to B. We have a homomorphism α_* from the group $Mor_\Sigma(D, A)$ to the group $Mor_\Sigma(D, B)$ defined by $\alpha_*(f) = \alpha f$. Clearly, $(I_A)_* = I_{Mor_\Sigma(D,A)}$ and $(\beta\alpha)_* = \beta_*\alpha_*$ for all morphism α and β for which $\beta\alpha$ is defined. This gives us a functor $Mor_\Sigma(D, -)$ from the category Σ to the category AB of abelian groups defined by $Mor_\Sigma(D, -)(A) = Mor_\Sigma(D, A)$ and $Mor_\Sigma(D, -)(\alpha) = \alpha_*$.

Similarly, we have a functor $Mor_\Sigma(-, D)$ from the category Σ^0 to the category AB of abelian groups defined by $Mor_\Sigma(-, D)(A) = Mor_\Sigma(A, D)$ and $Mor_\Sigma(-, D)(\alpha) = \alpha^*$, where $\alpha^*(f) = f\alpha$.

It is easily observed that the above two functors are additive functors.

Definition 1.2.16 An additive functor F from an abelian category Σ to an abelian category Ω is called a **left exact** functor if given any exact sequence of the type

$$0 \to A \overset{\alpha}{\to} B \overset{\beta}{\to} C \, in\, \Sigma,$$

the induced sequence

$$0 \to F(A) \overset{F(\alpha)}{\to} F(B) \overset{F(\beta)}{\to} F(C) \, in \, \Omega$$

is also exact. It is said to be a **right exact** functor if given any exact sequence of the type

$$A \overset{\alpha}{\to} B \overset{\beta}{\to} C \to 0 \, in \, \Sigma,$$

the induced sequence

$$F(A) \overset{F(\alpha)}{\to} F(B) \overset{F(\beta)}{\to} F(C) \to 0 \, in \, \Omega$$

is also exact. The functor F is said to be an **exact functor** if it is left as well as right exact functor.

Proposition 1.2.17 *Let Σ be an abelian category, and D be an object of Σ. Then the functor $Mor_\Sigma(D, -)$ is a left exact functor from Σ to AB, and the functor $Mor_\Sigma(-, D)$ is left exact from Σ^0 to AB.*

Proof We prove the left exactness of $Mor_\Sigma(D, -)$. The other part can be proved similarly. Let

$$0 \to A \xrightarrow{\alpha} B \xrightarrow{\beta} C$$

be an exact sequence in Σ. Then α is a monomorphism and it is the kernel of β. We have to show that the sequence

$$0 \to Mor_\Sigma(D, A) \xrightarrow{\alpha_\star} Mor_\Sigma(D, B) \xrightarrow{\beta_\star} Mor_\Sigma(D, C)$$

is an exact sequence of abelian groups. Suppose that $\alpha_\star(f) = \alpha f = 0_{D,B} = \alpha 0_{D,A}$. Since α is a monomorphism, $f = 0_{D,A}$. This shows that $\alpha_\star$ is injective. Further, $\beta_\star o \alpha_\star = (\beta\alpha)_\star = (0_{A,C})_\star$ is the zero map. Hence, $image \alpha_\star \subseteq ker \beta_\star$. Next, let $f \in ker \beta_\star$. Then $\beta f = 0_{D,C}$. Since $\alpha = ker \beta$, there is a unique morphism g from D to A such that $\alpha_\star(g) = \alpha g = f$. This shows that $image \alpha_\star \supseteq ker \beta_\star$. ♮

Now, we state (without proof) a very important and useful result known as the **full embedding theorem**. The proof can be found in the "Theory of categories" by B. Mitchell.

Theorem 1.2.18 (Freyd–Mitchell) *Let Σ be a small abelian category. Then there is a ring R with identity, and an exact faithful and full functor F from Σ to the category Mod-R of R-modules.* ♮

Evidently, under the embedding, all kernels, co-kernels, images, exact sequence, and commutative diagrams correspond. Indeed, all limits of finite directed systems, and inverse limits of finite inverse systems correspond. However, projective and injective objects need not correspond. Almost all important and useful diagram lemmas and theorems in abelian categories follow from the corresponding results in a category of modules. Diagram chasing is easier in the category of modules (see Chap. 7, Algebra 1) than arguing by using arrows in an abelian category. However, we illustrate a proof of the short five lemma by using arrows in an abelian category.

Proposition 1.2.19 *Consider the following commutative diagram:*

$$
\begin{array}{ccccccccc}
0 & \longrightarrow & A & \xrightarrow{\alpha} & B & \xrightarrow{\beta} & C & \longrightarrow & 0 \\
& & \downarrow{f} & & \downarrow{g} & & \downarrow{h} & & \\
0 & \longrightarrow & D & \xrightarrow{\gamma} & E & \xrightarrow{\delta} & F & \longrightarrow & 0
\end{array}
$$

where rows are exact, vertical arrows are morphisms, and the extreme vertical arrows f and h are isomorphisms in an abelian category Σ. Then the middle vertical arrow g is also an isomorphism. ♮

Proof Let $\eta : K \to B$ be a kernel of g. Then $h\beta\eta = \delta g\eta = 0_{K,F}$. Since h is a monomorphism, $\beta\eta = 0_{K,C}$. Since $\alpha = \ker\beta$, there is a unique morphism ρ from K to A such that $\alpha\rho = \eta$. In turn, $\gamma f\rho = g\alpha\rho = g\eta = 0_{K,E} = \gamma f 0_{K,A}$. Since γ and f are monomorphisms, $\rho = 0_{K,A}$. This means that $\eta = 0_{K,A}$. Since η is a kernel of g, K is a zero object and η is a zero morphism. This shows that g is a monomorphism. Similarly, looking at the co-kernel of g, we can show that g is an epimorphism. From Proposition 1.2.9, g is an isomorphism. ♯

Exercises

1.2.1 Show that the category of finite abelian groups is an abelian category. Show that the inclusion functor is exact faithful and full embedding into the category of $\mathbb{Z}$-modules. Show that the category of torsion abelian groups is also abelian, whereas the category of torsion-free abelian groups is not abelian.

1.2.2 Let R be a noetherian ring. Show that the category of finitely generated R-modules is an abelian category. What happens if R is not noetherian.

1.2.3 Prove the five lemma in an abelian category without using the full embedding theorem.

1.2.4 Show that an object P (I) in an abelian category Σ is projective (injective) object if and only if the functor $Mor_\Sigma(P, -)$ $(Mor_\Sigma(-, I))$ is an exact functor from Σ (Σ^0) to AB. Describe the projective and the injective objects, if any, in the categories of Exercise 1.2.1.

1.2.5 Show by means of an example that even under an exact faithful and full embedding projective and injective objects need not correspond.

1.3 Category of Chain Complexes and Homology

In the light of the full embedding theorem, in this section, we shall restrict our attention in the categories of right R-modules. Indeed, in the category of modules over rings, homological results are proved by chasing certain commutative diagrams. One picks up suitable elements at suitable places and then chase their images and pre-images to establish the desired result. This technique, however, does not make sense in an abelian category. The results can be established in an abelian category, otherwise by using the axioms of abelian category, but usually the proofs will be cumbersome. Using full embedding theorem, these results follow from the corresponding results in the category of modules. However, the existence and construction of certain types of objects such as free objects, projective, and injective objects cannot be deduced from their existence and constructions in the category of modules. Unless stated otherwise all the modules considered in this section are right modules. The basic theory of modules including the structure theory of finitely generated modules over a principal ideal domain, projective and injective modules, hom, tensor, and exterior powers has been studied in detail in Chap. 7 of Algebra 2. The reader is advised to visit this chapter of Algebra 2. A chain

$$X = \cdots \overset{d_{n+2}^X}{\to} X_{n+1} \overset{d_{n+1}^X}{\to} X_n \overset{d_n^X}{\to} X_{n-1} \overset{d_{n-1}^X}{\to} \cdots , n \in \mathbb{Z}$$

of R-module homomorphisms is called a **chain complex** in the category of R-modules if $d_n^X o d_{n+1}^X = 0$ (equivalently, $image d_{n+1}^X \subseteq ker d_n^X$) for all n. It is said to be a finite chain complex if $X_n = \{0\}$ for all but finitely many n. It is said to be positive chain complex if $X_n = \{0\}$ for all $n < 0$. A chain complex in the opposite category of Mod-R is called a **co-chain complex** of R-modules. Thus, a co-chain complex $X^\star$ of R-modules is a co-chain

$$X^\star = \cdots \overset{\delta^{n-2}}{\to} X^{n-1} \overset{\delta^{n-1}}{\to} X^n \overset{\delta^n}{\to} X^{n+1} \overset{\delta^{n+1}}{\to} \cdots , n \in \mathbb{Z}$$

of R-module homomorphisms such that $\delta^n o \delta^{n-1} = 0$ for all $n \in \mathbb{Z}$. A chain complex X can also be treated as a co-chain complex by putting $X^n = X_{-n}$, and $\delta^n = d_{-n}^X$ for all n.

Let

$$X = \cdots \overset{d_{n+2}^X}{\to} X_{n+1} \overset{d_{n+1}^X}{\to} X_n \overset{d_n^X}{\to} X_{n-1} \overset{d_{n-1}^X}{\to} \cdots , n \in \mathbb{Z},$$

and

$$Y = \cdots \overset{d_{n+2}^Y}{\to} Y_{n+1} \overset{d_{n+1}^Y}{\to} Y_n \overset{d_n^Y}{\to} Y_{n-1} \overset{d_{n-1}^Y}{\to} \cdots , n \in \mathbb{Z}$$

be chain complexes of R-modules. A set $f = \{f_n \in Hom_R(X_n, Y_n) \mid n \in \mathbb{Z}\}$ is called a **chain transformation** if the following diagram is commutative:

$$
\begin{array}{ccccc}
\longrightarrow & X_{n+1} & \overset{d_{n+1}^X}{\to} X_n & \overset{d_n^X}{\to} X_{n-1} & \longrightarrow \\
& \downarrow f_{n+1} & \downarrow f_n & \downarrow f_{n-1} & \\
\longrightarrow & Y_{n+1} & \overset{d_{n+1}^Y}{\to} Y_n & \overset{d_n^Y}{\to} Y_{n-1} & \longrightarrow
\end{array}
$$

More explicitly, $d_n^Y f_n = f_{n-1} d_n^X$ for all n.

If $f = \{f_n \in Hom_R(X_n, Y_n) \mid n \in \mathbb{Z}\}$ is a chain transformation from a chain complex X to a chain complex Y and $g = \{g_n \in Hom_R(Y_n, Z_n) \mid n \in \mathbb{Z}\}$ is a chain transformation from a chain complex Y to a chain complex Z, then $g \circ f = \{g_n \circ f_n \in Hom_R(X_n, Z_n) \mid n \in \mathbb{Z}\}$ is easily seen to be a chain transformation. $I_X = \{I_{X_n} \mid n \in \mathbb{Z}\}$ is the identity chain transformation on X. This gives us a category CH_R of chain complexes of R modules. It is easy to observe that CH_R is also an abelian category. Evidently, a chain transformation f from a chain complex X to a chain complex Y is a monomorphism (epimorphism) if and only if f_n is injective (surjective) for each n. A chain complex X is a subchain complex of Y if X_n is a submodule of Y_n for each n and the inclusion maps define a chain transformation. If X is a subchain complex of Y, then d_n^Y induces a homomorphism ν_n from Y_n/X_n to Y_{n-1}/X_{n-1} defined by $\nu_n(y + X_n) = d_n^Y(y) + X_{n-1}$. This gives us a chain complex

$$Y/X = \cdots \overset{\nu_{n+2}}{\to} Y_{n+1}/X_{n+1} \overset{\nu_{n+1}}{\to} Y_n/X_n \overset{\nu_n}{\to} Y_{n-1}/X_{n-1} \overset{\nu_{n-1}}{\to} \cdots, \quad n \in \mathbb{Z}.$$

The chain complex Y/X obtained above is called the quotient chain complex of Y modulo X. Let

$$X = \cdots \overset{d_{n+2}^X}{\to} X_{n+1} \overset{d_{n+1}^X}{\to} X_n \overset{d_n^X}{\to} X_{n-1} \overset{d_{n-1}^X}{\to} \cdots, \quad n \in \mathbb{Z}$$

be a chain complex of R modules. Then $d_{n+1}^X(X_{n+1}) \subseteq ker d_n^X$ for each n. We denote $d_{n+1}^X(X_{n+1})$ by $B_n(X)$, and it is called the module of **n-boundaries** of X. We also denote $ker d_n^X$ by $C_n(X)$, and call it the module of **n-cycles** of X. The quotient module $C_n(X)/B_n(X)$ is denoted by $H_n(X)$, and it is called the nth **homology** of the chain complex X. Let $f = \{f_n \mid n \in \mathbb{Z}\}$ is a chain transformation from X to Y. Then $d_n^Y f_n = f_{n-1} d_n^X$ for all n. This means that $f_n(C_n(X)) \subseteq C_n(Y)$, and $f_n(B_n(X)) \subseteq B_n(Y)$. Thus, f induces a homomorphism $H_n(f)$ from $H_n(X)$ to $H_n(Y)$ defined by $H_n(f)(a + B_n(X)) = f_n(a) + B_n(Y)$. If $a \in C_n(X)$, then the element $a + B_n(X)$ of $H_n(X)$ will be denoted by $[a]$. Thus, $H_n(f)[a] = [f_n(a)]$. It is easily observed that $H_n(g \circ f) = H_n(g) \circ H_n(f)$ and $H_n(I_X) = I_{H_n(X)}$ for all n. For each n, H_n defines a functor from CH_R to $Mod\text{-}R$. The chain complex X is exact if and only if $H_n(X) = \{0\}$ for all n.

Exact Homology Sequence
Let

$$E \equiv 0 \to X \overset{f}{\to} Y \overset{g}{\to} Z \to 0$$

be a short exact sequence of chain complexes. Then we have the commutative diagram

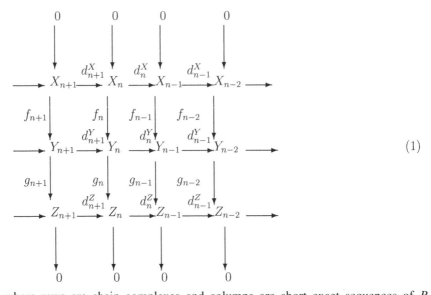

$$\tag{1}$$

where rows are chain complexes and columns are short exact sequences of R-modules. Let z be an element of $C_n(Z)$. Then $d_n^Z(z) = 0$. Since g_n is surjective, there is an element y in Y_n (not necessarily unique) such that $g_n(y) = z$. In turn, $g_{n-1}(d_n^Y(y)) = d_n^Z(g_n(y)) = d_n^Z(z) = 0$. Hence, $d_n^Y(y)$ belongs to ker g_{n-1}. Since the columns are exact, there is a unique element x in X_{n-1} such that $f_{n-1}(x) = d_n^Y(y)$. Further, $f_{n-2}(d_{n-1}^X(x)) = d_{n-1}^Y(f_{n-1}(x)) = d_{n-1}^Y(d_n^Y(y)) = 0$. Since f_{n-2} is injective, $d_{n-1}^X(x) = 0$. This shows that x is an element of $C_{n-1}(X)$. The choice of x depends upon the choice of y. Let y' be another element of Y_n such that $g_n(y') = z = g_n(y)$, and x' be an element of $C_{n-1}(X)$ such that $f_{n-1}(x') = d_n^Y(y')$. Since $g_n(y - y') = 0$, $y - y' = f_n(u)$ for some u in X_n. Now, $f_{n-1}(d_n^X(u)) = d_n^Y(f_n(u)) = d_n^Y(y - y') = f_{n-1}(x - x')$. Since f_{n-1} is injective, $x - x' = d_n^X(u)$ or equivalently, $x + B_{n-1}(X) = x' + B_{n-1}(X)$. This shows that $[x] = x + B_{n-1}(X)$ is independent of the choice of y, and it depends only on z in $C_n(Z)$. As such, we get a map η_n^E from $C_n(Z)$ to $H_{n-1}(X)$ defined by $\eta_n^E(z) = [x]$, where $f_{n-1}(x) = d_n^Y(y)$ with $g_n(y) = z$. It is easily observed that η_n^E is a homomorphism. Further, if z in $B_n(Z)$, then there is w in Z_{n+1} such that $d_{n+1}^Z(w) = z$. Since g_{n+1} is surjective, there is an element v in Y_{n+1} such that $g_{n+1}(v) = w$. Take $y = d_{n+1}^Y(v)$. Then $g_n(y) = z$, and $0 = d_n(y) = f_{n-1}(0)$. This shows that $B_n(Z) \subseteq ker\,\eta_n^E$. Hence, η_n^E induces a homomorphism ∂_n^E from $H_n(Z)$ to $H_{n-1}(X)$ given by $\partial_n^E[z] = [\eta_n^E(z)]$. The homomorphism ∂_n^E is called a **connecting homomorphism** also called the **Bockstein** homomorphism associated with the short exact sequence E of chain complexes.

Theorem 1.3.1 *Let*

$$E \equiv 0 \to X \xrightarrow{f} Y \xrightarrow{g} Z \to 0$$

be a short exact sequence of chain complexes of R-modules. Then we have the following long exact homology sequence:

$$\to H_{n+1}(Z) \xrightarrow{\partial_{n+1}^E} H_n(X) \xrightarrow{H_n(f)} H_n(Y) \xrightarrow{H_n(g)} H_n(Z) \xrightarrow{\partial_n^E} H_{n-1}(X) \to$$

of R-modules, where ∂_n^E is the connecting homomorphism. Further, the correspondence which associates the above long exact sequence to E is natural in the sense that if

$$E' \equiv 0 \to X' \xrightarrow{f'} Y' \xrightarrow{g'} Z' \to 0$$

is another short exact sequence of chain complexes together with a triple (α, β, γ) of chain transformations such that the diagram

$$
\begin{array}{ccccccccc}
0 & \longrightarrow & X & \xrightarrow{f} & Y & \xrightarrow{g} & Z & \longrightarrow & 0 \\
& & \downarrow{\alpha} & & \downarrow{\beta} & & \downarrow{\gamma} & & \\
0 & \longrightarrow & X' & \xrightarrow{f'} & Y' & \xrightarrow{g'} & Z' & \longrightarrow & 0
\end{array}
$$

then the following diagram is commutative:

$$
\begin{array}{ccccccccc}
\longrightarrow H_{n+1}(Z) & \xrightarrow{\partial_{n+1}^E} & H_n(X) & \xrightarrow{H_n(f)} & H_n(Y) & \xrightarrow{H_n(g)} & H_n(Z) & \xrightarrow{\partial_n^E} & H_{n-1}(X) \longrightarrow \\
\downarrow{H_{n+1}(\gamma)} & & \downarrow{H_n(\alpha)} & & \downarrow{H_n(\beta)} & & \downarrow{H_n(\gamma)} & & \downarrow{H_{n-1}(\alpha)} \\
\longrightarrow H_{n+1}(Z') & \xrightarrow{\partial_{n+1}^{E'}} & H_n(X') & \xrightarrow{H_n(f')} & H_n(Y') & \xrightarrow{H_n(g')} & H_n(Z') & \xrightarrow{\partial_n^{E'}} & H_{n-1}(X') \longrightarrow
\end{array}
$$

Proof The proof is by chasing the diagram given by (1). Let $[z] \in H_{n+1}(Z)$, where $z \in C_{n+1}(Z)$. We show that $H_n(f)(\partial_{n+1}^E([z])) = 0$. By definition $\partial_{n+1}^E([z]) = [x]$, where $f_n(x) = d_{n+1}^Y(y)$ for some y with $g_{n+1}(y) = z$. Evidently, $H_n(f)([x]) = [f_n(x)] = [d_{n+1}^Y(y)] = 0$. Thus, $image \partial_{n+1}^E \subseteq ker H_n(f)$. Let $[x]$ be an element $H_n(X)$ such that $H_n(f)[x] = [f_n(x)] = 0$. Then there is an element $y \in Y^{n+1}$ such that $d_{n+1}^Y(y) = f_n(x)$. Clearly, $z = g_{n+1}(y) \in C_{n+1}(Z)$, and $\partial_{n+1}^E([z]) = [x]$. This establishes the exactness at $H_n(X)$. Now, we prove the exactness at $H_n(Y)$. Since H_n is a functor and $g o f = 0$, $H_n(g) o H_n(f) = H_n(g o f) = H_n(0) = 0$. This shows that $image H_n(f) \subseteq ker H_n(g)$. Next, let $[y]$ be an element of $ker H_n(g)$, where $y \in Ker d_n^Y$. Then $g_n(y) = d_{n+1}^Z(z)$ for some $z \in Z_{n+1}$. Since g_{n+1} is surjective, there is an element $y' \in Y_{n+1}$ such that $g_{n+1}(y') = z$. In turn, $g_n(d_{n+1}^Y(y')) = d_{n+1}^Z(g_{n+1}(y')) = d_{n+1}(z) = g_n(y)$. This shows that $y - d_{n+1}^Y(y') \in ker g_n$. Since $ker g_n = image \ f_n$, there is an element $x \in X_n$ such that $f_n(x) = y - d_{n+1}^Y(y')$. Further, $f_{n-1}(d_n^X(x)) = d_n^Y(f_n(x)) = d_n^Y(y - d_{n+1}^Y(y')) = 0$. Hence $x \in C_n(X)$. Evidently, $H_n(f)[x] = [f_n(x)] = [y]$. This proves the exactness at $H_n(Y)$. Similarly, chasing the diagram, we can prove the exactness at $H_n(Z)$. The rest of the statement follows from the functoriality of the homology functors H_n and the natural construction of the connecting homomorphisms. ♯

Corollary 1.3.2 *Let*

$$E \equiv 0 \to X \xrightarrow{f} Y \xrightarrow{g} Z \to 0$$

be a short exact sequence of chain complexes of R-modules. Then we have the following:

   (i) If Z is exact, then $H_n(X)$ is isomorphic to $H_n(Y)$ for all n.
  (ii) If X is exact, then $H_n(Z)$ is isomorphic to $H_n(Y)$ for all n.
 (iii) If Y is exact, then $H_n(X)$ is isomorphic to $H_{n+1}(Z)$ for all n. ♯

Mapping Cone

Let f be a chain transformation from a chain complex X to a chain complex Y. We construct a chain complex $C(f)$ as follows: Take $C(f)_n = X_{n-1} \oplus Y_n$. Define $d_n^{C(f)}$ from $C(f)_n$ by $d_n^{C(f)}(x, y) = (-d_{n-1}^X(x), d_n^Y(y) + f_{n-1}(x))$. Clearly, $d_n^{C(f)} o d_{n+1}^{C(f)} = 0$. The chain complex $C(f)$, thus obtained, is called the **mapping cone** of f (the terminology has its justification in topology). Let X^+ denote the chain complex given by $X_n^+ = X_{n-1}$ and $d_n^{X^+} = d_{n-1}^X$. We have the short exact sequence

$$E(f) \equiv 0 \to Y \xrightarrow{i} C(f) \xrightarrow{j} X^+ \to 0,$$

$i_n(y) = (0, y)$ and $j_n(x, y) = x$. It is evident that $H_n(X^+) = H_{n-1}(X)$. Further, it is also easy to observe that $\partial_n^E = H_{n-1}(f)$. From Theorem 1.3.1, we obtain the following natural long exact homology sequence:

$$\to H_n(Y) \xrightarrow{H_n(i)} H_n(C(f)) \xrightarrow{H_n(j)} H_{n-1}(X) \xrightarrow{H_{n-1}(f)} H_{n-1}(Y) \to$$

Chain Homotopy

Let f and g be two chain transformations from a chain complex X to a chain complex Y. A family $s = \{s_n : X_n \to Y_{n+1} \mid n \in \mathbb{Z}\}$ of module homomorphisms is called a **chain homotopy** from f to g if

$$d_{n+1}^Y o s_n + s_{n-1} o d_n^X = f_n - g_n$$

for all $n \in \mathbb{Z}$. We say that f is chain homotopic to g if there is a chain homotopy from f to g.

Proposition 1.3.3 *If f and g are chain transformations from X to Y which are chain homotopic, then $H_n(f) = H_n(g)$ for all n.*

Proof Let s be a chain homotopy from f to g. Let $[a] \in H_n(X)$, where $a \in C_n(X)$. Then $d_n^X(a) = 0$. Hence $f_n(a) - g_n(a) = d_{n+1}^Y(s_n(a))$. This means that $H_n(f)([a]) = [f_n(a)] = [g_n(a)] = H_n(g)([a])$. ♯

Remark 1.3.4 The converse of the above proposition is not true, i.e., $H_n(f) = H_n(g)$ for all n need not imply that f is chain homotopic to g.

    We shall adopt the notation $f \sim g$ to say that f is chain homotopic to g. Clearly, $f \sim f$. Indeed, 0 is chain homotopy from f to f. If s is a chain homotopy from f to g, then $-s$ is a chain homotopy from g to f. Let s be a chain homotopy from f to g, and t a chain homotopy from g to h, then $s + t$ is a chain homotopy from f to h. Thus, $\sim$ defines an equivalence relation on the set $Hom(X, Y)$ of all chain transformations from X to Y. A chain transformation f from X to Y is said to be a chain equivalence if there is a chain transformation g from Y to X such that $gof \sim I_X$ and $fog \sim I_Y$. If f is a chain equivalence, then $H_n(f)$ is an isomorphism from $H_n(X)$ to $H_n(Y)$ for all n.

Proposition 1.3.5 *Let f and g be chain homotopic transformations from X to Y, and f' and g' be chain homotopic transformations from Y to Z. Then $f'of$ is chain homotopic to $g'og$.*

Proof Let s be a chain homotopy from f to g, and s' be a chain homotopy from f' to g'. Then

$$d_{n+1}^Y os_n + s_{n-1} od_n^X = f_n - g_n \tag{1.2}$$

and

$$d_{n+1}^Z os_n' + s_{n-1}' od_n^Y = f_n' - g_n' \tag{1.3}$$

for all n. Composing (1.2) with f_n' from left , and (1.3) with g_n from right, we obtain

$$f_n' o(d_{n+1}^Y os_n + s_{n-1} od_n^X) = f_n' of_n - f_n' og_n \tag{1.4}$$

and

$$(d_{n+1}^Z os_n' + s_{n-1}' od_n^Y)og_n = f_n' og_n - g_n' og_n \tag{1.5}$$

for all n. Adding (1.4) and (1.5), we get that

$$f_n' od_{n+1}^Y os_n + f_n' os_{n-1} od_n^X + d_{n+1}^Z os_n' og_n + s_{n-1}' od_n^Y og_n = f_n' of_n - g_n' og_n.$$

Using the fact that f' and g are chain transformations, we find that

$$d_{n+1}^Z o(f_{n+1}' os_n + s_n' og_n) + (f_n' os_{n-1} + s_{n-1}' og_{n-1})od_n^X = f_n' of_n - g_n' og_n$$

for all n. This shows that $t = \{t_n = (f_{n+1}' os_n + s_n' og_n) : X_n \to Z_{n+1}\}$ is a homotopy from $f'of$ to $g'og$. ♯

Let $[X, Y]$ denote the set of homotopy classes of chain transformations from X to Y. Let $[f]$ denote the homotopy class of chain transformations determined by f. Thus, $[f] = \{g \mid f \sim g\}$. The above proposition allows us to have a composition law $\cdot : [Y, Z] \times [X, Y] \to [X, Z]$ given by $[g] \cdot [f] = [gof]$. This also gives us a category $[CH]$ whose objects are chain complexes, and $Mor_{[CH]}(X, Y) = [X, Y]$. The composition law is given as above. The category $[CH]$ will be termed as the homotopy category of chain complexes.

Let X and Y be chain complexes. Denote the Cartesian product $\prod_{r=-\infty}^{\infty} Hom(X_r, Y_{r-n})$ by $Hom[X, Y]^n$. An element of $Hom[X, Y]^n$ is a family $\{f_r : X_r \to Y_{r-n} \mid r \in \mathbb{Z}\}$ of homomorphisms. Define a map δ^n from $Hom[X, Y]^n$

to $Hom[X, Y]^{n+1}$ by $\delta^n(f_r) = d^Y_{r-n} \circ f_r + (-1)^{n+1} f_{r-1} \circ d^X_r$. It is easily observed that $\delta^{n+1} \circ \delta^n = 0$. This gives us a co-chain complex $Hom[X, Y]$ of abelian groups described as above.

Every R-module A can be treated as a chain complex as follows: Define $A_0 = A$, and $A_n = \{0\}$ for all $n \neq 0$. Indeed, we can regard the category of R-modules as a subcategory of the category CH of chain complexes. Under this identification $Hom[X, A]^n = \prod_{r=-\infty}^{\infty} Hom(X_r, A_{r-n}) = Hom(X_n, A)$, and $\delta^n = (-1)^{n+1}(d^X_{n+1})^*$. Thus, $\delta^n(\alpha) = (-1)^{n+1}\alpha \circ d^X_{n+1}$. The justification for multiplying with $(-1)^{n+1}$ will follow later. The members of $ker\,\delta^n$ are called the n co-cycles of X with coefficient in A, and it is denoted by $C^n(X, A)$. The members of the image of δ^{n-1} are called the n co-boundaries of X with coefficient in A. The submodule of n co-boundaries is denoted by $B^n(X, A)$. The quotient module $C^n(X, A)/B^n(X, A)$ is denoted by $H^n(X, A)$, and it is called the nth **co-homology** of X with coefficient in A. Clearly, $H^n(-, A)$ defines a contra-variant functor from the category CH of chain complexes of R-modules to the category of abelian groups. Similarly, $H^n(X, -)$ defines a functor from the category of R-modules to the category of abelian groups. In a later section, we shall try to find relations between $H^n(X, A)$, $H_n(X)$, and A.

Now, we give some important examples having their applications in topology and geometry.

Simplicial Complex and Homology

An abstract **simplicial complex** is a pair (Σ, S), where Σ is a nonempty set, and S is a set of finite nonempty subsets of Σ such that every nonempty subset of a member of S is also a member of S. The members of Σ are called the vertices, and the members of S are called the faces of the simplicial complex. If $\sigma \in S$ and $\tau \subseteq \sigma$, then τ is called a face of σ. The members of S are also called the simplexes of the simplicial complex. An element σ of S containing $p + 1$ elements is called a p-simplex.

A simplicial map from a simplicial complex (Σ, S) to a simplicial complex (Σ', S') is a map f from Σ to Σ' such that $f(\sigma) \in S'$ for all $\sigma \in S$. Thus, we have a category SC of simplicial complexes, where morphisms are simplicial maps. A simplicial complex (Σ', S') is said to be a simplicial subcomplex of (Σ, S) if $\Sigma' \subseteq \Sigma$ and the inclusion map i from Σ to Σ' is a simplicial map. It is said to be a full simplicial subcomplex if $S \cap \wp(\Sigma') = S'$. More explicitly, a subsimplicial complex (Σ', S') of (Σ, S) will be termed as a full simplicial subcomplex if whenever $\sigma \in S$ and $\sigma \subseteq \Sigma'$, $\sigma \in S'$. Let $\bar{\sigma}$ denote the set of all faces of the simplex $\sigma \in S$. Then $(\sigma, \bar{\sigma})$ is a full subcomplex of (Σ, S). Similarly, let $\dot{\sigma}$ denote the set of all proper faces of σ. Then $(\sigma, \dot{\sigma})$ is a subcomplex of $(\sigma, \bar{\sigma})$. Obviously, it is not a full subcomplex.

Let (Σ_1, S_1) and (Σ_2, S_2) be subcomplexes of a simplicial complex (Σ, S). Then $(\Sigma_1 \cap \Sigma_2, S_1 \cap S_2)$ is also a subcomplex, and it is called the intersection (Σ_1, S_1) and (Σ_2, S_2). A simplicial complex (Σ, S) will be termed as direct sum of its subcomplexes (Σ_1, S_1) and (Σ_2, S_2) (in symbol $(\Sigma, S) = (\Sigma_1, S_1) \oplus (\Sigma_2, S_2)$) if $\Sigma_1 \cap \Sigma_2 = \emptyset$ and every simplex σ of (Σ, S) can be uniquely expressed as $\sigma = \rho \bigcup \tau$, where $\rho \in S_1$ and $\tau \in S_2$.

Proposition 1.3.6 *If* (Σ_1, S_1) *is a full simplicial subcomplex of* (Σ, S)*, then there is a simplicial subcomplex* (Σ_2, S_2) *of* (Σ, S) *such that* $(\Sigma, S) = (\Sigma_1, S_1) \oplus (\Sigma_2, S_2)$.

Proof Take $\Sigma_2 = \Sigma - \Sigma_1$ and $S_2 = \wp(\Sigma_2) \bigcap S$. Since (Σ_1, S_1) is a full simplicial subcomplex of (Σ, S), $(\Sigma, S) = (\Sigma_1, S_1) \oplus (\Sigma_2, S_2)$. ♯

An ordered p-simplex of (Σ, S) is a p-simplex σ together with a total order in σ. More explicitly, an ordered p-simplex in (Σ, S) is a pair (σ, α), where σ is a p-simplex and α is a bijective map from $\{0, 1, 2, \ldots, p\}$ to σ. An element $v \in \sigma$ is less than $w \in \sigma$ if $v = \alpha(i)$, $w = \alpha(j)$ for some $i < j$. Let (σ, α) be an ordered p-simplex in (Σ, S). Then for each $i, 0 \leq i \leq p$, we have an ordered $p - 1$-simplex (σ_i, α_i), where $\sigma_i = \sigma - \{\alpha(i)\}$ and the map α_i from $\{0, 1, \ldots, p - 1\}$ to σ_i is given by $\alpha_i(j) = \alpha(j)$ for $j \leq i - 1$ and $\alpha_i(j) = \alpha(j + 1)$ for $j \geq i$. The ordered $p - 1$ simplex (σ_i, α_i) defined above is called the ith face of the ordered p-simplex (σ, α).

Let $\Omega_p(\Sigma, S)$ denote the free abelian group on the set of all ordered p-simplexes of Σ. We have a homomorphism d_p from $\Omega_p(\Sigma, S)$ to $\Omega_{p-1}(\Sigma, S)$ defined by

$$d_p((\sigma, \alpha)) = \sum_{i=0}^{p} (-1)^i (\sigma_i, \alpha_i),$$

where (σ_i, α_i) is the ith face of the ordered p-simplex (σ, α). It is evident that $((\sigma_j)_i, (\alpha_j)_i) = ((\sigma_i)_{j-1}, (\alpha_i)_{j-1})$ for all $i < j$. Further,

$$(d_{p-1} o d_p)((\sigma, \alpha)) = d_{p-1}(\sum_{i=0}^{p}(-1)^i(\sigma_i, \alpha_i)) = \sum_{i=0}^{p}(-1)^i \sum_{j=0}^{p-1}(-1)^j((\sigma_i)_j, (\alpha_i)_j).$$

In turn,

$$(d_{p-1} o d_p)((\sigma, \alpha)) =$$
$$\sum_{0 \leq i \leq j \leq p-1}(-1)^{i+j}((\sigma_i)_j, (\alpha_i)_j) + \sum_{p \geq i > j \geq 0}(-1)^{i+j}((\sigma_i)_j, (\alpha_i)_j).$$

Using the identity $((\sigma_j)_i, (\alpha_j)_i) = ((\sigma_i)_{j-1}, (\alpha_i)_{j-1})$, it follows that $(d_{p-1} o d_p) = 0$. This gives us a chain complex $\Omega(\Sigma, S)$. This chain complex $\Omega(\Sigma, S)$ is called the the ordered simplicial chain complex associated with the simplicial complex (Σ, S). $\Omega(\Sigma, S)$ is a nonnegative chain complex with $\Omega_0(\Sigma, S)$ the free abelian group on the set $\dot{\Sigma}$ of 0-simplexes. Any element of $\Omega_0(\Sigma, S)$ is a finite sum $\sum_{v \in \Sigma} \alpha_v \{v\}$, where $\alpha_v \in \mathbb{Z}$ and all but finitely many α_v are 0. The map ϵ from $\Omega_0(\Sigma, S)$ to $\mathbb{Z}$ defined by $\epsilon(\sum_{v \in \Sigma} \alpha_v \{v\}) = \sum_{v \in \Sigma} \alpha_v$ is a surjective homomorphism. The map ϵ is called the **augmentation** map of the chain complex $\Omega(\Sigma, S)$. Let (σ, α) be an ordered 1-simplex in (Σ, S), where $\sigma = \{v, w\}$, $\alpha(0) = v$, and $\alpha(1) = w$. Then $d_1((\sigma, \alpha)) = \{w\} - \{v\}$. Hence $\epsilon d_1 = 0$. Thus, we have another chain complex $\tilde{\Omega}(\Sigma, S)$ given by

$$\tilde{\Omega}(\Sigma, S) \equiv \cdots \overset{d_{p+1}}{\to} \Omega_p(\Sigma, S) \overset{d_p}{\to} \cdots \overset{d_2}{\to} \Omega_1(\Sigma, S) \overset{d_1}{\to} Ker \, \epsilon \longrightarrow 0,$$

where $\tilde{d}_1$ is the homomorphism induced by d_1. The chain complex $\tilde{\Omega}(\Sigma, S)$ is called
the **reduced** order chain complex of (Σ, S). The qth homology $H_q(\Omega(\Sigma, S))$ of
$\Omega(\Sigma, S)$ is denoted by $H_q(\Sigma, S)$, and it is called the qth-order homology of (Σ, S).
The qth homology $H_q(\tilde{\Omega}(\Sigma, S))$ of $\tilde{\Omega}(\Sigma, S)$ is denoted by $\tilde{H}_q(\Sigma, S)$, and it is called
the reduced qth-order homology of (Σ, S). Evidently, $H_q(\Omega(\Sigma, S)) = \tilde{H}_q(\Sigma, S)$
for all $q \geq 1$. Further, since $\mathbb{Z}$ is free abelian, $\Omega_0(\Sigma, S) = Ker \, \epsilon \oplus \mathbb{Z}$, and hence
$H_0(\Omega(\Sigma, S)) = \tilde{H}_0(\Sigma, S) \oplus \mathbb{Z}$.

A simplicial complex (Σ, S) is said to be a **connected** simplicial complex if any
pair of distinct vertices belong to a simplex. Maximal connected simplicial subcomplexes of (Σ, S) are called the connected components of (Σ, S). It is clear that the
connected components of (Σ, S) form a partition of Σ. Let $\{(\Sigma_i, S_i) \mid i \in I\}$ be the
family of all connected components of (Σ, S). Evidently, $(\Sigma, S) = \oplus \sum_{i \in I} (\Sigma_i, S_i)$,
and $\Omega(\Sigma, S) = \oplus \sum_{i \in I} \Omega(\Sigma_i, S_i)$. In turn, $\Omega(\Sigma, S) = \oplus \sum_{i \in I} \Omega(\Sigma_i, S_i)$, and
$\tilde{\Omega}(\Sigma, S) = \oplus \sum_{i \in I} \tilde{\Omega}(\Sigma_i, S_i)$. Since the homology commutes with the direct sum,
$H_q(\Sigma, S) = \oplus \sum_{i \in I} H_q(\Sigma_i, S_i)$, and $\tilde{H}_q(\Sigma, S) = \oplus \sum_{i \in I} \tilde{H}_q(\Sigma_i, S_i)$.

Proposition 1.3.7 *If* (Σ, S) *is a connected simplicial complex, then* $H_0(\Sigma, S) \approx \mathbb{Z}$.

Proof Since ϵ is surjective, it is sufficient to show that $Image \, d_1 = Ker \, \epsilon$. Already,
$Image \, d_1 \subseteq Ker \, \epsilon$. Let $\sum_{i=1}^{n} \lambda_i \{v_i\}$ be a member of $Ker \, \epsilon$. Then $\sum_{i=1}^{n} \lambda_i = 0$.
Let v_0 be a member of Σ. Then

$$\sum_{i=1}^{n} \lambda_i \{v_i\} - (\sum_{i=1}^{n} \lambda_i) v_0 = \sum_{i=1}^{n} (v_i - v_0).$$

Without any loss, we may assume that $v_i \neq v_0$ for all i. Since (Σ, S) is connected,
for each i we have an ordered 2-simplex (σ_i, α_i), where $\sigma_i = \{v_0, v_1\}$ and α_i is
the map which takes 0 to v_0 and 1 to v_i. Evidently $\sum_{i=1}^{n} \lambda_i (\sigma_i, \alpha_i)$ is a member of
$\Omega_1(\Sigma, S)$ whose image under d_1 is $\sum_{i=1}^{n} \lambda_i \{v_i\}$. $\sharp$

The following corollary is immediate from the above result.

Corollary 1.3.8 $H_0(\Sigma, S)$ *is the direct sum of as many copies of* $\mathbb{Z}$ *as many connected components of* (Σ, S). $\sharp$

Let f be a simplicial map from a simplicial complex (Σ, S) to a simplicial complex (Σ', S'). For each $p \geq 0$, we have a homomorphism $\Omega_p(f)$ from $\Omega_p(\Sigma, S)$ to
$\Omega_p(\Sigma', S')$ given by $\Omega_p(f)(\sigma, \alpha) = 0$ if f is not injective on σ, and $\Omega_p(f)(\sigma, \alpha) =$
$(f(\sigma), f \circ \alpha)$ if f is injective. It is clear that $\Omega(f) = \{\Omega_p(f) \mid p \geq 0\}$ is a chain
transformation from $\Omega(\Sigma, S)$ to $\Omega(\Sigma', S')$. Evidently, Ω defines a functor from the
category SC of simplicial complexes to the category of chain complexes of abelian
groups. In turn, for each n, H_n defines a functor from the category of SC of simplicial
complexes to the category of abelian groups.

Let $A_p(\Sigma, S)$ denote the set of all ordered p-simplexes of (Σ, S). Define a relation
$\sim$ on $A_p(\Sigma, S)$ by $(\sigma, \alpha) \sim (\tau, \beta)$ if $\sigma = \tau$ and $\alpha \beta^{-1}$ is an even permutation in
S_{p+1}. Evidently, $\sim$ is an equivalence relation. The equivalence class determined by

(σ, α) is denoted by $[\sigma, \alpha]$, and it is called an **oriented** p-simplex in (Σ, S). For each p-simplex σ, there are exactly two equivalence classes. Let $\Lambda^p(\Sigma, S)$ denote the abelian group generated by the set of oriented p-simplexes $[\sigma, \alpha]$ subject to the relation $[\sigma, \alpha] + [\sigma, \beta] = 0$ whenever $[\sigma, \alpha] \neq [\sigma, \beta]$ (equivalently, whenever $\alpha\beta^{-1}$ is an odd permutation). Evidently, $\Lambda_p(\Sigma, S)$ is a free abelian of rank equal to the cardinality of the set of p-simplexes. Using the induction on p, the following proposition can be easily verified.

Proposition 1.3.9 *If* $[\sigma, \alpha] + [\sigma, \beta] = 0$, *then*

$$\sum_{i=0}^{p}(-1)^i[\sigma_i, \alpha_i] + \sum_{i=0}^{p}(-1)^i[\sigma_i, \beta_i] = 0.$$

♯

Thus, we have a homomorphism d_p from $\Lambda_p(\Sigma, S)$ to $\Lambda_{p-1}(\Sigma, S)$ given by

$$d_p([\sigma, \alpha]) = \sum_{i=0}^{p}(-1)^i[\sigma_i, \alpha_i].$$

As in case of order simplexes, it can be shown that $d_{p-1}d_p = 0$. This gives us a chain complex

$$\Lambda(\Sigma, S) \equiv \cdots \overset{d_{p+1}}{\to} \Lambda_p(\Sigma, S) \overset{d_{p-1}}{\to} \cdots \overset{d_2}{\to} \Lambda_1(\Sigma, S) \overset{d_1}{\to} \Lambda_0(\Sigma, S) \longrightarrow 0.$$

The simplicial complex $\Lambda(\Sigma, S)$ is called the **oriented chain complex** associated with the simplicial complex (Σ, S). Here also, we have an augmentation map ϵ from $\Lambda_0(\Sigma, S)$ to $\mathbb{Z}$ given by $\epsilon(\sum_{i=1}^{n} \lambda_i[v_i]) = \sum_{i=1}^{n} \lambda_i$. Clearly, ϵ is a surjective homomorphism. Again, we have a reduced oriented chain complex

$$\tilde{\Lambda}(\Sigma, S) \equiv \cdots \overset{d_{p+1}}{\to} \Lambda_p(\Sigma, S) \overset{d_p}{\to} \cdots \overset{d_2}{\to} \Lambda_1(\Sigma, S) \overset{\tilde{d}_1}{\to} Ker\ \epsilon \longrightarrow 0,$$

where $\tilde{d}_1$ is the homomorphism induced by d_1. Thus, Λ defines another functor from the category SC to the category of nonnegative chain complexes of abelian groups.

For each (Σ, S), we have a chain transformation $\mu_{(\Sigma, S)} = \{\mu_p \mid p \geq 0\}$ from $\Omega(\Sigma, S)$ to $\Lambda(\Sigma, S)$ given by $\mu_p((\sigma, \alpha)) = [\sigma, \alpha]$. We state the following theorem without proof. The reader is referred to the Algebraic topology by Spanier for the proof.

Theorem 1.3.10 $\mu_{(\Sigma, S)}$ *defined above is a chain equivalence, and μ is a natural equivalence between the functors Ω and Λ.* ♯

In turn, $H_q(\Omega(\Sigma, S)) \approx H_q(\Lambda(\Sigma, S))$ for each q. We denote $H_q(\Lambda(\Sigma, S))$ by $H_q(\Sigma, S)$, and call it the qth simplicial homology of the simplicial complex (Σ, S). We compute simplicial homologies of some simple simplicial complexes.

Example 1.3.11 Take $\Sigma = \{v_0, v_1, v_2\}$, and

$$S = \{\{v_0\}, \{v_1\}, \{v_2\}, \{v_0, v_1\}, \{v_0, v_2\}, \{v_1, v_2\}\}.$$

Clearly, the pair (Σ, S) defines a simplicial complex. The orientation (v_0, v_1, v_2) defines unique orientation of the simplicial complex. Clearly, $\bigwedge_p(\Sigma) = \{0\}$ for all $p < 0$. $\bigwedge_0(\Sigma)$ is the free abelian group generated by the vertices v_0, v_1, and v_2. Note that we identify the zero simplex $\{v_i\}$ by the vertex v_i itself. Thus, every element of $\bigwedge_0(\Sigma)$ is uniquely expressible as $\alpha_0 v_0 + \alpha_1 v_1 + \alpha_2 v_2$, where $\alpha_i \in \mathbb{Z}$. Next, $\bigwedge_1(\Sigma)$ is the free abelian group generated by the set $(v_0, v_1), (v_1, v_2), (v_0, v_2)$ of oriented 1-simplexes. Also $\bigwedge_p(\Sigma) = \{0\}$ for all $p \geq 2$. Evidently, $d_p = 0$ for all $p \neq 1$, and d_1 is given by

$$d_1(\alpha(v_0, v_1) + \beta(v_1, v_2) + \gamma(v_2, v_0)) = \alpha v_0 - \alpha v_1 + \beta v_1 - \beta v_2 + \gamma v_2 - \gamma v_0 = (\alpha - \gamma)v_0 + (\beta - \alpha)v_1 + (\gamma - \beta)v_2.$$

It follows that

$$B_0(\bigwedge(\Sigma)) = image\, d_1 = \{\alpha_0 v_0 + \alpha_1 v_1 + \alpha_2 v_2 \mid \alpha_0 + \alpha_1 + \alpha_2 = 0\},$$

and $Z_0(\bigwedge(\Sigma)) = ker\, d_0 = \bigwedge_0(\Sigma)$. The augmentation map ϵ from $\bigwedge_0(\Sigma)$ to $\mathbb{Z}$ given by $\epsilon(\alpha_0 v_0 + \alpha_1 v_1 + \alpha_2 v_2) = \alpha_0 + \alpha_1 + \alpha_2$ is a surjective homomorphism whose kernel is $B_0(\bigwedge(\Sigma))$. By the fundamental theorem of homomorphism, $H_0(\Sigma)$ is isomorphic to the additive group $\mathbb{Z}$. Also

$$Z_1(\bigwedge(\Sigma)) = ker\, d_1 = \{\alpha(v_0, v_1) + \beta(v_1, v_2) + \gamma(v_2, v_0)) \mid \alpha = \beta = \gamma\},$$

and $B_1(\bigwedge(\Sigma)) = \{0\}$. Clearly, $Z_1(\wedge(\Sigma))$ is isomorphic to $\mathbb{Z}$. Thus, $H_1(\Sigma)$ is also isomorphic to the additive group $\mathbb{Z}$.

We shall have more examples in Chap. 3.

Singular Chain Complex, and Singular Homology

The topological subspace

$$\Delta^q = \{\overline{\alpha} = (\alpha_0, \alpha_1, \ldots, \alpha_q) \in \mathbb{R}^{q+1} \mid \sum_{i=0}^{q} \alpha_i = 1\}\, of\, \mathbb{R}^{q+1}$$

is called the **standard q-simplex**. In particular, the standard 0-simplex Δ^0 is the single point $1 \in \mathbb{R}$. For $i, 0 \leq i \leq q$, the subspace

$$\Delta_i^q = \{\overline{\alpha} \in \Delta^q \mid \alpha_i = 0\}$$

is called the ith face of Δ^q. Further, for each i, the map ∂_i from Δ^{q-1} to Δ^q defined by

$$\partial_i(\alpha_0, \alpha_1, \ldots, \alpha_{q-1}) = (\alpha_0, \alpha_1, \ldots, \alpha_{i-1}, 0, \alpha_{i+1}, \ldots, \alpha_{q-1})$$

is a homeomorphism from Δ^{q-1} to Δ_i^q, and it is called the ith face map of Δ^q.

Let X be a topological space. A continuous map σ from Δ^q to X is called a **singular q-simplex** in X. Thus, a singular 0-simplex can be simply viewed as a point in X. Let $S_q(X)$ denote the free abelian group generated by the set of all singular q-simplexes in X. In particular, $S_0(X)$ is the free abelian group on X. Define a map d_q from $S_q(X)$ to $S_{q-1}(X)$ defined by

$$d_q(\sigma) = \sum_{i=0}^{q} (-1)^i \sigma o \partial_i.$$

It can be easily shown that $d_{q-1}od_q = 0$ for all q. This gives us a chain complex

$$S(X) \equiv \overset{d_{q+1}}{\to} S_q(X) \overset{d_q}{\to} S_{q-1}(X) \overset{d_{q-1}}{\to} \cdots \overset{d_1}{\to} S_0(X) \to 0.$$

This chain complex is called the **singular chain complex** of X. The qth homology of $S(X)$ is called the qth singular homology of X, and it is denoted by $H_q(X)$.

We have a surjective homomorphism ϵ from $S_0(X)$ to $\mathbb{Z}$ given by $\epsilon(\sum_{i=1}^{n} a_i x_i) = \sum_{i=1}^{n} a_i$. Evidently, $\epsilon od_1 = 0$. Thus, ϵ is an augmentation map of the singular chain complex. It is also clear that $S_0(X) = Ker\ \epsilon \oplus \mathbb{Z}$. We have the reduced singular chain complex $\tilde{S}(X)$ of X given by

$$\tilde{S}(X) \equiv \cdots \overset{d_{n+1}}{\to} S_n(X) \overset{d_n}{\to} \cdots \overset{d_2}{\to} S_1(X) \overset{\tilde{d}_1}{\to} Ker\ \epsilon \longrightarrow 0,$$

where $\tilde{d}_1$ is the homomorphism induced by d_1. The qth homology of $\tilde{S}(X)$ is called the reduced qth singular homology of X, and it is denoted by $\tilde{H}_q(X)$. Clearly, $H_q(X) = \tilde{H}_q(X)$ for all $q \geq 1$ and $H_0(X) = \tilde{H}_0(X) \oplus \mathbb{Z}$.

If f is a continuous map for a topological space X to a topological space Y, and σ is a singular q-simplex in X, then $fo\sigma$ is a singular q-simplex in Y. In turn, it induces a homomorphism $S_q(f)$ from $S_q(X)$ to $S_q(Y)$. Evidently, $S(f) = \{S_q(f)\}$ is a chain transformation from $S(X)$ to $S(Y)$ which respects the augmentation maps. Indeed, S is a functor from the category of topological spaces to the category of chain complexes of abelian groups. Consequently, for each n, we have the nth singular homology functor H_n from the category of topological spaces to the category of abelian groups.

Given a subspace Y of X, we get a short exact sequence

$$0 \longrightarrow S(Y) \overset{i}{\to} S(X) \overset{\nu}{\to} S(X)/S(Y) \longrightarrow 0$$

of chain complexes. The chain complex $S(X)/S(Y)$ is denoted by $S(X, Y)$, and it is called the singular chain complex of the pair (X, Y), where Y is a subspace of X. The nth homology of $S(X, Y)$ is denoted by $H_n(X, Y)$, and it is called the nth singular homology of the pair (X, Y). From Theorem 1.3.1, we have a long exact sequence

$$\cdots \overset{\partial_{n+1}}{\to} H_n(Y) \overset{H_n(i)}{\to} H_n(X) \overset{H_n(\nu)}{\to} H_n(X, Y) \overset{\partial_n}{\to} \cdots$$

associated with the pair (X, Y).

Let A be an abelian group. The nth co-homology group $H^n(X, A)$ of the co-chain complex $Hom(S(X), A)$ is called the nth **singular co-homology** of the space X with coefficient in A. Further, for an abelian group A we have a chain complex $S_n(X) \otimes A$. The nth homology $H_n(X, A) = H_n(S(X) \otimes A)$ is called the **singular homology** of X with coefficient in A.

Let X be a topological space, and $\{X_\alpha \mid \alpha \in \Gamma\}$ be the family of all path components of X. Then $X = \bigsqcup_{\alpha \in \Gamma} X_\alpha$ is disjoint topological sum of $\{X_\alpha \mid \alpha \in \Gamma\}$. Since Δ^q is path connected, and a continuous image of a path connected space is path connected, every singular q-simplex in X is a singular q-simplex in a unique X_α. As such, $S(X) = \oplus \sum_{\alpha \in \Gamma} S(X_\alpha)$. Since homology functor commutes with the direct sum functor, we have the following proposition.

Proposition 1.3.12 *Let $\{X_\alpha \mid \alpha \in \Gamma\}$ be the family of all path components of topological space X. Then $H_q(X) = \oplus \sum_{\alpha \in \Gamma} H_q(X_\alpha)$ for each q. ♯*

Proposition 1.3.13 *Let X be a path connected space. Then $H_0(X) \approx \mathbb{Z}$.*

Proof By definition $H_0(X) = S_0(X)/Image\ d_1$. Since the augmentation map ϵ is surjective, it is sufficient to show that $Image\ d_1 = Ker\ \epsilon$. Already $Image\ d_1 \subseteq Ker\ \epsilon$. Let $\sum_{i=1}^n a_i x_i$ be a member of $Ker\ \epsilon$. Then $\sum_{i=1}^n a_i = 0$. Let x_0 be a member of X. Then $\sum_{i=1}^n a_i x_i = \sum_{i=1}^n a_i(x_i - x_0)$. Since X is path connected, for each i, we have a continuous map σ_i from Δ^1 to X such that $\sigma_i((0, 1)) = x_i$ and $\sigma_i((1, 0)) = x_0$. Evidently, $d_1(\sum_{i=1}^n a_i \sigma_i) = \sum_{i=1}^n a_i x_i$. ♯

Corollary 1.3.14 *Let X be a topological space. Then the 0th singular homology $H_0(X)$ is direct sum of as many copies of $\mathbb{Z}$ as many path components of X. ♯*

In the third chapter, we shall further study simplicial and singular homologies together with some of their applications in topology and geometry.

Co-homology of Groups

Let G be a group, and A be a $\mathbb{Z}(G)$-module. Let $S^n(G, A)$ denote the set of all maps f from $G^n = \underbrace{G \times G \times \cdots \times G}_{n}$ to A for which $f(g_1, g_2, \ldots, g_n) = 0$ whenever $g_i = e$ for some i. We take $G^0 = \{e\}$. Clearly, $S^n(G, A)$ is an abelian group with respect to pointwise addition. Evidently $S^0(G, A)$ is the trivial group. Define a map d^n from $S^n(G, A)$ to $S^{n+1}(G, A)$ by

$$d^n(f)(g_1, g_1, \ldots, g_{n+1}) =$$
$$g_1 \cdot f(g_2, g_3, \ldots, g_{n+1}) + \sum_{i=1}^n (-1)^i f(g_1, g_2, \ldots, g_{i-1}, g_i g_{i+1}, g_{i+2}, \ldots, g_{n+1}) +$$
$$(-1)^{n+1} f(g_1, g_2, \ldots, g_n).$$

We may easily observe that $d^n(f)(g_1, g_1, \ldots, g_{n+1}) = 0$ whenever any $g_i = e$. Further, it is straightforward to verify that d^n is a homomorphism and $d^{n+1} o d^n = 0$ for all n. This gives us a co-chain complex

$$S(G, A) = \{0\} \to S^1(G, A) \xrightarrow{d^1} S^2(G, A) \xrightarrow{d_2} \cdots S^{n-1}(G, A) \xrightarrow{d_{n-1}} S^n(G, A) \xrightarrow{d_n} \cdots$$

of abelian groups. The nth co-homology of $S(G, A)$ is denoted by $H^n(G, A)$, and it is called the nth co-homology of the group G with coefficient in A.

Let us look at the low-dimensional co-homology groups. Clearly, $S^1(G, A)$ is the set of maps from G to A which preserve identity. If $f \in S^1(G, A)$, then $d_1(f)(g_1, g_2) = g_1 \cdot f(g_2) - f(g_1 g_2) + f(g_1)$. Thus, the group $Z^1(G, A) = ker d_1$ of all one co-cycles is the group of all crossed homomorphisms (see Chap. 10, Algebra 2) from G to A. Evidently, $B^1(G, A) = \{0\}$. Hence, $H^1(G, A)$ is the group of all crossed homomorphisms from G to A. In particular, if A is a trivial G-module, then $H^1(G, A) = Hom(G, A)$.

Let us describe $H^2(G, A)$. The map d^2 from $S^2(G, A)$ to $S^3(G, A)$ is given by

$$d^2(f)(g_1, g_2, g_3) = g_1 \cdot f(g_2, g_3) - f(g_1 g_2, g_3) + f(g_1, g_2 g_3) - f(g_1, g_2).$$

Thus, the group $Z^2(G, A) = ker d^2$ is precisely the group $FAC(G, A)$ of factor systems of central extensions of A by G (see Chap. 10, Algebra 2). Further, two factor systems are equivalent if and only if they differ by a boundary. It turns out that $H^2(G, A)$ is the group of equivalence classes of central extensions of A by G. The higher dimensional co-homology groups will be discussed in the next chapter.

Remark 1.3.15 The theory of co-homology groups arouse from topological considerations. To each group G, Eilenberg and MacLane associated a topological space (a cw-complex) $K(G, 1)$ whose first fundamental group is G and all higher homotopy groups are trivial. Indeed, such a cw-complex is unique up to homotopy type. The singular co-homology groups of $K(G, 1)$ with coefficient in an abelian group A are precisely the co-homology groups $H^n(G, A)$ of G with coefficient in the trivial G-module A.

Exercises

1.3.1 Find the simplicial homologies of the simplicial complex (Σ, S), where $\Sigma = \{v_0, v_1, v_2\}$ and S is the set of all nonempty subsets of Σ.

1.3.2 Find the simplicial homologies of the simplicial complex (Σ, S), where $\Sigma = \{v_0, v_1, v_2, v_3\}$ and S is the set of all proper nonempty subsets of Σ.

1.3.3 Let (Σ_1, S_1) and (Σ_2, S_2) be simplicial subcomplexes of a simplicial complex. Show that

$$\Omega(\Sigma_1 \cap \Sigma_2, S_1 \cap S_2) \xrightarrow{(i_1, -i_2)} \Omega(\Sigma_1, S_1) \oplus \Omega(\Sigma_2, S_2) \xrightarrow{j_1 + j_2}$$
$$\Omega(\Sigma_1 \cup \Sigma_2, S_1 \cup S_2) \longrightarrow 0$$

is a short exact sequence of chain complexes of abelian groups, where i_1, i_2, j_1, and j_2 are the corresponding inclusion simplicial maps. Establish the long exact sequence

$$\cdots \overset{\partial_{n+1}}{\to} H_n(\Sigma_1 \cap \Sigma_2, S_1 \cap S_2) \overset{(H_n(i_1), -H_n(i_2))}{\to} H_n(\Sigma_1, S_1) \oplus H_n(\Sigma_2, S_2) \overset{H_n(j_1)+H_n(j_2)}{\to} H_n(\Sigma_1 \cup \Sigma_2, S_1 \cup S_2) \overset{\partial_n}{\to} \cdots.$$

The above exact sequence is called the Mayer–Vietoris exact sequence.

1.3.4 Consider the simplicial complex (Σ, S), where $\Sigma = \{v_0, v_1, v_2, v_3, v_4\}$ and $S = \{\{v_0\}, \{v_1\}, \{v_1\}, \{v_2\}, \{v_3\}, \{v_4\}, \{v_0, v_1\}, \{v_0, v_2\}, \{v_1, v_2\}, \{v_0, v_3\}, \{v_0, v_4\}, \{v_3, v_4\}\}$. Express (Σ, S) as union of two proper subsimplicial complexes, and then use the above Mayer–Vietoris exact sequence to compute $H_n(\Sigma, S)$.

1.3.5 Compute $H^2(\mathbb{Z}_3, \mathbb{Z}_2)$ by treating $\mathbb{Z}_2$ as trivial $\mathbb{Z}_3$-module.

1.3.6 (**Snake lemma**) Consider the following commutative diagram:

$$
\begin{array}{ccccccccc}
0 & \longrightarrow & A & \overset{\alpha_1}{\longrightarrow} & M & \overset{\beta_1}{\longrightarrow} & B & \longrightarrow & 0 \\
 & & \downarrow f & & \downarrow g & & \downarrow h & & \\
0 & \longrightarrow & C & \underset{\alpha_2}{\longrightarrow} & N & \underset{\beta_2}{\longrightarrow} & D & \longrightarrow & 0
\end{array}
$$

where the rows are exact. Using Theorem 1.3.1 show that the induced sequence

$$0 \to kerf \to kerg \to kerh \to cokerf \to cokerg \to cokerh \to 0$$

is exact.

1.3.7 Show that the category of chain complexes in an abelian category is itself an abelian category.

1.3.8 Let X and Y be chain complexes in an abelian category Σ. Show that $H_n(X \oplus Y) \approx H_n(X) \oplus H_n(Y)$ for each n, where $\oplus$ denotes the co-products in the respective categories.

1.3.9 Let Σ be an abelian category, and $(D, \leq)$ be a directed set. Let $\{f_\alpha^\beta \in Mor_\Sigma(X^\alpha, X^\beta) \mid \alpha, \beta \in D, \ and \ \alpha \leq \beta\}$ be a directed system of chain complexes in Σ. Show that $H_n(\overset{Lim}{\to} X^\alpha) \approx \overset{Lim}{\to} H^n(X^\alpha)$ for each n.

1.3.10 Let R be a principal ideal domain and X be a chain complex of R-modules. For each n, show that there is a chain complex F^n of free R-modules and a chain transformation f^n from F^n to X such that (i) $H_m(F^n) = 0$ for $m \neq n$ and (ii) $H_n(f^n)$ is an isomorphism from $H_n(F^n)$ to $H_n(X)$.
Hint: Take F_n^n to be a free R-module on $C_n(X)$ with f_n^n the homomorphism from F_n^n to X_n with image $C_n(X)$, $F_{n+1}^n = (f_n^n)^{-1}(B_n(X))$, f_{n+1}^n the induced homomorphism from F_{n+1}^n to X_{n+1}, $d_{n+1}^{F^n}$ to be the inclusion map, and $F_m^n = 0$ for all m different from n and $n + 1$.

1.3.11 Let R be a principal ideal domain and X be a chain complex of R-modules. Use the above exercise to show the existence of a chain complex F of free R-modules together with a chain transformation f from F to X such that $H_n(f)$ is an isomorphism for each n.

1.3.12 Let Σ be a selective abelian category (in particular, category of modules). A pseudo-chain complex (see "Pseudo-co-homology of general extensions" by Lal and Sharma; Homology, Homotopy and Applications, vol 12, no 2) is a chain $(X, Y, d) = \{(X_n, Y_n, d_n) \mid n \in \mathbb{Z}\}$, where Y_n is a subobject X_n in Σ, and d_n is a morphism from X_n to X_{n-1} such that $d_{n-1}d_n/Y_n = 0$ for each n. $H_n(X, Y,) = Ker\, d_n/d_{n+1}(Y_{n+1})$ is called the nth pseudo-homology of (X, Y, d). A family $f = \{f_n \in Mor_\Sigma(X_n, X'_n) \mid n \in \mathbb{Z}\}$ is called a pseudo-chain transformation if (i) $f_n(Y_n) \subseteq Y'_n$ and (ii) $f_{n-1}d_n = d'_n f_n$ for all n. Let PC_Σ denote the category of all pseudo-chain complexes. Show that $F(X, Y) = \{(F(X, Y)_n, \tilde{d}^n) \mid n \in \mathbb{Z}\}$ is a chain complex, where $F(X, Y)_n = Y_n \cap d_n^{-1}(Y_{n-1})$ and $\tilde{d}^n$ is the restriction of d_n to $F(X, Y)_n$. Show further that F defines a reflector functor from PC_Σ to the category C_Σ of chain complexes in Σ such that $H_n(F(X, Y)) \subseteq H_n(X, Y)$.

1.3.13 Refer to the above exercise. Show that $G(X, Y) = \{(Y_n + Ker\, d_n, \tilde{d}^n) \mid n \in \mathbb{Z}\}$ is a chain complex in Σ. Show further that G is a functor such that $H_n(X, Y) = H_n(G(X, Y))$ for all n.

1.4 Extensions and the Functor EXT

In Chap. 10 of Algebra 2, we studied extensions of groups by groups, and described the group of equivalence classes of central extensions of an abelian group A by a group G as second co-homology group $H^2(G, A)$. Here, in this section, we shall study onefold extensions of an R-module A by another R-module B by imitating the corresponding description of extensions of a group by another group as described in Chap. 10, Algebra 2, and then describe the EXT functor. The functors $Ext_R^n(B, A)$ of equivalence classes of n-fold extensions of an R-module A by an R-module B as a co-homology group $H_R^n(B, A)$ will be discussed in the next chapter.

Let A and B be left R-modules. An exact sequence

$$0 \longrightarrow A \overset{\alpha}{\to} M_1 \overset{d_1}{\to} M_2 \overset{d_2}{\to} M_3 \cdots \overset{d_{n-1}}{\to} M_n \overset{\beta}{\to} B \longrightarrow 0$$

of R-modules is called an **n-fold extension** of A by B. In particular, a short exact sequence

$$0 \longrightarrow A \overset{\alpha}{\to} M \overset{\beta}{\to} B \to 0$$

is called a onefold extension of A by B. A onefold extension will be simply termed as an extension of A by B.

We say that an extension

$$E_1 \equiv 0 \longrightarrow A \overset{\alpha_1}{\to} M \overset{\beta_1}{\to} B \longrightarrow 0$$

of A by B is **equivalent** to an extension

$$E_2 \equiv 0 \longrightarrow A \overset{\alpha_2}{\to} N \overset{\beta_2}{\to} B \longrightarrow 0$$

of A by B if there is a homomorphism ϕ from M to N such that the diagram

$$
\begin{array}{ccccccccc}
0 & \longrightarrow & A & \overset{\alpha_1}{\longrightarrow} & M & \overset{\beta_1}{\longrightarrow} & B & \longrightarrow & 0 \\
& & \Big\downarrow I_A & & \Big\downarrow \phi & & \Big\downarrow I_B & & \\
0 & \longrightarrow & A & \overset{\alpha_2}{\longrightarrow} & N & \overset{\beta_2}{\longrightarrow} & B & \longrightarrow & 0
\end{array}
$$

is commutative. From the five lemma, ϕ is an isomorphism.

We use the notation $E_1 \equiv E_2$ to say that the extension E_1 is equivalent to the extension E_2. Evidently, the relation $\equiv$ is an equivalence relation on the class of all extensions of A by B.

Let

$$E \equiv 0 \longrightarrow A \overset{\alpha}{\to} M \overset{\beta}{\to} B \longrightarrow 0$$

be an extension of A by B. Let t be a section of the extension E. More explicitly, let t be a map from B to M such that $\beta o t = I_B$ with $t(0) = 0$. Then

$$\beta(t(x) + t(y)) = \beta(t(x)) + \beta(t(y)) = x + y = \beta(t(x + y)).$$

This shows that $t(x) + t(y) - t(x + y)$ belongs to the $ker\beta = imagea$. Hence, there is a unique member $f'(x, y)$ in A such that

$$t(x) + t(y) - t(x + y) = \alpha(f'(x, y)) \qquad (1.6)$$

Thus, for a given choice of the section t, we get a map f' from $B \times B$ to A given by Equation 1. Since $t(0) = 0$,

$$f'(0, x) = f'(x, 0) = 0 \qquad (1.7)$$

for all $x \in B$. Since $(t(x) + t(y)) + t(z) = t(x) + (t(y) + t(z))$ and $t(x) + t(y) = t(y) + t(x)$, we have

$$f'(y, z) - f'(x + y, z) + f'(x, y + z) - f'(x, y) = 0 \qquad (1.8)$$

and

$$f'(x, y) = f'(y, x) \qquad (1.9)$$

for all $x, y, z \in B$. Next, for $r \in R$ and $x \in B$, $\beta(t(rx)) = rx = r\beta(t(x)) = \beta(rt(x))$. This shows that $t(rx) - rt(x) \in ker\,\beta$. Since $ker\,\beta = image\,\alpha$ and α is injective, there is a unique element $g^t(r, x) \in A$ such that

$$\alpha(g^t(r, x)) = t(rx) - rt(x) \tag{1.10}$$

Using (1.6) and (1.10),

$$
\begin{aligned}
& t(rx + sx) \\
& = -\alpha(f^t(rx, sx)) + t(rx) + t(sx) \\
& = -\alpha(f^t(rx, sx)) + \alpha(g^t(s, x)) + \alpha(g^t(r, x)) + rt(x) + st(x)
\end{aligned} \tag{1.11}
$$

On the other hand,

$$t(rx + sx) = t((r + s)x) = \alpha(g^t(r + s, x)) + (r + s)t(x) \tag{1.12}$$

Equating (1.11) and (1.12), we get

$$f^t(rx, sx) = g^t(r, x) - g^t((r + s), x)) + g^t(s, x) \tag{1.13}$$

Again using (1.6) and (1.10),

$$
\begin{aligned}
& t(rx + ry) \\
& = -\alpha(f^t(rx, ry)) + t(rx) + t(ry) \\
& = -\alpha(f^t(rx, ry)) + \alpha(g^t(r, x)) + \alpha(g^t(r, y)) + rt(x) + rt(y)
\end{aligned} \tag{1.14}
$$

On the other hand,

$$
\begin{aligned}
t(rx + ry) = t(r(x + y)) & = \alpha(g^t(r, x + y)) + rt(x + y) \\
& = \alpha(g^t(r, x + y)) + -r\alpha(f^t(x, y)) + rt(x) + rt(y)
\end{aligned} \tag{1.15}
$$

Equating (1.14) and (1.15),

$$rf^t(x, y) - f^t(rx, ry) = g^t(r, x + y) - g^t(r, x) - g^t(r, y) \tag{1.16}$$

Further, using (1.10),

$$
\begin{aligned}
\alpha(g^t(rs, x)) + rst(x) = t((rs)x) = t(r(sx)) & = \alpha(g^t(r, sx)) + rt(sx) = \\
& \alpha(g^t(r, sx)) + r\alpha(g^t(s, x)) + rst(x).
\end{aligned}
$$

Hence,

$$rg^t(s, x) - g^t(rs, x) + g^t(r, sx) = 0 \qquad (1.17)$$

for all $r, s \in R$ and $x \in B$. The above discussion prompts us to have the following definition.

Definition 1.4.1 Let R be a ring with identity. A **factor system** over R is a quadruple (B, A, f, g), where B and A are R-modules, f is a map from $B \times B$ to A, and g is a map from $R \times B$ to A such that the following hold:

(i) $f(0, x) = f(x, 0) = 0$ for all $x \in B$.

(ii) $f(y, z) - f(x + y, z) + f(x, y + z) - f(x, y) = 0$ for all $x, y, z \in B$.

(iii) $f(x, y) = f(y, x)$ for all $x, y \in B$.

(iv) $f(rx, sx) = g(r, x) - g((r + s), x)) + g(s, x)$.

(v) $f(rx, ry) - rf(x, y) = g(r, x) - g(r, x + y) + g(r, y)$.

(vi) $rg(s, x) - g(rs, x) + g(r, sx) = 0$ for all $r, s \in R$ and $x \in B$.

In particular, it follows from (iv), (v), and (vi) that $g(0, x) = 0 = g(r, o) = g(1, x)$ for all $r \in R$ and $x \in B$.

Thus, for every pair (E, t), where E is an extension of an R-module A by an R-module B and t a section of E, we have an associated factor system $Fac(E, t) = (B, A, f^t, g^t)$ described as above. Conversely, to every factor system $\Sigma = (B, A, f, g)$, we associate the extension $Ext(\Sigma)$ as follows: Define a binary operation $+$ on $A \times B$ by

$$(u, x) + (v, y) = (u + v + f(x, y), x + y).$$

It is easily seen that $(A \times B, +)$ is an abelian group. Define the external multiplication $\cdot$ from $R \times (A \times B)$ to $A \times B$ by

$$r \cdot (u, x) = (ru + g(r, x), rx).$$

It can be further shown that $A \times B$ together with the above operations is an R-module, and we get an extension

$$Ext(\Sigma) \equiv 0 \longrightarrow A \overset{\alpha}{\to} A \times B \overset{\beta}{\to} B \longrightarrow 0$$

of A by B, where $\alpha(u) = (u, 0)$ and $\beta(u, x) = x$. The section t given by $t(x) = (0, x)$ is such that $Fac(Ext(\Sigma)) = \Sigma$. Further, given an extension

$$E \equiv 0 \longrightarrow A \overset{\alpha}{\to} M \overset{\beta}{\to} B \longrightarrow 0$$

of A by B, together with a section t of E, the map η from $A \times B$ to M defined by $\eta(a, x) = \alpha(a) + t(x)$ induces an equivalence from $Ext(Fac(E, t))$ to E.

Let EXT_R denote the category whose objects are extensions

$$1 \longrightarrow A \overset{\alpha}{\to} M \overset{\beta}{\to} B \longrightarrow 1$$

of R-modules, and a morphism between two extensions E_1 and E_2 given by the short exact sequences

$$0 \longrightarrow A_1 \xrightarrow{\alpha_1} M_1 \xrightarrow{\beta_1} B_1 \longrightarrow 0$$

and

$$0 \longrightarrow A_2 \xrightarrow{\alpha_2} M_2 \xrightarrow{\beta_2} B_2 \longrightarrow 0$$

is a triple (λ, μ, ν), where λ is a homomorphism from A_1 to A_2, μ is a homomorphism from M_1 to M_2, and ν is a homomorphism from B_1 to B_2 such that the following diagram is commutative:

$$
\begin{array}{ccccccccc}
0 & \longrightarrow & A_1 & \xrightarrow{\alpha_1} & M_1 & \xrightarrow{\beta_1} & B_1 & \longrightarrow & 0 \\
& & \downarrow{\lambda} & & \downarrow{\mu} & & \downarrow{\nu} & & \\
1 & \longrightarrow & A_2 & \xrightarrow{\alpha_2} & M_2 & \xrightarrow{\beta_2} & B_2 & \longrightarrow & 0
\end{array}
$$

The category EXT_R is called the category of **Schreier extensions** of R-modules. The isomorphisms in this category are called the **equivalences of extensions**.

Now, we describe the category EXT_R of extensions as a category of factor systems. Let (λ, μ, ν) be a morphism from the extension E_1 to the extension E_2 given by the commutative diagram

$$
\begin{array}{ccccccccc}
0 & \longrightarrow & A_1 & \xrightarrow{\alpha_1} & M_1 & \xrightarrow{\beta_1} & B_1 & \longrightarrow & 1 \\
& & \downarrow{\lambda} & & \downarrow{\mu} & & \downarrow{\nu} & & \\
1 & \longrightarrow & A_2 & \xrightarrow{\alpha_2} & M_2 & \xrightarrow{\beta_2} & B_2 & \longrightarrow & 1
\end{array}
$$

Let t_1 be a section of E_1, and t_2 be a section of E_2. Let $Fac(E_1, t_1) = (B_1, A_1, f^{t_1}, g^{t_1})$ and $Fac(E_2, t_2) = (B_2, A_2, f^{t_2}, g^{t_2})$ be the corresponding factor systems. Let $x \in B_1$. Then $\mu(t_1(x)) \in M_2$ and $\beta_2(\mu(t_1(x))) = \nu(\beta_1(t_1(x))) = \nu(x) = \beta_2(t_2(\nu(x)))$. Thus, $\mu(t_1(x)) - (t_2(\nu(x))) \in ker\beta_2 = image\alpha_2$. In turn, we have a unique $\phi(x) \in A_2$ such that $\alpha_2(\phi(x)) = \mu(t_1(x)) - (t_2(\nu(x)))$. Equivalently,

$$\mu(t_1(x)) = \alpha_2(\phi(x)) + t_2(\nu(x)) \tag{1.18}$$

Since $t_1(0) = 0 = t_2(0)$, it follows that

$$\phi(0) = 0 \tag{1.19}$$

Now, using the commutativity of the diagram and Eq. 1.18, we have

$$
\begin{aligned}
\mu(t_1(x) + t_1(y)) &= \mu(\alpha_1(f^{t_1}(x, y)) + t_1(x + y)) = \\
\alpha_2(\lambda(f^{t_1}(x, y)) + \mu(t_1(x + y)) &= \alpha_2(\lambda(f^{t_1}(x, y)) + \alpha_2(\phi(x + y)) + t_2(\nu(x + y)) = \\
\alpha_2(\lambda(f^{t_1}(x, y)) &+ \alpha_2(\phi(x + y)) + t_2(\nu(x) + \nu(y)).
\end{aligned}
$$

On the other hand, since μ is a homomorphism, using again Eq. 1.18,

$$\mu(t_1(x) + t_1(y)) \;=\; \mu(t_1(x)) + \mu(t_1(y)) \;=\; \alpha_2(\phi(x)) + t_2(\nu(x)) + \alpha_2(\phi(y)) +$$
$$t_2(\nu(y)) \;=\; \alpha_2(\phi(x)) + \alpha_2(\phi(y)) + t_2(\nu(x)) + t_2(\nu(y)) \;=\;$$
$$\alpha_2(\phi(x)) + \alpha_2(\phi(y)) + \alpha_2(f^{t_2}(\nu(x), \nu(y))) + t_2(\nu(x) + \nu(y)).$$

Equating the two expressions for $\mu(t_1(x) + t_1(y))$, and observing that α_2 is an injective homomorphism, we obtain the following identity:

$$\lambda(f^{t_1}(x, y)) + \phi(x + y) \;=\; \phi(x) + \phi(y) + f^{t_2}(\nu(x), \nu(y)) \qquad (1.20)$$

Next, from Eq. (1.10),

$$\alpha_1(g^{t_1}(r, x)) \;=\; t_1(rx) - rt_1(x).$$

Hence, $\alpha_2(\lambda(g^{t_1}(r, x))) \;=\; \mu(\alpha_1(g^{t_1}(r, x)))$
$= \; \mu(t_1(rx) - rt_1(x)) \;=\; \mu(t_1(rx)) - r\mu(t_1(x))$
$= \; \alpha_2(\phi(rx)) + t_2(\nu(rx)) - r\alpha_2(\phi(x)) - rt_2(\nu(x))$
$= \; \alpha_2(\phi(rx)) - r\alpha_2(\phi(x)) + t_2(r\nu(x)) - rt_2\nu(x)$
$= \; \alpha_2(\phi(rx) - r\phi(x)) + \alpha_2(g^{t_2}(r, x))$
$= \; \alpha_2(\phi(rx) - r\phi(x) + g^{t_2}(r, x)).$
Since α_2 is injective,

$$\lambda(g^{t_1}(r, x)) \;=\; \phi(rx) - r\phi(x) + g^{t_2}(r, x) \qquad (1.21)$$

Thus, a morphism (λ, μ, ν) between extensions E_1 and E_2 together with choices of sections t_1 and t_2 of the corresponding extensions induces a map ϕ from B_1 to A_2 such that the triple (ν, ϕ, λ) satisfies (1.19), (1.20), and (1.21). Such a triple may be viewed as a morphism from the factor system $(B_1, A_1, f^{t_1}, g^{t_1})$ to $(B_2, A_2, f^{t_2}, g^{t_2})$.

Let $(\lambda_1, \mu_1, \nu_1)$ be a morphism from an extension

$$E_1 \;\equiv\; 0 \longrightarrow A_1 \overset{\alpha_1}{\to} M_1 \overset{\beta_1}{\to} B_1 \longrightarrow 0$$

to an extension

$$E_2 \;\equiv\; 0 \longrightarrow A_2 \overset{\alpha_2}{\to} M_2 \overset{\beta_2}{\to} B_2 \longrightarrow 0,$$

and $(\lambda_2, \mu_2, \nu_2)$ be that from the extension E_2 to

$$E_3 \;\equiv\; 0 \longrightarrow A_3 \overset{\alpha_3}{\to} M_3 \overset{\beta_3}{\to} B_3 \longrightarrow 0.$$

Let t_1, t_2 and t_3 be corresponding choice of the sections. Then as in (1.14)

$$\mu_1(t_1(x)) \;=\; \alpha_2(\phi_1(x)) + t_2(\nu_1(x))$$

and

$$\mu_2(t_2(u)) = \alpha_3(\phi_2(u)) + t_3(\nu_2(u)),$$

where ϕ_1 is the uniquely determined map from B_1 to A_2, and ϕ_2 is that from B_2 to A_3. In turn,

$$\mu_2(\mu_1(t_1(x))) = \mu_2(\alpha_2(\phi_1(x))) + \mu_2(t_2(\nu_1(x))) =$$
$$\mu_2(\alpha_2(\phi_1(x))) + \alpha_3(\phi_2(\nu_1(x))) + t_3(\nu_2(\nu_1(x))) =$$
$$\alpha_3(\lambda_2(\phi_1(x))) + \alpha_3(\phi_2(\nu_1(x))) + t_3(\nu_2(\nu_1(x))) = \alpha_3(\phi_3(x)) + t_3(\nu_2(\nu_1(x))),$$

where $\phi_3(x) = \lambda_2(\phi_1(x)) + \phi_2(\nu_1(x))$. It follows that the composition $(\lambda_2 \circ \lambda_1, \mu_2 \circ \mu_1, \nu_2 \circ \nu_1)$ induces the triple $(\nu_2 \circ \nu_1, \phi_3, \lambda_2 \circ \lambda_1)$, where $\phi_3(x) = \lambda_2(\phi_1(x)) + \phi_2(\nu_1(x))$ for each $x \in B_1$.

Prompted by the above discussion, we introduce the category **FACS** whose objects are factor systems, and a morphism from (B_1, A_1, f^1, g^1) to (B_2, A_2, f^2, g^2) is a triple (ν, ϕ, λ), where ν is a homomorphism from B_1 to B_2, λ a homomorphism from A_1 to A_2, and ϕ a map from B_1 to A_2 such that

(i) $\phi(0) = 0$.
(ii) $\lambda(f^1(x, y)) + \phi(x + y) = \phi(x) + \phi(y) + f^2(\nu(x), \nu(y))$.
(iii) $\lambda(g^1(r, x)) = \phi(rx) - r\phi(x) + g^2(r, x)$.

The composition of morphisms $(\nu_1, \phi_1, \lambda_1)$ from (B_1, A_1, f^1, g^1) to (B_2, A_2, f^2, g^2) and the morphism $(\nu_2, \phi_2, \lambda_2)$ from (B_2, A_2, f^2, g^2) to (B_3, A_3, f^3, g^3) is the triple $(\nu_2 \circ \nu_1, \phi_3, \lambda_2 \circ \lambda_1)$, where ϕ_3 is given by $\phi_3(x) = \phi_2(\nu_1(x)) + \lambda_2(\phi_1(x))$ for each $x \in B_1$.

The following theorem is a consequence of the above discussion.

Theorem 1.4.2 *Let t_E be a choice of a section of an extension E of an R-module by another R-module (such a choice function t exists because of axiom of choice). Then for the association Fac which associates to each extension E the factor system $Fac(E, t_E)$ is an equivalence between the category $\mathbf{EXT}_R$ of extensions and the category $\mathbf{FACS}$ of factor systems. More explicitly, Fac which associates the factor system $Fac(E, t_E)$ to an extension E is a functor from EXT_R to $FACS$, and the association Ext which associates the extension $Ext(\Sigma)$ to a factor system Σ is a functor from the category $FACS$ to EXT_R such that $Ext \circ Fac$ and $Fac \circ Ext$ are naturally equivalent to the corresponding identity functors.* ♯

Fix a pair A and B of R-modules. We try to describe the equivalence classes of extensions of A by B. Let M be an extension of A by B given by the exact sequence

$$E \equiv 0 \longrightarrow A \xrightarrow{\alpha} M \xrightarrow{\beta} B \longrightarrow 0.$$

Let (λ, μ, ν) be an equivalence from this extension to another extension M' of A' by B' given by the exact sequence

$$E' \equiv 0 \longrightarrow A' \xrightarrow{\alpha'} M' \xrightarrow{\beta'} B' \longrightarrow 0.$$

Then it follows that M' is also an extension of A by B given by the exact sequence

$$E'' \equiv 0 \longrightarrow A \overset{\alpha' \circ \lambda}{\to} M' \overset{\nu^{-1} \circ \beta'}{\to} B \longrightarrow 0$$

such that (I_A, μ, I_B) is an equivalence from E to E''. Also, $(\lambda, I_{M'}, \nu)$ is an equivalence from E'' to E'.

As such, there is no loss of generality in restricting the concept of equivalence on the class $E_R(B, A)$ of all extensions of A by B by saying that two extensions

$$E_1 \equiv 0 \longrightarrow A \overset{\alpha_1}{\to} M_1 \overset{\beta_1}{\to} B \longrightarrow 0$$

and

$$E_2 \equiv 0 \longrightarrow A \overset{\alpha_2}{\to} M_2 \overset{\beta_2}{\to} B \longrightarrow 0$$

in $E(B, A)$ are equivalent if there is an isomorphism ϕ from M_1 to M_2 such that the diagram

$$
\begin{array}{ccccccccc}
0 & \longrightarrow & A & \overset{\alpha_1}{\longrightarrow} & M_1 & \overset{\beta_1}{\longrightarrow} & B & \longrightarrow & 0 \\
& & \downarrow{\scriptstyle I_A} & & \downarrow{\scriptstyle \phi} & & \downarrow{\scriptstyle I_B} & & \\
0 & \longrightarrow & A & \overset{\alpha_2}{\longrightarrow} & M_2 & \overset{\beta_2}{\longrightarrow} & B & \longrightarrow & 0
\end{array}
$$

is commutative. Indeed, for any extension E in $\mathbf{EXT_R}$ which is equivalent to a member E' of $E(B, A)$, there is a member E'' of $E(B, A)$ such that E is equivalent to E'' in the category $\mathbf{EXT_R}$ and E'' in $E_R(B, A)$ is equivalent to E' in the sense described above.

Let $FAC_R(B, A)$ denote the set $\{(f, g) \mid (B, A, f, g)\ is\ a\ factor\ system\}$. Let (f, g) and (f', g') be members of $FAC_R(B, A)$. It is easy to verify that $(f + f', g + g') \in FAC_R(B, A)$, and $FAC_R(B, A)$ is an abelian group with respect to this addition. Evidently, the extension $Ext(0)$ associated with the zero factor system $(0, 0)$ is the direct sum extension

$$0 \longrightarrow A \overset{i_1}{\to} A \times B \overset{p_2}{\to} B \longrightarrow 0.$$

The following proposition is an easy straightforward verification.

Proposition 1.4.3 *Let ϕ be a map from B to A such that $\phi(0) = 0$. Define a map μ^ϕ from $B \times B$ to A by*

$$\mu^\phi(x, y) = \phi(y) - \phi(x + y) + \phi(x),$$

and a map ν^ϕ from $R \times B$ to A by

$$\nu^\phi(r, x) = \phi(rx) - r\phi(x).$$

Then $(\mu^\phi, \nu^\phi) \in FAC_R(B, A)$, and we have an exact sequence

$$0 \longrightarrow Hom_R(B, A) \overset{i_{(B,A)}}{\rightarrow} Map(B, A) \overset{\partial_{(B,A)}}{\rightarrow} FAC_R(B, A), \qquad (1.22)$$

where $Map(B, A)$ is the group of zero-preserving maps from B to A, and $\partial_{(B,A)}$ is given by $\partial_{(B,A)}(\phi) = (\mu^\phi, \nu^\phi)$. ♯

Remark 1.4.4 The map ∂ can be viewed as an obstruction map in the sense that ϕ is a homomorphism if and only if $\partial(\phi) = (0, 0)$.

Let $EXT_R(B, A)$ denote the class of equivalence classes of extensions of A by B (a prior, there is no reason to believe that it is a set). The equivalence class determined by the extension E will be denoted by $[E]$. Let E be an extension of A by B and t a section of E. As observed earlier, E is equivalent to $Ext(Fac(E, t))$, and hence $[E] = [Ext(Fac(E, t))]$. Thus, we have a surjective map $\eta_{(B,A)}$ from $FAC_R(B, A)$ to $EXT_R(B, A)$ given by $\eta_{(B,A)}(\Sigma) = [Ext(\Sigma)]$ (in particular, it follows that $EXT_R(B, A)$ is a set for each pair of modules A and B).

Proposition 1.4.5 *Let Σ and Σ' be members of $FAC_R(B, A)$. Then $\eta_{(B,A)}(\Sigma) = \eta_{(B,A)}(\Sigma')$ if and only if $\Sigma + image\partial_{(B,A)} = \Sigma' + image\partial_{(B,A)}$. Consequently, $\eta_{(B,A)}$ induces a bijective map $\overline{\eta_{(B,A)}}$ from the group $FAC_R(B, A)/image\partial_{(B,A)}$ to $EXT_R(B, A)$ defined by*

$$\overline{\eta_{(B,A)}}(\Sigma + image\partial_{(B,A)}) = [Ext(\Sigma)].$$

Proof Suppose that $\Sigma = (f, g)$ and $\Sigma' = (f', g')$. Suppose further that $\Sigma + image\partial_{(B,A)} = \Sigma' + image\partial_{(B,A)}$. Then there is a map ϕ from B to A with $\phi(0) = 0$ such that $(f - f', g - g') = (\mu^\phi, \nu^\phi)$. Thus,

$$f(x, y) = f'(x, y) + \phi(y) - \phi(x + y) + \phi(x)$$

and

$$g(r, yx) = g'(r, x) + \phi(rx) - r\phi(x)$$

for all $x, y \in B$ and $r \in R$. The extension $Ext(\Sigma)$ is given by

$$Ext(\Sigma) \equiv 0 \longrightarrow A \overset{\alpha}{\rightarrow} M \overset{\beta}{\rightarrow} B \longrightarrow 0,$$

where $M = A \times B$ with module operations given by

$$(a, x) + (b, y) = (a + b + f(x, y), x + y)$$

and

$$r(a, x) = (ra + g(r, x), rx).$$

Similarly, the extension $Ext(\Sigma')$ is given by

$$Ext(\Sigma) \equiv 0 \longrightarrow A \overset{\alpha}{\to} M' \overset{\beta}{\to} B \longrightarrow 0,$$

where $M' = A \times B$ with module operations given by

$$(a, x) + (b, y) = (a + b + f'(x, y), x + y)$$

and

$$r(a, x) = (ra + g'(r, x), rx).$$

It is easily seen that the map ρ from M to M' given by $\rho(a, x) = (a + \phi(x), x)$ is a module homomorphism. Evidently, (I_A, ρ, I_B) is an equivalence between $Ext(\Sigma)$ and $Ext(\Sigma')$. This shows that $\Sigma + image\partial_{(B,A)} = \Sigma' + image\partial_{(B,A)}$ implies that $\eta_{(B,A)}(\Sigma) = \eta_{(B,A)}(\Sigma')$. Conversely, suppose that $\eta_{(B,A)}(\Sigma) = \eta_{(B,A)}(\Sigma')$. Then there is a homomorphism ρ from M to M' such that (I_A, ρ, I_B) is an equivalence between $Ext(\Sigma)$ and $Ext(\Sigma')$. Thus, there is a map ϕ from B to A such that $\rho(a, x) = (a + \phi(x), x)$. Evidently, $\phi(0) = 0$. Since ρ is a homomorphism, $(f, g) + \partial_{(B,A)}\phi = (f', g')$. The rest is evident. ♯

Clearly, Hom_R and Map are bi-functors from the category of R-modules to the category AB of abelian groups. Let (β, α) be a morphism from (B, A) to (D, C) in $(Mod - R)^o \times Mod - R$. More explicitly, let β be a module homomorphism from D to B and α a module homomorphism from A to C. Let (f, g) be a member of $FAC_R(B, A)$. Then it can be easily seen that $(\alpha o f o (\beta \times \beta), \alpha o g o (I_R \times \beta))$ is a member of $FAC_R(D, C)$. The map FAC_R from $FAC_R(B, A)$ to $FAC_R(D, C)$ given by $FAC_R(f, g) = (\alpha o f o (\beta \times \beta), \alpha o g o (I_R \times \beta))$ is a homomorphism of groups. This defines another bi-functor FAC_R from the category of R-modules to the category AB of abelian groups. Clearly, the inclusion i defines a natural transformation from Hom_R to Map, and ∂ defines a natural transformation from Map to FAC_R. In turn, we have a bi-functor $FAC_R/image\partial$ from the category of R-modules to the category AB of abelian groups defined by $FAC_R/image\partial(B, A) = FAC_R(B, A)/image\partial_{(B,A)}$. Further, for each object (B, A) of $(Mod - R)^o \times Mod - R$, the bijective map $\overline{\eta_{(B,A)}}$ introduced in Proposition 1.4.5 induces an abelian group structure on $EXT_R(B, A)$ such that EXT_R also becomes a bi-functor from Mod $-R$ to AB, and $\overline{\eta}$ defines a natural isomorphism from $FAC_R/image\partial$ to EXT_R. ♯

Now, we describe the abelian group structure of $EXT_R(B, A)$, and also the bi-functor EXT_R independently.

Proposition 1.4.6 *Let f be a homomorphism from an R-module D to an R-module B, and*

$$E \equiv 0 \longrightarrow A \overset{\alpha}{\to} M \overset{\beta}{\to} B \longrightarrow 0$$

be an extension of A by B. Then there is an extension $f^\star(E)$ of A by D together with a morphism (I_A, μ, f) from $f^\star(E)$ to E. Further, such an extension is unique up to equivalence.

Proof We have the pullback diagram

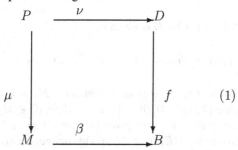

$$(1)$$

where $P = \{(m, d) \in M \times D \mid \beta(m) = f(d)\}$ is the submodule of $M \times D$, μ is the first projection, and ν is the second projection. Let χ be a homomorphism from A to P given by $\chi(a) = (\alpha(a), 0)$. Then we get an extension $f^*(E)$ of A by D given by

$$f^*(E) \equiv 0 \longrightarrow A \overset{\chi}{\to} P \overset{\nu}{\to} D \longrightarrow 0.$$

Evidently, (I_A, μ, f) is a morphism from $f^*(E)$ to E. Further, let

$$E' \equiv 0 \longrightarrow A \overset{\chi'}{\to} P' \overset{\nu'}{\to} D \longrightarrow 0$$

be an extension of A by D, and (I_A, μ', f) be a morphism from E' to E. Since

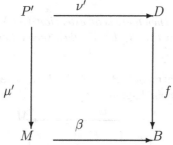

is a commutative diagram, and the diagram (1) is a pullback diagram, there is a homomorphism ϕ from P' to P such that (I_A, ϕ, I_D) is a morphism from E' to $f^*(E)$. By the five lemma this is an equivalence. ♯

Let

$$E_1 \equiv 0 \longrightarrow A \overset{\alpha_1}{\to} M_1 \overset{\beta_1}{\to} B \longrightarrow 0$$

and

$$E_2 \equiv 0 \longrightarrow A \overset{\alpha_2}{\to} M_2 \overset{\beta_2}{\to} B \longrightarrow 0$$

be equivalent extensions of A by B. Suppose that (I_A, ρ, I_B) is an equivalence from E_1 to E_2. Then the extension $f^*(E_1)$ of A by D is given by

$$f^\star(E_1) \equiv 0 \longrightarrow A \overset{\chi_1}{\to} P_1 \overset{p_2}{\to} D \longrightarrow 0,$$

and the extension $f^\star(E_2)$ of A by D is given by

$$f^\star(E_2) \equiv 0 \longrightarrow A \overset{\chi_2}{\to} P_2 \overset{p_2}{\to} D \longrightarrow 0,$$

where P_1 is the submodule of $M_1 \times D$ given by $\{(m, d) \in M_1 \times D \mid \beta_1(m) = f(d)\}$ and P_2 is the submodule of $M_2 \times D$ given by $P_2 = \{(m, d) \in M_2 \times D \mid \beta_2(m) = f(d)\}$. Clearly, the map $\overline{\rho}$ from P_1 to P_2 given by $\overline{\rho}(m, d) = (\rho(m), d)$ defines an equivalence $(I_A, \overline{\rho}, I_D)$ from $f^\star(E_1)$ to $f^\star(E_2)$. This defines a map $EXT_R(-, A)(f)$ from $EXT_R(D, A)$ to $EXT_R(B, A)$ which is given by $EXT_R(-, A)(f)([E]) = [f^\star(E)]$. It is straightforward to see that $EXT_R(-, A)(I_A) = I_{EXT_R(A,A)}$, and

$$EXT_R(-, A)(gof) = EXT_R(-, A)(f)oEXT_R(-, A)(g).$$

This ensures that $EXT_R(-, A)$ is a contra-variant functor from the category of R-modules to the category AB of abelian groups.

Dually, we have the following.

Proposition 1.4.7 *Let f be a homomorphism from an R-module A to an R-module C, and*

$$E \equiv 0 \longrightarrow A \overset{\alpha}{\to} M \overset{\beta}{\to} B \longrightarrow 0$$

be an extension of A by B. Then there is an extension $f_\star(E)$ of C by B together with a morphism (f, μ, I_B) from E to $f_\star(E)$. Further, such an extension is unique up to equivalence.

Proof Consider the quotient module $P = (C \times M)/L$, where $L = \{(f(a), -\alpha(a)) \mid a \in A\}$. We have the pushout diagram

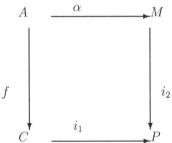

where $i_1(c) = (c, 0) + L$, and $i_2(m) = (0, m) + L$. Suppose that $(c, 0) \in L$. Then there exists an element $a \in A$ such that $c = f(a)$ and $0 = -\alpha(a)$. Since α is injective, $a = 0$. Hence $c = 0$. This shows that i_1 is injective homomorphism. Further, suppose that $(c, m) + L = (c', m') + L$. Then $m - m' = -\alpha(a) = f(a)$ for some $a \in A$. This means that $\beta(m) = \beta(m')$. In turn, we have a surjective homomorphism $\overline{\beta}$ from P to B given by $\overline{\beta}((c, m) + L) = \beta(m)$. Now,
$ker \overline{\beta} = \{(c, m) + L \mid \beta(m) = 0\}$

$$= \{(c, \alpha(a)) + L \mid c \in C, \ a \in A\}$$
$$= \{(c + f(a), 0) + L \mid c \in C, \ a \in A\}$$
$$= image \, i_1.$$

Thus, we get an extension $f_*(E)$ of C by B given by

$$f_*(E) \equiv 0 \longrightarrow C \overset{i_1}{\to} P \overset{\bar{\beta}}{\to} B \longrightarrow 0.$$

Evidently, (f, i_2, I_B) is a morphism from E to $f_*(E)$. Now, let

$$E' \equiv 0 \longrightarrow C \overset{\chi}{\to} P' \overset{\xi}{\to} B \longrightarrow 0$$

be an extension of C by B together with a morphism (f, ρ, I_B) from E to E'. We have the following commutative diagram:

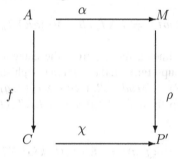

Since

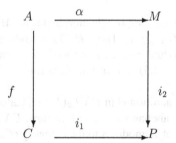

is a pushout diagram, we have homomorphism ϕ from P to P' such that (I_C, ϕ, I_B) is a morphism from $f_*(E)$ to E'. By the five lemma it is an equivalence. ♯

Let

$$E_1 \equiv 0 \longrightarrow A \overset{\alpha_1}{\to} M_1 \overset{\beta_1}{\to} B \longrightarrow 0$$

and

$$E_2 \equiv 0 \longrightarrow A \overset{\alpha_2}{\to} M_2 \overset{\beta_2}{\to} B \longrightarrow 0$$

be equivalent extensions of A by B. Suppose that (I_A, ρ, I_B) is an equivalence from E_1 to E_2. The extension $f_*(E_1)$ of C by B is given by

$$f_*(E_1) \equiv 0 \longrightarrow C \overset{i_1}{\to} P_1 \overset{\overline{\beta_1}}{\to} B \longrightarrow 0,$$

and the extension $f_*(E_2)$ of C by B is given by

$$f_*(E_2) \equiv 0 \longrightarrow C \overset{i_1}{\to} P_2 \overset{\overline{\beta_2}}{\to} B \longrightarrow 0.$$

Here P_1 is the quotient module $(C \times M_1)/L_1$, where $L_1 = \{(f(a), -\alpha_1(a)) \mid a \in A\}$, and P_2 is the quotient module $(C \times M_2)/L_2$, where $L_2 = \{(f(a), -\alpha_2(a)) \mid a \in A\}$. The maps $\overline{\beta_1}$ and $\overline{\beta_2}$ are given by $\overline{\beta_1}((c, m_1) + L_1) = \beta_1(m_1)$ and $\overline{\beta_2}((c, m_2) + L_2) = \beta_2(m_2)$ Clearly, the map $\overline{\rho}$ from P_1 to P_2 given by $\overline{\rho}((c, m) + L_1) = (c, \rho(m)) + L_2$ induces an equivalence $(I_A, \overline{\rho}, I_B)$ from $f_*(E_1)$ to $f_*(E_2)$. In turn, we have a map $EXT_R(B, -)(f)$ from $EXT_R(B, A)$ to $EXT_R(B, C)$ given by $EXT_R(B, -)(f)([E]) = [f_*(E)]$. Evidently, $EXT_R(B, -)(I_A) = I_{EXT_R(A,A)}$, and

$$EXT_R(B, -)(g \circ f) = EXT_R(B, -)(g) \circ EXT_R(B, -)(f).$$

As such, $EXT_R(B, -)$ defines a functor from the category of R-modules to the category AB of abelian groups. Further, if (f, g) is a morphism from (B, A) to (D, C) in the category $(Mod - R)^o \times Mod - R$, then $EXT_R(f, g) = EXT_R(D, -)(g) \circ EXT_R(-, A)(f)$ is a map from $EXT_R(B, A)$ to $EXT_R(D, C)$. It can be easily observed that

$$EXT_R((h, k) \circ (f, g)) = EXT_R(h, k) \circ EXT_R(f, g),$$

whenever (h, k) and (f, g) are composable morphisms in $(Mod - R)^o \times Mod - R$. Also $EXT_R(I_B, I_A) = I_{EXT_R(B,A)}$. Thus, EXT_R introduced above is a bi-functor from the category of R-modules to the category of sets. It can also be verified that $\overline{\eta}$ (introduced in Proposition 1.4.5) is a natural isomorphism from $FAC_R/image\partial$ to EXT_R.

Next, we introduce the addition $\uplus$ in $EXT_R(B, A)$, called the **Baer sum**, so that $EXT_R(B, A)$ becomes an abelian group, and, in turn, EXT_R turns out to be a bi-functor from the category of R-modules to the category of abelian groups. Let

$$E_1 \equiv 0 \longrightarrow A \overset{\alpha_1}{\to} M_1 \overset{\beta_1}{\to} B \longrightarrow 0$$

and

$$E_2 \equiv 0 \longrightarrow A \overset{\alpha_2}{\to} M_2 \overset{\beta_2}{\to} B \longrightarrow 0$$

be extensions of A by B. Consider the extension $E_1 \times E_2$ given by

$$E_1 \times E_2 \equiv 0 \longrightarrow A \times A \overset{\alpha_1 \times \alpha_2}{\to} M_1 \times M_2 \overset{\beta_1 \times \beta_2}{\to} B \times B \longrightarrow 0.$$

Let Δ_B denote the diagonal homomorphism from B to $B \times B$ given by $\Delta(b) = (b, b)$, and ∇_A denote the co-diagonal homomorphism from $A \times A$ to A given by $\nabla_A(a, b) = a + b$. This gives an extension $(\nabla_A)_\star((\Delta_B)^\star(E_1 \times E_2))$ of A by B. We denote this extension by $E_1 \uplus E_2$. Let (I_A, ϕ, I_B) be an equivalence from E_1 to E_1' and (I_A, ψ, I_B) be an equivalence from E_2 to E_2'. Clearly, $(I_{A \times A}, \phi \times \psi, I_{B \times B})$ is an equivalence from $E_1 \times E_2$ to $E_1' \times E_2'$. Hence, $(\nabla_A)_\star((\Delta_B)^\star(E_1 \times E_2))$ is equivalent to $(\nabla_A)_\star((\Delta_B)^\star(E_1' \times E_2'))$. This shows that $E_1 \uplus E_2$ is equivalent to $E_1' \uplus E_2'$. Thus, we have a binary operation $\uplus$ on $EXT_R(B, A)$ given by $[E_1] \uplus [E_2] = [E_1 \uplus E_2]$.

Finally, we show that the bijective map $\overline{\eta_{(B,A)}}$ from the group $FAC_R(B, A)/image\partial_{(B,A)}$ to the group $(EXT_R(B, A), \uplus)$ is an isomorphism. The reader may recall the earlier constructions. Let $\Sigma = (B, A, f, g)$ and $\Sigma' = (B, A, f', g')$ be members of $FAC_R(B, A)$. Then $\Sigma + \Sigma' = (B, A, f + f', g + g')$. We need to show that $[Ext(\Sigma + \Sigma')] = [Ext(\Sigma)] \uplus [Ext(\Sigma')]$. By the construction, there is an extension

$$E \equiv 0 \longrightarrow A \xrightarrow{\alpha} M \xrightarrow{\beta} B \longrightarrow 0$$

of A by B and a section t such that $(f, g) = (f^t, g^t)$, and also an extension

$$E' \equiv 0 \longrightarrow A \xrightarrow{\alpha'} M' \xrightarrow{\beta'} B \longrightarrow 0$$

of A by B and a section t' such that $(f', g') = (f^{t'}, g^{t'})$. Evidently, the commutative diagram

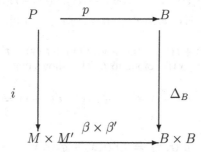

is a pullback diagram, where P is the submodule $\{(m, m') \in M \times M' \mid \beta(m) = \beta'(m')\}$ of $M \times M'$, the map p is given by $p(m, m') = \beta(m)$, and i is the inclusion map. In turn, $\Delta_B^\star(E \times E')$ is given by

$$\Delta_B^\star(E \times E') \equiv 0 \longrightarrow A \times A' \xrightarrow{\alpha \times \alpha'} P \xrightarrow{p} B \longrightarrow 0.$$

Let $\overline{P}$ denote the quotient module $(A \times P)/L$, where $L = \{(a + b, -((\alpha(a), 0) + (0, \alpha'(b)))) \mid (a, b) \in A \times A\}$. We have the following pushout diagram:

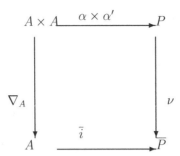

where $\bar{i}(a) = (a, (0, 0)) + L$ and $\nu((x, y)) = (0, (x, y)) + L$. In turn, $E \uplus E' = (\nabla_A)_*(\Delta_B^*(E \times E'))$ is given by

$$E \uplus E' \equiv 0 \longrightarrow A \xrightarrow{\bar{i}} \overline{P} \xrightarrow{\overline{p}} B \longrightarrow 0,$$

where $\overline{p}((a, (x, y)) + L) = p((x, y))$. We have a section s of $E \oplus E'$ which is given by $s(x) = (0, (t(x), t'(x))) + L$. Now,

$$(0, (t(x) + t(y), t'(x) + t'(y))) + L = s(x) + s(y) =$$
$$\bar{i}(f^s(x, y)) + s(x + y) = (f^s(x, y), (t(x + y), t'(x + y))) + L$$

and

$$(0, (g^t(r, x), g^{t'}(r, x))) = (0, (t(rx) - rt(x), t'(rx) - rt'(x))) + L =$$
$$s(rx) - rs(x) = \bar{i}(g^s(r, x)) = (g^s(r, x), (0, 0)) + L.$$

Hence $(-f^s(x, y), (t(x) + t(y) - t(x + y), t'(x) + t'(y) - t'(x + y)))$ and $(-g^s(r, x), (g^t(r, x), g^{t'}(r, x)))$ belong to L. This shows that

$$f^s(x, y) = t(x) + t(y) - t(x + y) + t'(x) + t'(y) - t'(x + y) = f^t(x, y) + f^{t'}(x, y),$$

and

$$g^s(r, x) = (g^t(r, x) + g^{t'}(r, x)$$

for all $x, y \in B$ and $r \in R$. It follows that $\eta_{(B,A)}(\Sigma + \Sigma') = \eta_{B,A}(\Sigma) \uplus \eta_{B,A}(\Sigma)'$. This shows that $(EXT^R(B, A), \uplus)$ is an abelian group, and $\overline{\eta_{(B,A)}}$ is natural isomorphism from $FAC_R(B, A)/image\partial_{(B,A)}$ to the group $EXT_R(B, A)$.

Since the extension associated with a trivial (zero) factor system is a split extension, it follows that $[E]$ represents 0 in $EXT_R(B, A)$ if and only if E is a split extension.

Recall that a module P over R is projective if any one (and hence all) of the following equivalent conditions is satisfied:

1. If β is a surjective homomorphism from B to C and f a homomorphism from P to C, then there exists a homomorphism $\overline{f}$ from P to B such that $\beta o \overline{f} = f$.
2. Every short exact sequence of the type

$$0 \longrightarrow A \xrightarrow{\bar{\alpha}} M \xrightarrow{\bar{\beta}} P \longrightarrow 0$$

splits.

3. P is direct summand of a free R-module.

In turn, we have the following corollary:

Corollary 1.4.8 *A module P over R is projective if and only if $EXT_R(P, A) = \{0\}$ for every module A.* ♯

Further, recall that a module I over R is injective if and only if very short exact sequence of the type

$$0 \longrightarrow I \xrightarrow{\bar{\alpha}} M \xrightarrow{\bar{\beta}} B \longrightarrow 0$$

splits.

Corollary 1.4.9 *A module I over R is injective if and only if $EXT_R(A, I) = \{0\}$ for every module A.* ♯

Also recall that every module is submodule of an injective module.

Obstructions and Extensions of Homomorphisms

Proposition 1.4.10 *Let A be a submodule of M and B be the quotient module M/A. Then a homomorphism f from A to C can be extended to a homomorphism $\bar{f}$ from M to C if and only if $f_\star(E)$ is a split exact, where E is the extension*

$$E \equiv 0 \longrightarrow A \xrightarrow{i} M \xrightarrow{\nu} B \longrightarrow 0.$$

Proof Suppose that f is extended to a homomorphism $\bar{f}$ from M to C. Then the homomorphism ϕ from M to $C \times B$ defined by $\phi(x) = (\bar{f}(x), \nu(x))$ is such that (f, ϕ, I_B) is a morphism from E to the direct sum extension

$$0 \longrightarrow C \xrightarrow{i_1} C \times B \xrightarrow{p_2} B \longrightarrow 0.$$

From Proposition 1.4.7, it follows that $f_\star(E)$ is a split extension. Conversely, suppose that we have a split extension

$$E' \equiv 0 \longrightarrow C \xrightarrow{\mu} M' \xrightarrow{\nu} B \longrightarrow 0$$

together with a morphism (f, ϕ, I_B) from E to E'. Then there is a homomorphism s from M' to C such that $s \circ \mu = I_C$. Clearly, $s \circ \phi$ is the extension of f. ♯

Dually, we have the following proposition.

Proposition 1.4.11 *Let β be a surjective homomorphism from M to B, and f be a homomorphism from D to B. Then f can be lifted to a homomorphism $\bar{f}$ from D to M in the sense that $\beta \circ \bar{f} = f$ if and only if $f^\star(E)$ is a split exact, where E is the extension*

$$E \equiv 0 \longrightarrow Ker\beta \xrightarrow{i} M \xrightarrow{\beta} B \longrightarrow 0.$$

Proof Suppose that there is a lifting $\overline{f}$ of f. Then we have a homomorphism $(I_{ker\beta}, \psi, f)$ from the split extension

$$0 \longrightarrow Ker\beta \overset{i_1}{\to} Ker\beta \times D \overset{p_2}{\to} D \longrightarrow 0$$

to E, where $\psi(a, x) = a + \overline{f}(x)$. From Proposition 1.4.6, it follows that $f^\star(E)$ is split extension. Conversely, suppose that $f^\star(E)$ splits. Let $(I_{ker\beta}, \rho, f)$ be a morphism from $f^\star(E)$ to E and t is a splitting of $f^\star(E)$. Then $\rho \circ t$ is the required lifting.   ♯

Theorem 1.4.12 *Let*

$$E \equiv 0 \longrightarrow A \overset{\alpha}{\to} B \overset{\beta}{\to} C \longrightarrow 0$$

be a short exact sequence of R-modules, and D be an R-module. Then we have the exact sequences

$$0 \longrightarrow Hom_R(C, D) \overset{Hom_R(-,D)(\beta)}{\longrightarrow} Hom_R(B, D) \overset{Hom_R(-,D)(\alpha)}{\longrightarrow}$$

$$Hom_R(A, D) \overset{\partial^E}{\longrightarrow} EXT_R(C, D) \overset{EXT_R(-,D)(\beta)}{\longrightarrow} EXT_R(B, D) \overset{EXT_R(-,D)(\alpha)}{\longrightarrow}$$

$$EXT_R(A, D)$$

and

$$0 \longrightarrow Hom_R(D, A) \overset{Hom_R(D,-)(\alpha)}{\longrightarrow} Hom_R(D, B) \overset{Hom_R(D,-)(\beta)}{\longrightarrow}$$

$$Hom_R(D, C) \overset{\partial_E}{\longrightarrow} EXT_R(D, A) \overset{EXT_R(D,-)(\alpha)}{\longrightarrow} EXT_R(D, B) \overset{EXT_R(D,-)(\beta)}{\longrightarrow}$$

$$EXT_R(D, C)$$

of abelian groups, where ∂^E is the natural connecting homomorphism given by $\partial^E(f) = [f_\star(E)]$ and ∂_E is the natural connecting homomorphism given by $\partial_E(f) = [f^\star(E)]$.

Proof We prove the exactness of the first sequence. Dual arguments will prove the exactness of the second sequence. The proofs of the exactness at $Hom_R(C, D)$ and also at $Hom_R(B, D)$ are already established in Algebra 2 (Theorem 7.2.11), and indeed, it also follows from Proposition 1.2.17 and the fact that the category of modules is an abelian category. However, for the sake of completeness, we prove it here also. Let f be a member of $ker Hom_R(-, D)(\beta)$. Then $f \circ \beta = Hom_R(-, D)(\beta)(f) = 0$. Since β is surjective, $f = 0$. This proves the exactness at $Hom_R(C, D)$. Since $Hom_R(-, D)$ is a contra-variant functor, $Hom_R(-, D)(\alpha) \circ Hom_R(-, D)(\beta) = Hom_R(-, D)(\beta \circ \alpha) = 0$. Thus, $image$ $Hom_R(-, D)(\beta)$ is contained in $ker Hom_R(-, D)(\alpha)$. Let $f \in ker Hom_R(-, D)(\alpha)$. Then $f \circ \alpha = 0$. Since $image \alpha = ker\beta$, f is zero on $ker\beta$. By the fundamental theorem of homomorphism, there is a homomorphism $\overline{f}$ from B to D such that $Hom_R(-, D)(\beta)(\overline{f}) = \overline{f} \circ \beta = f$. This proves the exactness at $Hom_R(B, D)$.

Now, we prove the exactness at $Hom_R(A, D)$. Let $f \in Hom_R(B, D)$. Then $Hom_R(-, D)(\alpha)(f) = f \circ \alpha$. Hence $\partial^E(Hom_R(-, D)(\alpha)(f)) = [(f \circ \alpha)_\star(E)] = [f_\star(\alpha_\star(E))]$. By Proposition 1.4.10, $\alpha_\star(E)$ is split exact. Hence $f_\star(\alpha_\star(E))$ is also split exact. This shows that $\partial^E(Hom_R(-, D)(\alpha)(f)) = 0$. Thus, $image\,Hom_R$ $(-, D)(\alpha)$ is contained in $ker\,\partial^E$. Let $f \in ker\,\partial^E$. Then $\partial^E(f) = [f_*(E)] = 0$. In other words, $f_*(E)$ is a split exact sequence. It follows from Proposition 1.4.10 that there is a homomorphism $\overline{f}$ from B to D such that $\overline{f}\,o\alpha = f$. Hence, $f \in image\,Hom_R(-, D)(\alpha)$. This proves the exactness at $Hom_R(A, D)$.

Next, we prove the exactness at $EXT_R(C, D)$. Let $f \in Hom_R(A, D)$. Then $\partial^E(f) = [f_*(E)]$, and $EXT_R(-, D)(\beta)([f_*(E)]) = [\beta^*(f_*(E))]$. We have a homomorphism (f, ρ, I_C) from E to $f_*(E)$. Clearly, ρ is a lifting of β. By Proposition 1.4.11, $\beta^*(f_*(E))$ is split exact. This means that $EXT_R(-, D)(\beta)(\partial^E(f)) = 0$. Thus, $image\,\partial^E$ is contained in $ker\,EXT_R(-, D)(\beta)$. Now, let $[E']$ be a member of $ker\,EXT_R(-, D)(\beta)$, where

$$E' \equiv 0 \longrightarrow D \overset{\alpha'}{\to} M \overset{\beta'}{\to} C \longrightarrow 0.$$

Then $\beta^*(E')$ is a split exact sequence. Suppose that

$$\beta^*(E') \equiv 0 \longrightarrow D \overset{\chi}{\to} L \overset{\eta}{\to} B \longrightarrow 0.$$

Let (I_D, ϕ, β) be the morphism from $\beta^*(E')$ to E', and t be a map from B to L which is a splitting of $\beta^*(E')$. Then $\beta'\phi t\alpha = \beta\alpha = 0$. From the exactness of E', we get a homomorphism f from A to D such that $\alpha'f = \phi t\alpha$. Evidently, $(f, \phi t, I_C)$ is a morphism from E to E'. By Proposition 1.4.7, $\partial^E(f) = [f_*(E)] = [E']$. This proves the exactness at $EXT_R(C, D)$.

Finally, we prove the exactness at $EXT_R(B, D)$. Since $EXT_R(-, D)$ is a contravariant functor, $EXT_R(-, D)(\alpha)o\,EXT_R(-, D)(\beta) = EXT_R(-, D)(\beta o\alpha) = 0$. This shows that $image\,EXT_R(-, D)(\beta) \subseteq ker\,EXT_R(-, D)(\alpha)$. Let $[\tilde{E}]$ be a member of $ker\,EXT_R(-, D)(\alpha)$, where

$$\tilde{E} \equiv 0 \longrightarrow D \overset{\phi}{\to} L \overset{\psi}{\to} B \longrightarrow 0.$$

Then

$$\alpha^*(\tilde{E}) \equiv 0 \longrightarrow D \overset{\phi'}{\to} L' \overset{\psi'}{\to} A \longrightarrow 0$$

splits. Let (I_D, ρ, α) be the morphism from $\alpha^*(\tilde{E})$ to $\tilde{E}$, and t be a splitting homomorphism of $\alpha^*(\tilde{E})$. Then $\psi o\rho o t = \alpha$. Let M denote the image of $\rho o t$. By the first isomorphism theorem, ψ induces a surjective homomorphism $\overline{\psi}$ from L/M to C which is given by $\overline{\psi}(x + M) = \beta(\psi(x))$. Since $\phi'(D) \cap t(A) = \{0\}$, it follows that $\phi(D) \cap M = \{0\}$. Thus, we have an injective homomorphism $\overline{\phi}$ from D to L/M given by $\overline{\phi}(y) = \phi(y) + M$, and the extension

$$\overline{E} \equiv 0 \longrightarrow D \overset{\overline{\phi}}{\to} L/M \overset{\overline{\psi}}{\to} C \longrightarrow 0$$

of D by C. Again, since (I_D, ν, β) is a morphism from $\tilde{E}$ to $\overline{E}$, we see that $EXT_R(-, D)(\beta)([\overline{E}]) = [\beta^*(\overline{E})] = [\tilde{E}]$. Evidently, $EXT_R(-, D)(\beta)([\overline{E}]) = [\tilde{E}]$. This completes the proof of the exactness of the first sequence. ♯

Proposition 1.4.13 *If R is a principal ideal domain, then the exact sequences in Theorem 1.4.12 remain exact if we add zero in the right. More explicitly, if α is an injective homomorphism from A to B and D is an R-module, then the homomorphism $EXT_R(-, D)(\beta)$ from $EXT_R(B, D)$ to $EXT_R(A, D)$ is surjective. Also if β is a surjective homomorphism from B to C and D is an R-module, then the homomorphism $EXT_R(D, -)(\alpha)$ from $EXT_R(D, B)$ to $EXT_R(D, C)$ is surjective.*

Proof Let α be an injective homomorphism from A to B and D be an R-module. Let F be a free R-module and β be a surjective homomorphism from F to B. Let $P = \beta^{-1}(\alpha(A))$. Then β induces a surjective homomorphism $\overline{\beta}$ from P to A given by $\overline{\beta}(x) = \alpha^{-1}(\beta(x))$. Evidently, $ker\beta = ker\overline{\beta}$. Put $K = ker\beta$. We have the following commutative diagram:

$$
\begin{array}{ccccccccc}
0 & \longrightarrow & K & \overset{i}{\longrightarrow} & P & \overset{\overline{\beta}}{\longrightarrow} & A & \longrightarrow & 0 \\
& & \downarrow{\scriptstyle I_K} & & \downarrow{\scriptstyle i} & & \downarrow{\scriptstyle \alpha} & & \\
0 & \longrightarrow & K & \overset{i}{\longrightarrow} & F & \overset{\beta}{\longrightarrow} & B & \longrightarrow & 0
\end{array}
$$

where rows are exact. Since submodule of a free module over a principal ideal domain is free, it follows that P and K are free. In turn, by Corollary 1.4.8, $EXT_R(P, D) = \{0\} = EXT_R(K, D)$. Using Theorem 1.4.12, we obtain the following commutative diagram:

$$
\begin{array}{ccccc}
Hom(K, D) & \overset{\partial}{\longrightarrow} & EXT_R(B, D) & \longrightarrow & 0 \\
\downarrow{\scriptstyle I_{Hom(K,D)}} & & \downarrow{\scriptstyle EXT_R(-, D)(\alpha)} & & \\
Hom(K, D) & \underset{\partial}{\longrightarrow} & EXT_R(A, D) & \longrightarrow & 0
\end{array}
$$

where the rows are exact. Evidently, both the connecting homomorphisms ∂ are surjective, and the left-hand vertical arrow is bijective. This shows that $EXT_R(-, D)(\alpha)$ is surjective. The second part can also be established similarly. ♯

Universal Coefficient Theorem for Co-homology

Let

$$X \equiv \overset{d_{n+2}}{\to} X_{n+1} \overset{d_{n+1}}{\to} X_n \overset{d_n}{\to} X_{n-1} \overset{d_{n-1}}{\to}$$

be a chain complex of R-modules, and A be an R-module. The co-chain complex $Hom(X, A)$ of the chain complex X with coefficient in A is given by

$$Hom(X, A) \equiv \overset{\delta^{n-2}}{\to} Hom(X_{n-1}, A) \overset{\delta^{n-1}}{\to} Hom(X_n, A) \overset{\delta^n}{\to} Hom(X_{n+1}, A) \overset{\delta^{n+1}}{\to},$$

where $\delta^n(f) = (-1)^{n+1}d_{n+1}^\star(f) = (-1)^{n+1} f o d_{n+1}$, $C_n(X)$ stands for the module of n-cycles of X, and $B_n(X)$ stands for the module of n-boundaries of X. Thus, the nth homology $H_n(X) = C_n(X)/B_n(X)$. Also $C^n(X, A)$ denotes the module of n co-cycles of $Hom(X, A)$, and $B^n(X, A)$ denotes the module of n co-boundaries of

$Hom(X, A)$. Thus, the nth co-homology $H^n(X, A)$ of X with coefficient in A is given by $H^n(X, A) = C^n(X, A)/B^n(X, A)$. Note that $C^n(-, -)$, $B^n(-, -)$, and $H^n(-, -)$ are all functors from $(CH_R)^o \times Mod - R)$ to the category AB of abelian groups (CH_R denotes the category of chain complexes of R-modules).

Let f be a member of $C^n(X, A)$. Then f is a homomorphism from X_n to A such that $0 = \delta^n(f) = (-1)^{n+1} f o d_{n+1}$. This means that $B_n(X) \subseteq ker f$. Hence, f induces a homomorphism $\overline{f}$ from $H_n(X)$ to A given by $\overline{f}(x + B_n(X)) = f(x)$. Further, if $f \in B^n(X, A)$, then $f = (-1)^n g o d_n$ for some $g \in Hom(X_{n-1}, A)$. But, then $f(x) = 0$ for all $x \in C_n(X)$. This gives us a *natural homomorphism* μ from $H^n(X, A)$ to $Hom(H_n(X), A)$ which is given by $\mu(f + B^n(X, A))(x + B_n(X)) = f(x)$.

Theorem 1.4.14 (Universal coefficient theorem) *Let X be a chain complex of free R-modules over a principal ideal domain R. Let A be an R-module. Then for each n, we have a split exact sequence*

$$0 \longrightarrow EXT_R(H_{n-1}(X), A) \xrightarrow{\rho} H^n(X, A) \xrightarrow{\mu} Hom(H_n(X), A) \longrightarrow 0,$$

where μ is the map given by $\mu(f + B_n(X, A))(x + B_n(X)) = f(x)$ and ρ is to be introduced during the proof. The splitting is natural in A.

Proof For simplicity, let us denote by C_n (instead of $C_n(X)$) the module of n-cycles, and by B_n the module of n-boundaries. Further, denote X_n/C_n by D_n, and $H_n(X)$ by H_n. Then D_n is isomorphic to B_{n-1}. Since X_{n-1} is free and submodule of a free module over a principal ideal domain is free, D_n is free for all n. In turn, $EXT_R(D_n, A) = 0$ for all n (Corollary 1.4.8). Also C_n is free, and so $EXT_R(C_n, A) = 0$. We have the following two short exact sequences:

$$E \equiv 0 \longrightarrow C_n \xrightarrow{i} X_n \xrightarrow{\nu} D_n \longrightarrow 0$$

and

$$E' \equiv 0 \longrightarrow D_{n+1} \xrightarrow{\widetilde{d_{n+1}}} C_n \xrightarrow{\widetilde{\nu}} H_n \longrightarrow 0,$$

where ν and $\widetilde{\nu}$ represent the respective quotient maps and $\widetilde{d_{n+1}}$ is the homomorphism induced by d_{n+1}. Applying Theorem 1.4.12 to these two exact sequences and fitting the co-chain complex $Hom(X, A)$ suitably in between, we get the following commutative diagram with exact rows and columns except the middle row:

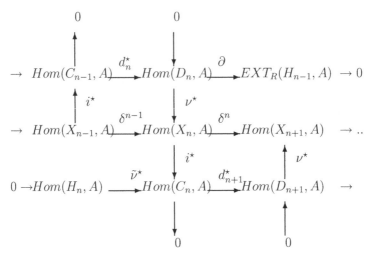

We chase the above diagram to establish the result. Let f be a homomorphism from $H_n = C_n/B_n$ to A. Then, we have a homomorphism h from C_n to A given by $h(x) = f(x + B_n)$, and $\tilde{\nu}^\star(f) = h$. Since $i^\star$ is surjective, there is a homomorphism g from X_n to A such that $i^\star(g) = g/C_n = h = \tilde{\nu}^\star(f)$. Now,

$$\delta^n(g) = \nu^\star d_{n+1}^\star i^\star(g) = \nu^\star d_{n+1}^\star \tilde{\nu}^\star(f) = 0.$$

Hence $g \in C^n(X, A)$. Clearly, $\mu(g + B^n(X, A))(x + B_n) = h(x) = f(x + B_n)$ for all $x \in C_n$. Thus, $\mu(g + B^n(X, A)) = f$. This shows that μ is surjective. Further, since the middle column is split exact, there is a homomorphism t from $Hom(C_n, A)$ to $Hom(X_n, A)$ such that $i^\star ot = I_{Hom(C_n, A)}$. In turn, we have a homomorphism s from $Hom(H_n, A)$ to $H^n(X, A)$ given by $s(f) = t\tilde{\nu}^\star(f) + B^n(X, A)$. Evidently, s is natural in second component and $\mu os = I_{Hom(H_n, A)}$.

Now, we introduce the homomorphism ρ and establish the exactness at EXT_R $(H_{n-1}(X), A)$ and also at $H^n(X, A)$. To define ρ, let $[E] \in EXT_R(H_{n-1}, A)$. Since ∂ is surjective, there is an element $f \in Hom(D_n, A)$ such that $\partial f = [E]$. Now, from the commutativity of the two bottom rows, $\delta^n = \nu^\star d_{n+1}^\star i^\star$. Hence $\delta^n \nu^\star(f) = \nu^\star d_{n+1}^\star i^\star \nu^\star(f) = 0$. This shows that $\nu^\star(f) \in C^n(X, A)$. Further, if g is another element of $Hom(D_n, A)$ such that $\partial(g) = [E] = \partial(f)$, then $\partial(g - f) = 0$. By the exactness of the top row, there is an element $h \in Hom(C_{n-1}, A)$ such that $d_n^\star(h) = g - f$. Also, since $i^\star$ is surjective, there is an element $k \in Hom(X_{n-1}, A)$ such that $i^\star(k) = h$. Hence $\nu^\star(g - f) = \nu^\star(d_n^\star i^\star(k)) = \delta^{n-1}(k)$. Thus, $\nu^\star(g) + B^n(X, A) = \nu^\star(f) + B^n(X, A)$. This ensures that we have a map ρ from $EXT_R(H_{n-1}, A)$ to $H^n(X, A)$ defined by $\rho([E]) = \nu^\star(f) + B^n(X, A)$, where $\partial(f) = [E]$. Evidently, ρ is a homomorphism. This introduces the homomorphism ρ.

Suppose that $\rho([E]) = 0$. Then there is an element $f \in Hom(D_n, A)$ such that $\partial(f) = [E]$ and $\nu^\star(f) \in B^n(X, A)$. Hence, there is an element $k \in Hom(X_{n-1}, A)$ such that $\delta^{n-1}(k) = \nu^\star(f)$. But, then $\nu^\star(d_n^\star i^\star(k)) = \nu^\star(f)$. Since $\nu^\star$ is injective,

$d_n^\star(i^\star(k)) = f$. In turn, it follows that $[E] = \partial f = \partial d_n^\star(i^\star(k)) = 0$. This shows that ρ is injective.

Finally, we prove the exactness at $H^n(X, A)$. Let $[E] \in EXT_R(H_{n-1}, A)$. Then by definition, $\rho([E]) = \nu^\star(f) + B^n(X, A)$, where $\partial(f) = [E]$. Again, by definition $\mu(\rho([E])) = h$, where $h \in Hom(H_n, A)$ such that $i^\star(\nu^\star(f)) = \nu^\star(h)$ (the reader may easily distinguish the two different $\nu^\star$). Since $i^\star \nu^\star = 0$, and $\tilde{\nu}^\star$ is injective, it follows that $\mu(\rho([E])) = h = 0$. This shows that $image\,\rho \subseteq ker\,\mu$. Next, let $g + B^n(X, A)$ be a member of $ker\,\mu$, where $g \in C^n(X, A)$. Then by definition of μ, $0 = \mu(g + B^n(X, A)) = i^\star(g)$. Since the middle column is exact, there is an element $f \in Hom(D_n, A)$ such that $\nu^\star(f) = g$. Take $[E] = \partial f$. By definition $\rho([E]) = g + B^n(X, A)$. This completes the proof of the universal coefficient theorem. ♯

Corollary 1.4.15 *If X is a chain complex of vector spaces over a field F and V a vector space over F, then $H^n(X, V)$ is isomorphic to $Hom(H_n(X), V)$.*

Proof The result follows, since every module over a field is free. ♯

Remark 1.4.16 We may observe that the proof of the universal coefficient theorem goes well if X is a chain complex of R-modules such that C_n and B_n are projective for all n.

Recall (see Algebra 2, Chap. 9) that a ring R is semi-simple ring if and only if all R-modules are projective (or equivalently, injective). For example, the group algebra $F(G)$, where G is a finite group such that characteristic of F does not divide $|G|$. Thus, we have the following corollary.

Corollary 1.4.17 *Let X be a chain complex of R-modules, where R is semi-simple. Let A be an R-module. Then $H^n(X, A)$ is isomorphic to $Hom(H_n(X), A)$.* ♯

Exercises

1.4.1 Compute $EXT_{\mathbb{Z}}(\mathbb{Z}, \mathbb{Z})$, $EXT_{\mathbb{Z}}(\mathbb{Z}_m, \mathbb{Z})$, $EXT_{\mathbb{Z}}(\mathbb{Z}, \mathbb{Z}_m)$, and $EXT_{\mathbb{Z}}(\mathbb{Z}_n, \mathbb{Z}_m)$.

1.4.2 Show that
(i) $EXT_R(\oplus \Sigma_{\alpha \in \Lambda} A_\alpha, C) \approx \prod_{\alpha \in \Lambda} EXT_R(A_\alpha, C)$ and
(ii) $EXT_R(A, \prod_{\alpha \in \Lambda} B_\alpha) \approx \prod_{\alpha \in \Lambda} EXT_R(A, B_\alpha)$.

1.4.3 Compute $EXT_{\mathbb{Z}}(\mathbb{Z}_m, A)$ and $EXT_{\mathbb{Z}}(A, \mathbb{Z}_m)$ for an abelian group A.

1.4.4 Show that $EXT_{\mathbb{Z}}(\mathbb{Q}/\mathbb{Z}, \mathbb{Z})$ is an extension of $\mathbb{Z}$ by $EXT_R(Q, \mathbb{Z})$.

1.4.5 Establish the exactness of the following sequence:

$$0 \longrightarrow Hom_{\mathbb{Z}}(\mathbb{Q}, \mathbb{Q}) \xrightarrow{\nu_*} Hom(\mathbb{Q}, \mathbb{Q}/\mathbb{Z}) \xrightarrow{\partial} EXT_{\mathbb{Z}}(Q, \mathbb{Z}) \longrightarrow 0.$$

Assuming that $EXT_{\mathbb{Z}}(\mathbb{Q}, \mathbb{Z}) \approx \mathbb{R}$, show that $Hom(\mathbb{Q}, \mathbb{Q}/\mathbb{Z}) \approx \mathbb{R}$.

1.4.6 Suppose that $Hom(A, \mathbb{Z}) = \{0\} = EXT_{\mathbb{Z}}(A, \mathbb{Z})$. Show that A is the trivial group.

1.4.7 Prove the second part of Theorem 1.4.12.

Chapter 2
Homological Algebra 2, Derived Functors

In this chapter, we introduce the concept of derived functors in an abelian category and develop its basic theory which is essential for the subsequent developments. The n-fold extension functors EXT_R^n and the functors Tor_n^R are introduced as derived co-homology and homology functors. We also establish the Kunneth formula and conclude the chapter with a basic introduction to spectral sequences.

2.1 Resolutions and Extensions

Let Σ be an abelian category. A nonnegative chain complex

$$P \equiv \cdots \overset{d_{n+1}}{\to} P_n \overset{d_n}{\to} P_{n-1} \overset{d_{n-1}}{\to} \cdots \overset{d_1}{\to} P_1 \overset{d_1}{\to} P_0 \longrightarrow 0$$

in Σ is called an **acyclic chain complex** if $H_n(P) = \{0\}$ for all $n \geq 1$. It is said to be a projective (free) chain complex if each P_i is projective (free). Dually, a nonnegative co-chain complex

$$I \equiv 0 \longrightarrow I^0 \overset{d^0}{\to} I^1 \overset{d^1}{\to} I^2 \overset{d^2}{\to} \cdots \overset{d^{n-1}}{\to} I^n \overset{d^n}{\to} \cdots$$

in Σ is called an **acyclic co-chain complex** if $H^n(I) = \{0\}$ for all $n \geq 1$. It is said to be an injective co-chain complex if each I^i is injective.

If P is an acyclic chain complex in Σ, then

$$P \equiv \cdots \overset{d_{n+1}}{\to} P_n \overset{d_n}{\to} P_{n-1} \overset{d_{n-1}}{\to} \cdots \overset{d_2}{\to} P_1 \overset{d_1}{\to} P_0 \overset{\nu}{\to} H_0(P) \longrightarrow 0$$

is exact. Further, if B is an object in Σ and there is an isomorphism η from $H_0(P)$ to B, then we have an exact sequence

© Springer Nature Singapore Pte Ltd. 2021
R. Lal, *Algebra 3*, Infosys Science Foundation Series,
https://doi.org/10.1007/978-981-33-6326-7_2

$$P \equiv \cdots \overset{d_{n+1}}{\to} P_n \overset{d_n}{\to} P_{n-1} \overset{d_{n-1}}{\to} \cdots \overset{d_2}{\to} P_1 \overset{d_1}{\to} P_0 \overset{\epsilon}{\to} B \longrightarrow 0,$$

where $\epsilon = \eta o \nu$. This exact sequence is called a **resolution** of B. If in addition each P_i is projective (free), then we say that P is a **projective (free) resolution** of B. We also express it by saying that P is an acyclic projective (free) chain complex over B or the acyclic projective (free) chain complex P is a projective resolution of B.

If I is an acyclic co-chain complex in Σ, then

$$I \equiv 0 \longrightarrow H^0(Q) \overset{i}{\to} I^0 \overset{d^0}{\to} I^1 \overset{d^1}{\to} \cdots I^2 \overset{d^2}{\to} \cdots \overset{d^{n-1}}{\to} I^n \overset{d^n}{\to} \cdots$$

is exact. Further, if A is an object in Σ and there is an isomorphism ρ from A to $H^0(I)$, then we have an exact sequence

$$I^A \equiv 0 \longrightarrow A \overset{\tau}{\to} I^0 \overset{d^0}{\to} I^1 \overset{d^1}{\to} \cdots I^2 \overset{d^2}{\to} \cdots \overset{d^{n-1}}{\to} I^n \overset{d^n}{\to} \cdots,$$

where $\tau = io\rho$. This exact sequence is called a **co-resolution** of A. If in addition each I^i is injective, then we say that I is an **injective co-resolution** of A. We also express it by saying that I is an acyclic injective co-chain complex over A or the acyclic injective co-chain complex I is an injective resolution of A.

An abelian category Σ is said to have enough projectives if for every object B in Σ, there is a projective object P in Σ together with an epimorphism β from P to B. Dually, Σ is said to have enough injectives if for every object A in Σ, there is an injective object I in Σ together with a monomorphism α from A to I. Thus, the category Mod-R of R-modules has enough projectives as well as enough injectives (see Sect. 7.2 of Algebra 2). An abelian category need not have enough projectives (injectives) (see Exercises 2.1.3–2.1.5).

Theorem 2.1.1 *Let Σ be an abelian category with enough projectives. Then for every object B in Σ, there is a projective resolution of B. Dually, if Σ has enough injectives, then for every object A in Σ, there is an injective co-resolution over A.*

Proof Suppose that σ has enough projectives, and B is an object of Σ. Using induction, we construct projective objects P_n together with morphism d_n from P_n to P_{n-1} so that

$$\cdots \overset{d_{n+1}}{\to} P_n \overset{d_n}{\to} P_{n-1} \overset{d_{n-1}}{\to} \cdots \overset{d_2}{\to} P_1 \overset{d_1}{\to} P_0 \overset{\epsilon}{\to} B \longrightarrow 0$$

is a projective resolution of B. Since Σ has enough projectives, there is a projective object P_0 together with an epimorphism ϵ from P_0 to B. Let $K_0 \overset{\epsilon_0}{\to} P_0$ be a kernel of ϵ. Again, since Σ has enough projectives, we have a projective object P_1 and an epimorphism ϵ_1 from P_1 to K_0. Take $d_1 = \epsilon_0\epsilon_1$. Evidently, $imaged_1 = ker\epsilon$, and

$$P_1 \overset{d_1}{\to} P_0 \overset{\epsilon}{\to} B \longrightarrow 0$$

is exact. Now, suppose that P_n and d_n have already been constructed. Let $K_n \xrightarrow{\epsilon_n} P_n$ be a kernel of d_n. Since Σ has enough projectives, there is a projective object P_{n+1} and an epimorphism ϵ_{n+1} from P_{n+1} to K_n. Take $d_{n+1} = \epsilon_n \epsilon_{n+1}$. This proves the existence of a projective resolution. Dually, we can prove the second statement. ♯

Theorem 2.1.2 *Let*

$$P \equiv \cdots \xrightarrow{d_{n+1}} P_n \xrightarrow{d_n} P_{n-1} \xrightarrow{d_{n-1}} \cdots \xrightarrow{d_2} P_1 \xrightarrow{d_1} P_0 \xrightarrow{\epsilon} B \longrightarrow 0$$

be a projective resolution of B. Let

$$X \equiv \cdots \xrightarrow{\alpha_{n+1}} X_n \xrightarrow{\alpha_n} X_{n-1} \xrightarrow{\alpha_{n-1}} \cdots \xrightarrow{\alpha_2} X_1 \xrightarrow{\alpha_1} X_0 \xrightarrow{\eta} C \longrightarrow 0$$

be a resolution of C. Let ϕ be a morphism from B to C. Then ϕ can be lifted to a chain transformation $f = \{f_n \mid n \in \mathbb{N} \bigcup \{0\}\}$ from P to X which is unique up to chain homotopy.

Proof We construct a chain transformation $f = \{f_n \mid n \in \mathbb{N} \bigcup \{0\}\}$ by induction on n. Since P_0 is projective, and η is an epimorphism from X_0 to C, there is a morphism f_0 from P_0 to X_0 such that $\phi\epsilon = \eta f_0$. By Proposition 1.2.7,

$$X_1 \xrightarrow{\alpha_1} X_0 = X_1 \xrightarrow{\rho_1} K_0 \xrightarrow{\eta_1} X_0,$$

where ρ_1 is an epimorphism and η_1 is the image of α_1. Since X is exact, $image \alpha_1$ is $ker \eta$. Thus, $K_0 \xrightarrow{\eta_1} X_0 = ker \eta$. Further, $\eta f_0 d_1 = \phi\epsilon d_1 = 0$. Hence, there is a morphism ξ_1 from P_1 to K_0 such that $\eta_1 \xi_1 = f_0 d_1$. Since P_1 is projective and $X_1 \xrightarrow{\rho_1} K_0$ is an epimorphism, there is a morphism f_1 from P_1 to X_1 such that $\rho_1 f_1 = \xi_1$. In turn, $\alpha_1 f_1 = \eta_1 \rho_1 f_1 = \eta_1 \xi_1 = f_0 d_1$. This completes the construction of f_1.

Assume that f_n has already been defined, $n \geq 1$. Again, by Proposition 1.2.7,

$$X_{n+1} \xrightarrow{\alpha_{n+1}} X_n = X_{n+1} \xrightarrow{\rho_{n+1}} K_n \xrightarrow{\eta_{n+1}} X_n,$$

where ρ_{n+1} is an epimorphism and η_{n+1} is the image of α_{n+1}. Since X is exact, $image \alpha_{n+1}$ is $ker \alpha_n$. Thus, $K_n \xrightarrow{\eta_{n+1}} X_n = ker \alpha_n$. Further, by the induction assumption, $\alpha_n f_n d_{n+1} = f_{n-1} d_n d_{n+1} = 0$. Hence, there is a morphism ξ_{n+1} from P_{n+1} to K_n such that $\eta_{n+1} \xi_{n+1} = f_n d_{n+1}$. Since $X_{n+1} \xrightarrow{\rho_{n+1}} K_n$ is an epimorphism and P_{n+1} is projective, there is a morphism f_{n+1} from P_{n+1} to X_{n+1} such that $\rho_{n+1} f_{n+1} = \xi_{n+1}$. In turn,

$$\alpha_{n+1} f_{n+1} = \eta_{n+1} \rho_{n+1} f_{n+1} = \eta_{n+1} \xi_{n+1} = f_n d_{n+1}.$$

This shows the existence of a required chain transformation.

Let $g = \{g_n \mid n \in \mathbb{N} \bigcup \{0\}\}$ be another chain transformation from P to X. We have to show the existence of a chain homotopy $s = \{s_n \mid n \in \mathbb{N} \bigcup \{0\}\}$ from f to g. Again, we do it by induction on n. Now, $\eta f_0 = \phi\epsilon = \eta g_0$. Thus, $\eta(f_0 - g_0) = 0$. Since

$K_0 \xrightarrow{\eta_1} X_0 = ker\eta$, there is a morphism t_0 from P_0 to K_0 such that $\eta_1 t_0 = f_0 - g_0$. Since ρ_1 is an epimorphism from X_1 to K_0 and P_0 is projective, there is a morphism s_0 from P_0 to X_1 such that $\rho_1 s_0 = t_0$. In turn,

$$\alpha_1 s_0 = \eta_1 \rho_1 s_0 = \eta_1 t_0 = f_0 - g_0.$$

This completes the construction of s_0. Further, $\alpha_1 f_1 = f_0 d_1$ and $\alpha_1 g_1 = g_0 d_1$. Hence $\alpha_1(f_1 - g_1) = (f_0 - g_0)d_1 = \alpha_1 s_0 d_1$. This shows that $\alpha_1(f_1 - g_1 - s_0 d_1) = 0$. Suppose that

$$X_2 \xrightarrow{\alpha_2} X_1 = X_2 \xrightarrow{\rho_2} K_1 \xrightarrow{\eta_2} X_1,$$

where ρ_2 is an epimorphism and $K_1 \xrightarrow{\eta_2} X_1 = ker\alpha_1$. Hence, there is a morphism t_1 from P_1 to K_1 such that $\eta_2 t_1 = f_1 - g_1 - s_0 d_1$. Since P_1 is projective and $X_2 \xrightarrow{\rho_2} K_1$ is an epimorphism, there is a morphism s_1 from P_1 to X_2 such that $\rho_2 s_1 = t_1$. In turn,

$$\alpha_2 s_1 = \eta_2 \rho_1 s_1 = \eta_2 t_1 = f_1 - g_1 - s_0 d_1.$$

Equivalently, $\alpha_2 s_1 + s_0 d_1 = f_1 - g_1$.

Assume that s_m, $m \leq n - 1$ has already been introduced with the required properties. Imitating the construction of s_1, we can construct s_n such that $\alpha_n s_n + s_{n-1} d_n = f_n - g_n$. This shows the existence of a chain homotopy from f to g. ♯

The following theorem is the dual of the above theorem, and it can be proved by using dual of the arguments used in the proof of the above theorem.

Theorem 2.1.3 *Let*

$$I \equiv 0 \longrightarrow A \xrightarrow{\tau} I^0 \xrightarrow{d^0} I^1 \xrightarrow{d^1} \cdots I^2 \xrightarrow{d^2} \cdots \xrightarrow{d^{n-1}} I^n \xrightarrow{d^n} \cdots$$

be an injective co-resolution of A. Let

$$Y \equiv 0 \longrightarrow D \xrightarrow{\eta} Y^0 \xrightarrow{\gamma^0} Y^1 \xrightarrow{\gamma^1} \cdots Y^2 \xrightarrow{\gamma^2} \cdots \xrightarrow{\gamma^{n-1}} Y^n \xrightarrow{\gamma^n} \cdots$$

be a co-resolution of D. Let ψ be a homomorphism from D to A. Then ψ can be extended to a co-chain transformation $f = \{f_n \mid n \in \mathbb{N} \bigcup \{0\}\}$ from Y to I which is unique up to co-chain homotopy. ♯

Corollary 2.1.4 *Any two projective (injective) resolutions (co-resolutions) of B (A) are chain (co-chain) equivalent.*

Proof Let P and P' (I and I') be two projective (injective) resolutions (co-resolutions) of B (A). From Theorem 2.1.2 (Theorem 2.1.3), there are chain (co-chain) transformations f and f' from P (I) to P' (I') and from P' (I') to P (I), respectively. Then $f' \circ f$ and I_P (I_I) are two chain (co-chain) transformations from P (I) to P (I). Again, from Theorem 2.1.2 (Theorem 2.1.3), $f' \circ f$ is chain (co-chain) homotopic to I_P (I_I). Similarly, $f \circ f'$ is chain (co-chain) homotopic to $I_{P'}$ ($I_{I'}$). ♯

Few remarks are in offing:

Remark 2.1.5 1. Let Σ be an abelian category with enough projectives. Let $\wp(\Sigma)$ denote the category whose objects are positive projective acyclic chain complexes in Σ, and morphisms are chain homotopy classes of chain transformations. Theorem 2.1.2 asserts that the functor H_0 from $\wp(\Sigma)$ to Σ gives an equivalence between the category $\wp(\Sigma)$ and Σ. More explicitly, using the axiom of choice, we have an association Γ from $Obj\,\Sigma$ to $Obj\,\wp(\Sigma)$ such that $\Gamma(B)$ is an acyclic projective chain complex over B. Then from Theorem 2.1.2, for any morphism β from B to D, we have a unique chain homotopy class $\Gamma(\beta)$ of chain transformations from $\Gamma(B)$ to $\Gamma(D)$ which lifts β. This defines a functor Γ from Σ to $\wp(\Sigma)$ such that $H_0 o \Gamma$ and $\Gamma o H_0$ are naturally equivalent to the corresponding identity functors.

2. We have the functor $Hom(-, -)$ from the category $\wp(\Sigma)^0 \times \Sigma$ to the category of positive co-chain complexes of abelian groups defined as follows:

$$Hom(P, A) \equiv 0 \longrightarrow Mor_\Sigma(P_0, A) \xrightarrow{d_1^\star} Mor_\Sigma(P_1, A) \xrightarrow{d_2^\star} \cdots \xrightarrow{d_n^\star}$$

$$Mor_\Sigma(P_n, A) \xrightarrow{d_{n+1}^\star} \cdots,$$

where

$$P \equiv \cdots \xrightarrow{d_{n+1}} P_n \xrightarrow{d_n} P_{n-1} \xrightarrow{d_{n-1}} \cdots \xrightarrow{d_2} P_1 \xrightarrow{d_1} P_0 \longrightarrow 0$$

is a positive acyclic projective chain complex in Σ. Next, if f is a chain transformation from P' to P, and α is a morphism from A to C, then $f^\star = \{f_n^\star \mid n \in \mathbb{N} \bigcup \{0\}\}$ is a co-chain transformation from $Hom(P, A)$ to $Hom(P', C)$.

3. In turn, for each $n \geq 0$, we have a functor $H^n(-, -)$ from $\wp_\Sigma^0 \times \Sigma$ to the category AB of abelian groups. Using the equivalence given in 1, we obtain a bi-functor $H^n(-, -)$ from the category Σ to AB. In particular, for each n, we have a bi-functor $H_\Sigma^n(-, -)$ from the category Σ to the category AB of abelian groups given by $H_\Sigma^n(B, A) = H^n(\Gamma(B), A)$, where Γ is as in 1. In particular, if we fix an object A in Σ, then we have contra-variant functors $H_\Sigma^n(-, A)$ from Σ to AB, and if we fix an object B in Σ, then we get a functor $H_\Sigma^n(B, -)$ from Σ to AB. If Σ is the category of R-modules, then we denote these functors by $H_R^n(B, A)$.

Dually,

4. Let Σ be an abelian category with enough injectives. Let $\Im(\Sigma)$ denote the category whose objects are positive injective acyclic co-chain complexes in Σ, and morphisms are homotopy classes of co-chain transformations. Theorem 2.1.3 asserts that the functor H^0 from $\Im(\Sigma)$ to Σ gives an equivalence between the category $\Im(\Sigma)$ and Σ. More explicitly, using the axiom of choice, we have an association Λ from $Obj\,\Sigma$ to $Obj\,\Im(\Sigma)$ such that $\Lambda(A)$ is an acyclic injective positive co-chain complex over A. Then from Theorem 2.1.3, for any morphism α from A to C, we have a unique homotopy class $\Lambda(\alpha)$ of co-chain transformations from $\Lambda(A)$ to $\Lambda(C)$ which extends α. This defines a functor Λ from Σ to $\Im(\Sigma)$ such that $H^0 o \Lambda$ and $\Lambda o H^0$ are naturally isomorphic to the corresponding identity functors.

5. Further, we have the functor $Hom(-, -)$ from the category $\Sigma^0 \times \Im(\Sigma)$ to the
category of positive co-chain complexes of abelian groups defined as follows:

$$0 \longrightarrow Hom(B, I) \equiv 0 \longrightarrow Mor_\Sigma(B, I^0) \overset{d_*^0}{\to} Mor_\Sigma(B, I^1) \overset{d_*^1}{\to} \cdots \overset{d_*^n}{\to}$$

$$Mor_\Sigma(B, I^{n+1}) \overset{d_*^{n+1}}{\to} \cdots ,$$

where

$$I \equiv 0 \longrightarrow I^0 \overset{d^0}{\to} I^1 \overset{d^1}{\to} \cdots \overset{d^{n-1}}{\to} I^n \overset{d^n}{\to} \cdots$$

is a positive acyclic injective co-chain complex in Σ. Next, if f is a co-chain transformation from I to I' and α is a morphism from A to C, then $f_* = \{f_*^n \mid n \in \mathbb{N} \bigcup \{0\}\}$
is a co-chain transformation from $Hom(B, I)$ to $Hom(B, I')$.

6. In turn, for each $n \geq 0$, we have a functor $H^n(-, -)$ from $\Sigma^0 \times \Im(\Sigma)$ to the
category AB of abelian groups. Using the equivalence given in 3, we obtain another
bi-functor $\overline{H^n}(-, -)$ from the category Σ to AB. In particular, for each n, we have
a bi-functor $\overline{H_\Sigma^n}(-, -)$ from the category Σ to the category AB of abelian groups
given by $\overline{H_\Sigma^n}(B, A) = H^n(B, \Lambda(A))$, where Λ is as in 3. In particular, if we fix an
object A in Σ, then we have a contra-variant functor $\overline{H_\Sigma^n}(-, A)$ from Σ to AB, and
if we fix an object B in Σ, then we get a functor $\overline{H_\Sigma^n}(B, -)$ from Σ to AB. If Σ is
the category of R-modules, then we denote these functors by $\overline{H_R^n}(B, A)$.

To compute $H_\Sigma^n(B, A)$, we need to take a suitable acyclic projective chain complex

$$P \equiv \cdots \overset{d_{n+1}}{\to} P_n \overset{d_n}{\to} P_{n-1} \overset{d_{n-1}}{\to} \cdots \overset{d_2}{\to} P_1 \overset{d_1}{\to} P_0 \longrightarrow 0$$

over B, and then compute $H^n(Hom(P, A))$. Dually, to compute $\overline{H_\Sigma^n}(B, A)$, we need
to take a suitable positive acyclic injective co-chain complex

$$I \equiv 0 \longrightarrow I^0 \overset{d^0}{\to} I^1 \overset{d^1}{\to} \cdots \overset{d^{n-1}}{\to} I^n \overset{d^n}{\to} \cdots$$

on A, and then compute $H^n(Hom(B, I))$. Soon we shall show that if Σ has enough
projective and also enough injectives, then the bi-functors $H_\Sigma^n(-, -)$ and $\overline{H_\Sigma^n}(-, -)$
are naturally isomorphic.

Proposition 2.1.6 *Suppose that the abelian category Σ has enough projectives.
Then $H_\Sigma^0(B, A)$ is naturally isomorphic to $Mor_\Sigma(B, A)$. If Σ has enough injectives, then $\overline{H_\Sigma^0}(B, A)$ is naturally isomorphic to $Mor_\Sigma(B, A)$. In particular, if Σ has
enough projective and also enough injectives, then $H_\Sigma^0(B, A)$ is naturally isomorphic
to $\overline{H_\Sigma^0}(B, A)$.*

Proof Suppose that the choice $\Gamma(B)$ of acyclic projective chain complex over B is

$$\Gamma(B) \equiv \cdots \overset{d_{n+1}}{\to} P_n \overset{d_n}{\to} P_{n-1} \overset{d_{n-1}}{\to} \cdots \overset{d_2}{\to} P_1 \overset{d_1}{\to} P_0 \longrightarrow 0.$$

Then

$$Hom(\Gamma(B), A) \equiv 0 \longrightarrow Mor_\Sigma(P_0, A) \overset{d_1^*}{\to} Mor_\Sigma(P_1, A) \overset{d_2^*}{\to} \cdots \overset{d_n^*}{\to}$$

$$Mor_\Sigma(P_n, A) \overset{d_{n+1}^*}{\to} \cdots .$$

Hence $H_\Sigma^0(B, A) = ker d_1^*$. Since the functor $Mor_\Sigma(-, A)$ is a left exact functor from Σ^o to AB, and

$$P_1 \overset{d_1}{\to} P_0 \overset{\epsilon}{\to} B \longrightarrow 0$$

is exact, we have the exact sequence

$$0 \longrightarrow Mor_\Sigma(B, A) \overset{\epsilon^*}{\to} Mor_\Sigma(P_0, A) \overset{d_1^*}{\to} Mor_\Sigma(P_1, A).$$

Hence, $ker d_1^*$ is naturally isomorphic to $Mor_\Sigma(B, A)$. The rest of the statements can be proved by using the dual arguments. ♯

Proposition 2.1.7 *Let P be a projective object and I be an injective object in Σ. Then $H_\Sigma^n(P, A) = 0 = H_\Sigma^n(B, I)$ for all objects A and B in Σ and $n \geq 1$. Dually, $\overline{H_\Sigma^n}(P, A) = 0 = \overline{H_\Sigma^n}(B, I)$ for all objects A and B in Σ and $n \geq 1$.*

Proof Let P be a projective object, then every projective acyclic chain complex over P is chain equivalent to

$$\cdots 0 \longrightarrow 0 \cdots \longrightarrow 0 \longrightarrow P \longrightarrow o.$$

Evidently, $H_\Sigma^n(P, A) = 0$ for all objects A and for all $n \geq 1$. Let I be an injective object in Σ. Then the functor $Mor_\Sigma(-, I)$ is an exact functor. Let

$$\Gamma(B) \equiv \cdots \overset{d_{n+1}}{\to} P_n \overset{d_n}{\to} P_{n-1} \overset{d_{n-1}}{\to} \cdots \overset{d_2}{\to} P_1 \overset{d_1}{\to} P_0 \longrightarrow 0$$

be an acyclic projective chain complex over B. Take any $n \geq 1$. Suppose that $f \in ker d_{n+1}^*$. Then $f d_{n+1} = 0$. From Proposition 1.2.7, we have the factorization

$$P_{n+1} \overset{d_{n+1}}{\to} P_n = P_{n+1} \overset{k_{n+1}}{\to} L_{n+1} \overset{h_{n+1}}{\to} P_n,$$

where $h_{n+1} = image d_{n+1} = ker d_n$ and k_{n+1} is an epimorphism (co-image of d_{n+1}). Similarly, we have the factorization

$$P_n \overset{d_n}{\to} P_{n-1} = P_n \overset{k_n}{\to} L_n \overset{h_n}{\to} P_{n-1},$$

where $h_n = image d_n = ker d_{n-1}$ and k_n is an epimorphism (co-image of d_n). Now, $f d_{n+1} = f h_{n+1} k_{n+1} = 0$. Since k_{n+1} is an epimorphism, $f h_{n+1} = 0$. In turn, there is a morphism ϕ from L_n to I such that $\phi k_n = f$. Since h_n is a monomorphism and

I is injective, ϕ can be extended to a morphism ψ from P_{n-1} to I such that $\psi h_n = \phi$. Thus, $\psi d_n = f$. This means that $f \in imaged_n^*$. This completes the proof of the fact that $H_\Sigma^n(B, I) = 0$. Dually, we can prove the rest of the statement. ♯

Proposition 2.1.8 *Let Σ be an abelian category with enough projectives. Let*

$$E \equiv 0 \longrightarrow B \overset{\beta}{\to} C \overset{\gamma}{\to} D \longrightarrow 0$$

be a short exact sequence in Σ. Then for each n, there is a natural connecting homomorphism ∂_E^n from $H_\Sigma^n(B, A)$ to $H_\Sigma^{n+1}(D, A)$ such that

$$\Omega(E, A) \equiv \cdots \overset{\partial_E^{n-1}}{\to} H_\Sigma^n(D, A) \overset{H^n(-,A)(\gamma)}{\to} H_\Sigma^n(C, A) \overset{H^n(-,A)(\beta)}{\to} H_\Sigma^n(B, A) \overset{\partial_E^n}{\to} \cdots$$

is a long exact sequence which is natural in E as well as in A. More explicitly, Ω given above defines a functor from the category $EXT_\Sigma^o \times \Sigma$ to the category of long exact sequences of abelian groups, where EXT_Σ denote the category of onefold extensions in Σ.

Proof Let

$$\Gamma(D) \overset{\epsilon_D}{\to} D \to 0 \equiv \cdots \overset{d_{n+1}^D}{\to} P_n^D \overset{d_n^D}{\to} P_{n-1}^D \overset{d_{n-1}^D}{\to} \cdots \overset{d_2^D}{\to} P_1^D \overset{d_1^D}{\to} P_0^D \overset{\epsilon_D}{\to} D \longrightarrow 0$$

be a projective resolution of D, and

$$\Gamma(B) \overset{\epsilon_B}{\to} B \to 0 \equiv \cdots \overset{d_{n+1}^B}{\to} P_n^B \overset{d_n^B}{\to} P_{n-1}^B \overset{d_{n-1}^B}{\to} \cdots \overset{d_2^B}{\to} P_1^B \overset{d_1^B}{\to} P_0^B \overset{\epsilon_B}{\to} B \longrightarrow 0$$

be that of B. Since P_0^D is a projective object and γ is an epimorphism, we have a morphism ϕ from P_0^D to C such that $\gamma\phi = \epsilon_D$. Let (P_0^C, i_1, i_2) denote the co-product of P_0^B and P_0^D, where i_1 is a morphism from P_0^B to P_0^C and i_2 is a morphism from P_0^D to P_0^C. In turn, from Proposition 1.2.2, we get morphisms p_1 from P_0^C to P_0^B and also a morphism p_2 from P_0^C to P_0^D such that

(i) $p_1 i_1 = I_{P_0^B}$, $p_2 i_2 = I_{P_0^D}$, and
(ii) $i_1 p_1 + i_2 p_2 = I_{P_0^C}$.

Further, from the definition of the co-product, there is a morphism ϵ_C from P_0^C to C such that $\epsilon_C i_1 = \beta$ and $\epsilon_C i_2 = \phi$. We have the commutative diagram

$$0 \longrightarrow P_0^B \overset{i_1}{\longrightarrow} P_0^C \overset{i_2}{\longrightarrow} P_0^D \longrightarrow 0$$

$$\epsilon_B \downarrow \quad \epsilon_C \downarrow \quad \epsilon_D \downarrow \qquad\qquad (1)$$

$$0 \longrightarrow B \overset{\beta}{\longrightarrow} C \overset{\gamma}{\longrightarrow} D \longrightarrow 0$$

where the rows are exact with top row split exact. Since ϵ_B and ϵ_D are epimorphisms, ϵ_C is also an epimorphism. Proceeding inductively, we get a projective resolution

$$\Gamma(C) \overset{\epsilon_C}{\to} C \to 0 \equiv \cdots \overset{d_{n+1}^C}{\to} P_n^C \overset{d_n^C}{\to} P_{n-1}^C \overset{d_{n-1}^C}{\to} \cdots \overset{d_2^C}{\to} P_1^C \overset{d_1^C}{\to} P_0^C \overset{\epsilon_C}{\to} C \longrightarrow 0$$

of C, where (P_n^C, i_1, i_2) is co-product of P_n^B and P_n^D, and d_n^C is the unique morphism from P_n^C to P_{n-1}^C defined by the properties $d_n^C i_1 = i_1 d_{n-1}^B$ and $d_n^C i_2 = i_2 d_{n-1}^D$. In turn, we have the split short exact sequence

$$0 \longrightarrow \Gamma(B) \overset{i_1}{\to} \Gamma(C) \overset{p_2}{\to} \Gamma(C) \longrightarrow 0$$

of projective positive acyclic chain complexes. Applying the contra-variant functor $Mor_\Sigma(-, A)$, we obtain a short exact sequence

$$0 \longrightarrow Hom(\Gamma(D), A) \overset{p_2^*}{\to} Hom(\Gamma(C), A) \overset{i_1^*}{\to} Hom(\Gamma(B), A) \longrightarrow 0$$

of co-chain complexes of abelian groups. Using Theorem 1.3.1 for short exact sequences of co-chain complexes, we obtain the desired long co-homology exact sequence. ♯

The following proposition is the dual of the above proposition, and it can be proved by using dual arguments.

Proposition 2.1.9 *Let Σ be an abelian category with enough injectives. Let*

$$E \equiv 0 \longrightarrow A \overset{\alpha}{\to} A' \overset{\alpha'}{\to} A'' \longrightarrow 0$$

be a short exact sequence in Σ. Then for each n, there is a natural connecting homomorphism δ_n^E from $\overline{H_\Sigma^n}(B, A'')$ to $\overline{H_\Sigma^{n+1}}(B, A)$ such that

$$\Delta(B, E) \equiv \cdots \overset{\delta_E^{n-1}}{\to} \overline{H_\Sigma^n}(B, A) \overset{\overline{H^n(-, A)}(\alpha)}{\to} \overline{H_\Sigma^n}(B, A') \overset{\overline{H^n(-, A')}(\beta)}{\to} \overline{H_\Sigma^n}(B, A'') \overset{\delta_E^n}{\to}$$

$$\cdots$$

is a long exact sequence which is natural in B as well as in E. More explicitly, Δ given above defines a functor from the category $\Sigma^o \times EXT_\Sigma$ to the category of long exact sequences of abelian groups. ♯

Proposition 2.1.10 *Let Σ be an abelian category with enough projectives. Let*

$$E \equiv 0 \longrightarrow A \overset{\alpha}{\to} A' \overset{\alpha'}{\to} A'' \longrightarrow 0$$

be a short exact sequence in Σ, and B be an object in Σ. Then for each n, we have a natural connecting homomorphism ∂_E^n from $H_\Sigma^n(B, A'')$ to $H_\Sigma^{n+1}(B, A)$ such that

$$\Omega^t(B, E) \equiv \cdots \overset{\partial_E^{n-1}}{\to} H_\Sigma^n(B, A) \overset{H^n(B,-)(\alpha)}{\to} H_\Sigma^n(B, A') \overset{H^n(B,-)(\alpha')}{\to} H_\Sigma^n(B, A'') \overset{\partial_E^n}{\to}$$
$$\cdots$$

is a long exact sequence which is natural in B as well as E. More explicitly, Ω^t given above defines a functor from the category $\Sigma^o \times EXT_\Sigma$ to the category of long exact sequences of abelian groups.

Proof Let

$$\Gamma(B) \overset{\epsilon_B}{\to} B \to 0 \equiv \cdots \overset{d_{n+1}^B}{\to} P_n^B \overset{d_n^B}{\to} P_{n-1}^B \overset{d_{n-1}^B}{\to} \cdots \overset{d_2^B}{\to} P_1^B \overset{d_1^B}{\to} P_0^B \overset{\epsilon_B}{\to} B \longrightarrow 0$$

be a projective resolution of B. Since P_n^B is projective for each n, the sequences

$$0 \longrightarrow Mor_\Sigma(P_n, A) \overset{\alpha_*}{\to} Mor_\Sigma(P_n, A') \overset{\alpha'_*}{\to} Mor_\Sigma(P_n, A'') \longrightarrow 0$$

are exact for each n. It follows that

$$0 \longrightarrow Hom(P, A) \overset{\alpha_*}{\to} Hom(P, A') \overset{\alpha'_*}{\to} Hom(P, A'') \longrightarrow 0$$

is a short exact sequence of co-chain complexes of abelian groups. The result follows from Theorem 1.3.1. ♯

Dually, we have the following proposition whose proof can be given by using dual arguments.

Proposition 2.1.11 *Let Σ be an abelian category with enough injectives. Let*

$$E \equiv 0 \longrightarrow B \overset{\beta}{\to} B' \overset{\beta'}{\to} B'' \longrightarrow 0$$

be a short exact sequence in Σ, and A be an object in Σ. Then for each n, we have a natural connecting homomorphism δ_E^n from $\overline{H_\Sigma^n}(B, A)$ to $\overline{H_\Sigma^{n+1}}(B'', A)$ such that

$$\Delta^t(E, A) \equiv \cdots \stackrel{\delta_E^{n-1}}{\to} \overline{H_\Sigma^n}(B'', A) \stackrel{\overline{H^n(-,A)}(\beta')}{\to} \overline{H_\Sigma^n}(B', A) \stackrel{\overline{H^n(-,A)}(\beta)}{\to} \overline{H_\Sigma^n}(B, A) \stackrel{\delta_E^n}{\to}$$

$$\cdots$$

is a long exact sequence which is natural in E as well as A. More explicitly, Δ^t given above defines a functor from the category $EXT_\Sigma^o \times \Sigma$ to the category of long exact sequences of abelian groups. ♯

Finally, we prove the following theorem.

Theorem 2.1.12 *Let Σ be an abelian category with enough projectives and also with enough injectives. Then the bi-functors $H_\Sigma^n(-, -)$ and $\overline{H_\Sigma^n}(-, -)$ from Σ to AB are naturally isomorphic for all n.*

Proof We define natural isomorphisms $\eta_{-,-}^n$ from the bi-functor $H_\Sigma^n(-, -)$ to $\overline{H_\Sigma^n}(-, -)$ by the induction on n. By Proposition 2.1.6, $H_\Sigma^0(-, -)$ and $\overline{H_\Sigma^0}(-, -)$ both are naturally isomorphic to the bi-functor $Mor_\Sigma(-, -)$. This gives us natural isomorphism $\eta_{-,-}^0$ from $H_\Sigma^0(-, -)$ to $\overline{H_\Sigma^0}(-, -)$. We construct $\eta_{-,-}^1$ as follows. Let B and A be objects in Σ. Since Σ has enough injectives, we have a short exact sequence

$$E \equiv 0 \longrightarrow A \stackrel{\alpha}{\to} I \stackrel{\beta}{\to} C \longrightarrow 0,$$

where I is an injective object. By Proposition 2.1.10, we have a long exact sequence

$$\Omega^t(B, E) \equiv o \to Hom(B, A) \stackrel{\alpha_*}{\to} Hom(B, I) \stackrel{\beta_*}{\to} Hom(B, C) \stackrel{\partial_E^0}{\to}$$

$$H_\Sigma^1(B, A) \stackrel{H^1(B,-)(\alpha)}{\to} H_\Sigma^1(B, I) \cdots \stackrel{\partial_E^{n-1}}{\to} H_\Sigma^n(B, A) \stackrel{H^n(B,-)(\alpha)}{\to}$$

$$H_\Sigma^n(B, A') \stackrel{H^n(B,-)(\alpha')}{\to} H_\Sigma^n(B, A'') \stackrel{\partial_E^n}{\to} \cdots,$$

and by Proposition 2.1.9, we have a long exact sequence

$$\Delta(B, E) \equiv o \to Hom(B, A) \stackrel{\alpha_*}{\to} Hom(B, I) \stackrel{\beta_*}{\to} Hom(B, C) \stackrel{\delta_E^0}{\to}$$

$$\overline{H_\Sigma^1}(B, A) \stackrel{\overline{H^1(B,-)}(\alpha)}{\to} \overline{H_\Sigma^1}(B, I) \cdots \stackrel{\delta_E^{n-1}}{\to} \overline{H_\Sigma^n}(B, A) \stackrel{\overline{H^n(B,-)}(\alpha)}{\to} \overline{H_\Sigma^n}(B, I) \stackrel{\overline{H^n(B,-)}(\beta)}{\to}$$

$$\overline{H_\Sigma^n}(B, C) \stackrel{\delta_E^n}{\to} \cdots$$

of abelian groups. By Proposition 2.1.7, $H_\Sigma^n(B, I) = 0 = \overline{H_\Sigma^n}(B, I)$ for all $n \geq 1$. Thus, we have the following commutative diagram:

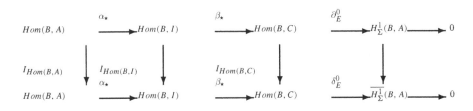

where rows are exact. Evidently, $\ker \partial_E^0 = \ker \delta_E^0$. This determines a unique morphism $\eta_{B,A}^1$ from $H_\Sigma^1(B, A)$ to $\overline{H_\Sigma^1}(B, A)$ so that the above diagram remains commutative with the last vertical arrow as $\eta_{B,A}^1$. Since the connecting homomorphisms ∂_E^0 and δ_E^0 are natural transformations between the corresponding functors, $\{\eta_{B,A}^1\}$ is a natural isomorphism. Now, assume that $\eta_{B,A}^n$ has already been defined for all B and A and $n \geq 1$. We need to construct $\eta_{B,A}^{n+1}$. Again, by Proposition 2.1.7, $H_\Sigma^k(B, I) = 0 = \overline{H_\Sigma^k}(B, I)$ for all $k \geq 1$, and we get the following commutative diagram:

$$
\begin{array}{ccccccccc}
 & & & & \partial_E^n & & & & \\
0 & \longrightarrow & H_\Sigma^n(B, C) & & \longrightarrow & H_\Sigma^{n+1}(B, A) & & \longrightarrow & 0 \\
 & & \downarrow{\scriptstyle \eta_{B,C}^n} & & & \downarrow & & & \\
0 & \longrightarrow & \overline{H_\Sigma^n}(B, C) & & \xrightarrow{\ \delta_E^n\ } & \overline{H_\Sigma^{n+1}}(B, A) & & \longrightarrow & 0
\end{array}
$$

where the rows are exact. We take $\eta_{B,A}^{n+1}$ to be the unique isomorphism which makes the above diagram commutative. Since the connecting homomorphisms are natural transformations between the corresponding functors, and $\eta_{B,A}^n$ is assumed to be a natural isomorphism, $\{\eta_{B,A}^{n+1}\}$ is a natural isomorphism. ♯

H_Σ^n and n-Fold Extensions

For the sake of simplicity, in rest of the section, we shall again restrict our self to the category $Mod - R$ of right R-modules. However, every thing can be done in any abelian category with enough projectives and enough injectives. Note that the category of modules has enough projectives and enough injectives.

Again, let Γ denote the choice functor from the category $Mod - R$ of modules over R to the category $\wp(R)$ of acyclic projective chain complexes in $Mod - R$. Indeed, in the category $Mod - R$, for each R-module B, we have the canonical choice of free acyclic chain complex $\Gamma(B)$ over B. Let

$$\Gamma(B) \xrightarrow{\epsilon} 0 \equiv \cdots \xrightarrow{d_{n+1}} P_n \xrightarrow{d_n} P_{n-1} \xrightarrow{d_{n-1}} \cdots \xrightarrow{d_2} P_1 \xrightarrow{d_1} P_0 \xrightarrow{\epsilon} B \longrightarrow 0$$

be the corresponding projective resolution of B. Further, let

$$E \equiv 0 \longrightarrow A \overset{\alpha}{\to} M \overset{\beta}{\to} B \longrightarrow 0$$

be an extension of A by B. From Theorem 2.1.2, we have a homomorphism f_0 from P_0 to M and a homomorphism f_1 from P_1 to A such that the following diagram is commutative:

$$
\begin{array}{ccccccccc}
& \overset{d_2}{\longrightarrow} & P_1 & \overset{d_1}{\longrightarrow} & P_0 & \overset{\epsilon}{\longrightarrow} & B & \longrightarrow & 0 \\
P_2 & & & & & & & & \\
\downarrow & & \downarrow f_1 & & \downarrow f_0 & & \downarrow I_B & & \\
0 & \longrightarrow & A & \underset{\alpha}{\longrightarrow} & M & \underset{\beta}{\longrightarrow} & B & \longrightarrow & 0
\end{array} \qquad (2)
$$

Consider the co-chain complex

$$Hom(\Gamma(B), A) \equiv 0 \longrightarrow Hom(P_0, A) \overset{d_1^*}{\to} Hom(P_1, A) \overset{d_2^*}{\to} Hom(P_2, A) \overset{d_3^*}{\to} \ldots.$$

Clearly, $H_R^0(B, A) = H^0(\Gamma(B), A) = ker d_1^*$. Again, since $Hom(-, A)$ is a left exact functor,

$$0 \longrightarrow Hom(B, A) \overset{\epsilon^*}{\to} Hom(P_0, A) \overset{d_1^*}{\to} Hom(P_1, A)$$

is exact. Hence, $H_R^0(B, A)$ is naturally isomorphic to $Hom(B, A)$. Evidently, f_1 appearing in the commutative diagram 2 is a member of $C^1(\Gamma(B), A)$. This gives us an association $\Lambda_{B,A}$ from the class $\mathbf{EXT_R(B, A)}$ of extensions of A by B to the group $H_R^1(B, A)$ given by $\Lambda_{B,A}(E) = f^1 + B^1(\Gamma(B), A)$, where E and f_1 are given as above.

Next, suppose that an extension

$$E' \equiv 0 \longrightarrow A \overset{\alpha'}{\to} M' \overset{\beta'}{\to} B \longrightarrow 0$$

is equivalent to the extension E. Without any loss of generality, we may assume that (I_A, μ, I_B) is an equivalence from E to E'. Then the diagram

$$
\begin{array}{ccccccccc}
 & \xrightarrow{d_2} & P_1 & \xrightarrow{d_1} & P_0 & \xrightarrow{\epsilon} & B & \longrightarrow & 0 \\
P_2 & & & & & & & & \\
\downarrow & & f_1 \downarrow & & \mu\circ f_0 \downarrow & & I_B \downarrow & & \\
 & & & & \alpha' & & \beta & & \\
0 & \longrightarrow & A & \longrightarrow & M' & \longrightarrow & B & \longrightarrow & 0
\end{array}
\qquad (3)
$$

is commutative. This shows that $\Lambda_{B,A}(E) = f_1 + B^1(\Gamma(B), A) = \Lambda_{B,A}(E')$. Thus, $\Lambda_{B,A}$ induces a map $\overline{\Lambda_{B,A}}$ from the group $EXT_R(B, A)$ to the group $H^1_R(B, A)$. Finally, we have the following theorem.

Theorem 2.1.13 *The map $\overline{\Lambda_{B,A}}$ from $EXT_R(B, A)$ to $H^1_R(B, A)$ defined above is an isomorphism.*

Proof Let

$$E \equiv 0 \longrightarrow A \xrightarrow{\alpha} M \xrightarrow{\beta} B \longrightarrow 0,$$

and

$$E' \equiv 0 \longrightarrow A \xrightarrow{\alpha'} M' \xrightarrow{\beta'} B \longrightarrow 0$$

be extensions of A by B such that $\overline{\Lambda_{B,A}}([E]) = f_1 + B^1(\Gamma(B), A)$ and $\overline{\Lambda_{B,A}}([E']) = f_1' + B^1(\Gamma(B), A)$, where the diagrams

$$
\begin{array}{ccccccccc}
 & \xrightarrow{d_2} & P_1 & \xrightarrow{d_1} & P_0 & \xrightarrow{\epsilon} & B & \longrightarrow & 0 \\
P_2 & & & & & & & & \\
\downarrow & & f_1 \downarrow & & f_0 \downarrow & & I_B \downarrow & & \\
 & & & & \alpha & & \beta & & \\
0 & \longrightarrow & A & \longrightarrow & M & \longrightarrow & B & \longrightarrow & 0
\end{array}
$$

and

$$
\begin{array}{ccccccccc}
 & \xrightarrow{d_2} & P_1 & \xrightarrow{d_1} & P_0 & \xrightarrow{\epsilon} & B & \longrightarrow & 0 \\
P_2 & & & & & & & & \\
\downarrow & & f_1' \downarrow & & f_0' \downarrow & & I_B \downarrow & & \\
 & & & & \alpha' & & \beta & & \\
0 & \longrightarrow & A' & \longrightarrow & M' & \longrightarrow & B & \longrightarrow & 0
\end{array}
$$

are commutative. Recall the construction of $E \uplus E'$. Let $T = \{(m, m') \in M \times M' \mid \beta(m) = \beta'(m')\}$, $L = \{(a + b, (-\alpha(a), \alpha'(b))) \mid (a, b) \in A \times A\}$. Then T is a submodule of $M \times M'$, and L is a submodule of $A \times T$. Take $\Omega = (A \times T)/L$. Let $\overline{i}$ be the map from A to Ω given by $\overline{i}(a) = (a, (0, 0)) + L$, and $\overline{p}$ be a map from Ω to B defined by $\overline{p}((a, (m, m')) + L) = \beta(m)$ (note that $\beta(m) = \beta'(m')$). Then

$$E \uplus E' \equiv 0 \longrightarrow A \xrightarrow{\overline{i}} \Omega \xrightarrow{\overline{\beta}} B \longrightarrow 0.$$

Let ϕ be the map from P_0 to Ω defined by $\phi(x) = (0, (f_0(x), f_0'(x))) + L$. Then the following diagram is commutative:

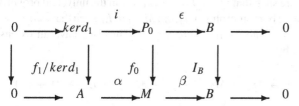

This shows that $\overline{\Lambda_{B,A}}([E] \uplus [E']) = f_1 + f_1' + B^1(\Gamma(B), A) = \overline{\Lambda_{B,A}}([E]) + \overline{\Lambda_{B,A}}([E'])$. It remains to show that $\overline{\Lambda_{B,A}}$ is bijective. Suppose that $\overline{\Lambda_{B,A}}([E]) = 0$. Then there is a map η from P_0 to A such that $\eta \circ d_1 = f_1$. In turn, we have the following commutative diagram:

$$
\begin{array}{ccccccccc}
0 & \longrightarrow & ker d_1 & \xrightarrow{i} & P_0 & \xrightarrow{\epsilon} & B & \longrightarrow & 0 \\
& & \downarrow{f_1/ker d_1} & & \downarrow{f_0}_{\alpha} & & \downarrow{I_B}_{\beta} & & \\
0 & \longrightarrow & A & \longrightarrow & M & \longrightarrow & B & \longrightarrow & 0
\end{array}
$$

We also have a homomorphism η from P_0 to A which is lifting of $f_1/ker d_1$. It follows from Proposition 1.4.10 that E is split exact. This shows that $[E] = 0$.

Finally, we prove that $\overline{\Lambda_{B,A}}$ is surjective. Let $f + B^1(\Gamma(B), A) \in H_R^1(B, A)$, where $f \in C^1(\Gamma(B), A)$. Then $f \circ d_2 = 0$. Thus, f can be treated as a map from $ker d_1$ to A. Take $E = f_*(\hat{E})$, where

$$\hat{E} \equiv 0 \longrightarrow ker d_1 \xrightarrow{i} P_0 \xrightarrow{\epsilon} B \longrightarrow 0.$$

Evidently, $\Lambda_{B,A}(E) = f + B^1(\Gamma(B), A)$. ♯

Corollary 2.1.14 *The family $\{\overline{\Lambda_{B,A}}\}$ defines a natural isomorphism between the bi-functors $EXT_R(-, -)$ and $H_R^1(-, -)$.* ♯

More generally, let $\Xi X T_R^n(B, A)$ denote the class of n-fold extensions of A by B. Thus, an arbitrary member of $\Xi X T_R^n(B, A)$ is an exact sequence

$$E \equiv 0 \longrightarrow A \xrightarrow{\alpha} M_1 \xrightarrow{\chi_1} M_2 \xrightarrow{\chi_2} \cdots \xrightarrow{\chi_{n-1}} M_n \xrightarrow{\beta} B \longrightarrow 0.$$

Let

$$E' \equiv 0 \longrightarrow A \xrightarrow{\alpha'} M_1' \xrightarrow{\chi_1'} M_2' \xrightarrow{\chi_2'} \cdots \xrightarrow{\chi_{n-1}'} M_n' \xrightarrow{\beta'} B \longrightarrow 0$$

be another n-fold extension of A by B. We say that $E \prec E'$ if for each k, $1 \leq k \leq n$, there is a homomorphism f_k from M_k to M_k' such that $f_1 \alpha = \alpha'$, $f_{k-1}\chi_k = \chi_k' f_k$, and $\beta' f_n = \beta$ (look at the corresponding commutative diagram). For $n = 1$, $\prec$ is an equivalence relation. However, for $n \geq 2$, $\prec$ is reflexive and transitive but it is not a symmetric relation. Let $\approx$ denote the equivalence relation generated by $\prec$. Let $EXT_R^n(B, A)$ denote the quotient $\Xi X T_R^n(B, A)/\approx$. It is easy to observe that $EXT_R^n(B, A)$ is a set. Indeed, by the induction on n, one can construct a set $X^n(B, A)$ of n-fold extensions of A by B so that each member of $\Xi X T_R^n$ is equivalent to a member of $X^n(B, A)$. The equivalence class determined by an n-fold extension E of A by B will be denoted by $[E]$.

Let E be an n-fold extension of A by B as given above. Let D be an R-module and γ be a homomorphism from D to B. Let the triple (L_n, p_n, χ) denote the pullback of the pair (β, γ) of homomorphisms, where p_n is a homomorphism from L_n to M_n and χ is a homomorphism from L_n to D. Since β is an epimorphism, χ is also an epimorphism. The zero homomorphism 0 from M_{n-1} to D and the homomorphism χ_{n-1} from M_{n-1} to M_n are such that $\gamma 0 = 0 = \beta \chi_{n-1}$. From the universal property of pullback, we have a unique homomorphism μ from M_{n-1} to L_n such that $p_n \mu = \chi_{n-1}$. Evidently, $ker \chi = image \mu$ and $ker \mu = image \chi_{n-2}$. Thus, we have an extension $\gamma^\star(E)$ of A by D given by

$$\gamma^\star(E) \equiv 0 \longrightarrow A \xrightarrow{\alpha} M_1 \xrightarrow{\chi_1} M_2 \xrightarrow{\chi_2} \cdots \xrightarrow{\chi_{n-2}} M_{n-1} \xrightarrow{\mu} L_n \xrightarrow{\chi} D \longrightarrow 0.$$

Suppose that $E \prec E'$, where E and E' are given as above. Then, for each k, $1 \leq k \leq n$, there is a homomorphism f_k from M_k to M_k' such that $f_1 \alpha = \alpha'$, $f_{k-1}\chi_k = \chi_k' f_k$, and $\beta' f_n = \beta$. Suppose that (L_n', p_n', χ') be the pullback of the pair (β', γ) of homomorphisms, where p_n' is a homomorphism from L_n' to M_n' and χ' is a homomorphism from L_n' to D, and

$$\gamma^\star(E') \equiv 0 \longrightarrow A \xrightarrow{\alpha'} M_1' \xrightarrow{\chi_1'} M_2' \xrightarrow{\chi_2'} \cdots \xrightarrow{\chi_{n-2}'} M_{n-1}' \xrightarrow{\mu'} L_n' \xrightarrow{\chi'} D \longrightarrow 0.$$

Since $\beta' f_n p_n = \beta p_n = \gamma \chi$ and (L_n', p_n', χ') is the pullback of the pair (β', γ), the universal property of a pullback ensures the existence of a unique homomorphism h_n from L_n to L_n' such that $p_n' h_n = f_n p_n$. Already, $f_1 \alpha = \alpha'$, $f_{k-1} d_k = d_k f_k$ for all $k \leq n - 1$. Also, $h_n \mu = \mu' f_{n-1}$, and $\chi' h_n = \chi = I_D \chi$. This shows that $\gamma^\star(E) \prec \gamma^\star(E')$. Since $\approx$ is the smallest equivalence relation containing $\prec$, we find that $E \approx E'$ implies that $\gamma^{\star}(E) \approx \gamma^{\star}(E')$. This induces a map $EXT_R^n(-, A)(\gamma)$ from $EXT_R^n(B, A)$ to $EXT_R^n(D, A)$ defined by $EXT_R^n(-, A)(\delta)([E]) = [\gamma^{\star}(E)]$. It is easy to observe that $EXT_R^n(-, A)(\delta\gamma) = EXT_R^n(-, A)(\gamma) o EXT_R^n(-, A)(\gamma)$, and $EXT_R^n(-, A)(I_B) = I_{EXT_R^n(B,A)}$. This defines a contra-variant functor $EXT_R^n(-, A)$ from the category of R-modules to the category SET of sets.

The dual considerations give a co-variant functor $EXT_R^n(B, -)$ from the category of R-modules to the category SET. These two functors can be easily seen to be compatible to give a bi-functor $EXT_R^n(-, -)$ from the category of R-modules to the category SET.

Let E_1 and E_2 be n-fold extensions of A by B. We have the direct sum n-fold extension $E_1 \times E_2$ of $A \times A$ by $B \times B$. As in case of $EXT_R(B, A)$, the n-fold extension $(\nabla_A)_\star(\Delta_B^\star(E_1 \times E_2))$ of A by B is denoted by $E \uplus E'$. If $E_1 \approx E_1'$ and $E_2 \approx E_2'$, then as above it can be shown that $E_1 \uplus E_2 \approx E_1' \uplus E_2'$. This defines the sum $\uplus$ on $EXT_R^n(B, A)$ by $[E] \uplus [E'] = [E \uplus E']$. The sum $\uplus$ defined is again called the Baer sum.

Theorem 2.1.15 $EXT_R^n(B, A)$ is a group with respect to the Baer sum $\uplus$, and the bi-functors $EXT_R^n(-, -)$ and $H_R^n(-, -)$ from $Mod - R$ to AB are naturally isomorphic.

Proof Let

$$E \equiv 0 \longrightarrow A \xrightarrow{\alpha} M_1 \xrightarrow{\chi_1} M_2 \xrightarrow{\chi_2} \cdots \xrightarrow{\chi_{n-1}} M_n \xrightarrow{\beta} B \longrightarrow 0$$

be a member of $\Xi XT_R^n(B, A)$. Let

$$\Gamma(B) \xrightarrow{\epsilon} 0 \equiv \cdots \xrightarrow{d_{n+1}} P_n \xrightarrow{d_n} P_{n-1} \xrightarrow{d_{n-1}} \cdots \xrightarrow{d_2} P_1 \xrightarrow{d_1} P_0 \xrightarrow{\epsilon} B \longrightarrow 0$$

be the choice of projective resolution of B. From Theorem 2.1.2, we have a chain transformation $f = \{f_k \mid k \geq 0\}$ from $\Gamma(B) \xrightarrow{\epsilon} 0$ to E which is unique up to chain homotopy. More explicitly, we have the commutative diagram

$$
\begin{array}{ccccccccccccc}
P_{n+1} & \xrightarrow{d_{n+1}} & P_n & \xrightarrow{d_n} & P_{n-1} & \xrightarrow{d_{n-1}} & \cdots & \xrightarrow{d_2} & P_1 & \xrightarrow{d_1} & P_0 & \xrightarrow{\epsilon} & B & \longrightarrow & 0 \\
\downarrow & & \downarrow f_n & & \downarrow f_{n-1} & & & & \downarrow f_1 & & \downarrow f_0 & & \downarrow I_B & & \\
0 & \longrightarrow & A & \xrightarrow{\alpha} & M_1 & \xrightarrow{\chi_1} & \cdots & \xrightarrow{\chi_2} & M_{n-1} & \xrightarrow{\chi_{n-1}} & M_n & \xrightarrow{\beta} & B & \longrightarrow & 0
\end{array}
$$

where the rows are exact. Consider the co-chain complex $Hom(\Gamma(B), A)$. Since $f_n d_{n+1} = 0$, $f_n \in C^n(\Gamma(B), A)$. If $g = \{g_k \mid k \geq 0\}$ is another chain transformation from $\Gamma(B) \xrightarrow{\epsilon} 0$ to E, then there is a chain homotopy $s = \{s_k, k \geq 0\}$ from f to

g. Thus, there is a homomorphism s_{n-1} from P_{n-1} to A such that $s_{n-1}d_n = f_n - g_n$. This shows that $f_n + B^n(\Gamma(B), A) = g_n + B^n(\Gamma(B), A)$. Thus, we have a map $\Lambda_{B,A}$ from $\Xi X T^n(B, A)$ to $H^n(\Gamma(B), A) = H_R^n(B, A)$ defined by $\Lambda_{B,A}(E) = f_n + B^n(\Gamma(B), A)$. Let

$$E' \equiv 0 \longrightarrow A \xrightarrow{\alpha} M_1' \xrightarrow{\chi_1'} M_2' \xrightarrow{\chi_2'} \cdots \xrightarrow{chi_{n-1}'} M_n' \xrightarrow{\beta'} B \longrightarrow 0$$

be another member of $\Xi X T^n(B, A)$ such that $E \prec E'$. Then for each k, $1 \le k \le n$, there is a homomorphism ϕ_k from M_k to M_k' such that the diagram

is commutative. In turn, the diagram

is commutative, where $\psi_k = \phi_k f_{n-k}$. This shows that $\Lambda_{B,A}(E') = f_n + B^n(\Gamma(B), A) = \Lambda_{B,A}(E)$. Since the equivalence relation $\approx$ is generated by $\prec$, it follows that $\Lambda_{B,A}(E) = \Lambda_{B,A}(E')$ whenever $E \approx E'$. Thus, $\Lambda_{B,A}$ induces a map $\overline{\Lambda_{B,A}}$ from $EXT_R^n(B, A)$ to $H_R^n(B, A)$ defined by $\overline{\Lambda_{B,A}}([E]) = [\Lambda_{B,A}(E)]$. As in the proof of Theorem 2.1.13, it follows that $\overline{\Lambda_{B,A}}([E] \uplus [E']) = \overline{\Lambda_{B,A}}([E]) + \overline{\Lambda_{B,A}}([E'])$. To prove that $\overline{\Lambda_{B,A}}$ is bijective, we construct its natural inverse $\Theta_{B,A}$. Let $f_n + B^n(\Gamma(B), A)$ be a member of $H_R^n(B, A)$, where $f_n \in C^n(\Gamma(B), A)$. Since $f_n d_{n+1} = 0$ and $image\, d_{n+1} = ker\, d_n$, f_n is zero on $ker\, d_n$. Also $image\, d_n = ker\, d_{n-1}$. We may treat d_n as a map from P_n to $ker\, d_{n-1}$. From the fundamental theorem of homomorphism, we have a unique homomorphism $\overline{f_n}$ from $ker\, d_{n-1}$ to A such that $\overline{f_n} o d_n = f_n$. We have an n-fold extension

$$\Re(\Lambda(B)) \equiv 0 \longrightarrow ker\, d_{n-1} \xrightarrow{i} P_{n-1} \xrightarrow{d_{n-1}} \cdots \xrightarrow{d_2} P_1 \xrightarrow{d_1} P_0 \xrightarrow{\epsilon} B \longrightarrow 0$$

of $ker d_{n-1}$ by B, and $\overline{f_n}$ is a homomorphism from $ker d_{n-1}$ to A. Let the triple (M_1, i_1, i_2) denote the pushout of the triple $(ker d_{n-1}, \overline{f_n}, i)$. Let $\widehat{d_{n-1}}$ denote the obvious map from M_1 to P_{n-2}. We have the n-fold extension

$$\overline{f_n}_\star(\Re(\Lambda(B))) \equiv 0 \longrightarrow A \overset{i_1}{\to} M_1 \overset{\widehat{d_{n-1}}}{\to} P_{n-2} \overset{d_{n-2}}{\to} \cdots \overset{d_1}{\to} P_0 \overset{\epsilon}{\to} B \longrightarrow 0$$

of A by B. Let f_n' be another member of $C^n(\Gamma(B), A)$ such that $f_n' + B^n(\Gamma(B), A) = f_n + B^n(\Gamma(B), A)$. Then $f_n - f_n'$ belong to $B^n(\Gamma(B), A)$. Hence, there is a homomorphism s_{n-1} from P_{n-1} to A such that $f_n - f_n' = s_{n-1}d_n$. Thus, $\overline{f_n} = \overline{f_n'} + t_{n-1}$, where t_{n-1} is the restriction of s_{n-1} to $ker d_{n-1}$. Again, we have an n-fold extension

$$\overline{f_n'}_\star(\Re(\Lambda(B))) \equiv 0 \longrightarrow A \overset{i_1'}{\to} M_1' \overset{\widehat{d_{n-1}'}}{\to} P_{n-2} \overset{d_{n-2}}{\to} \cdots \overset{d_1}{\to} P_0 \overset{\epsilon}{\to} B \longrightarrow 0$$

of A by B. Clearly, $i_1' \circ \overline{f_n} = \eta \circ i$, where η is the homomorphism from P_{n-1} to M_1' given by $\eta(x) = i_2'(x) + i_1'(s_{n-1}(x))$. Since the triple (M_1, i_1, i_2) is a pushout of the triple $(ker d_{n-1}, \overline{f_n}, i)$, there is a unique homomorphism ρ from M_1 to M_1' such that $\rho \circ i_1 = i_1'$ and $\rho \circ i_2 = i_2'$. In turn, we have the commutative diagram

This means that $\overline{f_n}^\star[(\Re(\Lambda(B)))] = \overline{f_n'}^\star[(\Re(\Lambda(B)))]$. Thus, we have a map $\Theta_{B,A}$ from $H_R^n(B, A)$ to $EXT_R^n(B, A)$ given by $\Theta_{B,A}(f^n + B^n(\Gamma(B), A)) = [\overline{f_n}^\star(\Re(\Lambda(B)))]$. Evidently, $\Theta_{B,A}$ is the natural inverse of $\Lambda_{B,A}$. In turn, it also follows that $EXT_R^n(B, A)$ is an abelian group with respect to $\uplus$. ♯

The above theorem justifies the notation $EXT_R^n(B, A)$ for $H_R^n(B, A)$ which is adopted in the literature. Thus, the easiest way to compute the group $EXT_R^n(B, A)$ of n-fold extensions of A by B is to start with a projective resolution $\Gamma(B) \overset{\epsilon}{\to} B \to 0$ of B, and then find the nth co-homology $H_R^n(Hom(\Gamma(B), A))$ or to start with an injective resolution $0 \to A \overset{\eta}{\to} \Lambda(A)$ of A, and then find $H_R^n(B, \Lambda(A))$ (see Theorem 2.1.12).

Corollary 2.1.16 *Let*

$$E \equiv 0 \longrightarrow B \overset{\beta}{\to} C \overset{\gamma}{\to} D \longrightarrow 0$$

be a short exact sequence in R-modules, and A be an R-module. Then for each n, there is a natural connecting homomorphism ∂_E^n from $EXT_R^n(A, D)$ to $EXT_R^{n+1}(A, B)$ such that

$$\equiv \cdots \xrightarrow{\partial_E^{n-1}} EXT_R^n(A,B) \xrightarrow{EXT_R^n(-,A)(\beta)} EXT_R^n(A, C) \xrightarrow{EXT_R^n(-,A)(\gamma)}$$

$$EXT_R^n(A, D) \xrightarrow{\partial_E^n} \cdots$$

is a long exact sequence which is natural in E as well as in A. ♯

Corollary 2.1.17 *Let*

$$E \equiv 0 \longrightarrow B \xrightarrow{\beta} C \xrightarrow{\gamma} D \longrightarrow 0$$

be a short exact sequence of R-modules. Then for each n, there is a natural connecting homomorphism ∂_E^n from $EXT_R^n(B, A)$ to $EXT_R^{n+1}(D, A)$ such that

$$\equiv \cdots \xrightarrow{\partial_E^{n-1}} EXT_R^n(D,A) \xrightarrow{EXT_R^n(-,A)(\gamma)} EXT_R^n(C, A) \xrightarrow{EXT_R^n(-,A)(\beta)}$$

$$EXT_R^n(B, A) \xrightarrow{\partial_E^n} \cdots$$

is a long exact sequence which is natural in E as well as in A. ♯

Exercises

2.1.1 Let
$$0 \to A \to M \to B \to 0$$

be a short exact sequence in an abelian category Σ. Suppose that A and B both have finite projective resolution. Show that M also has a finite projective resolution.

2.1.2 Let R be a principal ideal domain. Show that $EXT_R^n(B, A) = 0$ for all $n \geq 2$. In particular, $EXT_{K[X]}^n(B, A) = 0$ for all $n \geq 2$, where K is a field. Show that $EXT_{K[X,Y]}^2(B, A)$ need not be 0. Is $EXT_{K[X,Y]}^n(B, A) = 0$ for all $n \geq 3$?

2.1.3 What are projective (injective) objects in the category of finite abelian groups? Does it have enough projectives (injectives). Support.

2.1.4 Show that the category of finitely generated $\mathbb{Z}$-modules have enough projectives. Does it have nontrivial injectives?

2.1.5 Show that the category of torsion $\mathbb{Z}$-modules have enough injective. Does it have nontrivial projectives?

2.1.6 Show that

(i) $EXT_R^n(B \oplus C, A) \approx EXT_R^n(B, A) \oplus EXT_R^n(C, A)$ and

(ii) $EXT_R^n(B, A \oplus C) \approx EXT_R^n(B, A) \oplus EXT_R^n(B, C)$.

2.1.7 Show that P is projective if and only if $EXT_R(P, A) = 0$ for all R-modules A.

2.1.8 Show that I is injective if and only if $EXT_R(B, I) = 0$ for all R-modules B.

2.1.9 Show that a right R-module I is injective if and only if $EXT_R(R/A, I) = 0$ for all right ideals A of R.
Hint: Use suitably the Zorn's lemma.

2.1.10 Compute (i) $EXT_{\mathbb{Z}}(\mathbb{Z}_m, A)$ and (ii) $EXT_{\mathbb{Z}}(\mathbb{Z}_m, \mathbb{Z}_n)$.

2.1.11 Let B be an R-module. We say that the projective dimension $Pd_R(B)$ of B is $\leq n$ if $EXT_R^m(B, A) = 0$ for all R-modules A and for all $m \geq n + 1$. Show that the following conditions are equivalent:

(i) $Pd_R(B) \leq n$.
(ii) $EXT_R^{n+1}(B, A) = 0$ for all modules A.
(iii) There is a projective resolution of B of length n.
(iv) For all projective resolution

$$\cdots \to P_m \overset{d_m}{\to} P_{m-1} \overset{d_{m-1}}{\to} \cdots \overset{d_2}{\to} P_1 \overset{d_1}{\to} P_0 \overset{\epsilon}{\to} B \to 0$$

of B, $image d_n$ is projective.

2.1.12 State and prove the dual of the above exercise.

2.1.13 The smallest m (if exists) such that $Pd_R(B) \leq m$ is called the projective dimension of B. Similarly, we define the injective dimension $Id_R(A)$ of A. Show that $Pd_R(B \oplus C) = Max(Pd_R(B), Pd_R(C))$ and $Id_R(B \oplus C) = Max(Id_R(B), Id_R(C))$.

2.1.14 Let R be a ring and B be an R-module. Then $B \otimes_R R[X]$ is a right $R[X]$-module which we denote by $B[X]$. Show that $Pd_R(B)$ is finite if and only if $Pd_{R[X]}(B[X])$ is finite, and then $Pd_R(B) = Pd_{R[X]}(B[X])$.

2.1.15 Let R be a ring. Define the right projective dimension $rPd(R)$ of R to be $sup\{Pd(B) \mid B \ is \ a \ right \ R - module\}$, and the right injective dimension $rId(R)$ to be $sup\{Id(B) \mid B \ is \ a \ right \ R - module\}$. Show that $rPd(R) = rId(R)$. This common number is called the **right global dimension** of R, and it is denoted by $rgd(R)$. Similarly, one can define left global dimension by considering left R-modules. In general, $rgd(R)$ may be different from $lgd(R)$. However, for rings which are left as well as right noetherian the equality holds.

2.2 Tensor and Tor Functors

In the previous section, we introduced the functors $EXT_R^n(-, -)$ as derived functors of the $Hom(-, -)$ functor. In this section, we introduce the derived functors $Tor_n^R(-, -)$ of the tensor functor.

Torsion Products

Before introducing the concept of torsion product, we recall the concept of tensor product and also some of its basic properties (see Chap. 7, Algebra 2). Let M be a right R-module, and N be a left R-module. Recall that the tensor product $M \otimes_R N$ of M and N is an abelian group generated by the set $\{m \otimes n \mid m \in M, n \in N\}$ of formal symbols subject to the following relations: (i) $(m + m') \otimes n = m \otimes n + m' \otimes n$, (ii) $(m \otimes (n + n') = m \otimes n + m \otimes n'$, and (iii) $m\alpha \otimes n = m \otimes \alpha n$ for all m, m' in $M; n, n'$ in N, and $\alpha \in R$. Recall further that a map f from $M \times N$ to an abelian group A is called a balanced map if the following three conditions hold:

(i) $f(m + m', n) = f(m, n) + f(m', n)$;
(ii) $f(m, n + n') = f(m, n) + f(m, n')$; and
(iii) $f(m\alpha, n) = f(m, \alpha n)$ for all m, m' in $M; n, n'$ in N, and $\alpha \in R$.

Thus, the map η from $M \times N$ to $M \otimes_R N$ defined by $\eta(m, n) = m \otimes n$ is a balanced map. The tensor product $M \otimes_R N$ is completely described by the following universal property:

"If f is balanced map from $M \times N$ to an abelian group L, then there is a unique homomorphism $\overline{f}$ from $M \otimes_R N$ to L subject to $\overline{f}(m \otimes n) = f(m, n)$."

Definition 2.2.1 Let R and S be rings with identities. An abelian group M which is a left R-module, and also a right S-module is called a **Bi** $-$ (**R, S**) module if $(a \cdot x) \cdot b = a \cdot (x \cdot b)$ for all $x \in M, a \in R$, and $b \in S$.

Observe that if R is a commutative ring with identity, then a left R-module M is also a right R-module (define $x \cdot a = a \cdot x$). In fact, it is a bi-(R, R) module.

Proposition 2.2.2 *Let M be a right R-module and N a bi-(R, S) module. Then $M \otimes_R N$ has unique right S-module structure defined by $(x \otimes u) \cdot b = (x \otimes (u \cdot b))$. If M is bi-(S, R) module and N a left R-module, then $M \otimes_R N$ is a left S-module.*

Proof Let M be a right R-module and N be a bi-(R, S)-module. Let $b \in S$. Define a map f_b from $M \times N$ to $M \otimes N$ by $f_b(x, u) = x \otimes ub$. It is easy to observe (using the fact that N is a bi-(R, S)-module) that f_b is a balanced map. From the universal property of the tensor product, we have a unique homomorphism ϕ_b from $M \otimes_R N$ to itself defined by the property $\phi_b(x \otimes u) = x \otimes ub$. Define an external multiplication on $M \otimes_R N$ by elements of S from right by $z \cdot b = \phi_b(z)$ for all $z \in M \otimes_R N$, and $b \in S$. Since ϕ_b is a homomorphism for all $b \in S$, and $\phi_{b_1 b_2} = \phi_{b_2} o \phi_{b_1}$ for all $b_1, b_2 \in S$, it follows that $M \otimes_R N$ is a right S-module with respect to the external multiplication defined above. The rest can be proved similarly. ♯

In particular, we have the following corollary.

Corollary 2.2.3 *If R is a commutative ring, then $M \otimes_R N$ is a both-sided R-module.* ♯

Proposition 2.2.4 *Let M and N be bi-(R, R) modules. Then we have a unique isomorphism f from $M \otimes_R N$ to $N \otimes_R M$ such that $f(x \otimes y) = y \otimes x$.*

Proof The map ϕ from $M \times N$ to $N \otimes_R M$ defined by $\phi(x, y) = y \otimes x$ is a balanced (in fact bilinear) map. From the universal property of the tensor product, we have a unique homomorphism f subject to the condition $f(x \otimes y) = y \otimes x$. Similarly, we have a unique homomorphism g from $N \otimes_R M$ to $M \otimes N$ subject to the condition $g(y \otimes x) = x \otimes y$. Clearly, $gof(x \otimes y) = x \otimes y$ for all $x \in M$ and $y \in N$. Since $\{x \otimes y \mid x \in M, y \in N\}$ is a set of generators of $M \otimes_R N$, it follows that $gof = I_{M \otimes_R N}$. Similarly, fog is also the identity map. This shows that f is an isomorphism. ♯

In particular, we have the following corollary.

Corollary 2.2.5 *Let R be a commutative ring, and M and N be R-modules. Then $M \otimes_R N$ is isomorphic to $N \otimes_R M$.* ♯

Proposition 2.2.6 *Let M be a right R-module, N a bi-(R, S)-module, and L a left S-module. Then there is a unique isomorphism ϕ from $(M \otimes_R N) \otimes_S L$ to $M \otimes_R (N \otimes_S L)$ subject to the condition $\phi((x \times y) \otimes z) = x \otimes (y \otimes z)$ for all $x \in M$, $y \in N$, and $z \in L$.*

Proof Let $x \in M$. The map $(y, z) \rightsquigarrow (x \otimes y) \otimes z$ defines a balanced map from $N \times L$ to $(M \otimes_R N) \otimes_S L$. Hence, there is a unique homomorphism ϕ_x from $N \otimes_S L$ to $(M \otimes_R N) \otimes_S L$ subject to the condition $\phi_x(y \otimes z) = (x \otimes y) \otimes z$. The map $(x, u) \rightsquigarrow \phi_x(u)$, where $u \in N \otimes_S L$, is also a balanced map from $M \times (N \otimes_S L)$ to $(M \otimes_R N) \otimes_S L$. Thus, there is a unique homomorphism ϕ from $M \otimes_R (N \otimes_S L)$ to $(M \otimes_R N) \otimes_S L$ subject to the condition $\phi(x \otimes (y \otimes z)) = (x \otimes y) \otimes z$. Similarly, we have a unique homomorphism ψ from $(M \otimes_R N) \otimes_S L$ to $M \otimes_R (N \otimes_S L)$ subject to the condition $\psi((x \otimes y) \otimes z) = x \otimes (y \otimes z)$. It is clear that ϕ and ψ are inverses of each other. ♯

Remark 2.2.7 The above result, in particular, says that if R is a commutative ring with identity and $M_1, M_2, \cdots, M_n$ are R-modules, then the tensor product of $M_1, M_2, \cdots, M_n$ taken in same order with respect to any two bracket arrangements is naturally isomorphic. Thus, we can define the tensor product $M_1 \otimes_R M_2 \otimes_R \cdots \otimes_R M_n$ unambiguously. It is universal with respect to n-linear maps in the sense that if ϕ is an n-linear map from $M_1 \times M_2 \times \cdots \times M_n$ to an R-module L, then there is a unique homomorphism ψ from $M_1 \otimes_R M_2 \otimes_R \cdots \otimes_R M_n$ to L subject to $\psi(x_1 \otimes x_2 \otimes \cdots \otimes x_n) = \phi(x_1, x_2, \cdots, x_n)$.

Proposition 2.2.8 *Let R be a ring with identity and A be a left (right) R-module. Then there is an R-isomorphism ρ_A from $R \otimes_R A$ ($A \otimes_R R$) to A defined by $\rho_A(a \otimes x) = ax$ ($\rho_A(x \otimes a) = xa$). Further, if f is a homomorphism from A to B, then the diagram*

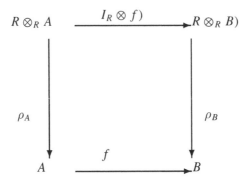

is commutative.

Proof The map $(a, x) \rightsquigarrow ax$ is clearly a balance map from $R \times A$ to A. Hence, there is a unique homomorphism ρ_A from $R \otimes_R A$ to A such that $\rho_A(a \otimes x) = ax$. Indeed, ρ_A is a R-homomorphism. Also the map ξ_A from A to $R \otimes_R A$ defined by $\xi_A(x) = 1 \otimes x$ is a homomorphism. Now, $(\xi_A o \rho_A)(a \otimes x) = \xi_A(ax) = 1 \otimes ax = 1a \otimes x = a \otimes x$. Thus, $\xi_A o \rho_A = I_{R \otimes_R A}$. Similarly, $\rho_A o \xi_A = I_A$. This shows that ρ_A is an R-isomorphism. The commutativity of the diagram is evident. ♯

Proposition 2.2.9 *Let $\{M_\alpha \mid \alpha \in \Lambda\}$ be a family of right R-modules and N be a left R-module. Then there is a unique isomorphism $\overline{\phi}$ from $(\oplus \Sigma_{\alpha \in \Lambda} M_\alpha) \otimes_R N$ to $\oplus \Sigma_{\alpha \in \Lambda} (M_\alpha \otimes_R N)$ such that $\overline{\phi}(f \otimes n)(\alpha) = f(\alpha) \otimes n$. Similar result holds if N is a right R-module and M_α is left R-module for each α.*

Proof The map ϕ from $(\oplus \Sigma_{\alpha \in \Lambda} M_\alpha) \times N$ to $\oplus \Sigma_{\alpha \in \Lambda} (M_\alpha \otimes_R N)$ defined by $\phi((f, n))(\alpha) = f(\alpha) \otimes n$ is easily seen to be a balanced map. Hence, there is a unique homomorphism $\overline{\phi}$ such that $\overline{\phi}(f \otimes n)(\alpha) = f(\alpha) \otimes n$. The inverse map is an obvious map. The proof of the second part is similar. ♯

Remark 2.2.10 A free left R-module is isomorphic to direct sum of several copies of R, which are also bi-(R, R) modules. Thus, a free left (right) R-module is also a free bi-(R, R) module.

Corollary 2.2.11 *Tensor product of free left R-modules is a free left R-module. In turn, the tensor product $P \otimes Q$ of a projective right R-module P with a projective left R-module Q is a projective bi-(R, R)-module.*

Proof Since a free left R-module is direct sum of so many copies of R, and since $R \otimes R$ is isomorphic to R, the first part of the result follows from the above proposition. Further, let P be a projective bi-(R, R) module and Q a projective left R-module. Then there exists a right R-module L and a left R-module M such that $P \oplus L$ is a free R-module, and $Q \oplus M$ is also a free R-module. Since tensor products of free R-modules are free R-modules, $(P \oplus L) \otimes (Q \oplus M)$ is a free R-module. From the previous proposition, $(P \otimes Q) \oplus U$ is free, where $U = (P \otimes M) \oplus (L \otimes Q) \oplus (L \otimes M)$. Hence, $P \otimes Q$ is a projective module. ♯

Let f be a homomorphism from a right R-module M to a right R-module M', and g be a homomorphism from a left R-module N to a module N'. Then the map $f \times g$ from $M \times N$ to $M' \otimes_R N'$ defined by $(f \times g)(m, n) = f(m) \otimes g(n)$ is easily observed to be a balanced map. In turn, it induces a homomorphism $f \otimes g$ from $M \otimes_R N$ to $M' \otimes_R N'$ subject to $(f \otimes g)(m \otimes n) = f(m) \otimes g(n)$. This homomorphism $f \otimes g$ is called the tensor product of f and g. If further f' is a homomorphism from the right R-module M' to another right R-module M'' and g' is a homomorphism from the left R-module N' to another left R-module N'', then $(f' of) \otimes (g' og) = (f' \otimes g') o (f \otimes g)$. Also $I_M \otimes I_N = I_{M \otimes N}$. Thus, we have a functor $- \otimes -$ from the product category $(Mod - R) \times (R - Mod)$ to the category AB of abelian groups which is given by $(- \otimes -)(M, N) = M \otimes_R N$, and $(- \otimes -)(f, g) = f \otimes g$. This functor is called **tensor functor**. In particular, for a fixed right R-module M, we have a functor $M \otimes -$ from the category of left R-modules to the category AB of abelian groups given by $(M \otimes -)(N) = M \otimes_R N$ and $(M \otimes -)(g) = I_M \otimes g$, and for a fixed left R-module N, we have a functor $- \otimes N$ from the category of left R-modules to the category AB of abelian groups given by $(- \otimes N)(M) = M \otimes_R N$ and $(- \otimes N)(f) = f \otimes I_N$.

Evidently, tensor products of epimorphisms are epimorphisms. However, tensor products of two monomorphisms need not be a monomorphism.

Let

$$X \equiv \cdots \xrightarrow{d_{n+1}^X} X_n \xrightarrow{d_n^X} X_{n-1} \xrightarrow{d_{n-1}^X} \cdots$$

be a chain complex of right R-modules and

$$Y \equiv \cdots \xrightarrow{d_{n+1}^Y} Y_n \xrightarrow{d_n^Y} Y_{n-1} \xrightarrow{d_{n-1}^Y} \cdots$$

be a chain complexes of left R-modules. We define the tensor product $X \otimes_R Y$ of the chain complexes X and Y as follows: put $(X \otimes_R Y)_n = \oplus \Sigma_{p+q=n} X_p \otimes_R Y_q$. Define a homomorphism $d_n^{X \otimes_R Y}$ from $(X \otimes_R Y)_n$ to $(X \otimes_R Y)_{n-1}$ by defining

$$d_n^{X \otimes_R Y}(x \otimes y) = d_p^X(x) \otimes y + (-1)^p x \otimes d_q^Y(y),$$

where $x \in X_p$, $y \in Y_q$, $p + q = n$. It is easily seen that $X \otimes_R Y$ is a chain complex of abelian groups. This chain complex is called the tensor product of X and Y. Further, let $f = \{f_p \mid p \in \mathbb{Z}\}$ be a chain transformation from a chain complex X of right R-modules to a chain complex X' of right R-modules, and $g = \{g_q \mid q \in \mathbb{Z}\}$ be a chain transformation from a chain complex Y of left R-modules to a chain complex Y' of left R-modules. Then for each n, we have a homomorphism $(f \otimes g)_n$ from $(X \otimes_R Y)_n$ to $(X' \otimes_R Y')_n$ given by $(f \otimes g)_n(x \otimes y) = f_p(x) \otimes g_q(y)$, where $x \in X_p$ and $y \in Y_q$, $p + q = n$. It can be checked easily that $f \otimes g = \{(f \otimes g)_n \mid n \in \mathbb{Z}\}$ is a chain transformation from $X \otimes Y$ to $X' \otimes Y'$. Thus, tensor product defines a functor from the product category $CH_R \times_R CH$ to the category $CH_{\mathbb{Z}}$, CH_R denotes the category of chain complexes of right R-modules, $_RCH$ denotes the category of chain complexes of left R-modules, and $CH_{\mathbb{Z}}$ denotes the category of chain complexes of abelian groups. In a latter section, we shall discuss the relationship between the homologies of X, the homologies of Y, and the homologies of $X \otimes_R Y$.

Treating a left R-module A as a chain complex $(A_0 = A, \ and \ A_n = \{0\}$ for all $n \neq 0)$, we get the chain complex $X \otimes_R A$. Here $(X \otimes_R A)_n = X_n \otimes_R A$, and $d_n^{X \otimes_R A}$ is given by

$$d_n^{X \otimes_R A}(x \otimes a) = d_n^X(x) \otimes a + (-1)^n x \otimes d_n^A(a) = d_n^X(x) \otimes a.$$

The nth homology $H_n(X \otimes A)$ of $X \otimes A$ is denoted by $H_n(X, A)$, and it is termed as the nth homology of X with coefficient in A.

Tensoring is a right exact functor in the sense of the following theorem.

Theorem 2.2.12 *Let*

$$A \xrightarrow{\quad \alpha \quad} B \xrightarrow{\quad \beta \quad} C \longrightarrow 0$$

be an exact sequence of right R-modules and D be a left R-module. Then the sequence

$$A \otimes_R D \xrightarrow{\quad \alpha \otimes I_D \quad} B \otimes_R D \xrightarrow{\quad \beta \otimes I_D \quad} C \otimes_R D \longrightarrow 0$$

is exact. Also if

$$A \xrightarrow{\quad \alpha \quad} B \xrightarrow{\quad \beta \quad} C \longrightarrow 0$$

is an exact sequence of left R-modules and D is a right R-module. Then the sequence is

$$D \otimes_R A \xrightarrow{\quad I_D \otimes \alpha \quad} D \otimes_R B \xrightarrow{\quad I_D \otimes \beta \quad} D \otimes_R C \longrightarrow 0$$

Proof Since β is surjective, and the tensor products of surjective homomorphisms are surjective, $\beta \otimes I_D$ is surjective. Thus, we need to show that $ker \beta \otimes I_D = image \alpha \otimes I_D$. Again, since $\beta o \alpha = 0$, we have $0 = (\beta \otimes I_D o \alpha \otimes I_D) = (\beta o \alpha) \otimes I_D$. Hence, $image \alpha \otimes I_D \subseteq ker \beta \times I_D$. Put $image \alpha \otimes I_D = L$. By the fundamental theorem of homomorphism, we have a unique homomorphism ϕ from $(B \otimes_R D)/L$ to $C \otimes_R D$ defined by $\phi((b \otimes d) + L) = \beta(b) \otimes d$. It is sufficient to show that ϕ is an isomorphism. We construct its inverse. If $\beta(b) = \beta(b')$, then $b - b'$ belongs to $ker \beta = image \alpha$. Hence, $(b \otimes d) - (b' \otimes d) = (b - b') \otimes d$ is in L. This ensures that we have a map $(c, d) \rightsquigarrow b \otimes d + L$ from $C \times D$ to $(B \otimes_R D)/L$ where $\beta(b) = c$. This is a balanced map (verify). Hence, we have a unique homomorphism ψ from $C \otimes_R D$ to $(B \otimes_R D)/L$ such that $\psi(c \otimes d) = b \otimes d + L$, where $\beta(b) = c$. Now $(\phi o \psi)(c \otimes d) = \phi(b \otimes d + L) = \beta(b) \otimes d = c \otimes d$. This shows that $\phi o \psi$ is the identity map. Similarly, $\psi o \phi$ is also the identity map. This proves that ϕ is an isomorphism, and so $L = ker\beta \otimes I_D$. Similarly, the second statement follows. ♯

Consider the homomorphism f from $\mathbb{Z}$ to $\mathbb{Z}$ defined by $f(a) = 5a$. Then f is injective but $f \otimes I_{\mathbb{Z}_5}$ from $\mathbb{Z} \otimes_{\mathbb{Z}} \mathbb{Z}_5$ to itself is the zero map, for $f \otimes I_{\mathbb{Z}_5}(m \otimes \bar{a}) = f(m) \otimes \bar{a} = 5m \otimes \bar{a} = m \otimes 5\bar{a} = 0$. Since $\mathbb{Z} \otimes_{\mathbb{Z}} \mathbb{Z}_5 \approx \mathbb{Z}_5$ is nontrivial, $f \otimes I_{\mathbb{Z}_5}$ is not injective. This shows that tensoring is not left exact. However, we have the following proposition.

Proposition 2.2.13 *If*

$$0 \longrightarrow A \xrightarrow{\alpha} B \xrightarrow{\beta} C \longrightarrow 0$$

is a split exact sequence, then

$$0 \longrightarrow A \otimes_R D \xrightarrow{\alpha \otimes I_D} B \otimes_R D \xrightarrow{\beta \otimes I_D} C \otimes_R D \longrightarrow 0$$

is also a split exact sequence.

Proof If s is a homomorphism from B to A such that $s o \alpha = I_A$, then $(s \otimes I_D) o (\alpha \otimes I_D) = I_A \otimes I_D = I_{A \otimes D}$. This shows that $\alpha \otimes I_D$ is injective, and the result follows. ♯

Example 2.2.14 Let A be an abelian group. Then $\mathbb{Z}_m \otimes_{\mathbb{Z}} A$ is isomorphic to A/mA: Consider the exact sequence

$$0 \longrightarrow \mathbb{Z} \xrightarrow{\alpha} \mathbb{Z} \xrightarrow{\nu} \mathbb{Z}_m \longrightarrow 0$$

where α is the multiplication by m. Taking tensor product with A, and observing the fact that tensoring is right exact, we see that $\mathbb{Z}_m \otimes_{\mathbb{Z}} A$ is isomorphic to $(\mathbb{Z} \otimes_{\mathbb{Z}} A)/ker\nu_\star$. Again, $ker\nu_\star = image\alpha_\star$. The isomorphism f from $\mathbb{Z} \otimes_{\mathbb{Z}} A$ to A given by $f(n \otimes a) = na$ takes $image\alpha_\star$ to mA. The assertion follows from the fundamental theorem of homomorphism. In particular, $\mathbb{Z}_m \otimes \mathbb{Z}_n$ is isomorphic to $\mathbb{Z}_d$, where d is g.c.d of m and n.

Definition 2.2.15 A left (right) R-module D is called a **flat module** if $- \otimes_R D$ $(D \otimes_R -)$ is also left exact functor. More explicitly, D is said to be a flat module if $f \otimes I_D$ $(I_D \otimes f)$ is injective whenever f is an injective homomorphism.

Proposition 2.2.16 *Let P be a projective left (right) R-module. Then P is a flat module.*

Proof From Proposition 2.2.8, it follows that R is a flat left (right) R-module. Further, since a free R-module is direct sum of several copies of R and since the tensor product respects direct sum, it follows that every free R-module is flat. Let P be a projective left R-module. Then there is a projective left R-module Q such that $P \oplus Q$ is free. Let f be an injective homomorphism from a right R-module A to a right R-module B. Then we have the following commutative diagram:

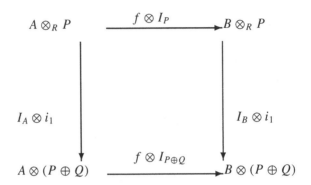

Since the vertical arrows and the bottom arrow are injective, the top row is also injective. This proves that P is flat. ♯

Let

$$0 \longrightarrow A \xrightarrow{\ \alpha\ } B \xrightarrow{\ \beta\ } C \longrightarrow 0$$

be an exact sequence of abelian groups and D be an abelian group. Let us examine the $\ker \alpha \otimes I_D$, where A, B, C, and D are abelian groups ($\mathbb{Z}$-modules). As observed above, $\alpha \otimes I_D$ need not be injective. Indeed, if $a \in A$ is such that $\alpha(a) = mb$ for some $m \in \mathbb{N}$ and $b \in B$ and $d \in D$ is such that $md = 0$, then $(\alpha \otimes I_D)(a \otimes d) = mb \otimes d = b \otimes md = 0$, whereas $a \otimes d$ may not be zero in $A \otimes D$. Further, if $\beta(b) = \beta(b')$, then $b' - b = \alpha(a')$ for a unique $a' \in A$. In turn, $\alpha(a + ma') = mb + m\alpha(a') = mb'$. Also $(a + ma') \otimes d = a \otimes d$. Thus, the above kernel element $a \otimes d$ with $md = 0$ depends only on the element $\beta(b) = c \in C$ with $\alpha(a) = mb$. Evidently, $mc = 0$. More explicitly, a triple (c, m, d), where $c \in C$, $d \in D$, and $mc = 0 = md$, determines an element $a \otimes d$ of the $\ker \alpha \otimes I_D$, where $\alpha(a) = mt(c)$, t being a section of the extension. Let us denote the set $\{(c, m, d) \in C \times \mathbb{Z} \times D \mid mc = 0 = md\}$ by X. Define the map ϕ from X to the $\ker \alpha \otimes I_D$ by $\phi((c, m, d)) = a \otimes d$, where $\alpha(a) = mt(c)$.

Let (c, m, d) and (c, m, d') be two members of X. Then $(c, m, d + d')$ also belongs to X. By definition of ϕ, $\phi((c, m, d)) = a \otimes d$, $\phi((c, m, d')) = a \otimes d'$, and $\phi((c, m, d + d')) = a \otimes (d + d') = (a \otimes d) + (a \otimes d')$, where $\alpha(a) = mt(c)$. This shows that

$$\phi((c, m, d + d')) = \phi((c, m, d)) + \phi((c, m, d')). \tag{2.1}$$

Suppose that (c, m, d) and (c', m, d) are members of X. Then $(c + c', m, d)$ also belongs to X, and $\phi((c + c', m, d) = u \otimes d$, where $\alpha(u) = mt(c + c')$. Further, $\phi((c, m, d)) = a \otimes d$, where $\alpha(a) = mt(c)$, and $\phi((c', m, d)) = a' \otimes d$, where $\alpha(a') = mt(c')$. Since $\beta(t(c) + t(c')) = \beta(t(c + c'))$, there is an element $v \in A$ such that $t(c) + t(c') = \alpha(v) + t(c + c')$. Hence, if we take $u = a + a' - mv$, then $\alpha(u) = mt(c + c'')$. This shows that $\phi((c + c', m, d) = u \otimes d = (a + a' - mv) \otimes d = a \otimes d + a' \otimes d$. This shows that

$$\phi((c + c', m, d)) = \phi((c, m, d)) + \phi((c', m, d)). \tag{2.2}$$

Next, suppose that $mnc = 0 = nd$, then it can be easily seen that

$$\phi(c, mn, d) = \phi(mc, n, d). \tag{2.3}$$

Finally, if $mc = 0 = mnd$, then again it follows that

$$\phi(c, mn, d) = \phi(c, m, nd). \tag{2.4}$$

The above discussion prompts us to have the following definition.

Definition 2.2.17 Let C and D be abelian groups. Then the abelian group generated by the set $X = \{(c, m, d) \in C \times \mathbb{N} \times D \mid mc = 0 = md\}$ of symbols subject to the relations

  (i) $(c, m, d + d') = (c, m, d) + (c, m, d')$,
 (ii) $(c + c', m, d) = (c, m, d) + (c', m, d)$,
(iii) $(c, mn, d) = (mc, n, d)$ whenever $mnc = 0 = nd$, and
 (iv) $(c, mn, d) = (c, m, nd)$ whenever $mc = 0 = mnd$

is called the **torsion product** of C and D, and it is denoted by $Tor(C, D)$.

Proposition 2.2.18 *Let C and D be abelian groups. Then*

 (i) $(0, m, d) = 0 = (c, m, 0)$ *for all $m \in \mathbb{Z}$, $c \in C$, and $d \in D$.*
(ii) $(c, 0, d) = 0$ *for all $c \in C$ and $d \in D$.*

Proof (i) Since $(0, m, d) = (0 + 0, m, d) = (0, m, d) + (0, m, d)$, $(0, m, d) = 0$ for all $m \in \mathbb{Z}$ and $d \in D$. Similarly, $(c, m, 0) = 0$ for all $m \in \mathbb{Z}$ and $c \in C$.
   (ii) $(c, 0, d) = (c, 0 \cdot 0, d) = (0c, 0, d) = (0, 0, d) = 0$. ♯

Corollary 2.2.19 *If C is torsion free or D is torsion free, then $Tor(C, D) = 0$.*

Proof Suppose that C is torsion free. Then the generating set X of $Tor(C, D)$ is $\{(0, m, 0) \mid m \in \mathbb{Z}\} \bigcup \{(c, 0, d) \mid c \in C \text{ and } d \in D\}$. From the previous proposition $X = \{0\}$. ♯

Proposition 2.2.20 $Tor(C, D)$ *is naturally isomorphic to* $Tor(D, C)$.

Proof The map τ from the generating set X of $Tor(C, D)$ to $Tor(D, C)$ given by $\tau(c, m, d) = (d, m, c)$ respects the defining relations. Hence, it induces a homomorphism $\bar{\tau}$ from $Tor(C, D)$ to $Tor(D, C)$ whose restriction to X is τ. Similarly, we have map ρ from the generating set $Y = \{(d, m, c) \mid d \in D, \ m \in \mathbb{Z}, c \in C, \ and \ md = 0 = mc\}$ of $Tor(D, C)$ to $Tor(C, D)$ which respects the defining relations. In turn, ρ induces a homomorphism $\bar{\rho}$ from $Tor(D, C)$ to $Tor(C, D)$ whose restriction to Y is ρ. Clearly $\bar{\rho}o\bar{\tau}$ is identity map on X. Hence, $\bar{\rho}o\bar{\tau} = I_{Tor(C,D)}$. Similarly, $\bar{\tau}o\bar{\rho} = I_{Tor(D,C)}$. ♯

Proposition 2.2.21 *(i)* $Tor(C \oplus A, D) \approx Tor(C, D) \oplus Tor(A, D)$ *and* *(ii)* $Tor(C, A \oplus D) \approx Tor(C, A) \oplus Tor(C, D)$.

Proof (i) It is easily seen that the association $(c \oplus a, m, d) \rightarrow (c, m, d) \oplus (a, m, d)$ induces the required isomorphism. Similarly, (ii) follows. ♯

Thus, to compute $Tor(C, D)$ for finitely generated abelian groups, it is sufficient to compute $Tor(\mathbb{Z}_m, \mathbb{Z})$ and $Tor(\mathbb{Z}_m, \mathbb{Z}_n)$. From Corollary 2.2.19, $Tor(\mathbb{Z}_m, \mathbb{Z}) = 0$.

If f is a homomorphism from C to A and g a homomorphism from D to B, then the map (f, g) from the generating set X of $Tor(C, D)$ to $Tor(A, B)$ given by $(f, g)(c, m, d) = (f(c), m, g(d))$ respects the defining relations for $Tor(C, D)$. As such, it induces a unique homomorphism $Tor(f, g)$ from $Tor(C, D)$ to $Tor(A, B)$ whose restriction to X is (f, g). Evidently, $Tor(I_C, I_D) = I_{Tor(C,D)}$, and Tor $(f', g')oTor(f, g) = Tor(f'of, g'og)$, where f' is a homomorphism from A to A' and g' a homomorphism from B to B'. Thus, we have a functor $Tor(-, -)$ from $AB \times AB$ to AB which is given by $Tor(-, -)(C, D) = Tor(C, D)$ and $Tor(-, -)(f, g) = Tor(f, g)$.

Proposition 2.2.22 *Let A be an abelian group. Then there is an isomorphism ϕ_A from $Tor(\mathbb{Z}_m, A)$ to the m-torsion $A_m = \{a \in A \mid ma = 0\}$ of A. Further, ϕ_A is natural in A in the sense that if f is a homomorphism from A to B, then the following diagram is commutative:*

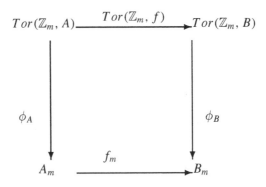

where f_m is the restriction of f to A_m.

Proof Define the map ϕ_A from A_m to $Tor(\mathbb{Z}_m, A)$ by $\phi_A(a) = (\overline{1}, m, a)$. Evidently, ϕ_A is a homomorphism from A_m to $Tor(\mathbb{Z}_m, A)$. We construct its inverse ψ_A. The generating set X of $Tor(\mathbb{Z}_m, A)$ is

$$\{(\overline{r}, t, a) \mid t\overline{r} = \overline{0} \in \mathbb{Z}_m, \text{ and } ta = 0\}.$$

Suppose that $(\overline{r}, t, a) \in X$. Then $t\overline{r} = \overline{0} \in \mathbb{Z}_m$. This means m divides tr. Also, $\{(\overline{r}, t, a) = (\overline{1}, rt, a) = (\overline{1}, m, \frac{rt}{m}a)$. Define a map ξ_A from X to A_m by $\xi_A((\overline{r}, t, a)) = \frac{tr}{m}\overline{a}$. Clearly, ξ_A respects the defining relation, and so it defines a homomorphism ψ_A from $Tor(\mathbb{Z}_m, A)$ to A_m whose restriction to X is ξ_A. Evidently, ψ_A is the inverse of ϕ_A. The commutativity of the diagram in the statement of the theorem follows from the definitions of ϕ_A and of ϕ_B. ♯

Corollary 2.2.23 *$Tor(\mathbb{Z}_m, \mathbb{Z}_n)$ is isomorphic to $\mathbb{Z}_d$, where d is the greatest common divisor of m and n.*

Proof Clearly, $(\mathbb{Z}_n)_m = \{\overline{a} \in \mathbb{Z}_n \mid m\overline{a} = \overline{0}\}$. The map η from $\mathbb{Z}_d$ to $(\mathbb{Z}_n)_m$ defined by $\eta(\overline{r}) = \overline{\frac{nr}{d}}$ is an isomorphism. ♯

Corollary 2.2.24 *$Tor(C, D) = 0$ for all abelian group C if and only if D is torsion free. In particular, $Tor(C, D) = 0$ if C or D is projective $\mathbb{Z}$-module.* ♯

Let

$$E \equiv 0 \longrightarrow A \overset{\alpha}{\to} B \overset{\beta}{\to} C \longrightarrow 0$$

be an exact sequence of abelian groups, and D be an abelian group. Let t be a section of the above short exact sequence. As described in Definition 2.2.17, we have a map ϕ from the generating set $X = \{(c, m, d) \in C \times \mathbb{Z} \times D \mid mc = 0 = md\}$ of $Tor(C, D)$ to $A \otimes D$ defined by $\phi((c, m, d)) = a \otimes d$, where $\alpha(a) = mt(c)$. Further, it respects the defining relations. Hence, it defines a homomorphism ∂_E from $Tor(C, D)$ to $A \otimes D$ whose restriction to X is ϕ. This homomorphism is again called a connecting homomorphism. The functoriality of the Tor and the tensor functors implies that the connecting homomorphism is natural in E and also in D. The following proposition is an easy verification.

Proposition 2.2.25 *Let E be the short exact sequence of abelian groups as given above, and m be a natural number. Let η_A denote the isomorphism from $A \otimes \mathbb{Z}_m$ to A/mA given by $\eta_A(a \otimes \overline{r}) = rA + mA$. Let t be a section of E. Then we have a map χ_E from C_m to A/mA given by $\chi(c) = a + mA$, where $mt(x) = \alpha(a)$ such that following commutative diagram is commutative:*

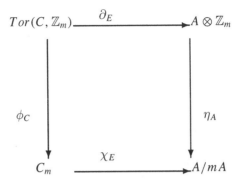

♯

Theorem 2.2.26 *Let*

$$E \equiv 0 \longrightarrow A \overset{\alpha}{\to} B \overset{\beta}{\to} C \longrightarrow 0$$

be an exact sequence of abelian groups, and D be an abelian group. Then we have the following natural (in E and in D) exact sequence of abelian groups:

$$0 \longrightarrow Tor(A, D) \overset{Tor(\alpha, I_D)}{\to} Tor(B, D) \overset{Tor(\beta, I_D)}{\to} Tor(C, D) \overset{\partial_E}{\to} A \otimes D \overset{\alpha \otimes I_D}{\to}$$

$$B \otimes D \overset{\beta \otimes I_D}{\to} C \otimes D \longrightarrow 0.$$

Proof In the light of Theorem 2.2.12, we need to show the exactness at $A \otimes D$, $Tor(C, D)$, $Tor(B, D)$, and at $Tor(A, D)$. Suppose that $D = \mathbb{Z}$. Then we have the following commutative diagram:

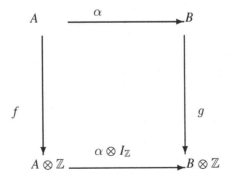

where f is the isomorphism given by $f(a) = 1 \otimes a$ and g is the isomorphism given by $f(b) = 1 \otimes b$. It follows that $\alpha \otimes I_{\mathbb{Z}}$ is injective. Hence, the sequence

$$0 \longrightarrow A \otimes \mathbb{Z} \overset{\alpha \otimes I_\mathbb{Z}}{\rightarrow} B \otimes \mathbb{Z} \overset{\beta \otimes I_\mathbb{Z}}{\rightarrow} C \otimes \mathbb{Z} \longrightarrow 0$$

is exact. Since $Tor(A, \mathbb{Z}) = Tor(B, \mathbb{Z}) = Tor(C, \mathbb{Z}) = 0$, the result is verified for $D = \mathbb{Z}$.

Next, suppose that $D = \mathbb{Z}_m$. From Proposition 2.2.22, we have $\phi_B \circ Tor(\alpha, I_{\mathbb{Z}_m}) = \alpha_m \circ \phi_A$ and $\phi_C \circ Tor(\beta, I_{\mathbb{Z}_m}) = \beta_m \circ \phi_B$. From Proposition 2.2.25, $\eta_A \circ \partial_E = \chi_E \circ \phi_C$. Again, $\eta_B \circ \alpha \otimes I_{\mathbb{Z}_m} = \alpha^m \circ \eta_A$, where α^m is the homomorphism from A/mA to B/mB defined by $\alpha^m(a + mA) = \alpha(a) + mB$. Since ϕ_A, ϕ_B, ϕ_C, η_A, and η_B are isomorphisms, it is sufficient to observe that the following sequence is exact:

$$0 \to A_m \overset{\alpha_m}{\rightarrow} B_m \overset{\beta_m}{\rightarrow} C_m \overset{\chi_E}{\rightarrow} A/mA \overset{\alpha^m}{\rightarrow} B/mB.$$

This verifies the result for $D = \mathbb{Z}_m$. Since tensor functor $\otimes$ and the torsion functor respect the direct sum, the result follows for all finitely generated abelian groups. The result follows if we observe that torsion and the tensor functors commute with direct limit, homology functor commutes with direct limit, and every abelian group D is direct limit of its finitely generated subgroups. ♯

Remark 2.2.27 The reader can easily observe that the above discussions (including the definitions) and the results hold good for modules over principal ideal domains.

Let C and D be a $\mathbb{Z}$-modules. Since submodule of a projective $\mathbb{Z}$-module is a projective, we have a projective presentation

$$\Gamma(C) \overset{\epsilon}{\to} C \to 0 = E \equiv 0 \longrightarrow P_1 \overset{i}{\to} P_0 \overset{\epsilon}{\to} C \longrightarrow 0$$

of C, where $P_1 = ker\,\epsilon$. Further, every projective resolution of C is chain equivalent to the above projective resolution. From Theorem 2.2.26, and the fact that $Tor(P, D) = 0$ whenever P is projective, we get the following exact sequence:

$$0 \longrightarrow Tor(C, D) \overset{\partial_E}{\to} P_1 \otimes D \overset{i \otimes I_D}{\rightarrow} P_0 \otimes D \overset{\epsilon \otimes I_D}{\rightarrow} C \otimes D \to 0.$$

Also

$$\Gamma(C) \otimes D \equiv 0 \longrightarrow P_1 \otimes D \overset{i \otimes I_D}{\rightarrow} P_0 \otimes D \longrightarrow 0.$$

Thus,

$$H_0(\Gamma(C), D) = (P_0 \otimes D)/image\, i \otimes I_D = (P_0 \otimes D)/ker\,\epsilon \otimes I_D \approx C \otimes D,$$

and

$$H_1(\Gamma(C), D) = ker\,\alpha \otimes I_D \approx Tor(C, D).$$

This prompts us to define the nth torsion product $Tor_n^R(C, D)$ of a right R-module C with a left R-module D as follows.

Consider the category $\wp(Mod - R)$ whose objects are positive projective acyclic chain complexes of right R-modules and morphisms are homotopy classes of chain transformations. Let Γ be the choice functor from $Mod - R$ to $\wp(Mod - R)$ which is an equivalence (see Remark 2.1.5). Suppose that

$$\Gamma(C) \overset{\epsilon}{\to} C \to 0 \equiv \cdots \overset{d_{n+i}}{\to} P_n \overset{d_{n-1}}{\to} P_{n-1} \cdots \overset{d_2}{\to} P_1 \overset{d_1}{\to} P_0 \overset{\epsilon}{\to} C \longrightarrow 0,$$

and D is a left R-module. The nth homology $H_n(\Gamma(C), D)$ of $\Gamma(C)$ with coefficient in D is denoted by $Tor_n^R(C, D)$, and it is called the **n^{th} torsion product** of C with D. In the light of Remark 2.1.5, $Tor_n^R(-, -)$ is a functor from the product category $Mod\text{-}R \times R\text{-}Mod$ to the category AB of abelian groups. It follows from the earlier discussion that $Tor_0^{\mathbb{Z}}(C, D)$ is naturally isomorphic to $C \otimes_{\mathbb{Z}} D$, and $Tor_1^{\mathbb{Z}}(C, D)$ is naturally isomorphic to $Tor(C, D)$.

Proposition 2.2.28 *The functor $Tor_0^R(-, -)$ is naturally isomorphic to the functor* $- \otimes_R -$.

Proof Let C be a right R-module and D be a left R-module. Consider the choice projective resolution

$$\Gamma(C) \overset{\epsilon}{\to} C \to o \equiv \cdots \overset{d_3}{\to} P_2 \overset{d_2}{\to} P_1 \overset{d_1}{\to} P_0 \overset{\epsilon}{\to} C \longrightarrow 0$$

of C. Since tensoring is a right exact functor, we have the exact sequence

$$P_1 \otimes_R D \overset{d_1 \otimes I_D}{\to} P_0 \otimes_R D \overset{\epsilon \otimes I_D}{\to} C \otimes_R D \longrightarrow 0.$$

In turn, $ker(\epsilon I_D) = image(d_1 \otimes I_D)$. Thus, we have an isomorphism $\mu_{C,D}$ from $Tor_0^R(C, D) = (P_0 \otimes D)/image d_1 \otimes I_D$ to $C \otimes_R D$ given by $\mu_{C,D}(x \otimes d + image d_1 \otimes I_D) = \epsilon(x) \otimes d$. We show that $\mu_{C,D}$ is natural in C and D. Let ϕ be a homomorphism from C to C' and ψ be a homomorphism from D to D'. Let

$$\Gamma(C') \overset{\epsilon'}{\to} C' \to o \equiv \cdots \overset{d_3'}{\to} P_2' \overset{d_2'}{\to} P_1' \overset{d_1'}{\to} P_0' \overset{\epsilon'}{\to} C' \longrightarrow 0$$

be the choice projective resolution of C', and $\Gamma(\phi) = [f]$, where $f = \{f_n : P_n \to P_n' \mid n \geq 0\}$ is a chain transformation from $\Gamma(C)$ to $\Gamma(C')$ which lifts ϕ. Evidently, $(\phi \otimes I_D) o \mu_{C,D} = \mu_{C',D'} o (f_0 \otimes \psi)$. Thus, $\mu_{C',D'} o Tor(\phi, \psi) = (\phi \otimes \psi) o \mu_{C,D}$. This proves the naturality of μ. ♯

Proposition 2.2.29 *If D is a flat left R-module, then $Tor_n^R(C, D) = 0$ for all $n \geq 1$.*

Proof Consider the choice projective resolution

$$\Gamma(C) \overset{\epsilon}{\to} C \to o \equiv \cdots \overset{d_3}{\to} P_2 \overset{d_2}{\to} P_1 \overset{d_1}{\to} P_0 \overset{\epsilon}{\to} C \longrightarrow 0$$

of C. Suppose that $n \geq 1$. We have the following sequence:

$$0 \longrightarrow ker d_n \overset{i}{\to} P_n \overset{d_n}{\to} P_{n-1}.$$

Further, since $ker d_n = image d_{n+1}$, d_{n+1} may be treated as a surjective homomorphism from P_{n+1} to $ker d_n$. Since D is flat, we have the exact sequence

$$0 \longrightarrow ker d_n \otimes_R D \overset{i \otimes I_D}{\to} P_n \otimes D \overset{d_n \otimes I_D}{\to} P_{n-1} \otimes D.$$

Since tensoring is right exact, $image d_{n+1} \otimes I_D = (i \otimes I_D)(ker d_n \otimes_R D) = ker(d_n \otimes I_D)$. By definition,

$$Tor_n^R(C, D) = (Ker d_n \otimes I_D)/(image d_{n+1} \otimes I_D) = H_n(\Gamma(C), D) = 0.$$

♯

Corollary 2.2.30 *If C is projective right R-module or if D is a projective left R-module, then $Tor_n^R(C, D) = 0$ for all $n \geq 1$.*

Proof Suppose that D is a projective left R-module. By Proposition 2.2.16, D is flat. From the above proposition, $Tor_n^R(C, D) = 0$ for all $n \geq 1$. Suppose that C is projective, then

$$0 \longrightarrow 0 \longrightarrow \cdots \longrightarrow 0 \longrightarrow C \overset{I_C}{\to} C \longrightarrow 0$$

is a projective resolution of C. Evidently, $Tor_n^R(C, D) = 0$ for all $n \geq 1$. ♯

Proposition 2.2.31 *Let*

$$E \equiv 0 \longrightarrow A \overset{\alpha}{\to} B \overset{\beta}{\to} C \longrightarrow 0$$

be a short exact sequence of left R-modules, and D be a right R-module. Then for each n we have a natural (in E and D) connecting homomorphisms $\partial_n(E)$ from $Tor_n^R(D, C)$ to $Tor_{n-1}^R(D, A)$ such that the following sequence is exact.

$$\cdots \overset{\partial_{n+1}(E)}{\to} Tor_n^R(D, A) \overset{Tor_n^R(I_D, \alpha)}{\to} Tor_n^R(D, B) \overset{Tor_n^R(I_D, \beta)}{\to} Tor_n^R(D, C) \overset{\partial_n(E)}{\to} \cdots.$$

Proof Consider the choice projective acyclic chain complex $\Gamma(D)$ over D. Since every projective module is flat, we have the short exact sequence

$$0 \longrightarrow \Gamma(D) \otimes_R A \overset{I_{\Gamma(D)} \otimes \alpha}{\to} \Gamma(D) \otimes_R B \overset{I_{\Gamma(D)} \otimes \beta}{\to} \Gamma(D) \otimes_R C \longrightarrow 0$$

of chain complexes. The result follows if we apply Theorem 1.3.1 to the above short exact sequence of chain complexes. ♯

Proposition 2.2.32 *Let*

$$E \equiv 0 \longrightarrow A \overset{\alpha}{\to} B \overset{\beta}{\to} C \longrightarrow 0$$

be a short exact sequence of right R-modules, and D be a left R-module. Then for each n we have a natural (in E and D) connecting homomorphisms $\partial_n(E)$ from $Tor_n^R(C, D)$ to $Tor_{n-1}^R(A, D)$ such that the following sequence is exact:

$$\cdots \overset{\partial_{n+1}(E)}{\to} Tor_n^R(A, D) \overset{Tor_n^R(\alpha, I_D)}{\to} Tor_n^R(B, D) \overset{Tor_n^R(\beta, I_D)}{\to} Tor_n^R(C, D) \overset{\partial_n(E)}{\to} \cdots .$$

Proof As in the proof of Proposition 2.1.8, we have a split short exact sequence

$$0 \longrightarrow \Gamma(A) \overset{i_1}{\to} \Gamma(B) \overset{p_2}{\to} \Gamma(C) \longrightarrow 0$$

of positive acyclic projective chain complexes of right R-modules, where $\Gamma(A)$ is projective chain complex over A, $\Gamma(B)$ is projective chain complex over B, and $\Gamma(C)$ is projective chain complex over C. In turn, we have the short exact sequence

$$0 \longrightarrow \Gamma(A) \otimes_R D \overset{i_1 \otimes I_D}{\to} \Gamma(B) \otimes_R D \overset{p_2 \otimes I_D}{\to} \Gamma(C) \otimes_R D \longrightarrow 0$$

of chain complexes of abelian groups. The result follows if we apply Theorem 1.3.1 to the above short exact sequence of chain complexes. ♯

Next, consider the category $\wp(R\text{-}Mod)$ (see Remark 2.1.5) whose objects are positive acyclic projective chain complexes of left R-modules, and morphisms are homotopy classes of chain transformations. Again, let $\tilde{\Gamma}$ denote the choice functor from $R\text{-}Mod$ to $\wp(R\text{-}Mod)$. Let D be a left R-module, and C be a right R-module. Then, as above, we have a functor $\overline{Tor_n^R}(-, -)$ from the product category $(Mod\text{-}R) \times (R\text{-}Mod)$ to AB which is given by $\overline{Tor_n^R}(C, D) = H_n(C, \tilde{\Gamma}(D)) = H_n(C \otimes \tilde{\Gamma}(D))$.

Imitating the proofs of Proposition 2.2.28, Proposition 2.2.29, Corollary 2.2.30, Proposition 2.2.31, and Proposition 2.2.32, respectively, we can establish the following five results.

Proposition 2.2.33 *The functor $\overline{Tor_0^R}(-, -)$ is naturally isomorphic to the functor $- \otimes_R -$.*

Proposition 2.2.34 *If D is a flat left R-module, then $\overline{Tor_n^R}(C, D) = 0$ for all $n \geq 1$.* ♯

Corollary 2.2.35 *If C is projective right R-module or if D is a projective left R-module, then $\overline{Tor_n^R}(C, D) = 0$ for all $n \geq 1$.* ♯

Proposition 2.2.36 *Let*

$$E \equiv 0 \longrightarrow A \xrightarrow{\alpha} B \xrightarrow{\beta} C \longrightarrow 0$$

be a short exact sequence of right R-modules, and D be a left R-module. Then for each n we have a natural (in E and D) connecting homomorphisms $\overline{\partial}_n(E)$ from $\overline{Tor}_n^R(C, D)$ to $\overline{Tor}_{n-1}^R(A, D)$ such that the following sequence is exact.

$$\cdots \xrightarrow{\overline{\partial}_{n+1}(E)} \overline{Tor}_n^R(A, D) \xrightarrow{\overline{Tor}_n^R(\alpha, I_D)} \overline{Tor}_n^R(B, D) \xrightarrow{\overline{Tor}_n^R(\beta, I_D)} \overline{Tor}_n^R(C, D) \xrightarrow{\overline{\partial}_n(E)} \cdots .$$

♯

Proposition 2.2.37 Let

$$E \equiv 0 \longrightarrow A \xrightarrow{\alpha} B \xrightarrow{\beta} C \longrightarrow 0$$

be a short exact sequence of left R-modules, and D be a right R-module. Then for each n we have a natural (in E and D) connecting homomorphisms $\overline{\partial}_n(E)$ from $\overline{Tor}_n^R(D, C)$ to $\overline{Tor}_{n-1}^R(D, A)$ such that the following sequence is exact.

$$\cdots \xrightarrow{\overline{\partial}_{n+1}(E)} \overline{Tor}_n^R(D, A) \xrightarrow{\overline{Tor}_n^R(I_D, \alpha)} \overline{Tor}_n^R(D, B) \xrightarrow{\overline{Tor}_n^R(I_D, \beta)} \overline{Tor}_n^R(D, C) \xrightarrow{\overline{\partial}_n(E)} \cdots .$$

♯

Theorem 2.2.38 The functors $Tor_n^R(-, -)$ and $\overline{Tor}_n^R(-, -)$ are naturally isomorphic.

Proof The proof is by the induction on n. For $n = 0$, it follows from Proposition 2.2.28 and Proposition 2.2.33 that $Tor_0^R(-, -)$ and $\overline{Tor}_0^R(-, -)$ are naturally isomorphic. Let us again denote this natural isomorphism by $\mu_0(-, -)$. Consider the choice projective resolution

$$\Gamma(C) \xrightarrow{\epsilon} C \to o \equiv \cdots \xrightarrow{d_3} P_2 \xrightarrow{d_2} P_1 \xrightarrow{d_1} P_0 \xrightarrow{\epsilon} C \longrightarrow 0$$

of C. We have the short exact sequence

$$E \equiv 0 \longrightarrow \ker \epsilon \xrightarrow{i} P_0 \xrightarrow{\epsilon} C \longrightarrow 0.$$

From Corollaries 2.2.29 and 2.2.34, and the fact that P_0 is projective, it follows that $Tor_1^R(P_0, D) = 0 = \overline{Tor}_1^R(P_0, D)$. Further, from Propositions 2.2.32 and 2.2.36, we have the following commutative diagram:

$$
\begin{array}{ccccccc}
0 & \longrightarrow & Tor_1^R(C,D) & \xrightarrow{\;\partial_1(E)\;} & Tor_0^R(ker\,\epsilon, D) & \xrightarrow{\;Tor_0^R(i,\,I_D)\;} & Tor_0^R(P_0, D) \\
& & \Big\downarrow \mu_0(ker\,\epsilon, D) & & \Big\downarrow & & \Big\downarrow \mu_0(P_0, D) \\
0 & \longrightarrow & \overline{Tor}_1^R(C,D) & \xrightarrow{\;\overline{\partial}_1(E)\;} & \overline{Tor}_0^R(ker\,\epsilon, D) & \xrightarrow{\;\overline{Tor}_0^R(i,\,I_D)\;} & \overline{Tor}_0^R(P_0, D)
\end{array}
$$

Since the connecting homomorphisms $\partial_1(E)$ and $\overline{\partial}_1(E)$ are natural injections, and the vertical map $\mu_0(ker\,\epsilon, D)$ is a natural isomorphism, we have a unique natural isomorphism $\mu_1(C, D)$ so that the above diagram remains commutative with the first vertical arrow as $\mu_1(C, D)$. Assume that $\mu_n(-, -)$ is a natural isomorphism from $Tor_n^R(-, -)$ to $\overline{Tor}_n^R(-, -)$. Since $Tor_n^R(P_0, D) = 0$ for all $n \geq 1$, using Propositions 2.2.32 and 2.2.36, we have the following commutative diagram:

$$
\begin{array}{ccccccc}
0 & \longrightarrow & Tor_{n+1}^R(C,D) & \xrightarrow{\;\partial_{n+1}(E)\;} & Tor_n^R(ker\,\epsilon, D) & \longrightarrow & 0 \\
& & \Big\downarrow & & \Big\downarrow \mu_n(ker\,\epsilon, D) & & \\
0 & \longrightarrow & \overline{Tor}_{n+1}^R(C,D) & \xrightarrow{\;\overline{\partial}_{n+1}(E)\;} & \overline{Tor}_n^R(ker\,\epsilon, D) & \longrightarrow & 0
\end{array}
$$

The connecting homomorphisms are natural isomorphisms, and by the assumption, $\mu_n(ker\,\epsilon, D)$ is also a natural isomorphism. Evidently, we have a unique natural isomorphism $\mu_{n+1}(C, D)$ from $Tor_{n+1}^R(C, D)$ to $\overline{Tor}_{n+1}^R(C, D)$ which makes the above diagram commutative. This completes the proof. ♯

Exercises

2.2.1 Show that $A \otimes_{\mathbb{Z}} A = 0$ implies that $A = 0$.

2.2.2 Show that $Tor_{\mathbb{Z}}(\mathbb{Q}/\mathbb{Z}, A)$ is isomorphic to the torsion part of A.

2.2.3 Describe the torsion product $Tor_{\mathbb{Z}}(\mathbb{Z}_{(p)}, A)$, where p is a prime.

2.2.4 Describe $Tor_{\mathbb{Z}}(\mathbb{Q}, A)$ for a finite abelian group A.

2.2.5 Characterize an abelian group A such that $Tor_{\mathbb{Z}}(A, B) = 0$ for all abelian groups B.

2.2.6 Show that $Tor_{\mathbb{Z}}(A, B)$ is isomorphic to $Tor_{\mathbb{Z}}(Tor(A), Tor(B))$.

2.2.7 Show that $EXT_R^n(A, -) = 0$ implies that $Tor_n^R(A, -) = 0$.

2.2.8 Describe natural transformations from $EXT_{\mathbb{Z}}(\mathbb{Z}_m, -)$ to $EXT_{\mathbb{Z}}(\mathbb{Z}_n, -)$.

2.2.9 Describe natural transformations from $- \otimes_R B$ to $- \otimes_R C$.

2.2.10 Let R be a principal ideal domain. Show that $EXT_R(A, -)$ is a right exact functor.

2.2.11 Check if $\eta = \{\eta_{A,F} \mid A \in \Sigma \ and \ F \in SET^{\Sigma}\}$ is a natural equivalence.

2.2.12 Show that the Tor functor preserves direct limits.

2.3 Abstract Theory of Derived Functors

So far, we introduced and studied the bi-functors $EXT_R^n(-, -)$ and Tor_n^R from the category of R-modules to the category AB of abelian groups as derived functors of Hom and tensor functors. There are many other important abelian categories with enough injectives or with enough projectives such as the category of sheaves of modules over ringed spaces, category of vector bundles, and also additive functors such as section functors. The corresponding derived functors have very significant applications in geometry, topology, and representation theory. Some of these applications may also appear in the following chapters of this book. We shall quickly develop the general theory of abstract-derived functors. The proofs of the results are mostly the imitations of the corresponding results for $EXT_R^n(-, -)$ and $Tor_R^R(-, -)$ functors.

Let Σ be an abelian category with enough projectives. Consider the category $\wp(\Sigma)$ of positive projective acyclic chain complexes, where morphisms are homotopy classes of chain transformations, and also a choice functor Γ from the category Σ to $\wp(\Sigma)$ (see Remark 2.1.5). Note that Γ is an equivalence. Let Ω be abelian category. Let F be an additive functor from Σ to Ω. Thus, the map $f \mapsto F(f)$ from $Mor_{\Sigma}(A, B)$ to $Mor_{\Omega}(F(A), F(B))$ is a homomorphism of groups. Let C be an object of Σ, and

$$\Gamma(C) \overset{\epsilon}{\to} C \to 0 \equiv \cdots \overset{d_{n+1}}{\to} P_n \overset{d_n}{\to} \cdots \overset{d_1}{\to} P_0 \overset{\epsilon}{\to} C \longrightarrow 0$$

be the choice projective resolution of C. Then we have the chain complex $F(\Gamma(C))$ in Ω given by

$$F(\Gamma(C)) \equiv \cdots \overset{F(d_{n+1})}{\to} F(P_n) \overset{F(d_n)}{\to} \cdots \overset{F(d_1)}{\to} F(P_0) \longrightarrow 0.$$

This gives us a functor $L_n F$ from Σ to Ω defined by $L_n F(C) = H_n(F(\Gamma(C)))$. The functor $L_n F$ is called the **n^{th} left $-$ derived functor** of F.

Thus, $EXT_R^n(A, -)$ is the nth left-derived functor of $Hom_R(A, -)$, and Tor_n^R $(A, -)$ is the nth left-derived functor of $A \otimes_R -$.

Suppose that Σ and Ω are small abelian categories with Σ having enough projectives. Let Ω^{Σ} denote the category whose objects are additive functors from Σ to Ω

and the morphisms between two functors are natural transformations. We have taken Σ and Ω to be small categories so that the class of natural transformation between two functors becomes a set (in general, Ω^{Σ} need not be a category). We define the evaluation functor E_v from $\Omega^{\Sigma} \times \Sigma$ to Ω as follows: define $E_v(F, A) = F(A)$, and $E_v(\eta, f) = G(f) o \eta_A = \eta_B o F(f)$, where (η, f) is a morphism from (F, A) to (G, B) in the category $\Omega^{\Sigma} \times \Sigma$.

Now, we define the left-derived functors of the evaluation functor E_v as follows: let F and G be two objects in Ω^{Σ}. Let η be a natural transformation from F to G. Then for each $A \in Obj \Sigma$, η induces a chain transformation $\overline{\eta_A}$ from $F(\Gamma(A))$ to $G(\Gamma(A))$ such that the diagram

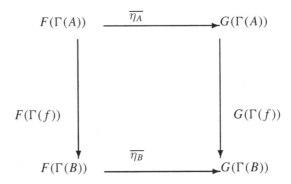

is commutative, where $f \in Mor_{\Sigma}(A, B)$. This gives us a morphism

$$L_n(\eta, f) = H_n(\overline{\eta_B} o F(\Gamma(f)) = H_n(G(\Gamma(f) o \overline{\eta_A})$$

from $L_n F(A) = H_n(F(\Gamma(A)))$ to $L_n G(B) = H_n(G(\Gamma(B)))$. Thus, we have a functor $L_n(-, -)$ from $\Omega^{\Sigma} \times \Sigma$ to Ω defined by $L_n(-, -)(F, A) = L_n F(A)$, and $L_n(-, -)(\eta, f) = L_n(\eta, f)$. The functor $L_n(-, -)$ is called the nth left-derived functor of the evaluation functor E_v. Fixing F, we get the left-derived functor $L_n F = L_n(F, -)$ of F.

Proposition 2.3.1 *Let P be a projective object in Σ. Then $L_n F(P) = 0$ for all $n \geq 1$, and $L_0 F(P) = P$ for all additive functors F.*

Proof The result follows if we observe that

$$\cdots 0 \longrightarrow 0 \longrightarrow \cdots \longrightarrow 0 \longrightarrow P \overset{I_P}{\to} P \longrightarrow 0$$

is a projective resolution of P, and as such, all projective resolutions of P are chain equivalent to the above projective resolution. ♯

Recall that a functor F is said to be a right exact functor if given any short exact sequence

$$0 \longrightarrow A \xrightarrow{\alpha} B \xrightarrow{\beta} C \longrightarrow 0$$

in Σ, the sequence

$$F(A) \xrightarrow{F(\alpha)} F(B) \xrightarrow{F(\beta)} F(C) \longrightarrow 0$$

is also exact in Ω. It is said be a left exact functor if

$$0 \longrightarrow F(A) \xrightarrow{F(\alpha)} F(B) \xrightarrow{F(\beta)} F(C)$$

is exact. It is said to be an exact functor if

$$0 \longrightarrow F(A) \xrightarrow{F(\alpha)} F(B) \xrightarrow{F(\beta)} F(C) \longrightarrow 0$$

is exact.

Proposition 2.3.2 *Let F be a right exact additive functor from Σ to Ω, where Σ and Ω are as above. Then the functor L_0F is naturally equivalent to F. Indeed, if we denote the subcategory of Ω^Σ consisting of right exact functors by $R\Omega^\Sigma$, then the functor $L_0(-, -)$ restricted to $R\Omega^\Sigma \times \Sigma$ is naturally equivalent to the evaluation functor E_v restricted to $R\Omega^\Sigma \times \Sigma$.*

Proof Let

$$\Gamma(A) \xrightarrow{\epsilon} A \to 0 \equiv \cdots \xrightarrow{d_{n+1}} P_n \xrightarrow{d_n} \cdots \xrightarrow{d_1} P_0 \xrightarrow{\epsilon} A \longrightarrow 0$$

be the choice projective resolution of A. Since F is right exact,

$$F(P_1) \xrightarrow{F(d_1)} F(P_0) \xrightarrow{F(\epsilon)} F(A) \longrightarrow 0$$

is an exact sequence. Evidently, $L_0F(A) = H_0(F(\Gamma(A))) = F(P_0)/image F(d_1) = F(P_0)/ker\epsilon$ is isomorphic to $F(A)$, and the isomorphism is natural in A. It is also easy to observe that the isomorphism is natural in F also. ♯

Proposition 2.3.3 *Let F be a right exact additive functor from Σ to Ω, where Σ and Ω are as above. Let*

$$\Gamma(A) \xrightarrow{\epsilon} A \to 0 \equiv \cdots \xrightarrow{d_{n+1}} P_n \xrightarrow{d_n} \cdots \xrightarrow{d_1} P_0 \xrightarrow{\epsilon} A \longrightarrow 0$$

be the choice projective resolution of A. Then for each n, we have a natural morphism ϕ_n^F from $L_nF(A)$ to $F(Kerd_{n-1})$ such that the sequence

$$0 \longrightarrow L_nF(A) \xrightarrow{\phi_n^F} F(Kerd_{n-1}) \xrightarrow{F(i)} F(P_{n-1})$$

is exact.

Proof We have the following commutative diagram:

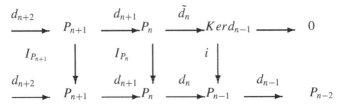

where the rows are exact. Here, $\tilde{d}_n$ is the obvious morphism induced by d_n. Applying the functor F, we get the following commutative diagram:

$$
\begin{array}{ccccccc}
F(P_{n+1}) & \xrightarrow{F(d_{n+1})} & F(P_n) & \xrightarrow{F(\tilde{d}_n)} & F(Ker d_{n-1}) & \longrightarrow & 0 \\
\downarrow{I_{F(P_{n+1})}} & & \downarrow{I_{F(P_n)}} & & \downarrow{F(i)} & & \\
F(P_{n+1}) & \xrightarrow{F(d_{n+1})} & F(P_n) & \xrightarrow{F(d_n)} & F(P_{n-1}) & \xrightarrow{F(d_{n-1})} & F(P_{n-2})
\end{array}
$$

Since the functor F is right exact, the top row is exact, whereas the bottom row is a chain complex. By definition $L_n F(A) = Ker F(d_n)/Image F(d_{n+1})$. Using the full embedding theorem, it is sufficient to prove the result by assuming that the above diagram is that of R-modules for some R. Now, we chase the diagram. Since the top row is exact, we get an injective homomorphism ϕ_n^F from $L_n F(A) = Ker F(d_n)/Image F(d_{n+1})$ to $F(Ker d_{n-1})$ defined by $\phi_n^F(x + Image F(d_{n+1})) = F(\tilde{d}_n)(x)$. Since $F(i) \circ F(\tilde{d}_n) = F(d_n)$, it follows that $Ker F(i) = Image \phi_n^F$. This proves the exactness of

$$
0 \longrightarrow L_n F(A) \xrightarrow{\phi_n^F} F(Ker d_{n-1}) \xrightarrow{F(i)} F(P_{n-1}).
$$

♯

In particular, applying the above result to the right exact functor $- \otimes_R D$ from the category Mod-R of right R-modules to the category AB of abelian groups, we get the following corollary.

Corollary 2.3.4 *Given a projective resolution*

$$
\cdots \xrightarrow{d_{n+1}} P_n \xrightarrow{d_n} \cdots \xrightarrow{d_1} P_0 \xrightarrow{\epsilon} A \longrightarrow 0
$$

of a right R-module A, we have natural injective homomorphism ϕ_n^D from Tor_n^R (A, D) to $Ker d_{n-1} \otimes_R D$ such that the sequence

$$0 \longrightarrow Tor_n^R(A, D) \overset{\phi_n^D}{\to} Ker d_{n-1} \otimes_R D \overset{i \otimes I_D}{\to} P_{n-1} \otimes_R D$$

is exact.

Similarly, if F is a left exact functor, then we have the following.

Proposition 2.3.5 *Let F be a left exact additive functor from Σ to Ω, where Σ and Ω are as above. Let*

$$\Gamma(A) \overset{\epsilon}{\to} A \to 0 \equiv \cdots \overset{d_{n+1}}{\to} P_n \overset{d_n}{\to} \cdots \overset{d_1}{\to} P_0 \overset{\epsilon}{\to} A \longrightarrow 0$$

be the choice projective resolution of A. Then for each n, we have a morphism ψ_{n-1}^F from $F(Ker d_{n-1})$ to $L_{n-1}F(A)$ such that the sequence

$$F(P_n) \overset{F(\tilde{d}_n)}{\to} F(Ker d_{n-1}) \overset{\psi_{n-1}^F}{\to} L_{n-1}F(A) \longrightarrow 0$$

is exact, where $\tilde{d}_n$ is the obvious morphism induced by d_n. ♯

In particular, we have the following corollary.

Corollary 2.3.6 *Let A and D be right R-modules, and*

$$\cdots \overset{d_{n+1}}{\to} P_n \overset{d_n}{\to} \cdots \overset{d_1}{\to} P_0 \overset{\epsilon}{\to} A \longrightarrow 0$$

be a projective resolution of A. Then for each n we have a surjective homomorphism ψ_{n-1}^D from $Hom(A, Ker d_{n-1})$ to $EXT_R^{n-1}(A, D)$ such that the sequence

$$Hom(A, P_n) \overset{\tilde{d}_n}{\to} Hom(A, Ker d_{n-1}) \overset{\psi_{n-1}^D}{\to} EXT_R^{n-1}(A, D) \longrightarrow 0$$

is exact, where $\tilde{d}_n$ is the obvious homomorphism induced by d_n. ♯

Imitating the proofs of Propositions 2.2.31 and 2.2.32, we can establish the following two propositions giving the two long exact sequences of left-derived functors of an additive functor F.

Proposition 2.3.7 *Let*

$$E \equiv 0 \longrightarrow A \overset{\alpha}{\to} B \overset{\beta}{\to} C \longrightarrow 0$$

be a short exact sequence in Σ, and F be an additive functor from Σ to Ω. Then for each n, we have a natural (in E and D) connecting homomorphism $\partial_n(E)$ from $L_nF(C)$ to $L_{n-1}F(A)$ such that the following sequence is exact.

$$\cdots \overset{\partial_{n+1}(E)}{\to} L_nF(A) \overset{L_n(I_F, \alpha)}{\to} L_nF(B) \overset{L_n(I_F, \beta)}{\to} L_nF(C) \overset{\partial_n(E)}{\to} \cdots.$$

♯

Proposition 2.3.8 *Let*

$$E \equiv 0 \longrightarrow F \overset{\mu}{\to} G \overset{\nu}{\to} H \longrightarrow 0$$

be a short exact sequence in the category $R(\Omega^{\Sigma})$ of right exact functors from Σ to Ω, and A be an object in Σ. Then for each n we have a natural (in E and A) connecting homomorphisms $\partial_n(E)$ from $L_n H(A)$ to $L_{n-1} F(A)$ such that the following sequence is exact.

$$\cdots \overset{\partial_{n+1}(E)}{\to} L_n F(A) \overset{L_n(\mu, I_A)}{\to} L_n G(A) \overset{L_n(\nu, I_A)}{\to} L_n H(A) \overset{\partial_n(E)}{\to} \cdots .$$

♯

Dually, we introduce the right-derived functors $R^n F$ of an additive functor F from Σ to Ω. We assume that Σ has enough injectives. Let A be an object of Σ, and

$$0 \longrightarrow A \overset{\eta}{\to} \Lambda(A) \equiv 0 \longrightarrow A \overset{\eta}{\to} I_0 \overset{d^0}{\to} I_1 \overset{d^1}{\to} \cdots \overset{d^{n-1}}{\to} I_n \overset{d^n}{\to} \cdots$$

be the choice injective co-resolution of A. We get a co-chain complex

$$0 \longrightarrow F(I_0) \overset{F(d^0)}{\to} F(I_1) \overset{F(d^1)}{\to} \cdots \overset{F(d^{n-1})}{\to} F(I_n) \overset{F(d^n)}{\to} \cdots .$$

In turn, we get a functor $R^n F$ from Σ to Ω defined by

$$R^n F(A) = H^n(F(\Lambda(A))) = Ker F(d^n)/Image F(d^{n-1}).$$

The functor $R^n F$, thus obtained, is called the **n^{th} right − derived functor** of F. Indeed, if Σ and Ω are small categories, then the functor $R^n(-, -)$ from $\Omega^{\Sigma} \times \Sigma$ to Ω is called the right-derived functor of the evaluation functor E_v. In addition to that, if we denote by $L\Omega^{\Sigma}$ the category of left exact functors from Σ to Ω, then as in the case of left-derived functors, $R^0(-, -)$ is naturally equivalent to E_v when restricted to $L\Omega^{\Sigma} \times \Sigma$.

Suppose, now that F is an additive contra-variant functor from Σ to Ω. Then F is an additive functor (co-variant) from Σ^o to Ω. The left-derived $L_n F$ of F is defined to be the nth left-derived functor of F considered as a functor from Σ^o to Ω. Thus, $L_n F$ is also a contra-variant functor from Σ to Ω. Now, a projective resolution of A in Σ^o is in fact an injective resolution of A in Σ. Let

$$0 \longrightarrow A \overset{\eta}{\to} \Lambda(A) \equiv 0 \longrightarrow A \overset{\eta}{\to} I_0 \overset{d^0}{\to} I_1 \overset{d^1}{\to} \cdots \overset{d^{n-1}}{\to} I_n \overset{d^n}{\to} \cdots$$

be the choice injective co-resolution of A in Σ (and so a projective resolution Σ^o). We get a chain complex

$$F(\Lambda(A)) \equiv \cdots \overset{F(d^n)}{\to} F(I_n) \overset{F(d^{n-1})}{\to} F(I_{n-1}) \overset{F(d^{n-2})}{\to} \cdots \overset{F(d^1)}{\to} F(I_1) \overset{F(d^0)}{\to} F(I_0) \longrightarrow 0$$

in Σ. In turn,

$$L_n F(A) = H_n(F(\Lambda(A))) = Ker\, F(d^{n-1})/Image\, F(d^n).$$

If f is a morphism from A to B, then $F(\Lambda(f))$ is a chain transformation from $F(\Lambda(B))$ to $F(\Lambda(A))$ in the category Ω, and so we get a morphism $L_n F(f)$ from $L_n F(B)$ to $L_n F(A)$.

Similarly, we can define the right-derived functors of a contra-variant functor F from Σ to Ω. Again, F is a co-variant functor from Σ^o to Ω. An injective resolution of an object A in Σ^o is in fact a projective resolution of A in Σ. Let

$$\Gamma(A) \overset{\epsilon}{\to} A \longrightarrow 0 \equiv \overset{d_{n+1}}{\to} P_n \overset{d_n}{\to} \cdots \overset{d_1}{\to} P_0 \overset{\epsilon}{\to} A \longrightarrow 0$$

be the choice projective resolution of A. Then the co-chain complex $F(\Gamma(A))$ is given by

$$0 \longrightarrow F(P_0) \overset{F(d_1)}{\to} F(P_1) \overset{F(P_2)}{\to} \cdots \overset{F(d_n)}{\to} F(P_n) \overset{F(d_{n+1})}{\to} F(P_{n+1}) \overset{F(d_{n+2})}{\to} \cdots.$$

Thus, the nth right-derived $R^n F$ of the contra-variant functor F is given by $R^n F(A) = Ker\, F(d_{n+1})/Image\, F(d_n)$. Evidently, $R^n F$ is also a contra-variant functor.

We prove the analogue of Proposition 2.3.3.

Proposition 2.3.9 *Let F be a contra-variant left exact functor from Σ to Ω, where Σ has enough projectives (and so Σ^o has enough injectives). Let*

$$\overset{d_{n+1}}{\to} P_n \overset{d_n}{\to} \cdots \overset{d_1}{\to} P_0 \overset{\epsilon}{\to} A \longrightarrow 0$$

be a projective resolution of A. Then for each n there is a morphism ϕ_n^F from $F(Ker\, d_{n-1})$ to $R^n F(A)$ such that the sequence

$$F(P_{n-1}) \overset{F(i)}{\to} F(Ker\, d_{n-1}) \overset{\phi_n^F}{\to} R^n F(A) \longrightarrow 0$$

is exact.

Proof We have the following commutative diagram:

$$\xrightarrow{d_{n+2}} P_{n+1} \xrightarrow{d_{n+1}} P_n \xrightarrow{\tilde{d}_n} Ker\,d_{n-1} \longrightarrow 0$$

$$I_{P_{n+1}} \downarrow \qquad I_{P_n} \downarrow \qquad i \downarrow$$

$$\xrightarrow{d_{n+2}} P_{n+1} \xrightarrow{d_{n+1}} P_n \xrightarrow{d_n} P_{n-1} \xrightarrow{d_{n-1}} P_{n-2}$$

where the rows are exact, and $\tilde{d}_n$ is the obvious morphism induced by d_n. Since F is a left exact contra-variant functor, applying F to the above diagram, we get the following commutative diagram:

$$F(P_{n-2}) \xrightarrow{F(d_{n-1})} F(P_{n-1}) \xrightarrow{F(d_n)} F(P_n) \xrightarrow{F(d_{n+1})} F(P_{n+1})$$

$$\downarrow \qquad F(i) \downarrow \qquad I_{F(P_n)} \downarrow \qquad I_{F(P_{n+1})} \downarrow$$

$$0 \longrightarrow F(Ker\,d_{n-1}) \xrightarrow{F(\tilde{d}_n)} F(P_n) \xrightarrow{F(d_{n+1})} F(P_{n+1})$$

where the bottom row is exact, and the top row is a co-chain complex. Clearly, $F(\tilde{d}_n)$ induces an epimorphism ϕ_n^F from $F(Ker\,d_{n-1})$ to
$Ker\,F(d_{n+1})/Image\,F(d_n) = R^n F(A)$. It also follows from the commutativity of the diagram that $Ker\,\phi_n = Image\,F(i)$. ♯

In particular, we have the following corollary.

Corollary 2.3.10 *Let Σ be an abelian category with enough projectives. Let A and B be objects in Σ. Let*

$$\cdots \xrightarrow{d_{n+1}} P_n \xrightarrow{d_n} P_{n-1} \xrightarrow{d_{n-1}} \cdots \xrightarrow{d_2} P_1 \xrightarrow{d_1} P_0 \xrightarrow{\epsilon} A \longrightarrow 0$$

be a projective resolution of A. Then we have a natural surjective homomorphism ϕ_n^B from $Mor_\Sigma(Ker\,d_{n-1}, B)$ to $EXT_\Sigma^n(A, B)$ such that

$$Mor_\Sigma(P_{n-1}, B) \xrightarrow{Mor_\Sigma(i, I_B)} Mor_\Sigma(Ker\,d_{n-1}, B) \xrightarrow{\phi_n^B} EXT_\Sigma^n(A, B) \longrightarrow 0$$

is exact sequence of abelian groups.

Proof The result follows immediately if we apply the above proposition to the left exact contra-variant functor $Mor_\Sigma(-, B)$. ♯

Derived Functors and EXT^n_Σ

Let Σ be an abelian category with enough projectives. Let $\hat{P}(\Sigma)$ denote the category of positive projective acyclic chain complexes with morphisms as chain transformations (note the difference between $\hat{P}(\Sigma)$ and $\wp(\Sigma)$). Let F be an additive functor from Σ to Ω. Let

$$P \equiv \cdots \overset{d_{n+1}}{\to} P_n \overset{d_n}{\to} P_{n-1} \overset{d_{n-1}}{\to} \cdots \overset{d_2}{\to} P_1 \overset{d_1}{\to} P_0 \longrightarrow 0$$

be an object in $\hat{P}(\Sigma)$. For $n \geq 1$, put $\tilde{L}_n F(P) = Ker F(i)$, where i represents the kernel of d_{n-1}. For $n = 0$, put $\tilde{L}_0 F(P) = H_0(P)$. If $f = \{f_n \mid n \geq 0\}$ is a chain transformation from P to P', where

$$P' \equiv \cdots \overset{d'_{n+1}}{\to} P'_n \overset{d'_n}{\to} P'_{n-1} \overset{d'_{n-1}}{\to} \cdots \overset{d'_2}{\to} P'_1 \overset{d'_1}{\to} P'_0 \longrightarrow 0,$$

then f_{n-1} induces a morphism $\tilde{f}_{n-1}$ from $Ker d_{n-1}$ to $Ker d'_{n-1}$ such that $i f_{n-1} = \tilde{f}_{n-1} i'$. In turn, $F(i)F(f_{n-1}) = F(\tilde{f}_{n-1})F(i')$, and so $F(\tilde{f}_{n-1})$ induces a morphism from $\tilde{L}_n F(P) = Ker F(i)$ to $\tilde{L}_n F(P') = Ker F(i')$. We denote this induced homomorphism by $\tilde{L}_n F(f)$. Thus, $\tilde{L}_n F$ defines a functor from $\hat{P}(\Sigma)$ to Ω.

In addition, suppose that F is right exact. Then from Proposition 2.3.3, it follows that $L_n F(A)$ is naturally isomorphic to $\tilde{L}_n F\Gamma(A)$ for all choice functors Γ from Σ to $\wp(\Sigma)$, and for all objects A. In turn, it also follows that if f is chain homotopic to f', then $\tilde{L}_n F(f) = \tilde{L}_n F(f')$. In this case, $\tilde{L}_n F$ can be treated as a functor from Σ to AB which is same as $L_n F$.

In general, it is not so clear that the functor $\tilde{L}_n F$ from $\hat{P}(\Sigma)$ to Ω induces a functor Σ to Ω. However, in what follows, we shall show that $\tilde{L}_n F$ induces a functor from Σ to AB whenever F is an additive functor from Σ to AB. More explicitly, if $\Omega = AB$, then whenever f and g are homotopic chain transformations from P to P' in $\hat{P}(\Sigma)$, $\tilde{L}_n F(f) = \tilde{L}_n F(g)$.

Theorem 2.3.11 *Let Σ be an abelian category with enough projectives. Let Γ be the choice equivalence from Σ to $\wp(\Sigma)$. Let F be an additive functor from Σ to AB. Then for each n, there is an isomorphism $\eta_{A,F}$ from $Mor_{AB^\Sigma}(EXT^n_\Sigma(A, -), F)$ to $\tilde{L}_n F\Gamma(A)$, where AB^Σ denotes the category of additive functors from Σ to AB.*

Proof Let

$$\Gamma(A) \overset{\epsilon}{\to} A \longrightarrow 0 \equiv \cdots \overset{d_{n+1}}{\to} P_n \overset{d_n}{\to} P_{n-1} \overset{d_{n-1}}{\to} \cdots \overset{d_2}{\to} P_1 \overset{d_1}{\to} P_0 \overset{\epsilon}{\to} A \longrightarrow 0$$

be the choice projective resolution of A. In what follows, $Ker d_{n-1}$ will be denoted by K_{n-1}. Let A be an object in Σ, and F be an additive functor from Σ to AB. Let $\Theta = \{\Theta_B \in Hom(EXT^n_\Sigma(A, B), F(B)) \mid B \in Obj\Sigma\}$ be a natural transformation from $EXT^n_\Sigma(A, -)$ to F. Then the following diagram:

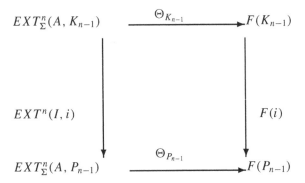

is commutative. Here i denotes the corresponding monomorphism and I denotes the corresponding identity morphism. For each object D in Σ, $Mor_\Sigma(-, D)$ is a left exact contra-variant additive functor from the category Σ to the category AB. Using Corollary 2.3.10, we obtain the following commutative diagram:

$$Mor_\Sigma(P_{n-1}, K_{n-1}) \xrightarrow{\ Mor(i, I)\ } Mor_\Sigma(K_{n-1}, K_{n-1}) \xrightarrow{\ \phi_n^{K_{n-1}}\ } EXT_\Sigma^n(A, K_{n-1}) \longrightarrow 0$$

$$\Big\downarrow Mor(I, i) \qquad\qquad \Big\downarrow Mor(I, i) \qquad\qquad \Big\downarrow EXT^n(I, i)$$

$$Mor_\Sigma(P_{n-1}, P_{n-1}) \xrightarrow{\ Mor(i, I)\ } Mor_\Sigma(K_{n-1}, P_{n-1}) \xrightarrow{\ \phi_n^{P_{n-1}}\ } EXT_\Sigma^n(A, P_{n-1}) \longrightarrow 0$$

with exact rows. From the above commutative diagram, it follows that

$$EXT^n(I, i)\phi_n^{K_{n-1}}(I_{K_{n-1}}) = \phi_n^{P_{n-1}} Mor(I, i)(I_{K_{n-1}}) =$$
$$\phi_n^{P_{n-1}} Mor(i, I)(I_{P_{n-1}}) = 0.$$

Thus, $\Theta_{P_{n-1}}(EXT^n(I, i))(\phi_n^{K_{n-1}}(I_{K_{n-1}}) = 0$. From the commutativity of the earlier diagram $F(i)(\Theta_{K_{n-1}}(\phi_n^{K_{n-1}}(I_{K_{n-1}}))) = 0$. This shows that $\Theta_{K_{n-1}}(\phi_n^{K_{n-1}}(I_{K_{n-1}}))$ belongs to $\tilde{L}_n F(\Gamma(A))$. Define a map $\eta_{A,F}$ from $Mor_{AB^\Sigma}(EXT_\Sigma^n(A, -), F)$ to $\tilde{L}_n F\Gamma(A)$ by

$$\eta_{A,F}(\Theta) = \Theta_{K_{n-1}}(\phi_n^{K_{n-1}}(I_{K_{n-1}})).$$

It is easily observed that $\eta_{A,F}$ is a homomorphism.

Now, we construct the natural inverse of $\eta_{A,F}$. Let χ be an element of $\tilde{L}_n F\Gamma(A)$. We show the existence of a unique Θ in $Mor_{AB^\Sigma}(EXT_\Sigma^n(A, -), F)$ such that $\eta_{A,F}(\Theta) = \chi$. Now, $\chi \in F(K_{n-1})$ such that $F(i)(\chi) = 0$. For each $B \in Obj\Sigma$, we need to define a homomorphism Θ_B from $EXT_\Sigma^n(A, B)$ to $F(B)$ so that $\Theta = \{\Theta_B \mid B \in Obj\,\Sigma\}$ becomes a natural transformation from $EXT_\Sigma^n(A, -)$ to F with $\Theta_{K_{n-1}}(\phi_n^{K_{n-1}}(I_{K_{n-1}})) = \chi$. Again, from Corollary 2.3.10, for each object B in Σ, we have a functorial exact sequence

$$Mor_\Sigma(P_{n-1}, B) \xrightarrow{Mor_\Sigma(i, I_B)} Mor_\Sigma(K_{n-1}, B) \xrightarrow{\phi_n^B} EXT_\Sigma^n(A, B) \longrightarrow 0.$$

Let $h \in EXT_\Sigma^n(A, B)$. Since ϕ_n^B is surjective, there is an element $\sigma \in Mor_\Sigma(K_{n-1}, B)$ such that $\phi_n^B(\sigma) = h$. Further, σ gives the following commutative diagram:

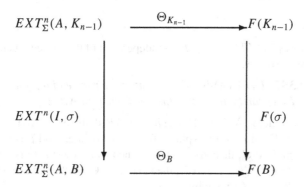

with exact rows. Clearly, $\sigma = Mor_\Sigma(I, \sigma)(I_{K_{n-1}})$. Using the commutativity of the above diagram, we see that $h = EXT_\Sigma^n(I, \sigma)(\phi_n^{K_{n-1}}(I_{K_{n-1}}))$. Again, the requirement for Θ to be a natural transformation forces that the diagram

$$
\begin{array}{ccc}
EXT_\Sigma^n(A, K_{n-1}) & \xrightarrow{\Theta_{K_{n-1}}} & F(K_{n-1}) \\
\Big\downarrow{EXT^n(I, \sigma)} & & \Big\downarrow{F(\sigma)} \\
EXT_\Sigma^n(A, B) & \xrightarrow{\Theta_B} & F(B)
\end{array}
$$

should be commutative. Thus, we are forced to define

$$\Theta_B(h) = \Theta_B(EXT_\Sigma^n(I, \sigma)(\phi_n^{K_{n-1}}(I_{K_{n-1}}))) =$$
$$F(\sigma)(\Theta_{K_{n-1}}(\phi_n^{K_{n-1}}(I_{K_{n-1}}))) = F(\sigma)(\chi).$$

We need to show that the definition of $\Theta_B(h)$ is independent of the choice of σ. Suppose that $\phi_n^B(\tau) = h = \phi_n^B(\sigma)$. Then $\phi_n^B(\sigma - \tau) = 0$. This means that $(\sigma - \tau) \in$

$ker \phi_n^B$. From the exactness, we see that there is morphism ψ from P_{n-1} to B such that $\psi o i = (\sigma - \tau)$. But then, $F(\sigma)(\chi) - F(\tau)(\chi) = F(\psi)(F(i)(\chi))$. Since $\chi \in \tilde{L}_n F(\Gamma(A))$ and $\tilde{L}_n F(\Gamma(A)) = ker F(i)$, it follows that $F(\sigma)(\chi) = F(\tau)(\chi)$. This completes the definition of Θ.

Now, it remains to show that Θ is a natural transformation. Let f be a morphism from B to C. We need to show that $F(f)F(\Theta_B(h)) = \Theta_C(EXT^n(I, f)(h))$ for all $h \in EXT_\Sigma^n(A, B)$. By definition $\Theta_B(h) = F(\sigma)(\chi)$, where $\phi_n^B(\sigma) = h$. From the commutativity of the diagram

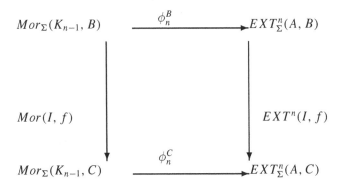

it follows that $EXT_\Sigma^n(I, f)(h) = \phi_n^C(f o \sigma)$. Hence, by definition of Θ_C,

$$\Theta_C(EXT^n(I, f)(h)) = F(f o \sigma)(\chi) = F(f)(F(\sigma)(\chi)) = F(f)(\Theta_B(h)).$$

♯

Since $Mor_{AB^\Sigma}(EXT^n(A, -), F)$ is independent of the choice functor Γ, we have the following corollary.

Corollary 2.3.12 $\tilde{L}_n F\Gamma(A)$ *is independent of the choice of a projective resolution of* A, *and* $\tilde{L}_n F(A)$ *can be treated as functor from* Σ *to* AB. ♯

We have a functor Λ from $\Sigma \times AB^\Sigma$ to AB defined by $\Lambda(A, F) = Mor_{AB^\Sigma}(EXT_\Sigma^n(A, -), F)$. If α is a morphism from A to B in Σ, and μ is a natural transformation from F to G, then $\Lambda(\alpha, \mu)$ is defined as follows: if Θ is a natural transformation from $EXT_\Sigma^n(A, -)$ to F, then $\Lambda(\alpha, \mu)(\Theta)$ is the natural transformation from $EXT_\Sigma^n(B, -)$ to G given by

$$\Lambda(\alpha, \mu)(\Theta) = \{\Lambda(\alpha, \mu)(\Theta)(C) = \mu_C o \Theta_C o EXT^n(\alpha, I_C) \mid C \in Obj\Sigma\}.$$

We have another functor Δ from $\Sigma \times AB^\Sigma$ to AB defined by $\Delta(A, F) = \tilde{L}_n F(A)$. The following corollary is a straightforward (some what painstaking) verification, and it is left as an exercise.

Corollary 2.3.13 *The family* $\{\eta_{A,F} \mid (A, F) \in \Sigma \times AB^\Sigma\}$ *defines a natural equivalence between* Λ *and* Δ, *where* $\eta_{A,F}$ *is as given in Theorem 2.3.11.* ♯

Corollary 2.3.14 *If F is an additive right exact functor from Σ to AB, then $Mor_{AB^\Sigma}(EXT_\Sigma^n(A, -), F)$ is naturally equivalent to $L_n F$.*

Proof Follows from Corollary 2.3.10 and Corollary 2.3.12. ♯

Corollary 2.3.15 *Let B be a right R-module, and A be a left R-module. Then the group of natural transformations from $EXT_R^n(A, -)$ to $B \otimes_R -$ is naturally isomorphic to $Tor_n^R(B, A)$.*

Proof Take F to be the right exact additive functor $B \otimes_R -$, and apply the above corollary. ♯

Proposition 2.3.16 *Let Σ be an abelian category with enough projectives. Then for any pair of objects (A, B) in Σ, we have a surjective map $\Lambda_{A,B}$ from $Mor_\Sigma(B, A)$ to $Mor_{Set^\Sigma}(EXT_\Sigma(A, -), EXT_\Sigma(B, -)$ defined by $\Lambda_{A,B}(\alpha) = EXT_\Sigma(\alpha, -)$.*

Proof Let $\alpha \in Mor_\Sigma(B, A)$. Evidently, for each morphism h from C to D, the diagram

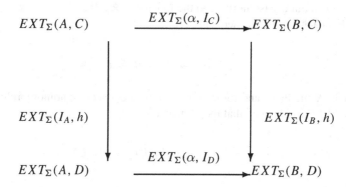

is commutative. This shows that

$$EXT_\Sigma(\alpha, -) = \{EXT_\Sigma(\alpha, I_C) \mid C \in Obj\Sigma\}$$

is a natural transformation from $EXT_\Sigma(A, -)$ to $EXT_\Sigma(B, -)$. Let

$$\cdots \xrightarrow{d_2} P_1 \xrightarrow{d_1} P_0 \xrightarrow{\epsilon} A \longrightarrow 0$$

be the choice projective resolution of A. We have the short exact sequence

$$E \equiv 0 \longrightarrow K_0 \xrightarrow{i} P_0 \xrightarrow{\epsilon} A \longrightarrow 0,$$

where K_0 is the kernel of ϵ. From Corollary 2.1.16, we have the exact sequence

$$\cdots \overset{Mor(I,i)}{\to} Mor_\Sigma(B, P_0) \overset{Mor(I,\epsilon)}{\to} Mor_\sigma(B, A) \overset{\partial_E^1}{\to} EXT_\Sigma(B, K_0) \overset{EXT_\Sigma(I,i)}{\to}$$

$$EXT_\Sigma(B, P_0) \overset{EXT(I,\epsilon)}{\to} \cdots .$$

Again, by definition $\tilde{L}_1 EXT_\Sigma(B, -)(A) = Ker EXT_\Sigma(I, i) = Image \partial_E^1$. Thus, ∂_E^1 can be treated as a surjective map from $Mor_\Sigma(B, A)$ to $\tilde{L}_1 EXT_\Sigma(B, -)(A)$. From Theorem 2.3.11, the map $\eta_{A, EXT_\Sigma(B, -)}$ is a bijective map from $Mor_{Set^\Sigma}(EXT_\Sigma(A, -), EXT_\Sigma(B, -))$ to $\tilde{L}_1 EXT_\Sigma(B, -)(A)$. In turn, we get a surjective map $\eta_{A, EXT_\Sigma(B, -)}^{-1} o \partial_E^1$ from $Mor_\Sigma(B, A)$ to $Mor_{Set^\Sigma}(EXT_\Sigma(A, -), EXT_\Sigma(B, -))$. Indeed, it can be checked that $\Lambda_{A, B} = \eta_{A, EXT_\Sigma(B, -)}^{-1} o \partial_E^1$. ♯

Exercises

2.3.1 Let $\{F_n \mid n \geq 0\}$ be a sequence of additive functors from the category $Mod\text{-}R$ of right R-modules to the category AB of abelian groups, and D be a left R-module such that

(i) F_0 is naturally isomorphic to the functor $- \otimes_R D$.
(ii) For each short exact sequence

$$E \equiv 0 \longrightarrow A \overset{\alpha}{\to} B \overset{\beta}{\to} C \longrightarrow o$$

of right R-modules, and for each n, there is a connecting homomorphism ∂_n^E from $F_n(C)$ to $F_{n-1}(A)$ such that the sequence

$$\cdots \overset{\partial_{n+1}^E}{\to} F_n(A) \overset{F_n(\alpha)}{\to} F_n(B) \overset{F_n(\beta)}{\to} F_n(C) \overset{\partial_n^E}{\to} \cdots$$

is exact.

(iii) $F_n(P) = 0$ for all projective modules P.

Use induction on n to show that F_n is naturally isomorphic to $Tor_n^R(-, D)$ for all n.

2.3.2 Let $\{F_n \mid n \geq 0\}$ be a sequence of additive functors from the category $Mod\text{-}R$ of right R-modules to the category AB of abelian groups, and A be a right R-module such that

(i) F_0 is naturally isomorphic to the functor $Hom_R(A, -)$.
(ii) For each short exact sequence

$$E \equiv 0 \longrightarrow B \overset{\beta}{\to} C \overset{\gamma}{\to} D \longrightarrow 0$$

of right R-modules, and for each n, there is a connecting homomorphism ∂_n^E from $F_n(D)$ to $F_{n-1}(B)$ such that the sequence

$$\cdots \overset{\partial_{n+1}^E}{\to} F_n(B) \overset{F_n(\beta)}{\to} F_n(C) \overset{F_n(\gamma)}{\to} F_n(D) \overset{\partial_n^E}{\to} \cdots$$

is exact.

(iii) $F_n(I) = 0$ for all injective modules P.

Use induction on n to show that F_n is naturally isomorphic to $EXT_R^n(A, -)$ for all n.

2.3.3 Let f and g be homotopic chain transformations from a chain complex X of right R-modules to a chain complex X' of right R-modules. Let f' and g' be homotopic chain transformations from a chain complex Y of left R-modules to a chain complex Y' of left R-modules. Show that $f \otimes g$ is chain homotopic to $f' \otimes g'$. Deduce that the tensor products of chain equivalences are chain equivalences.

2.3.4 Let

$$\Gamma(A) \overset{\epsilon^A}{\to} 0 \equiv \cdots \overset{d_{n+1}^A}{\to} P_n^A \overset{d_n^A}{\to} \cdots \overset{d_1^A}{\to} P_0^A \overset{\epsilon^A}{\to} A \longrightarrow 0$$

and

$$\Gamma(B) \overset{\epsilon^B}{\to} 0 \equiv \cdots \overset{d_{n+1}^B}{\to} P_n^B \overset{d_n^B}{\to} \cdots \overset{d_1^B}{\to} P_0^B \overset{\epsilon^B}{\to} B \longrightarrow 0$$

be choice projective resolutions of A and B, respectively. Treating B as a chain complex and ϵ^B as a chain transformation, describe the chain transformation $I_{\Gamma(A)} \otimes \epsilon^B$ from $\Gamma(A) \otimes \Gamma(B)$ to $\Gamma(A) \otimes B$.

The following set of exercises are designed to show that $H_n(\Gamma(A) \otimes \Gamma(B))$ is isomorphic to $Tor_n^R(A, B)$.

2.3.5 For each $m \geq 0$, show that we have a truncated chain subcomplex X^m of $\Gamma(A) \otimes \Gamma(B)$ defined by

$$X_n^m = \oplus \sum_{k=0}^{m} P_k^A \otimes P_{n-k}^B.$$

Further, show that

$$X_n^m = X_n^{m-1} \oplus P_m^A \oplus P_{n-m}^B.$$

Show that $H_n(X^m)$ is isomorphic to $H_n((\Gamma(A) \otimes \Gamma(B))$ for all $m \geq n + 1$.

2.3.6 Show that the quotient chain complex X^m / X^{m-1} is given by

$$P_m^A \otimes P_n^B \overset{I \otimes d_n^B}{\to} P_m^A \otimes P_{n-1}^B \overset{I \otimes d_{n-1}^B}{\to} \cdots \overset{I \otimes d_1^B}{\to} P_m^A \otimes P_0^B.$$

2.3.7 For each $m \geq 0$, again we have a truncated chain subcomplex Y^m of $\Gamma(A) \otimes B$ given by $Y_n^m = P_n \otimes B$ for $n \leq m$, and $Y_n^m = 0$ for $n \geq m + 1$. Show that $I_{\Gamma(A)} \otimes \epsilon^B$ restricted to X^m is a chain transformation from X^m to Y^m. Deduce that $I_{\Gamma(A)} \otimes \epsilon^B$ induces the commutative diagram

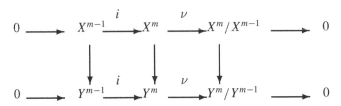

of chain complexes, where the rows are exact. Further, use Theorem 1.3.1 to get the following commutative diagram:

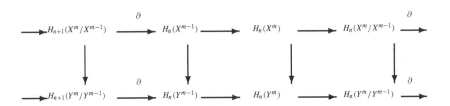

where the rows are exact.

2.3.8 Show that $H_n(Y^m)$ is isomorphic to $H_n((\Gamma(A) \otimes B))$ for all $m \geq n + 1$.

2.3.9 Show that the chain complex Y^m / Y^{m-1} is the chain complex which is zero in all dimensions except in the dimension m in which it is $P_m^A \otimes B$.

2.3.10 Using the fact that P_m^A is projective show that the chain transformation $I_{\Gamma(A)} \otimes \epsilon^B$ induces chain transformation from X^m / X^{m-1} to Y^m / Y^{m-1} which, in turn, induces isomorphisms between their homologies.

2.3.11 Apply the induction, and use Exercises 2.3.10 and 2.3.7 to deduce that $I_{\Gamma(A)} \otimes \epsilon^B$ induces isomorphism from $H_n(X^m)$ to $H_n(Y^m)$ for all n and m. Further, use Exercise 2.3.8 to deduce that $H_n(\Gamma(A) \otimes \Gamma(B))$ is naturally isomorphic to $Tor_n^R(A, B)$.

2.4 Kunneth Formula

The Eilenberg–Zilber theorem, an important result in algebraic topology, asserts that the singular chain complex of the product of two topological spaces is chain equivalent to the tensor product of their singular chain complexes. This prompts us to discuss the problem of describing homologies (and co-homologies) of the tensor product $X \otimes Y$ of two chain complexes X and Y in terms of the homologies (and co-homologies) of X and the homologies (co-homologies) of Y. This section mainly concerns to this problem.

Let X be a chain complex of right R-modules and Y be a chain complex of left R-modules. Recall that $C_n(X)$ denotes the module of n-cycles, and $B_n(X)$ denotes the module of n-boundaries of X. Usually, we shall denote the element $x + B_n(X)$ of $H_n(X)$ by $[x]$. Let $(x, y) \in C_p(X) \times C_q(Y)$. Since $d_{p+q}(x \otimes y) = d_p^X(x) \otimes y + (-1)^p x \otimes d_q^Y(y) = 0, x \otimes y \in C_{p+q}(X \otimes Y)$. If $x \in B_p(X)$, then there is an element $x' \in X_{p+1}$ such that $d_{p+1}^X(x') = x$. But then, $x \otimes y = d_{p+1}^X(x') \otimes y = d_{p+1+q}^{X \otimes Y}(x' \otimes y)$. This means that $x \otimes y \in B_{p+q}(X \otimes Y)$. Similarly, $x \in C_p(X)$ and $y \in B_q(Y)$ imply that $x \otimes y \in B_{p+q}(X \otimes Y)$. We have a map $\lambda_{p,q}$ from $H_p(X) \times H_q(Y)$ to $H_{p+q}(X \otimes Y)$ given by $\lambda_{p,q}([x], [y]) = [x \otimes y]$. It can be seen easily that $\lambda_{p,q}$ is a balanced map. Thus, $\lambda_{p,q}$ induces a homomorphism $\bar{\lambda}_{p,q}$ from $H_p(X) \otimes H_q(Y)$ to $H_{p+q}(X \otimes Y)$ given by $\bar{\lambda}_{p,q}([x] \otimes [y]) = [x \otimes y]$. In turn, we get a homomorphism λ_n from $\oplus \sum_{p+q=n} H_p(X) \otimes H_q(Y)$ to $H_n(X \otimes Y)$ whose restriction on $H_p(X) \otimes H_q(Y)$ is $\bar{\lambda}_{p,q}$. The homomorphism λ_n is called the **homology product** at the dimension n. In particular, if A is a right R-module treated as a chain complex, then $d_n^{A \otimes_R Y}(a \otimes y) = (-1)^n a \otimes d_n^Y(y)$, and the homology product λ_n is a homomorphism from $A \otimes_R H_n(Y)$ to $H_n(A \otimes_R Y)$ given by $\lambda_n(a \otimes [x]) = [a \otimes x]$. More generally, let X be a chain complex of right R-modules such that the boundary map d_n^X is zero for all n, and let Y be a chain complex of left R-modules. Then, for each n, $(X \otimes_R Y)_n = \oplus \sum_{p \in \mathbb{Z}} X_p \otimes Y_{n-p}$. Since $d_p^X = 0$ for all p, $d_n^{X \otimes_R Y}$ is a homomorphism from $\oplus \sum_{p \in \mathbb{Z}} X_p \otimes Y_{n-p}$ to $\oplus \sum_{p \in \mathbb{Z}} X_p \otimes Y_{n-1-p}$ given by $d_n^{X \otimes_R Y} = \oplus \sum_{p \in \mathbb{Z}} (-1)^p I_{X_p} \otimes d_{n-p}^Y$. Note that $H_p(X) = X_p$ for each p. The homology product λ_n is given by $\lambda_n(x \otimes [y]) = [x \otimes y]$, where $x \in X_p = H_p(X)$ and $y \in C_{n-p}(Y)$. Under some very stringent restrictions, λ_n will turn out to be an isomorphism.

Lemma 2.4.1 *Let X be a chain complex of flat right R-modules such that the boundary map d_n^X is zero for all n. Let Y be a chain complex. Then the homology product λ_n is an isomorphism for all n.*

Proof We have the following commutative diagram:

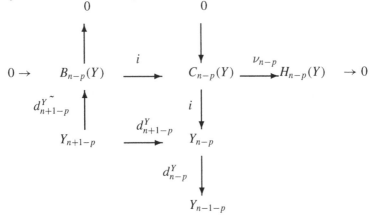

where rows and columns are exact, and $d^{Y\;\widetilde{}}_{n+1-p}$ is the map induced by d^Y_{n+1-p}. Since X_p is flat right R-module, taking the tensor product with X_p, we get the following commutative diagram:

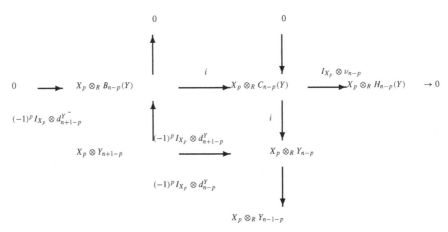

where the rows and the columns are exact. Summing over $p \in \mathbb{Z}$, we obtain the following commutative diagram:

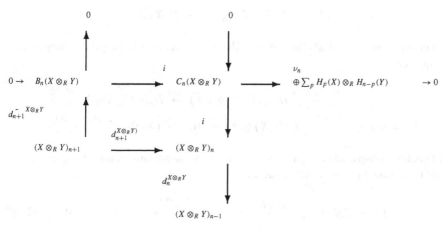

where rows and the columns are exact. Evidently, ν_n induces an isomorphism $\overline{\nu_n}$ from $H_n(X \otimes_R Y)$ to $\oplus \sum_p H_p(X) \otimes_R H_{n-p}(Y)$ which is, indeed, the inverse of the homology product λ_n. ♯

Treating a right R-module A as a complex X such that $X_0 = A$, and $X_n = 0$ for $n \neq 0$, we obtain the following corollary.

Corollary 2.4.2 *Let A be a flat right R-module. Then the homology product λ_n from $A \otimes_R H_n(Y)$ to $H_n(A \otimes Y)$ is an isomorphism.* ♯

More generally, we have the following Kunneth formula.

Theorem 2.4.3 *(Kunneth Formula). Let X be a chain complex of right R-modules such that $C_n(X)$ and $B_n(X)$ are flat modules for all n. Then for each n, we have a short exact sequence*

$$0 \longrightarrow \oplus \sum_{p+q=n} H_p(X) \otimes H_q(Y) \xrightarrow{\lambda_n} H_n(X \otimes Y) \xrightarrow{\mu_n}$$

$$\oplus \sum_{p+q=n-1} Tor_1^R(H_p(X), H_q(Y)) \longrightarrow 0,$$

where λ_n is the homology product and μ_n is a natural homomorphism.

Proof Let $C(X)$ denote the subchain complex of X, where $C(X)_n = C_n(X)$ and the boundary maps are 0 maps. Let $D(X)$ denote the chain complex $D(X)_n = D_n(X) = X_n/C_n(X) \approx B_{n-1}(X)$ with zero boundary maps. We have the obvious quotient chain transformation ν from X to $D(X)$ giving the short exact sequence

$$0 \longrightarrow C(X) \xrightarrow{i} X \xrightarrow{\nu} D(X) \longrightarrow 0$$

of chain complexes. Since $D(X)$ is chain complex of flat modules, taking the tensor product with the chain complex Y, we obtain the following short exact sequence:

$$0 \longrightarrow C(X) \otimes_R Y \overset{i \otimes I_Y}{\to} X \otimes_R Y \overset{\nu \otimes I_Y}{\to} D(X) \otimes_R Y \longrightarrow 0$$

of chain complexes of abelian groups. Using Theorem 1.3.1, we get the long exact sequence

$$\cdots \overset{H_{n+1}(\nu \otimes I_Y)}{\to} H_{n+1}(D(X) \otimes_R Y) \overset{\partial_{n+1}}{\to} H_n(C(X) \otimes_R Y) \overset{H_n(i \otimes I_Y)}{\to}$$

$$H_n(X \otimes_R Y) \overset{H_n(\nu \otimes I_Y)}{\to} H_n(D(X) \otimes Y) \overset{\partial_n}{\to} H_{n-1}(C(X) \otimes_R Y) \overset{H_{n-1}(i \otimes I_Y)}{\to} \cdots$$

of abelian groups, where ∂_n is a connecting homomorphism. In turn, for each n, we get the following short exact sequence:

$$0 \longrightarrow Coker \partial_{n+1} \overset{\overline{H_n(i \otimes I_Y)}}{\to} H_n(X \otimes_R Y) \overset{\overline{H_n(\nu \otimes I_Y)}}{\to} Ker \partial_n \longrightarrow 0 \qquad (2.5)$$

of abelian groups, where $\overline{H_n(i \otimes I_Y)}$ is the homomorphism induced by $H_n(i \otimes I_Y)$ and $\overline{H_n(\nu \otimes I_Y)}$ is the homomorphism induced by $H_n(\nu \otimes I_Y)$. Now, we view the connecting homomorphism ∂_{n+1} in the above long exact sequence from a different angle. For each $p \in \mathbb{Z}$, we have the following short exact sequence:

$$E_p \equiv 0 \longrightarrow D_{p+1}(X) \overset{\overline{d_{p+1}^X}}{\to} C_p(X) \overset{\nu_p}{\to} H_p(X) \longrightarrow 0,$$

where $\overline{d_{p+1}^X}$ is the map induced by the boundary map d_{p+1}^X, and ν_p is the quotient map. Since $C_p(X)$ is flat module for all p, $Tor_1^R(C_p(X), H_{n-p}(Y)) = 0$. Thus, for each p and for each n, the long exact sequence of Tor functors gives the exact sequence

$$0 \longrightarrow Tor_1^R(H_p(X), H_{n-p}(Y)) \overset{\partial_1^{E_p}}{\to} D_{p+1}(X) \otimes_R H_{n-p}(Y) \overset{\overline{d_{p+1}^X} \otimes I}{\to} \qquad (2.6)$$

$$C_p(X) \otimes_R H_{n-p}(Y) \overset{\nu_p \otimes I}{\to} H_p(X) \otimes_R H_{n-p}(Y) \longrightarrow 0,$$

where I denotes the corresponding identity maps (observe that $H_{p+1}(D(X)) = D_{p+1}(X)$ and $H_p(C(X)) = C_p(X)$). Taking the sum over $p \in \mathbb{Z}$, we obtain the exact sequence

$$0 \longrightarrow \oplus \sum_{p \in \mathbb{Z}} Tor_1^R(H_p(X), H_{n-p}(Y)) \overset{\oplus \sum_p \partial_1^{E_p}}{\to} \qquad (2.7)$$

$$\oplus \sum_p H_{p+1}(D(X)) \otimes_R H_{n-p}(Y) \overset{\oplus \sum_p \overline{d_{p+1}^X} \otimes I}{\to}$$

$$\oplus \sum_p H_p(C(X)) \otimes_R H_{n-p}(Y) \overset{\oplus \sum_p \nu_p \otimes I}{\to} \oplus \sum_p H_p(X) \otimes_R H_{n-p}(Y) \longrightarrow 0.$$

Further, looking at the definition of connecting homomorphism, we show that the diagram

$$\oplus \sum_p H_{p+1}(D(X)) \otimes_R H_{n-p}(Y) \xrightarrow{\ \oplus \sum_p \overline{d^X_{p+1}} \otimes I\ } \oplus \sum_p H_p(C(X)) \otimes_R H_{n-p}(Y)$$

$$\Big\downarrow{\lambda_{n+1}} \qquad\qquad\qquad\qquad\qquad \Big\downarrow{\lambda_n}$$

$$H_{n+1}(D(X) \otimes_R Y) \xrightarrow{\qquad\qquad \partial_{n+1} \qquad\qquad} H_n(C(X) \otimes_R Y)$$

is commutative, where λ represents the corresponding homology products. Let us denote the element $x + C_{p+1}(X)$ in $H_{p+1}(D(X))$ by $\bar{x}$ (note that $H_{p+1}(D(X)) = X_{p+1}/C_{p+1}(X)$). Let $\sum_{i=1}^r \bar{x}_i \otimes [y_i]$ be a member of $H_{p+1}(D(X)) \otimes_R H_{n-p}(Y)$, where $x_i \in X_{p+1}$ and $y_i \in C_{n-p}(Y)$. Then

$$\lambda_{n+1}\left(\sum_{i=1}^r \bar{x}_i \otimes [y_i]\right) = \left[\sum_{i=1}^r x_i \otimes y_i\right].$$

Now, the element $\sum_{i=1}^r x_i \otimes y_i \in X_{p+1} \otimes_R Y_{n-p}$ is such that $(\nu \otimes I)(\sum_{i=1}^r x_i \otimes y_i) = [\sum_{i=1}^r x_i \otimes y_i]$. Next,

$$d_{n+1}^{X \otimes_R Y}\left(\sum_{i=1}^r x_i \otimes y_i\right) = \sum_{i=1}^r d_{p+1}^X(x_i) \otimes y_i = i \otimes I_Y\left(\sum_{i=1}^r d_{p+1}^X(x_i) \otimes y_i\right).$$

Thus, from the definition of the connecting homomorphism (see Theorem 1.3.1),

$$\partial_{n+1}\left(\left[\sum_{i=1}^r x_i \otimes y_i\right]\right) = \left[\sum_{i=1}^r d_{p+1}^X(x_i) \otimes y_i\right].$$

Again, by definition,

$$\lambda_n\left(\sum_{i=1}^r [d_{p+1}^X(x_i)] \otimes [y_i]\right) = \left[\sum_{i=1}^r d_{p+1}^X(x_i) \otimes y_i\right].$$

This shows that the diagram is commutative. It also follows from Lemma 2.4.1 that λ_{n+1} and λ_n are isomorphisms. Thus, we have an isomorphism $\overline{\lambda_n}$ from $Coker(\oplus \sum_p \overline{d_{p+1}^X} \otimes I)$ to $Coker \partial_{n+1}$ given by $\overline{\lambda_n}(x \otimes [y] + Image \oplus \sum_p \overline{d_{p+1}^X} \otimes I) = [x \otimes y] + Image \partial_{n+1}$, $x \in C_p(X)$, and $y \in C_{n-p}(Y)$. From Eq. (2.6), we have an isomorphism χ_n from $\oplus \sum_p H_p(X) \otimes_R H_{n-p}(Y)$ to $Coker(\oplus \sum_p \overline{d_{p+1}^X} \otimes I)$

given by $\chi_n([x] \otimes [y]) = x \otimes [y] + Image(\oplus \sum_p \overline{d_{p+1}^X} \otimes I)$. Evidently, $H_n(i \otimes I_Y) o (\overline{\lambda_n})^{-1} o \chi_n$ is the homology product λ_n in the statement of the theorem.

Replacing n by $n-1$, the homomorphism λ_n from $\oplus \sum_p H_{p+1}(D(X)) \otimes_R H_{n-1-p}(Y)$ to $H_n(D(X) \otimes_R Y)$ induces an isomorphism $\overline{\lambda_n}$ from $Ker(\oplus \sum_p \overline{d_{p+1}^X} \otimes I)$ to $Ker \partial_n$. Again, from (2.6), we have an isomorphism ρ_{n-1} from $Ker(\oplus \sum_p \overline{d_{p+1}^X} \otimes I)$ to $\oplus \sum_p Tor_1^R(H_p(X), H_{n-1-p}(Y))$. Taking $\mu_n = \rho_{n-1} o(\overline{\lambda_n})^{-1} o \overline{H_n(\nu \otimes I_Y)}$, we obtain the desired exact sequence of the Kunneth formula. Also, note that μ_n is natural. ♯

Corollary 2.4.4 *Let X be a chain complex of right R-modules such that $H_n(X)$, and $C_n(X)$ are projective modules for all n. Let Y be a chain complex of left R-modules. Then the homology product is an isomorphism.*

Proof Since $H_n(X)$ is projective, the sequence

$$0 \longrightarrow B_n(X) \overset{i}{\to} C_n(X) \overset{\nu}{\to} H_n(X) \longrightarrow 0$$

is split exact. Since $C_n(X)$ is projective, $B_n(X)$ is also projective. Since a projective module is flat, $Tor_1^R(H_p(X), H_{n-p}(Y)) = 0$. The result follows from the above Kunneth formula. ♯

Theorem 2.4.5 *Let R be a principal ideal domain and X be a chain complex of torsion-free R-modules. Let Y be a chain complex of R-modules. Then we have split short exact sequence*

$$0 \longrightarrow \oplus \sum_{p+q=n} H_p(X) \otimes H_q(Y) \overset{\lambda_n}{\to} H_n(X \otimes Y) \overset{\mu_n}{\to}$$

$$\oplus \sum_{p+q=n-1} Tor_1^R(H_p(X), H_q(Y)) \longrightarrow 0,$$

where λ_n is the homology product and μ_n is a natural homomorphism.

Proof Since X_n is torsion free, $C_n(X)$ and $B_n(X)$ are also torsion free. In particular, $C_n(X)$ and $B_n(X)$ are flat modules. The exactness of the short exact sequence follows from Theorem 2.4.3. We need to show that the sequence splits. From Exercise 1.3.11, there are chain complexes $\tilde{X}$ and $\tilde{Y}$ of free R-modules together with a chain transformation f from $\tilde{X}$ to X and g from $\tilde{Y}$ to Y such that $H_n(f)$ and $H_n(g)$ are isomorphisms for each n. We have two extensions $\tilde{E}$ and E of abelian groups given by

$$\tilde{E} \approx 0 \longrightarrow \oplus \sum_p H_p(\tilde{X}) \otimes_R H_{n-p}(\tilde{Y}) \overset{\lambda_n}{\to} H_n(\tilde{X} \otimes_R \tilde{Y}) \overset{\nu_n}{\to}$$

$$\oplus \sum_p Tor_1^R(H_p(\tilde{X}), H_{n-1-p}(\tilde{Y})) \longrightarrow 0$$

and

$$E \approx 0 \longrightarrow \oplus \sum_p H_p(X) \otimes_R H_{n-p}(Y) \overset{\lambda_n}{\to} H_n(X \otimes_R Y) \overset{\nu_n}{\to}$$

$$\oplus \sum_p Tor_1^R(H_p(X), H_{n-1-p}(Y)) \longrightarrow 0.$$

Using the naturality of the homology product λ_n and of ν_n, we obtain a morphism $(\oplus \sum_p H_p(f) \otimes H_{n-p}(g), H_n(f \otimes g), \oplus \sum_p Tor(H_p(f), H_{n-1-p}(g)))$ from $\tilde{E}$ to E. Since $H_n(f)$ and $H_n(g)$ are isomorphisms for all n, the extreme two maps are isomorphisms. This means that $\tilde{E}$ and E are equivalent. Thus, it is sufficient to assume that X and Y are chain complexes of free R-modules. Since submodule of a free R-module over a principal ideal domain is free, $C_n(X)$ and $D_n(X)$ are also free. Thus, $X_n = C_n(X) \oplus D_n(X)$. We have a map ϕ_n from X_n to $H_n(X)$ given by $\phi_n((c, d)) = [c]$ which vanishes on the boundaries. Similarly, we have a homomorphism ψ_n from Y_n to $H_n(Y)$ which extends the map $y \mapsto [y]$ and it vanishes on the boundaries. This gives us a homomorphism $(\phi \otimes \psi)_n$ from $(X \otimes Y)_n$ to $\oplus \sum_p H_p(X) \otimes_R H_{n-p}(Y)$ given by $(\phi \otimes \psi)_n(x \otimes y) = \phi_p(x) \otimes \psi_{n-p}(y)$, $x \in X_p$, and $y \in Y_{n-p}$. Since ϕ_n and ψ_n vanish on the boundaries, $(\phi \otimes \psi)_n$ is zero on $B_n(X \otimes_R Y)$. Thus, $(\phi \otimes \psi)_n$ induces a homomorphism $\overline{(\phi \otimes \psi)_n}$ from $H_n(X \otimes_R Y)$ to $\oplus \sum_p H_p(X) \otimes_R H_{n-p}(Y)$ which is the left inverse of the homology product λ_n. This shows that the sequence splits. ♯

Earlier (Theorem 1.4.14) in Sect. 1.4, we established the universal coefficient theorem for co-homology. Following is the universal coefficient theorem for homology, and it follows immediately from the Kunneth formula.

Corollary 2.4.6 (Universal coefficient theorem for homology) *Let R be a principal ideal domain. Let X be a chain complex of torsion-free R-modules, and A be an R-module. Then for each n, we have a split exact sequence*

$$0 \longrightarrow H_n(X) \otimes_R A \overset{\lambda_n}{\to} H_n(X \otimes A) \overset{\nu_n}{\to} Tor^R(H_{n-1}(X), A) \longrightarrow 0.$$

♯

Exercises

2.4.1 Let X and Y be chain complexes of torsion-free abelian groups, and f be a chain transformation from X to Y such that $H_n(f)$ is an isomorphism for all n. Show that $H_n(f \otimes I_A)$ is also an isomorphism for all n and for all abelian groups A.

2.4.2 Let X be a chain complex of finitely generated free abelian groups, and A be an abelian group. Show that $Hom(X, \mathbb{Z}) \otimes_{\mathbb{Z}} A$ is chain equivalent to $Hom(X, A)$. In turn, deduce the existence of the natural short exact sequence given below:

$$0 \longrightarrow H^n(X, \mathbb{Z}) \otimes_{\mathbb{Z}} A \overset{\lambda^n}{\to} H^n(X, A) \overset{\nu_n}{\to} Tor(H^{n+1}(X, \mathbb{Z}), A) \longrightarrow 0.$$

2.4.3 Let X be a chain complex of finitely generated abelian groups. Show that the Betti number of $H_n(X)$ is the dimension of the vector space $H_n(X \otimes_{\mathbb{Z}} \mathbb{Q})$.

2.4.4 Let X and Y be a chain complex of vector spaces over a field F. Show that $H_n(X \otimes Y)$ is isomorphic to $\oplus \sum_p H_p(X) \otimes H_{n-p}(Y)$.

2.4.5 Let (X, Y) and (X', Y') be pseudo-chain complexes (see Exercise 1.3.12) of R-modules, where R is a principal ideal domain and X_n is torsion free for all n. Take $(X \otimes_R X')_n = \oplus \sum_{p+q=n} X_p \otimes X'_q$ and $(Y \otimes_R Y')_n$ to be the submodule of $(X \otimes_R X')_n$ generated by $Ker\, d_p \otimes_R Y'_q \cup Y_p \otimes_R X'_q \cup Y_p \otimes_R Y'_q$, $p + q = n$ (since X_n is torsion-free $Ker\, d_p \otimes_R Y'_q \cup Y_p \otimes_R X'_q \cup Y_p \otimes_R Y'_q$, $p + q = n$ can be treated as a subset of $X_p \otimes_R X'_q$). Define $\tilde{d}_n(x \otimes y) = d_p(x) \otimes y + (-1)^p x \otimes d_q(y)$, $p + q = n$. Show that $((X \otimes_R X')_n, (Y \otimes_R Y')_n)$ together with $\tilde{d}_n$ defined above is a pseudo-chain complex. Define the pseudo-homology product λ_n from $\oplus \sum_{p+q=n} H_p(X, Y) \otimes_R H_q(X', Y')$ to $H_n((X, Y) \otimes_R (X', Y'))$ by $\lambda_n([x] \otimes [y]) = [x \otimes y]$. Use the functor G introduced in Exercise 1.3.13, and the Kunneth formula to show that the pseudo-homology product is a monomorphism. Can we interpret the co-kernel?

2.5 Spectral Sequences

Spectral sequence is a tool for computations of homologies and co-homologies. For example, it enables us to approximate the homology of group with the help of homologies of a normal subgroup H and the homologies of the corresponding quotient group. This section is devoted to a very brief introduction to the theory of spectral sequences.

Recall that a $\mathbb{Z}$ bi-graded R-module E is a family $\{E_{p,q} \mid p, q \in \mathbb{Z}\}$ of R-modules. A morphism from a $\mathbb{Z}$ bi-graded R-module E to a $\mathbb{Z}$ bi-graded R-module E' is a family $f = \{f_{p,q} \in Hom_R(E_{p,q}, E'_{p,q}) \mid p, q \in \mathbb{Z}\}$ of homomorphisms. Evidently, we have an abelian category of $\mathbb{Z}$ bi-graded R-modules. A boundary operator on E of bi-degree $(-r, r - 1)$ is a family $d^E = \{d^E_{p,q} \in Hom_R(E_{p,q}, E_{p-r,q+r-1}) \mid p, q \in \mathbb{Z}\}$ of homomorphisms such that $d^E_{p,q} od^E_{p+r,q-r+1} = 0$ for all $p, q \in \mathbb{Z}$. The homology $H(E)$ of E is a $\mathbb{Z}$ bi-graded R-module $\{H_{p,q}(E) \mid p, q \in \mathbb{Z}\}$, where $H_{p,q} = Ker d^E_{p,q} / Image d^E_{p+r,q-r+1}$. To the pair (E, d^E), we can associate a chain complex $(T(E), d^{T(E)})$, where $T(E)_n = \oplus \sum_{p+q=n} E_{p,q}$ and for $x \in E_{p,q}$, $d_n^{T(E)}(x) = d^E_{p,q}(x)$. Evidently, $H_n(T(E)) = \oplus \sum_{p+q=n} H_{p,q}(E)$.

Definition 2.5.1 A **spectral sequence** is a triple (E^r, d^r, λ_r), $r \geq 2$, where $E^r = \{E^r_{p,q} \mid p, q \in \mathbb{Z}\}$ is a $\mathbb{Z}$ bi-graded R-module, d^r is boundary operator on E^r of bi-degree $(-r, r - 1)$, and λ_r is a $\mathbb{Z}$ bi-graded isomorphism from $H(E^r)$ to E^{r+1}. E^2 is called the initial term of the spectral sequence.

Remark 2.5.2 Note that d^{r+1} has nothing to do with λ_r.

A morphism from an spectral sequence $E = \{(E^r, d^r, \lambda_r) \mid r \geq 2\}$ to an spectral sequence $E' = \{(E'^r, d'^r, \lambda'_r) \mid r \geq 2\}$ is a sequence $f^r = \{f^r_{p,q} \in Hom_R(E_{p,q}, E'_{p,q}) \mid p, q \in \mathbb{Z}\}$, $r \geq 2$ of bi-graded homomorphisms of bi-degrees $(0, 0)$ such that $d'_{p,q} f^r_{p,q} = f^r_{p-r,q+r-1} d^r_{p,q}$, $\lambda'_r H(f^r) = f^{r+1} \lambda_r$ for all $r \geq 2$ and $p, q \in \mathbb{Z}$.

Without any loss of generality, we may assume that $E^{r+1} = H(E^r)$ and λ_r is the identity map for each r. Thus, λ_r may be omitted from the definition. In turn, a morphism $f = \{f^r \mid r \geq 2\}$ is uniquely determined by f^2.

Spectral Sequences and Filtrations

Let (E^r, d^r), $r \geq 2$ be a spectral sequence of R-modules. Then $E^3 = H(E^2)$. For each $p, q \in \mathbb{Z}$, $E^3_{p,q} = Ker\, d^2_{p,q}/Image\, d^2_{p+2,q-1}$. Denote $Ker\, d^2_{p,q}$ by $C^2_{p,q}$, and $Image\, d^2_{p+2,q-1}$ by $B^2_{p,q}$. Thus, $E^3_{p,q} = C^2_{p,q}/B^2_{p,q}$. This gives us a chain

$$0 = B^1 \subseteq B^2 \subseteq C^2 \subseteq C^1 = E^2$$

of $\mathbb{Z}$ bi-graded R-modules such that $E^2 = C^1/B^1$, d^2 is such that and $E^3 = C^2/B^2$. The boundary operator d^3 on $E^3 = C^2/B^2$ which is of bi-degree $(-3, 2)$. Further, $E^4 = H(E^3)$. Thus, $E^4_{p,q} = Ker\, d^3_{p,q}/Image\, d^3_{p+3,q-2}$. Suppose that $Ker\, d^3_{p,q} = C^3_{p,q}/B^2_{p,q}$ and $Image\, d^3_{p+3,q-2} = B^3_{p,q}/B^2_{p,q}$. Again, we have enlarged chain

$$0 = B^1 \subseteq B^2 \subseteq B^3 \subseteq C^3 \subseteq C^2 \subseteq C^1 = E^2$$

of $\mathbb{Z}$ bi-graded R-modules such that $E^4 = (C^3/B^2)/(B^3/B^2)$ is naturally isomorphic to C^3/B^3. Without any loss, we may take $E^4 = C^3/B^3$. Proceeding inductively, we get a chain

$$0 = B^1 \subseteq B^2 \subseteq B^3 \subseteq \cdots B^n \subseteq B^{n+1} \subseteq \cdots \cdots \subseteq \cdots C^{n+1} \subseteq C^n \subseteq \cdots \subseteq \quad (2.8)$$
$$C^3 \subseteq C^2 \subseteq C^1 = E^2$$

of $\mathbb{Z}$ bi-graded R-modules such that $E^r = C^{r-1}/B^{r-1}$ and d^r is a homomorphism from C^{r-1}/B^{r-1} to itself of bi-degree $(-r, r-1)$ with $Ker\, d^r = C^r/B^{r-1}$ and $Image\, d^r = B^r/B^{r-1}$. Since $Ker\, d^r = C^r/B^{r-1}$, $Image\, d^r$ is canonically isomorphic to C^{r-1}/C^r. In turn, we have a canonical isomorphism ϕ^r from C^{r-1}/C^r to B^r/B^{r-1} such that

$$d^r = C^{r-1}/B^{r-1} \xrightarrow{\nu^r} C^{r-1}/C^r \xrightarrow{\phi^r} B^r/B^{r-1} \xrightarrow{i^r} C^{r-1}/B^{r-1},$$

where ν^r is the quotient map and i^r is the inclusion map.

Conversely, suppose that we have a chain (2.8) together with isomorphisms ϕ^r from C^{r-1}/C^r to B^r/B^{r-1}. Then we have a spectral sequence $\{(E^r, d^r) \mid r \geq 1\}$, where $d^r = i^r \circ \phi^r \circ \nu^r$.

Further, it is evident that $B^\infty = \bigcup_{r=1}^\infty B^r$ is a submodule of $C^\infty = \bigcap_{r=1}^\infty C^r$. The $\mathbb{Z}$ bi-graded module $E^\infty = C^\infty/B^\infty$ is called the term of the spectral sequence at infinity. We also write that $Lim\, E^r = E^\infty$. We treat E^r as approximations of E^∞.

Let E and E' be spectral sequences, and f be a morphism from E to E'. Let

$$0 = B^1 \subseteq B^2 \subseteq B^3 \subseteq \cdots B^n \subseteq B^{n+1} \subseteq \cdots \subseteq \cdots C^{n+1} \subseteq C^n \subseteq \cdots \subseteq$$
$$C^3 \subseteq C^2 \subseteq C^1 = E^2$$

and

$$0 = B'^1 \subseteq B'^2 \subseteq B'^3 \subseteq \cdots B'^n \subseteq B'^{n+1} \subseteq \cdots \subseteq \cdots C'^{n+1} \subseteq C'^n \subseteq \cdots \subseteq$$
$$C'^3 \subseteq C'^2 \subseteq C'^1 = E'^2$$

be the corresponding filtrations. Then f^2 is a $\mathbb{Z}$ bi-graded homomorphism from $E^2 = C^1$ to $E'^2 = C'^1$ of bi-degree $(0, 0)$ such that $f^2(C^r) \subseteq C'^r$ and $f^2(B^r) \subseteq B'^r$, and it is such that $d'^r f^r = f^r d^r$ for all r, where f^r is the induced homomorphism from C^{r-1}/B^{r-1} to C'^{r-1}/B'^{r-1}. In turn, f induces a $\mathbb{Z}$ bi-graded homomorphism f^∞ of bi-degree $(0, 0)$ from E^∞ to E'^∞.

We say that a spectral sequence $E = \{E^r \mid r \geq 2\}$ **bounded below** if for each $n \geq 0$, $E^2_{-p,n+p}$ is eventually 0. More explicitly, for each n there is an integer $\lambda(n)$ such that $E^2_{-p,n+p} = 0$ for all $p \geq \lambda(n)$. Thus, a first quadrant spectral sequence (a spectral sequence $\{E^r \mid r \geq 2\}$ with $E^r_{p,q} = 0$ whenever $p < 0$ or $q < 0$) is a sequence which is bounded below.

Theorem 2.5.3 (Mapping theorem) *Let f be a morphism from a spectral sequence E to a spectral sequence E' such that f^{k_0} is an isomorphism from E^{k_0} to E'^{k_0}. Then f^r is an isomorphism from E^r to E'^r for all $r \geq k_0$. If in addition to that, E and E' are bounded below, then f^∞ is an isomorphism from E^∞ to E'^∞.*

Proof Since $d'^r f^r = f^r d^r$ for all r, it follows that $f^{k_0+1} = H(f^{k_0})$ is also an isomorphism. By the induction, it follows that f^r is an isomorphism for all $r \geq k_0$. Now, suppose that E and E' are bounded below. Let $a' + B'^\infty_{p,q}$ be a member of $E'^\infty_{p,q} = C'^\infty/B'^\infty_{p,q}$, where $C'^\infty_{p,q} = \bigcap_{r=2}^\infty C'^r_{p,q}$ and $B'^\infty = \bigcup_{r=2}^\infty B'^r_{p,q}$. Since E and E' are bounded below, then for $p, q \in \mathbb{Z}$, there exists $k_1 \geq k_0$ such that $d^r_{p,q} = 0 = d'^r_{p,q}$ for all $r \geq k_1$. This means that $C^{k_1}_{p,q} = C^\infty_{p,q}$ and $C'^{k_1}_{p,q} = C'^\infty_{p,q}$. Let $a' \in C'^\infty_{p,q} = C'^{k_1}_{p,q}$. Then $a' + B'^{k_1}$ belongs to $E'^{k_1}_{p,q} = C'^{k_1}_{p,q}/B'^{k_1}_{p,q}$. Since f^{k_1+1} is an isomorphism, there is an element $a \in C^{k_1}_{p,q}$ such that $f^{k_1+1}(a + B^{k_1}_{p,q}) = a' + B'^{k_1}_{p,q}$. It follows that $f^\infty(a + B^\infty_{p,q}) = a' + B'^\infty_{p,q}$. This shows that f^∞ is surjective. Suppose that $f^\infty(a + B^\infty_{p,q}) = 0$, where $a \in C^\infty_{p,q} = C^{k_1}_{p,q}$. Then $f^2(a) \in B'^\infty_{p,q}$. This means that $f^2(a) \in B'^{k_2}_{p,q}$ for some $k_2 \geq k_1$, and so $f^{k_2}(a + B^\infty_{p,q}) = 0$. Since f^{k_2} is an isomorphism, $a \in B^{k_2}_{p,q}$. Hence $a + B^\infty_{p,q} = 0$. It follows that f^∞ is injective. ♯

Let Σ be a selective abelian category. Let A be an object in Σ. A **filtration** F of an object A is a family $\{F_p(A) \mid p \in \mathbb{Z}\}$ of subobjects of A such that $F_p(A) \subseteq F_{p+1}(A)$ for all $p \in \mathbb{Z}$. The pair (F, A) is called a $\mathbb{Z}$-filtered object in Σ. We have a category $Fill\Sigma$ whose objects are $\mathbb{Z}$-filtered objects (F, A), and a morphism from (F, A) to (F', A') is a morphism f from A to A' such that

$f(F_p(A)) \subseteq F'_p(A')$ for each p. To each filtered object (F, A) of Σ, we can associate a $\mathbb{Z}$-graded object $GF(A) = \{GF(A)_p = F_p(A)/F_{p-1}(A) \mid p \in \mathbb{Z}\}$ in Σ. Indeed, this defines a functor from the category $Fill\Sigma$ to the category of $\mathbb{Z}$-graded objects in Σ. If in addition $A = \{A_n \mid n \in \mathbb{Z}\}$ is also a $\mathbb{Z}$-graded object in Σ and F is a filtration on $\mathbb{Z}$-graded object A, then we have an associated $\mathbb{Z}$ bi-graded module $GF(A) = \{GF(A)_{p,q} = F_p(A_{p+q})/F_{p-1}(A_{p+q}) \mid p, q \in \mathbb{Z}\}$. Here p is called the filtration degree and q is called the complimentary degree.

In particular, let us take Σ to be the category CH_R of chain complexes of right R-modules. Let

$$(F, A) = \cdots \overset{i_{p-2}}{\to} F_{p-1}(A) \overset{i_{p-1}}{\to} F_p(A) \overset{i_p}{\to} F_{p+1}(A) \overset{i_{p+1}}{\to} \cdots$$

be a $\mathbb{Z}$-filtered chain complex of right R-modules, where

$$A \equiv \cdots \overset{d_{n+1}}{\to} A_n \overset{d_n}{\to} A_{n-1} \overset{d_{n-1}}{\to} \cdots$$

is a chain complex of right R-modules, and i_p denotes the inclusion chain transformation from $F_p(A)$ to $F_{p+1}(A)$. Let $H(A)$ denote the $\mathbb{Z}$-graded homology module $\{H_n(A) = Z_n(A)/B_n(A) \mid n \in \mathbb{Z}\}$ of A. We can also treat i_p as an inclusion chain transformation from $F_p(A)$ to A. Let $F_p(H(A))$ denote the image of $H(i_p)$. Thus, $F_p(H(A))$ is the $\mathbb{Z}$-graded submodule $\{(Z_n(A) \cap F_p(A_n) + B_n(A))/B_n(A) \mid n \in \mathbb{Z}\}$ of $H(A)$ (note that $(Z_n(A) \cap F_p(A_n) + B_n(A))/B_n(A) \approx (Z_n(A) \cap F_p(A_n))/(B_n(A) \cap F_p(A_n))$. This gives us a filtration

$$\cdots \subseteq F_{p-1}(H(A)) \subseteq F_p(H(A)) \subseteq F_{p+1}(H(A)) \subseteq \cdots$$

of $\mathbb{Z}$-graded homology module $H(A)$ of A. In particular, we have the $\mathbb{Z}$ bi-graded module $\Lambda(F, A) = \{\Lambda_{p,q} = F_p(H_{p+q}(A)) \mid p, q \in \mathbb{Z}\}$ associated with the filtered chain complex (F, A).

A filtration F of a chain complex A is said to be a bounded filtration if for each n, there are integers λ_n and μ_n with $\lambda_n < \mu_n$ such that $F_{\lambda_n}(A_n) = 0$ and $F_{\mu_n}(A_n) = A_n$.

Theorem 2.5.4 *To every filtered chain complex (F, A), we can associate a spectral sequence $\{(E^r, d^r) \mid r \geq 1\}$ such that $E^1_{p,q} \approx H_{p+q}(F_p(A)/F_{p-1}(A))$. Further, if the filtration F is bounded, then the spectral sequence converges to the $\mathbb{Z}$-graded homology module $H(A)$ of A in the sense that $E^\infty_{p,q} = F_p(H_{p+q}(A))/F_{p-1}(H_{p+q}(A))$.*

Proof Put

$$Z^r_{p,q} = F_p(A_{p+q}) \cap d^{-1}_{p+q}(F_{p-r}(A_{p+q-1})),$$

and

$$B^r_{p,q} = d_{p+q+1}(Z^{r-1}_{p-1,q+1}), r \geq 1.$$

Evidently, $B^r_{p,q} \subseteq Z^r_{p,q}$. Let E^r denote the $\mathbb{Z}$ bi-graded module

$$\{E^r_{p,q} = (Z^r_{p,q} + F_{p-1}(A_{p+q}))/(B^r_{p,q} + F_{p-1}(A_{p+q})) \mid p, q \in \mathbb{Z}\}.$$

The boundary operator d on A induces a boundary d^r on E^r of bi-degree $(-r, r - 1)$ given by

$$d^r_{p,q}(a + B^r_{p,q} + F_{p-1}(A_{p+q})) = d_{p+q}(a) + B^r_{p-r,q+r-1} + F_{p-r-1}(A_{p+q-1}).$$

Evidently, $E^1_{p,q} = H_{p+q}(F_p(A)/F_{p-1}(A))$. It is straightforward, of course a little painstaking, to verify that $\{(E^r, d^r) \mid r \geq 1\}$ is a spectral sequence with the desired property. I leave the details. ♯

Exact Couples

An **exact couple** in the category of right R-modules is a quintuple $E_C = \{D, E; \alpha, \beta, \gamma\}$, where D and E are R-modules, α is a homomorphism from D to D, β is a homomorphism from D to E, and γ is a homomorphism from E to D such that the triangle

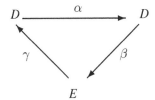

is exact in the sense that $Ker\ \beta = Image\ \alpha$, $Ker\ \gamma = Image\ \beta$, and $Ker\ \alpha = Image\ \gamma$.

A morphism from an exact couple from $E_C = \{D, E; \alpha, \beta, \gamma\}$ to an exact couple $E'_C = \{D', E'; \alpha', \beta', \gamma'\}$ is a pair (ϕ, ψ), where ϕ is a homomorphism from D to D', and ψ is a homomorphism from E to E' such that $\phi\alpha = \alpha'\phi$, $\beta'\phi = \psi\beta$, and $\gamma'\psi = \phi\gamma$. This gives us a category ECR of exact couples of right R-modules.

Given an exact couple E_C as above, take $d = \beta\gamma$. Then d is a homomorphism from E to E. Since $\gamma\beta = 0$, $d^2 = 0$. Let E^1 denote the homology module $H(E) = (Ker\ d)/(Image\ d)$ of E. Take $D^1 = \alpha(D)$, and α^1 the homomorphism from D^1 to D^1 which is restriction of α. Now, we define a homomorphism β^1 from D^1 to E^1 and a homomorphism γ^1 from E^1 to D^1 as follows. Suppose that $\alpha(x) = \alpha(x')$. Then $x - x'$ belongs to $Ker\ \alpha = Image\ \gamma$. Hence, $\beta(x) - \beta(x') = \beta(\gamma(u))$ for some u in E. This means that $\beta(x) + Image\ d = \beta(x') + Image\ d$. In turn, we get a homomorphism β^1 from D^1 to E^1 given by $\beta^1(\alpha(x)) = \beta(x) + Image\ d$. Next, let u be a member of $Ker\ d$. Then $\beta(\gamma(u)) = 0$, and so $\gamma(u)$ is a member of $Ker\ \beta = Image\ \alpha = D^1$. Further, if $u \in Image\ d$, then $u = \beta(\gamma(v))$ for some $v \in E$. But, then $\gamma(u) = 0$. Thus, we have a homomorphism γ^1 from E^1 to D^1 given by $\gamma^1(u + Image\ d) = \gamma(u)$.

Theorem 2.5.5 *The quintuple $E_C^1 = \{D^1, E^1; \alpha^1, \beta^1, \gamma^1\}$ defined above is an exact couple of R-modules.*

Proof Exactness at the right-hand D^1: Let $x = \alpha(d)$ be a member of D^1. Then, by definition, $\beta^1(\alpha^1(x)) = \beta^1(\alpha(\alpha(d))) = \beta(\alpha(d)) + Image\ d = 0$. This shows that $\beta^1\alpha^1 = 0$. Thus, $Image\ \alpha^1 \subseteq Ker\ \beta^1$. Let $x = \alpha(y)$ belongs to $Ker\ \beta^1$, where $y \in D$. By definition, $\beta^1(x) = \beta(y) + Image\ d = 0$. This means that $\beta(y) \in Image\ \beta\gamma$. Thus, $\beta(y) = \beta(\gamma(u))$ for some $u \in E$. Hence $y - \gamma(u)$ is a member of $Ker\ \beta = Image\ \alpha$. Suppose that $y - \gamma(u) = \alpha(v)$, where $v \in D$. Then $x = \alpha(y) = \alpha(y) - \alpha(\gamma(u)) = \alpha^2(v) = \alpha^1(\alpha(v))$ is a member of $Image\ \alpha^1$. This proves the exactness at right-hand D^1.

Exactness at E^1: Let $x = \alpha(y)$ be a member of D^1, where $y \in D$. Then, by definition, $\beta^1(x) = \beta(y) + Image\ d$ (note that $\beta(y) \in Ker\ d$). Again, by definition, $\gamma^1(\beta^1(x)) = \gamma(\beta(y)) = 0$. This shows that $Image\ \beta^1 \subseteq Ker\ \gamma^1$. Let $u + Image\ d$ be an element of $Ker\ \gamma^1$, where $u \in Ker\ d$. Then $\gamma(u) = \gamma^1(u + Image\ d) = 0$. By the exactness, there is an element $x \in D$ such that $u = \beta(x)$. By definition $u + Image\ d = \beta^1(\alpha(x))$. This shows that $u + Image\ d$ is a member of $Image\ \beta^1$.

Exactness at the left-hand D^1: Let $u + Image\ d$ be a member of E^1, where $u \in Ker\ d$. By definition, $\alpha^1(\gamma^1(u + Image\ d)) = \alpha^1(\gamma(u)) = \alpha(\gamma(u)) = 0$. Hence $Image\ \gamma^1 \subseteq Ker\ \alpha^1$. Now, suppose that $\alpha(x) \in Ker\ \alpha^1$. Then $\alpha(\alpha(x)) = \alpha^1(\alpha(x)) = 0$. From the exactness in E_C, $\alpha(x) = \gamma(u)$ for some $u \in E$. Since $\beta(\alpha(x)) = 0$, $\beta(\gamma(u)) = 0$. This means that $u \in Ker\ d$. In turn, $\gamma^1(u + Image\ d) = \alpha(x)$. This shows that $Ker\ \alpha^1 \subseteq Image\ \gamma^1$. ♯

The exact couple E_C^1 is called the **derived exact couple** of E_C.
The following corollary is evident.

Corollary 2.5.6 *Let E_C be an exact couple as given above. Iterating, we get a sequence $\{(E^n, d^n) \mid n \geq 1\}$ of modules E^n with boundary operators d^n on E^n such that $H(E^n, d^n) = E^{n+1}$ for all n.* ♯

More generally, an exact couple of $\mathbb{Z}$ bi-graded R-modules is a quintuple $E^C = \{D, E; \alpha, \beta, \gamma\}$, where D and E are $\mathbb{Z}$ bi-graded R-modules, α is a $\mathbb{Z}$ bi-graded homomorphism from D to D of bi-degree (m, n), β is a $\mathbb{Z}$ bi-graded homomorphism from D to E of bi-degree (p, q), and γ is a $\mathbb{Z}$ bi-graded homomorphism from E to D of bi-degree (r, s) such that the triangle

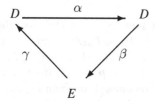

is exact in the sense that $Ker\ \beta_{j,k} = Image\ \alpha_{j-m,k-n}$, $Ker\ \gamma_{j,k} = Image\ \beta_{j-p,k-q}$, and $Ker\ \alpha_{j,k} = Image\ \gamma_{j-r,k-s}$ for all $j, k \in \mathbb{Z}$. As in the above theorem, we have the derived exact couple

$$E_C^1 = \{D^1, E^1; \alpha^1, \beta^1, \gamma^1\}$$

of $\mathbb{Z}$ bi-graded R-modules, where $D^1 = \alpha(D)$, α^1 is the restriction of α to D^1, E^1 is the homology module $H(E) = Ker\ \beta\gamma/Image\ \beta\gamma$, β^1 is given by $\beta^1(\alpha(x)) = \beta(x) + Image\ \beta\gamma$ (note that $\alpha(x) \in D_{j,k}^1$ implies that $x \in D_{j-m,k-n}$), and γ^1 is given by $\gamma^1(u + Image\ \beta\gamma) = \gamma(u)$. Evidently, α^1 is a $\mathbb{Z}$ bi-graded homomorphism of bi-degree (m, n), β^1 is a $\mathbb{Z}$ bi-graded homomorphism of bi-degree $(p - m, q - n)$, γ^1 is of bi-degree (r, s), and the boundary operator $d = \beta\gamma$ on E^1 is of bi-degree $(p + r, q + s)$. Iterating, we get a sequence

$$\{E_C^k = \{D^k, E^k; \alpha^k, \beta^k, \gamma^k\}, k \geq 1\}$$

of derived couples of $\mathbb{Z}$ bi-graded modules, where α^k is of bi-degree (m, n), β^k is of bi-degree $(p - km, q - kn)$, and γ_k is of bi-degree (r, s). Further, $d^k = \beta^{k-1}\gamma^{k-1}$ is a boundary operator on E^k of bi-degree $(p - (k - 1)m + r, q - (k - 1)n + s)$. Also $H(E^k, d^k) = E^{k+1}$. In particular, if $(m, n) = (1, -1)$, $(p, q) = (0, 0)$, and $(r, s) = (-1, 0)$, then

$$E_C^k = \{D^k, E^k; \alpha^k, \beta_k, \gamma_k\}$$

is an exact couple of $\mathbb{Z}$ bi-graded modules, where α^k is of bi-degree $(1, -1)$, β^k is of bi-degree $(-k, k)$, γ^k is of bi-degree $(-1, 0)$, and the boundary operator d^k is of degree $(-k, k - 1)$. Also $H(E^k) = E^{k+1}$. In turn, it gives us an spectral sequence $\{(E^k, d^k) \mid k \geq 2\}$.

Let A be a chain complex of right R-modules and F be a filtration on A. Then for each $p \in \mathbb{Z}$, we have a short exact sequence

$$0 \longrightarrow F_{p-1}(A) \overset{i_{p-1}}{\to} F_p(A) \overset{\nu_p}{\to} F_p(A)/F_{p-1}(A) \longrightarrow 0$$

of chain complexes. In turn, we have a long exact homology sequence

$$\cdots \overset{\partial_{n+1}}{\to} H_n(F_{p-1}(A)) \overset{H_n(i_{p-1})}{\to} H_n(F_p(A)) \overset{H_n(\nu_p)}{\to} H_n(F_p(A)/F_{p-1}(A)) \overset{\partial_n}{\to} \cdots,$$

where ∂_n is a natural connecting homomorphism. Let D denote the $\mathbb{Z}$ bi-graded module $\{D_{p,q} \mid p, q \in \mathbb{Z}\}$, where $D_{p,q} = H_{p+q}(F_p(A))$, and E denote the $\mathbb{Z}$ bi-graded module $\{E_{p,q} \mid p, q \in \mathbb{Z}\}$, where $E_{p,q} = H_{p+q}(F_p(A)/F_{p-1}(A))$. Let α denote the $\mathbb{Z}$ bi-graded homomorphism $\{\alpha_{p,q} \mid p, q \in \mathbb{Z}\}$ from D to D of degree $(1, -1)$, where $\alpha_{p,q} = H_{p+q}(i_p)$ is the homomorphism from $D_{p,q} = H_{p+q}(F_p(A))$ to $D_{p-1,q+1} = H_{p+q}(F_{p+1})$. Let β denote the $\mathbb{Z}$ bi-graded homomorphism $\{\beta_{p,q} \mid p, q \in \mathbb{Z}\}$ from D to E of degree $(0, 0)$, where $\beta_{p,q} = H_{p+q}(\nu_p)$ is a homomorphism from $D_{p,q} = H_{p+q}(F_p(A))$ to $E_{p,q} = H_{p+q}(F_p(A)/F_{p-1}(A))$. Let γ denote the $\mathbb{Z}$ bi-graded homomorphism $\{\gamma_{p,q} \mid p, q \in \mathbb{Z}\}$ from E to D of degree $(-1, 0)$, where $\gamma_{p,q} = \partial_{p+q}$ is a homomorphism from $E_{p,q} = H_{p+q}(F_p(A)/F_{p-1}(A))$ to $D_{p-1,q} = H_{p-1+q}(F_{p-1}(A))$. The above long exact homology sequence ensures that $\{D, E; \alpha, \beta, \gamma\}$ is an exact couple. In turn, it induces, as above, a spectral sequence which can be easily seen to be the spectral sequence introduced in Theorem 2.5.4.

In Sect. 3.2, we shall further use the spectral sequence arguments to establish an important theorem of Hurewicz.

Exercises

2.5.1 Let $E = \{(E^r, d^r) \mid r \geq 1\}$ and $E' = \{(E'^r, d'^r) \mid r \geq 1\}$ be spectral sequences. Let $\tilde{E} = \{(\tilde{E}^r, \tilde{d}^r \mid r \geq 1\}$, where $\tilde{E}^r_{p,q} = \oplus \sum_{k+m=p,l+n=q} E^r_{k,l} \otimes E'_{m,n}$ and $\tilde{d}^r$ is the usual tensor product of d^r and d'^r. Show that $\tilde{E}$ is a spectral sequence.

2.5.2 Develop the theory of spectral sequences, as above, in a selective complete and co-complete abelian category.

2.5.3 Let Σ be a selective complete and co-complete abelian category. Let (S, T) be a pair of $\mathbb{Z}$ bi-graded objects in Σ, where T is a $\mathbb{Z}$ bi-graded subobject of S. A pseudo-boundary operator on (S, T) of bi-degree $(-r, r - 1)$ is a $\mathbb{Z}$ bi-graded morphism $d = \{d_{p,q} \in Mor_{\Sigma}(S_{p,q}, S_{p-r,q+r-1}) \mid d_{p,q}d_{p+r,q-r+1}(T_{p+r,q-r+1}) = 0, p, q \in \mathbb{Z}\}$. Let $H(S, T)$ denote the $\mathbb{Z}$ bi-graded homology object $\{H_{p,q}(S, T) = Ker d_{p,q}/ d_{p+r,q-r+1}(T_{p+r,q-r+1}) \mid p, q \in \mathbb{Z}\}$ of (S, T). Let

$$F(S, T) = \{F_{p,q}(S, T) = d^{-1}_{p,q}(T_{p-r,q+r-1}) \bigcap T_{p,q} \mid p, q \in \mathbb{Z}\}.$$

Show that d induces a $\mathbb{Z}$ bi-graded boundary operator on $F(S, T)$ of bi-degree $(-r, r - 1)$ such that $H(F(S, T))$ is naturally isomorphic to a $\mathbb{Z}$ bi-graded subobject of $H(S, T)$.

2.5.4 A sequence $E = \{(E^r, d^r, \lambda_r) \mid r \geq 1\}$, where $E^r = (S^r, T^r)$ is a pair of $\mathbb{Z}$ bi-graded object in Σ with T^r as a graded subobject, d^r as a pseudo-boundary operator on E^r of bi-degree $(-r, r - 1)$, λ_r as an isomorphism from $H((S^r, T^r))$ inducing isomorphism from $H(F(S^r, T^r))$ will be termed as a pseudo-spectral sequence. Show that a pseudo-spectral sequence determines the following data:

1. Towers

$$0 = B^0 \subseteq B^1 \subseteq B^2 \subseteq \cdots B^n \subseteq B^{n+1} \subseteq \cdots \cdots \subseteq \cdots C^{n+1} \subseteq C^n \subseteq \cdots \subseteq$$
$$C^3 \subseteq C^2 \subseteq C^1 \subseteq = E^1$$

and

$$0 = V^0 \subseteq V^1 \subseteq V^2 \subseteq \cdots V^n \subseteq V^{n+1} \subseteq \cdots \cdots \subseteq \cdots U^{n+1} \subseteq U^n \subseteq \cdots \subseteq$$
$$U^3 \subseteq U^2 \subseteq U^1 \subseteq U^0 = E^1,$$

with the condition $U^i \subseteq C^i$, $V^i \subseteq B^i$ and $V^i = U^i \bigcap B^i = V_i$, $U^{r-1} \bigcap C^r = U^r$ for each i.

2. Isomorphisms ϕ^r from C^{r-1}/C^r to B^r/B^{r-1} such that $\phi^r i^r (U^{r-1}/U^r) = j^r (V^r/V^{r-1})$ for each r, where i^r is the obvious injective homomorphisms from U^{r-1}/U^r to C^{r-1}/C^r and j^r is the obvious injective homomorphisms from V^r/V^{r-1} to B^r/B^{r-1}.

Conversely, show that the above data determines a pseudo-spectral sequence which in turn determines this data.

2.5.5 Establish the mapping theorem for pseudo-spectral sequences.

Chapter 3
Homological Algebra 3: Examples and Applications

This chapter is devoted to associate homological invariants with mathematical structures, and to have some important applications in topology, geometry, and group theory. We assume that the reader is familiar with the basics in metric spaces and topology together with some amount of calculus .

3.1 Polyhedrons and Simplicial Homology

In this section, we associate simplicial homology with polyhedrons.

Simplicial Complex and Polyhedrons

Recall the basic definitions and terminologies regarding simplicial complexes from Sect. 1.3 of Chap. 1. Throughout this section, we shall restrict our attention to those simplicial complexes (Σ, S) which are **locally finite** in the sense that for each vertex $v \in \Sigma$, there are only finitely many simplexes $\sigma \in S$ such that $v \in \sigma$. Indeed, mostly, we shall be interested in finite simplicial complexes (complexes for which Σ is finite). Thus, SC stands for the category of locally finite simplicial complexes. A simplicial complex (Σ, S) is called a finite simplicial complex if Σ is finite.

We define a functor from the category SC of simplicial complexes to the category TOP of topological spaces as follows. Let (Σ, S) be a simplicial complex. Let $\mid (\Sigma, S) \mid$ denote the set consisting of all maps α from Σ to $[0, 1]$ such that (i) $\{v \in \Sigma \mid \alpha(v) \neq 0\}$ is a simplex and (ii) $\sum_{v \in \Sigma} \alpha(v) = 1$. The number $\alpha(v)$ is called the v_{th} **barycentric coordinate** of α. Evidently, any $\alpha \in \mid (\Sigma, S) \mid$ is 0 at all but finitely many vertices in Σ. We have a metric d on $\mid (\Sigma, S) \mid$ defined by

$$d(\alpha, \beta) = +\sqrt{\sum_{v} (\alpha(v) - \beta(v))^2}.$$

© Springer Nature Singapore Pte Ltd. 2021
R. Lal, *Algebra 3*, Infosys Science Foundation Series,
https://doi.org/10.1007/978-981-33-6326-7_3

If σ is a simplex of (Σ, S), then $\mid (\sigma, \bar{\sigma}) \mid$ can be identified with the subspace

$$\{\alpha \in \mid (\Sigma, S) \mid \mid \alpha(v) \neq 0 \Rightarrow v \in \sigma\}$$

of $\mid (\Sigma, S) \mid$. This subspace is denoted by $\mid \sigma \mid$, and it is termed as a closed simplex. Similarly, $\mid (\sigma, \dot{\sigma}) \mid$ is denoted by $\mid \dot{\sigma} \mid$. Evidently, $\mid \dot{\sigma} \mid$ is a boundary of $\mid \sigma \mid$. The subset $< \sigma > = \mid \sigma \mid - \mid \dot{\sigma} \mid$ is an open subspace of $\mid \sigma \mid$, and it is called an open simplex. Recall that the standard q-simplex Δ^q is the convex hull of the standard basis $\{\overline{e_0}, \overline{e_1}, \ldots, \overline{e_q}\}$ of $\mathbb{R}^{q+1}$. More explicitly, Δ^q is the subspace

$$\Delta^q = \left\{ \bar{a} = \sum_{i=0}^{q} a_i \overline{e_i} \in \mathbb{R}^{q+1} \mid a_i \geq 0 \text{ and } \sum a_i = 1 \right\}$$

of $\mathbb{R}^{q+1}$ with the Euclidean metric. If $\sigma = \{v_0, v_1, \ldots, v_q\}$ is a q-simplex in (Σ, S), then we have an isometry ϕ_q from standard q-simplex Δ^q to $\mid \sigma \mid$ given by $\phi_q(\bar{a})(v_i) = a_i, 0 \leq i \leq q$, and $\phi_q(\bar{a})(v) = 0$ if $v \notin \sigma$.

Thus, $\mid \sigma \mid$ is a compact (and so also a closed) subset of $\mid (\Sigma, S) \mid$. Since (Σ, S) is locally finite, it follows easily that a subset A of $\mid (\Sigma, S) \mid$ is a closed subset if and only if $A \bigcap \mid \sigma \mid$ is a closed subset of $\mid \sigma \mid$ for all simplexes σ of (Σ, S). In particular, a map f from $\mid (\Sigma, S) \mid$ to a topological subspace X is continuous if and only if f restricted to each closed simplex is continuous.

For a fixed $v \in \Sigma$, consider the map ϕ_v from $\mid (\Sigma, S) \mid$ to $[0, 1]$ defined by $\phi(\alpha) = \alpha(v)$. Evidently, ϕ_v is a continuous map. Hence $\phi_v^{-1}((0, 1]) = \{\alpha \in \mid (\Sigma, S) \mid \mid \alpha(v) \neq 0\}$ is an open subset of $\mid (\Sigma, S) \mid$. This set is called the **star** of v, and it is denoted by $St(v)$. Thus,

$$St(v) = \{\alpha \in \mid (\Sigma, S) \mid \mid \alpha(v) \neq 0\}.$$

Proposition 3.1.1 *Let (Σ, S) be a locally finite simplicial complex. Then $\mid (\Sigma, S) \mid$ is locally compact.*

Proof Let α be a member of $\mid (\Sigma, S) \mid$. Then there is a $v \in \Sigma$ such that $\alpha \in St(v)$. Let $F = \{\sigma \in S \mid v \in \sigma\}$. Since (Σ, S) is locally finite, F is a finite set. Evidently, $St(v) \subseteq \bigcup_{\sigma \in F} \mid \sigma \mid$. Since $St(v)$ is open, and the finite union of compact sets is compact, it follows that $\bigcup_{\sigma \in F} \mid \sigma \mid$ is a compact neighborhood of α. ♯

Proposition 3.1.2 $\mid (\Sigma, S) \mid$ *is compact if and only if Σ is finite.*

Proof If Σ is finite, then S is also finite. In turn, $\mid (\Sigma, S) \mid = \bigcup_{\sigma \in S} \mid \sigma \mid$ is compact. Conversely, suppose that $\mid (\Sigma, S) \mid$ is compact. The family $\{< \sigma > = \mid \sigma \mid - \mid \dot{\sigma} \mid \mid \sigma \in S\}$ is a family of nonempty sets. Using the axiom of choice, we obtain a map c from S to $\mid (\Sigma, S) \mid$ such that $c(\sigma) \in < \sigma >$. Since $< \sigma > \bigcap < \tau > \neq \emptyset$ implies that $\sigma = \tau$, it follows that c is an injective map. Put $X = \{c(\sigma) \mid \sigma \in S\}$. Clearly, $A \bigcap \mid \sigma \mid$ is finite for all subsets A of X, and for all $\sigma \in S$. This means that every subset of X is closed in $\mid (\Sigma, S) \mid$. Thus, X is a discrete subset of $\mid (\Sigma, S) \mid$. Since $\mid (\Sigma, S) \mid$ is compact, X is finite. Hence S is finite. This shows that Σ is finite. ♯

Example 3.1.3 Consider the simplicial complex (Σ, S) of Example 1.3.11. The space $|(\Sigma, S)|$ can be identified to the subspace of $\mathbb{R}^3$ consisting of the points on the boundary of the triangle $\{(x, y, z) \in \mathbb{R}^3 \mid x, y, z \geq 0 \text{ and } x + y + z = 1\}$ of $\mathbb{R}^3$.

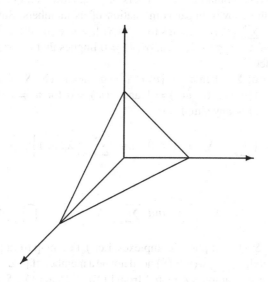

Example 3.1.4 Consider the simplicial complex (Σ, S), where $\Sigma = \{v_0, v_1, v_2\}$ and S is the set of all nonempty subsets of Σ. Then $|(\Sigma, S)|$ can naturally be identified with the triangle described in the above example.

Remark 3.1.5 For an abstract simplicial complex (Σ, S), not necessarily locally finite, we have a coherent topology T_c on $|(\Sigma, S)|$, where a subset F of $|(\Sigma, S)|$ is closed if and only if $F \cap |\sigma|$ is closed in $|\sigma|$ for all simplexes in (Σ, S). Equivalently, a map f from $|(\Sigma, S)|$ to a topological space X is continuous with respect to the coherent topology T_c if and only if f restricted to $|\sigma|$ is continuous for all simplexes σ. Thus the topology T_c is finer than the metric topology T_d defined above. In fact, T_c is a metric topology if and only if (Σ, S) is locally finite, and then $T_c = T_d$, where d is the above metric on $|(\Sigma, S)|$.

Let f be a simplicial map from (Σ, S) to (Σ', S'). Consider the map $|f|$ from $|(\Sigma, S)|$ to $|(\Sigma', S')|$ defined by $|f|(\alpha)(v') = \sum_{v \in f^{-1}(v')} \alpha(v)$ if $v' \in f(\Sigma)$ and 0 otherwise. Clearly, $|f|$ restricted to each closed simplex is continuous. In turn, $|f|$ is a continuous map from $|(\Sigma, S)|$ to $|(\Sigma', S')|$. Thus, $|-|$ defines a functor from the category SC to the category $LCTOP$ of locally compact Hausdorff topological spaces.

Linear Structure in Simplicial Complexes

Let (Σ, S) be a simplicial complex. For any $\alpha \in \mid (\Sigma, S) \mid$, there is a simplex σ such that $\alpha \in \mid \sigma \mid$. The smallest σ such that $\alpha \in \mid \sigma \mid$ is called the **carrier** of α. Clearly, σ is the carrier of α if and only if $\alpha \in < \sigma >$. Let $\alpha_1, \alpha_2, \ldots, \alpha_n$ be members of a closed simplex $\mid \sigma \mid$, and $\lambda_1, \lambda_2, \ldots, \lambda_n$ be nonnegative real numbers such that $\sum_{i=1}^{n} \lambda_i = 1$. Then $\sum_{i=1}^{n} \lambda_i \alpha_i$ belongs to $\mid \sigma \mid$. Thus, a closed simplex $\mid \sigma \mid$ is closed under the convex linear combination of members of $\mid \sigma \mid$. However, $\mid (\Sigma, S) \mid$ need not be closed under the convex linear combination of its members. Suppose that $\sum_{i=1}^{n} \lambda_i = 1$ and $\alpha = \sum_{i=1}^{n} \lambda_i \alpha_i$ belongs to $\mid (\Sigma, S) \mid$, where $\alpha_i \in \mid (\Sigma, S) \mid$. Let σ be a carrier of α. Then $\alpha \in < \sigma >$. In turn, $\alpha_i(v) \neq 0$ implies that $v \in \sigma$ for each i. Thus, $\alpha_i \in \mid \sigma \mid$ for each i.

Let v be a member of Σ. Then $\sigma_v = \{v\}$ is a 0-simplex in (Σ, S). Consider the map $\hat{v}$ from Σ to $[0, 1]$ given by $\hat{v}(v) = 1$ and $\hat{v}(w) = 0$ for $w \neq v$. Evidently, $\mid \sigma_v \mid = \{\hat{v}\} = < \sigma_v >$. For any simplex σ,

$$\mid \sigma \mid = \left\{ \sum_{v \in \sigma} \lambda_v \hat{v} \mid \lambda_v \geq 0, \ and \ \sum_{v \in \sigma} \lambda_v = 1 \right\},$$

and

$$< \sigma > = \left\{ \sum_{v \in \sigma} \lambda_v \hat{v} \mid \lambda_v > 0, \ and \ \sum_{v \in \sigma} \lambda_v = 1 \right\} = \bigcap_{v \in \sigma} St(v).$$

Let (Σ_1, S_1) and (Σ_2, S_2) be simplicial complexes. Let f be a map from $\mid (\Sigma_1, S_1) \mid$ to $\mid (\Sigma_2, S_2) \mid$. In general, $\sum_{v \in \Sigma_1} \alpha(v) f(\hat{v})$ need not be a member of $\mid (\Sigma_2, S_2) \mid$ even if $\alpha \in \mid (\Sigma_1, S_1) \mid$ (give an example). A map f from $\mid (\Sigma_1, S_1) \mid$ to $\mid (\Sigma, S_2) \mid$ is called a **linear** map if $\sum_{v \in \Sigma_1} \alpha(v) f(\hat{v})$ is a member of $\mid (\Sigma_2, S_2) \mid$ for each $\alpha \in \mid (\Sigma_1, S_1) \mid$, and then $f(\alpha) = \sum_{v \in \Sigma_1} \alpha(v) f(\hat{v})$. Evidently, if ϕ is a simplicial map from (Σ_1, S_1) to (Σ_2, S_2), then $\mid \phi \mid$ is a linear map.

Subdivision and Simplicial Approximation

Let (Σ_1, S_1) and (Σ_2, S_2) be simplicial complexes. A continuous map f from $\mid (\Sigma_1, S_1) \mid$ to $\mid (\Sigma_2, S_2) \mid$ may not be induced by a simplicial map. A simplicial map ϕ from (Σ_1, S_1) to (Σ_2, S_2) is called a **simplicial approximation** to f if $f^{-1}(< \sigma_2 >) \subseteq \mid \phi \mid^{-1} (\mid \sigma_2 \mid)$ for all $\sigma_2 \in S_2$. More precisely, if $f(\alpha) \in < \sigma_2 >$, then $\mid \phi \mid (\alpha) \in \mid \sigma_2 \mid$ for all $\sigma_2 \in S_2$. In particular, if v is a vertex in Σ_1 such that $f(\hat{v}) = \hat{w}$ for some $w \in \Sigma_2$, then $\phi(v) = w$. Thus, if ϕ is a simplicial map, then ϕ is the only simplicial approximation of $\mid \phi \mid$.

Theorem 3.1.6 *Let ϕ be a map from Σ_1 to Σ_2, where (Σ_1, S_1) and (Σ_2, S_2) are simplicial complexes. Then ϕ is a simplicial approximation of f if and only if $f(St(v)) \subseteq St(\phi(v))$ for each $v \in \Sigma_1$.*

Proof Suppose that ϕ is a simplicial approximation of f. Let α be a member of $St(v)$. Then $\alpha(v) \neq 0$. There is a unique simplex $\sigma_2 \in S_2$ such that $f(\alpha) \in < \sigma_2 >$. Since ϕ is a simplicial approximation of f, $\mid \phi \mid (\alpha)$ belongs to $\mid \sigma_2 \mid$. Again, since ϕ is a simplicial map and $\alpha(v) \neq 0$, $\mid \phi \mid (\alpha)(\phi(v)) \neq 0$. This means that $\phi(v)$ is a vertex of σ_2. Since $f(\alpha) \in< \sigma_2 >$, $f(\alpha)(\phi(v)) \neq 0$. This shows that $f(\alpha) \in St(\phi(v))$.

Conversely, suppose that ϕ is a map from Σ_1 to Σ_2 such that $f(St(v)) \subseteq St(\phi(v))$ for each $v \in \Sigma_1$. We first show that ϕ is a simplicial map. Let $\sigma = \{v_0, v_1, \ldots, v_q\}$ be a simplex of (Σ_1, S_1). Clearly, $< \sigma >\subseteq St(v_i)$ for each i. Under the assumption, $f(< \sigma >) \subseteq St(\phi(v_i))$ for each i. This means that there is a simplex $\tau \in S_2$ such that $\phi(v_i) \in \tau$ for each i. Since the subset of a simplex is a simplex, $\phi(\sigma)$ is a simplex in (Σ_2, S_2). This shows that ϕ is a simplicial map. Finally, we show that ϕ is a simplicial approximation of f. Let σ_2 be a simplex of (Σ_2, S_2), and $\alpha \in f^{-1}(< \sigma_2 >)$. Then $f(\alpha) \in< \sigma_2 >$. Suppose that $\alpha \in< \sigma_1 >$. Then for each $v \in \sigma_1$, $\alpha \in St(v)$. Since $f(St(v)) \subseteq St(\phi(v))$, $f(\alpha) \in St(\phi(v))$ for each $v \in \sigma_1$. Since $f(\alpha) \in< \sigma_2 >$, $\phi(v) \in \sigma_2$ for each $v \in \sigma_1$. Since ϕ is already seen to be a simplicial map, $\mid \phi \mid (\mid \sigma_1 \mid) \subseteq \mid \sigma_2 \mid$. In particular, $\mid \phi \mid (\alpha)$ belongs to $\mid \sigma_2 \mid$. Thus, $f^{-1}(< \sigma_2 >) \subseteq \mid \phi \mid^{-1} (\mid \sigma_2 \mid)$ for each $\sigma_2 \in S_2$. This shows that ϕ is a simplicial approximation of f. ♯

In general a continuous map f from $\mid (\Sigma_1, S_1) \mid$ to $\mid (\Sigma_2, S_2) \mid$ need not have any simplicial approximation from (Σ_1, S_1) to (Σ_2, S_2). However, if (Σ_1, S_1) is finite, we shall construct a simplicial complex (Σ_1', S_1') together with a linear homeomorphism ρ from $\mid (\Sigma_1', S_1') \mid$ to $\mid (\Sigma_1, S_1) \mid$ and a simplicial map ψ from (Σ_1', S_1') to (Σ_2, S_2) which is a simplicial approximation of $f \circ \rho$.

Definition 3.1.7 Let (Σ, S) be a simplicial complex. A simplicial complex (Σ', S') is called a **subdivision** of (Σ, S) if the following hold.

(i) $\Sigma' \subseteq \mid (\Sigma, S) \mid$.
(ii) For all $\sigma' \in S'$, there is a simplex $\sigma \in S$ such that $\sigma' \subseteq \mid \sigma \mid$.
(iii) The linear map λ from $\mid (\Sigma', S') \mid$ to $\mid (\Sigma, S) \mid$ given by

$$\lambda\left(\sum\nolimits_{v' \in \sigma'} \alpha_{v'} \hat{v}'\right) = \sum\nolimits_{v' \in \sigma'} \alpha_{v'} v'$$

is a homeomorphism .

Example 3.1.8 Consider the simplicial complex (Σ, S), where $\Sigma = \{v_0, v_1, v_2\}$ and S is the set of all nonempty subsets of Σ. Let Σ' be the subset $\{\hat{v}_0, \hat{v}_1, \hat{v}_2, \frac{1}{4}\hat{v}_0 + \frac{1}{4}\hat{v}_1 + \frac{1}{2}\hat{v}_2\}$ of $\mid (\Sigma, S) \mid$. Let us denote $\frac{1}{4}\hat{v}_0 + \frac{1}{4}\hat{v}_1 + \frac{1}{2}\hat{v}_2$ by α_3. Take $S' = \{\{\hat{v}_0\}, \{\hat{v}_1\}, \{\hat{v}_2\}, \{\alpha_3\}, \{\hat{v}_0, \hat{v}_1\}, \{\hat{v}_0, \hat{v}_2\}, \{\hat{v}_1, \hat{v}_2\}, \{\hat{v}_0, \alpha_3\}, \{\hat{v}_1, \alpha_3\}, \{\hat{v}_2, \alpha_3\}, \{\hat{v}_0, \hat{v}_1, \alpha_3\}, \{\hat{v}_0, \hat{v}_2, \alpha_3\}, \{\hat{v}_1, \hat{v}_2, \alpha_3\}\}$. It is easily seen that (Σ', S') is a subdivision of (Σ, S). See the figure below.

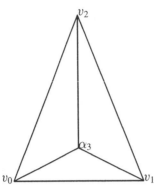

It is evident that a subdivision of a subdivision of (Σ, S) is a subdivision of (Σ, S). We are interested in a uniform and special type of subdivisions termed as barycentric subdivisions. Let (Σ, S) be a simplicial complex and $\sigma = \{v_0, v_2, \ldots, v_q\}$ be a q-simplex. Recall that the element $b(\sigma) = \frac{1}{q+1} \sum_{i=0}^{q} \hat{v}_i$ is called the **barycenter** of the simplex σ. Let Σ_{bd} denote the set of all the barycenters of the simplexes in (Σ, S). Let

$$S_{bd} = \{\{b(\sigma_0), b(\sigma_1), \ldots, b(\sigma_q)\} \mid \sigma_i \text{ is a } face \text{ of } \sigma_{i+1}\}.$$

More explicitly, to each ordered simplex $\sigma = \{v_0, v_1, \ldots, v_q\}$ in (Σ, S), there is a unique q-simplex

$$\left\{ \hat{v}_0, \frac{1}{2}(\hat{v}_0 + \hat{v}_1), \frac{1}{3}(\hat{v}_0 + \hat{v}_1 + \hat{v}_2), \ldots, \frac{1}{q+1}(\hat{v}_0 + \hat{v}_1 + \cdots + \hat{v}_q) \right\}$$

in S_{bd}. Then (Σ_{bd}, S_{bd}) is simplicial complex. For example, consider the simplicial complex (Σ, S), where $\Sigma = \{v_0, v_1, v_2\}$ and S is the set of nonempty subsets of Σ. Denote $b(\{v_0, v_1\}) = \frac{1}{2}(\hat{v}_0 + \hat{v}_1)$ by $\alpha_{0,1}$, $b(\{v_0, v_2\}) = \frac{1}{2}(\hat{v}_0 + \hat{v}_2)$ by $\alpha_{0,2}$, $b(\{v_1, v_2\}) = \frac{1}{2}(\hat{v}_1 + \hat{v}_2)$ by $\alpha_{1,2}$, and $b(\{v_0, v_1, v_2\}) = \frac{1}{3}(\hat{v}_0 + \hat{v}_1 + \hat{v}_2)$ by $\alpha_{0,1,2}$. Then $\Sigma_{bd} = \{\hat{v}_0, \hat{v}_1, \hat{v}_2, \alpha_{0,1}, \alpha_{0,2}, \alpha_{1,2}, \alpha_{0,1,2}\}$, and $S_{bd} = \{\{\hat{v}_0\}, \{\hat{v}_1\}, \{\hat{v}_2\}, \{\alpha_{0,1}\}, \{\alpha_{0,2}\}, \{\alpha_{1,2}\}, \{\alpha_{0,1,2}\}, \{\hat{v}_0, \alpha_{0,1}\}, \{v_0, \alpha_{0,1,2}\},$
$\{\alpha_{0,1}, \alpha_{0,1,2}\}, \{v_0, \alpha_{0,1}, \alpha_{0,1,2}\}, \{\alpha_{0,1}, v_1\}, \{v_1, \alpha_{0,1,2}\}, \{\alpha_{0,1}, v_1, \alpha_{0,1,2}\}, \{v_1, \alpha_{1,2}\},$
$\{\alpha_{1,2}, \alpha_{0,1,2}\}, \{v_1, \alpha_{1,2}, \alpha_{0,1,2}\}, \{\alpha_{1,2}, v_2\}, \{v_2, \alpha_{0,1,2}\}, \{\alpha_{1,2}, v_2, \alpha_{0,1,2}\}, \{v_2, \alpha_{0,2}\},$
$\{\alpha_{0,2}, \alpha_{0,1,2}\}, \{v_2, \alpha_{0,2}, \alpha_{0,1,2}\}, \{v_0, \alpha_{0,2}\}, \{v_0, \alpha_{0,2}, \alpha_{0,1,2}\}\}.$

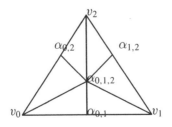

The linear map λ from $| (\Sigma_{bd}, S_{bd}) |$ to $| (\Sigma, S) |$ induced by the inclusion map from (Σ_{bd}, S_{bd}) to $| (\Sigma, S) |$ can easily be seen to be a homeomorphism. Thus, we have the following theorem.

Theorem 3.1.9 (Σ_{bd}, S_{bd}) *is a subdivision of* (Σ, S). ♮

The subdivision (Σ_{bd}, S_{bd}) of (Σ, S) is called the **barycentric subdivision** of (Σ, S). The nth iterated barycentric subdivision of (Σ, S) will be denoted by $(\Sigma_{bd}^n, S_{bd}^n)$. Since the composite of linear homeomorphisms are linear homeomorphisms, for each n, we have a linear homeomorphism from $| (\Sigma_{bd}^n, S_{bd}^n) |$ to $| (\Sigma, S) |$ which is induced by a map from Σ_{bd}^n to $| (\Sigma, S) |$.

Recall the metric d on $| (\Sigma, S) |$ defined by

$$d(\alpha, \beta) = +\sqrt{\sum_{v \in \Sigma} (\alpha(v) - \beta(v))^2}.$$

Evidently, $d(\hat{v}, \hat{w}) = \sqrt{2}$ for any distinct pair of vertices in Σ. Further, since $\alpha(v)$ and $\beta(v)$ are nonnegative,

$$\sum_{v \in \Sigma} (\alpha(v) - \beta(v))^2 \leq \sum_{v \in \Sigma} \alpha(v)^2 + \sum_{v \in \Sigma} \beta(v)^2.$$

Again, since $\sum_{v \in \Sigma} \alpha(v) = 1 = \sum_{v \in \Sigma} \beta(v)$,

$$\sum_{v \in \Sigma} \alpha(v)^2 \leq 1 \geq \sum_{v \in \Sigma} \beta(v)^2.$$

It follows that $d(\alpha, \beta) \leq \sqrt{2}$ for all $\alpha, \beta \in | (\Sigma, S) |$. Thus, the diameter $diam(| (\Sigma, S) |, d) = \sqrt{2}$ provided that Σ contains more than one elements. In fact, $diam | \sigma | = \sqrt{2}$ for all q-simplexes $\sigma \in S, q \geq 1$. Given an equivalent metric $\tilde{d}$ (a metric $\tilde{d}$ on $| (\Sigma, S) |$ which induces the same topology as the topology induced by the usual metric on $| (\Sigma, S) |$) on $| (\Sigma, S) |$, $sup\{diam | \sigma | \mid \sigma \in S\}$ is called the **mesh** of (Σ, S) relative to the metric $\tilde{d}$. Thus, $mesh(\Sigma, S)$ relative to the usual metric d on $| (\Sigma, S) |$ as given above is $\sqrt{2}$. If (Σ', S') is a subdivision of (Σ, S) and λ is the corresponding linear homeomorphism from $| (\Sigma', S') |$ to $| (\Sigma, S) |$, then the usual metric d on $| (\Sigma, S) |$ induces the metric d_λ on $| (\Sigma', S') |$ (equivalent to the usual metric d' on $| (\Sigma', S') |$) through the map λ. More explicitly,

$$d_\lambda \Big(\sum_{v' \in \Sigma'} \alpha_{v'} \hat{v}', \sum_{v' \in \Sigma'} \beta_{v'} \hat{v}' \Big) = d \Big(\sum_{v' \in \Sigma'} \alpha_{v'} v', \sum_{v' \in \Sigma'} \beta_{v'} v' \Big),$$

where $\alpha_{v'}, \beta_{v'} \geq 0$, the subsets $\{v' \in \Sigma' \mid \alpha_{v'} \neq 0\}$ and $\{v' \in \Sigma' \mid \beta_{v'} \neq 0\}$ of Σ' are members of S', and also $\sum_{v'} \alpha_{v'} = 1 = \sum_{v'} \beta_{v'}$. From now onward, without any reference, $mesh(\Sigma', S')$ of a subdivision of (Σ, S) will always mean the *mesh* of (Σ', S') relative to d_λ. Observe that the metric d_λ and the usual metric d on $| (\Sigma', S') |$ are different, and of course, they induce the same topology on $| (\Sigma', S') |$. Evidently, $diam(| (\Sigma', S') |, d_\lambda) = diam(| (\Sigma, S) |, d) = \sqrt{2}$.

If $\sigma' \in S'$, then there is a simplex $\sigma \in S$ such that $\sigma' \subseteq |\sigma|$. In turn, $\lambda(|\sigma'|) \subseteq |\sigma|$, and hence $mesh(\Sigma', S') \leq mesh(\Sigma, S) = \sqrt{2}$.

Recall that a subset X of $\mathbb{R}^n$ is called a convex subset if $t\overline{x} + (1 - t)\overline{y}$ belongs to X for all $\overline{x}, \overline{y} \in \mathbb{R}^n$ and $t \in [0, 1]$. Clearly, intersections of a family of convex sets are convex sets. Thus, for any subset S of $\mathbb{R}^n$, we have the smallest convex set containing S. This is called the **convex hull** of S.

Proposition 3.1.10 *Let X denote the convex hull of the set $\{\overline{w_0}, \overline{w_1}, \ldots, \overline{w_q}\}$ of points in $\mathbb{R}^{q+1}$. Let $\overline{x}$ be a member of $\mathbb{R}^{q+1}$. Then*

$$sup\{\|\overline{x} - \overline{u}\| \mid \overline{u} \in X\} = sup\{\|\overline{x} - \overline{w_i}\| \mid 0 \leq i \leq q\}.$$

Proof Suppose that $\sum_{i=0}^{q} \lambda_i = 1$, $\lambda_i \geq 0$. Then

$$\|\overline{x} - \sum_{i=0}^{q} \lambda_i \overline{w_i}\| = \|\sum_{i=0}^{q} \lambda_i(\overline{x} - \overline{w_i})\| \leq \sum_{i=0}^{q} \lambda_i(sup\{\|\overline{x} - \overline{w_i}\| \mid 0 \leq i \leq q\}) \leq sup\{\|\overline{x} - \overline{w_i}\| \mid 0 \leq i \leq q\}. \qquad \sharp$$

Corollary 3.1.11 *If X is the convex hull of the set $\{\overline{w_0}, \overline{w_1}, \ldots, \overline{w_q}\}$ of points in $\mathbb{R}^{q+1}$, then*
$$diam X = sup\{\|w_i - w_j\| \mid 0 \leq i, j \leq q\}.$$

Proof From the above proposition, it follows that

$$sup\{\|\sum_{i=0}^{q} \lambda_i w_i - \sum_{i=0}^{q} \mu_i w_i\| \mid \lambda_i, \mu_i \geq 0, \sum_{i=0}^{q} \lambda_i = 1 = \sum_{i=0}^{q} \mu_i\} = sup\{\|\sum_{i=0}^{q} \lambda_i w_i - w_j\| \mid 0 \leq j \leq q\} = sup\{\|w_i - w_j\|, 0 \leq i, j \leq q\}. \qquad \sharp$$

A simplicial complex (Σ, S) is said to be of **infinite dimension** if for each q there is a q-simplex. It is said to be of **finite dimension m** if there is an m-simplex but there is no $m + 1$-simplex.

Proposition 3.1.12 *Let (Σ, S) be a simplicial complex of dimension m. Then (Σ_{bd}, S_{bd}) is also of dimension m, and*

$$mesh(\Sigma_{bd}, S_{bd}) \leq \frac{m}{m + 1}\sqrt{2}.$$

In turn,

$$mesh(\Sigma_{bd}^n, S_{bd}^n) \leq \left(\frac{m}{m + 1}\right)^n \sqrt{2}.$$

Proof By definition, a simplex σ' in (Σ_{bd}, S_{bd}) is of the form

$$\left\{\hat{v_0}, \frac{1}{2}(\hat{v_0} + \hat{v_1}), \ldots, \frac{1}{q + 1}(\hat{v_0} + \hat{v_1} + \cdots + \hat{v_q})\right\},$$

where $\{v_0, v_1, \ldots, v_q\}$ is a q-simplex of (Σ, S). This shows that (Σ_{bd}, S_{bd}) is of dimension m. Indeed, $(\sigma', \bar{\sigma}')$ is the barycentric subdivision $(\sigma_{bd}, \bar{\sigma}_{bd})$ of the simplicial subcomplex $(\sigma, \bar{\sigma})$ of (Σ, S). We have the linear isometric isomorphism ψ_σ from $(|(\sigma, \bar{\sigma})|, d)$ to Δ^q given by $\psi_\sigma(\sum_{i=0}^{q} \lambda_i \hat{v}_i) = \sum_{i=0}^{q} \lambda_i \overline{e}_i$. Thus, the diameter of $(|(\sigma', \bar{\sigma}')|, d_\lambda)$ is the same as the diameter of the image of $\lambda o \psi_\sigma$ in Δ^q. Evidently, the image of $\lambda o \psi_\sigma$ is the convex hull of

$$\left\{ \overline{e}_0, \frac{1}{2}(\overline{e}_0 + \overline{e}_1), \ldots, \frac{1}{q+1}(\overline{e}_0 + \overline{e}_1 + \cdots + \overline{e}_q) \right\}.$$

From the above corollary,

$$diam((|(\sigma', \bar{\sigma}')|, d_\lambda)) = sup \left\{ \| \frac{1}{r+1}(\sum_{i=o}^{r} \overline{e}_i) - \frac{1}{s+1}(\sum_{i=o}^{s} \overline{e}_i) \| \mid 0 \leq r < s \leq q \right\}.$$

Again, applying the above corollary,

$$sup \left\{ \| \frac{1}{r+1}(\sum_{i=o}^{r} \overline{e}_i) - \frac{1}{s+1}(\sum_{i=o}^{s} \overline{e}_i) \| \mid 0 \leq r < s \leq q \right\}$$

$$= sup \left\{ \| e_j - \frac{1}{s+1}(\sum_{i=o}^{s} \overline{e}_i) \| \mid 0 \leq j \leq q \right\}$$

$$= sup \left\{ \frac{1}{s+1} sum_{i=0}^{s} \| \overline{e}_j - \overline{e}_i \| \mid 0 \leq j \leq q \right\}$$

$$\leq \frac{s}{s+1} sup\{\| \overline{e}_j - \overline{e}_i \| \mid 0 \leq i, j \leq q\}$$

$$= \frac{s}{s+1} \sqrt{2}.$$

Since $s \leq m$ implies that $\frac{s}{s+1} \leq \frac{m}{m+1}$, it follows that $mesh(\Sigma_{bd}, S_{bd}) \leq \frac{m}{m+1}\sqrt{2}$. The rest of the assertion follows by the induction. ♯

From now onward, the space $|(\Sigma', S')|$ of a subdivision (Σ', S') of (Σ, S) will be identified with the space $|(\Sigma, S)|$ through the linear homeomorphism induced by the inclusion map i from (Σ', S') to $|(\Sigma, S)|$.

Recall that an open cover $\{U_\alpha \mid \alpha \in \Lambda\}$ of a topological space X is said to be a refinement of an open cover $\{V_\gamma \mid \gamma \in \Gamma\}$ if for each $\alpha \in \Lambda$, there is a $\gamma \in \Gamma$ such that $U_\alpha \subseteq V_\gamma$. Also observe that if (Σ, S) is a simplicial complex, then $\{St(v) \mid v \in \Sigma\}$ is an open cover of $|(\Sigma, S)|$.

Proposition 3.1.13 *Let (Σ, S) be a finite simplicial complex, and (Σ', S') be a simplicial complex. Let f be a continuous map from $|(\Sigma, S)|$ to $|(\Sigma', S')|$, and $\{U_\alpha \mid \alpha \in \Lambda\}$ be an open cover of $|(\Sigma', S')|$. Then there is a natural number N such that $\{St(w) \mid w \in \Sigma_{bd}^N\}$ is a refinement of $\{f^{-1}(U_\alpha) \mid \alpha \in \Lambda\}$.*

Proof Since (Σ, S) is finite, $| (\Sigma, S) |$ is compact metric space. Hence, the open cover $\{f^{-1}(U_\alpha) \mid \alpha \in \Lambda\}$ of $| (\Sigma, S) |$ has a Lebesgue number δ (a number δ such that $d(x, y) < \delta$ implies that $x, y \in f^{-1}(U_\alpha)$ for some $\alpha \in \Lambda$ (see "Topology and Modern Analysis" by G. F. Simons for its existence)). Suppose that the dimension of (Σ, S) is m. Choose N sufficiently large so that $(\frac{m}{m+1})^N < \frac{\delta}{2}$. Then from the previous proposition, $mesh(\Sigma_{bd}^N, S_{bd}^N) < \frac{\delta}{2}$. Let w be a member of Σ_{bd}^N. Let μ and ν be members of $St(w)$. Since $mesh(\Sigma_{bd}^N, S_{bd}^N) < \frac{\delta}{2}, d(\mu, \hat{w}) < \frac{\delta}{2} > d(\hat{w}, \nu)$. In turn, $d(\mu, \nu) < \delta$. This shows that $diam(St(w)) < \delta$ for all $w \in \Sigma_{bd}^N$. Hence for each $w \in \Sigma_{bd}^N$, there is an $\alpha \in \Lambda$ such that $St(w) \subseteq U_\alpha$.   ♯

Corollary 3.1.14 (Simplicial approximation theorem) *Let (Σ, S) be a finite simplicial complex, and (Σ', S') be a simplicial complex. Let f be a continuous map from $| (\Sigma, S) |$ to $| (\Sigma', S') |$. Then there is a natural number N and a simplicial approximation ϕ from Σ_{bd}^N to Σ' of f from $| (\Sigma_{bd}^N, S_{bd}^N) | = | (\Sigma, S) |$ to $| (\Sigma', S') |$.*

Proof Since $\{St(u) \mid u \in \Sigma'\}$ is an open cover of $| (\Sigma', S') |$, from the above proposition, there is a natural number N such that $\{St(w) \mid w \in \Sigma_{bd}^N\}$ is a refinement of $\{f^{-1}(St(u)) \mid u \in \Sigma'\}$. Thus, for each $w \in \Sigma_{bd}^N$, there is a $\phi(w) \in \Sigma'$ such that $St(w) \subseteq f^{-1}(St(\phi(w)))$. This gives us a map ϕ from Σ_{bd}^N to Σ' such that $f(St(w)) \subseteq St(\phi(w))$. From Theorem 3.1.6, it follows that ϕ is a simplicial approximation of f.   ♯

Subdivision Chain Map

Let (Σ, S) be a simplicial complex. For each $p \geq 0$, we shall define a homomorphism sd_p from $\Lambda_p(\Sigma, S)$ to $\Lambda_p(\Sigma_{bd}, S_{bd})$ such that $sd = \{sd_p \mid p \geq 0\}$ is an augmentation-preserving chain transformation from $\Lambda(\Sigma, S)$ to $\Lambda(\Sigma_{bd}, S_{bd})$. The chain map sd will be termed as a **subdivision chain map**. This we do by induction on p. By definition, $\Lambda_0(\Sigma, S)$ is the free abelian group on the set $\{\{v\} \mid v \in \Sigma\}$ of oriented 0-simplexes. We define sd_0 to be the unique homomorphism from $\Lambda_0(\Sigma, S)$ to $\Lambda_0(\Sigma_{bd}, S_{bd})$ which maps $\{v\}$ to $\{\hat{v}\}$. Clearly, sd_0 respects the augmentation maps. Given a 1-simplex $\sigma = \{v_0, v_1\}$, we have two ordered 1-simplexes (v_0, v_1) and (v_1, v_0) associated with σ. Indeed, they have different orientations also, and so $[v_0, v_1] \neq [v_1, v_0]$. Define a map ϕ_1 from $A_1(\Sigma, S)$ to $\Lambda_1(\Sigma_{bd}, S_{bd})$ by

$$\phi_1(v, w) = \left[\frac{1}{2}(\hat{v} + \hat{w}), \hat{w} \right] - \left[\frac{1}{2}(\hat{w} + \hat{v}), \hat{v} \right].$$

Evidently, $\phi_1(v, w) + \phi_1(w, v) = 0$. Thus, ϕ_1 induces a homomorphism sd_1 from $\Lambda_1(\Sigma, S)$ to $\Lambda_1(\Sigma_{bd}, S_{bd})$ defined by

$$sd_1[v, w] = \left[\frac{1}{2}(\hat{v} + \hat{w}), \hat{w} \right] - \left[\frac{1}{2}(\hat{w} + \hat{v}), \hat{v} \right].$$

Further,

$$d_1 sd_1[v, w] = d_1\left(\left[\frac{1}{2}(\hat{v} + \hat{w}), \hat{w}\right]\right) - d_1\left(\left[\frac{1}{2}(\hat{w} + \hat{v}), \hat{v}\right]\right) = [\hat{w}] - [\hat{v}] = sd_0 d_1[v, w]$$

for all $v, w \in \Sigma$. This shows that $d_1 sd_1 = sd_0 d_1$. For convenience, $[v_0, v_1, \ldots, v_q]$ is also denoted by $v_0 \cdot [v_1, \ldots, v_q]$. More generally, if $\sum_i n_i[\sigma_i, \alpha_i]$ is an element of $\Lambda_q(\Sigma, S)$, and v is a vertex such that $v \cdot [\sigma_i, \alpha_i]$ is defined for all i, then $\sum_i n_i(v \cdot [\sigma_i, \alpha_i])$ is denoted by $v \cdot \left(\sum_i n_i[\sigma_i, \alpha_i]\right)$. Thus,

$$sd_1[v, w] = \frac{1}{2}(\hat{v} + \hat{w}) \cdot \{\hat{w}\} - \frac{1}{2}(\hat{w} + \hat{v}) \cdot \{\hat{v}\}.$$

Suppose that sd_q is already defined for all $q < p$ satisfying the condition $d_q sd_q = sd_{q-1} d_q$ for all $q < p$. Let $[\sigma, \alpha]$ be an oriented p-simplex. Suppose that $[\sigma, \alpha] \neq [\sigma, \beta]$. It can be verified that

$$b(\sigma) \cdot \sum_{i=0}^{p} (-1)^i sd_{p-1}([\sigma_i, \alpha_i]) + b(\sigma) \cdot \sum_{i=0}^{p} (-1)^i sd_{p-1}([\sigma_i, \beta_i]) = 0.$$

This ensures the existence of a unique homomorphism sd_p from $\Lambda_p(\Sigma, S)$ to $\Lambda_p(\Sigma_{bd}, S_{bd})$ subject to

$$sd_p([\sigma, \alpha]) = b(\sigma) \cdot \sum_{i=0}^{p} (-1)^i sd_{p-1}([\sigma_i, \alpha_i]) = b(\sigma) \cdot sd_{p-1}(d_p([\sigma, \alpha])).$$

In turn,

$$\begin{aligned} d_p sd_p([\sigma, \alpha]) &= d_p(b(\sigma) \cdot sd_{p-1}(d_p([\sigma, \alpha]))) \\ &= sd_{p-1}(d_p([\sigma, \alpha])) - b(\sigma) \cdot d_{p-1}(sd_{p-1}(d_p([\sigma, \alpha]))) \\ &= sd_{p-1}(d_p([\sigma, \alpha])) - b(\sigma) \cdot s_{p-2}(d_{p-1}(d_p(\sigma, \alpha))) = sd_{p-1}(d_p([\sigma, \alpha])) \end{aligned}$$

for all oriented p-simplex $[\sigma, \alpha]$. This shows that $d_p sd_p = sd_{p-1} d_p$. Thus, sd is a chain transformation and it is called the **subdivision chain map**.

We shall show that sd induces isomorphisms on the corresponding simplicial homology groups. For the purpose, we define simplicial map χ from (Σ_{bd}, S_{bd}) to (Σ, S) such that $\Lambda(\chi)osd$ and $sdo\Lambda(\chi)$ are chain equivalent to the corresponding identity chain transformations. By definition, $\Sigma_{bd} = \{b(\sigma) \mid \sigma \in S\}$. The axiom of choice gives us a map χ from Σ_{bd} to Σ such that $\chi(b(\sigma)) \in \sigma$. Let σ' be a simplex of (Σ_{bd}, S_{bd}). By definition, there is an ordered simplex $(\sigma, \alpha) = (v_0, v_1, \ldots, v_q)$ in (Σ, S) such that $\sigma' = \{\hat{v}_0, \frac{1}{2}(\hat{v}_0 + \hat{v}_1), \ldots, \frac{1}{q+1}(\hat{v}_0 + \hat{v}_1 + \cdots + \hat{v}_q)\}$. Indeed, there is a natural ordering in σ' induced by the order α on σ. Clearly, $\chi(\sigma') \subseteq \sigma$, and hence $\chi(\sigma')$ is a simplex in (Σ, S). This shows that χ is a simplicial map. Indeed, χ is a simplicial approximation of the tautological identity map on $|(\Sigma_{bd}, S_{bd})| = |(\Sigma, S)|$. Let σ be a q-simplex in S. By the induction, define elements $\chi^i(\sigma) \in \sigma$ for each $i, 0 \leq i \leq q$ as follows: Take $\chi^0(\sigma) = \chi(b(\sigma))$, and $\chi^1(\sigma) = \chi(b(\sigma - \{\chi^0(\sigma)\}))$. Assume that $\chi^i(\sigma), i < q$, has already been defined. Define $\chi^{i+1}(\sigma) = \chi(b(\sigma - \bigcup_{j=1}^{i}\{\chi^j(\sigma)\}))$. This gives us an oriented q-simplex $[\sigma, \overline{\chi}]$, where $\overline{\chi}(i) = \chi^i(\sigma)$.

Proposition 3.1.15 $\Lambda(\chi)osd$ and $sdo\Lambda(\chi)$ are chain homotopic to the corresponding identity chain transformations.

Proof Consider $\Lambda(\chi)osd$. We show that $\Lambda_q(\chi)osd_q = I_{\Lambda_q(\Sigma, S)}$ for all q. By definition, $sd_0(\{v\}) = \{\hat{v}\}$ and $\Lambda_0(\chi)(\{\hat{v}\}) = \Lambda_0(\chi)(\{b(\{v\})\}) = \{\chi(b(\{v\}))\} = \{v\}$. This shows that $\Lambda_0(\chi)osd_0 = I_{\Lambda_0(\Sigma, S)}$. Any oriented p-simplex $[\sigma, \alpha]$ can be represented by $[v_0, v_1, \ldots, v_p]$, where $\alpha(i) = v_i$. Let $[v_0, v_1]$ be an oriented 1-simplex. By definition, $sd_1([v_0, v_1]) = [b(\{v_0, v_1\}), b(\{v_1\})] - [b(\{v_0, v_1\}), b(\{v_0\})]$. Thus,

$\Lambda_1(\chi)(sd_1([v_0, v_1]))$
$= \Lambda_1(\chi)([b(\{v_0, v_1\}), b(\{v_1\})] - [b(\{v_0, v_1\}), b(\{v_0\})])$
$= \Lambda_1(\chi)([b(\{v_0, v_1\}), b(\{v_1\})]) - \Lambda_1(\chi)([b(\{v_0, v_1\}), b(\{v_0\})])$
$= [\chi(b(\{v_0, v_1\})), \chi(b(\{v_1\}))] - [\chi(b(\{v_0, v_1\})), \chi(b(\{v_0\}))]$
$= [\chi(b(\{v_0, v_1\})), v_1] - [\chi(b(\{v_0, v_1\})), v_0]$
$= [v_0, v_1]$

for all choices of χ (observe that $[v_1, v_0] = -[v_0, v_1]$). This shows that $\Lambda_1(\chi)osd_1 = I_{\Lambda_1(\Sigma, S)}$.

Assume that $\Lambda_q osd_q = I_{\Lambda_q(\Sigma, S)}$ for all $q < p$. We prove it for p. Let $[\sigma, \alpha] = [v_0, v_1, \ldots, v_p]$ be an oriented p-simplex. Suppose that $b(\sigma) = v_k$. Then

$\Lambda_p(\chi)(sd_p([\sigma, \alpha]))$
$= \Lambda_p(\chi)(b(\sigma) \cdot \sum_{i=0}^{p}(-1)^i sd_{p-1}([\sigma_i, \alpha_i]))$
$= \Lambda_p(\chi)(b(\sigma) \cdot (-1)^k sd_{p-1}([\sigma_k, \alpha_k]))$
$= (v_k \cdot (-1)^k \Lambda_{p-1}(sd_{p-1}([\sigma_k, \alpha_k]))$
$= (-1)^k v_k \cdot [\sigma_k, \alpha_k]$ (by the induction hypothesis)
$= [\sigma, \alpha]$.

This shows that $\Lambda_p(\chi)osd_p = I_{\Lambda_p(\Sigma, S)}$ for all p.

Finally, we show that $sdo\Lambda(\chi)$ is chain homotopic to $I_{\Lambda(\Sigma_{sd}, S_{bd})}$. By the induction on q, we define chain homotopy $t = \{t_q, q \geq 0\}$ from $I_{\Lambda(\Sigma_{bd}, S_{bd})}$ to $sdo\Lambda(\chi)$. By definition, $\Lambda_0(\Sigma_{bd}, S_{bd})$ is the free abelian group on $\{\{b(\sigma)\} \mid \sigma \in S\}$, and

$$(sd_0 o\Lambda_0(\chi))(\{b(\sigma)\}) = sd_0(\{\chi(b(\sigma))\}) = \{\chi(\hat{b}(\sigma))\}.$$

We have a unique homomorphism t_0 from $\Lambda_0(\Sigma_{bd}, S_{bd})$ to $\Lambda_1(\Sigma_{bd}, S_{bd})$ given by $t_0[(\{b(\sigma)\})] = [\chi(\hat{b}(\sigma)), b(\sigma)]$. Hence

$$d_1(t_0([\{b(\sigma)\}]) = d_1([\chi(\hat{b}(\sigma)), b(\sigma)]) = [\{b(\sigma)\}] - [\chi(\hat{b}(\sigma))] =$$
$$I_{\Lambda_0(\Sigma_{bd}, S_{bd})}([\{b(\sigma)\}]) - (sd_0 o\Lambda_0(\chi))([\{b(\sigma)\}]).$$

Assume that t_q has already been defined for all $q < p$ satisfying the condition

$$d_{q+1}t_q + t_{q-1}d_q = I_{\Lambda_q(\Sigma_{bd}, S_{bd})} - sd_q o\Lambda_q(\chi).$$

We define t_p such that the above identity is satisfied for all $q \leq p$. Let $[\sigma', \alpha']$ be an oriented p-simplex of (Σ_{bd}, S_{bd}). Then, using the induction assumption,

$d_p(I_{\Lambda_p(\Sigma_{bd}, S_{bd})} - sd_p o\Lambda_p(\chi) - t_{p-1}d_p([\sigma', \alpha']))$
$= d_p([\sigma', \alpha']) - d_p(sd_p(\Lambda_p(\chi)([\sigma', \alpha']))) - (d_p o t_{p-1} o d_p)([\sigma', \alpha'])$

$$= d_p([\sigma', \alpha']) - d_p(sd_p(\Lambda_p(\chi)([\sigma', \alpha']))) - ((I_{\Lambda_{p-1}(\Sigma_{bd}, S_{bd})} - sd_{p-1}o\Lambda_{p-1}(\chi) -$$
$$t_{p-2}od_{p-1})od_p)([\sigma', \alpha'])$$
$$= d_p([\sigma', \alpha']) - d_p(sd_p(\Lambda_p(\chi)([\sigma', \alpha']))) - ((I_{\Lambda_{p-1}(\Sigma_{bd}, S_{bd})} - sd_{p-1}o\Lambda_{p-1}(\chi)$$
$$(d_p([\sigma', \alpha']))))$$

$= 0$, since $sd_po\Lambda_p(\chi)$ is a chain transformation. This means that $I_{\Lambda_p(\Sigma_{bd}, S_{bd})} - sd_po\Lambda_p(\chi) - t_{p-1}d_p([\sigma', \alpha'])$ is a cycle in the acyclic chain complex $\Lambda(\sigma', \mid \sigma'\mid)$. Define $t_p([\sigma', \alpha']))$ by the requirement that $d_{p+1}(t_p([\sigma', \alpha'])) = I_{\Lambda_p(\Sigma_{bd}, S_{bd})} - sd_po\Lambda_p(\chi) - t_{p-1}d_p([\sigma', \alpha'])$. Extend t_p, by linearity, to a homomorphism from $\Lambda_p(\Sigma_{bd}, S_{bd})$ to itself. We denote this extended homomorphism by t_p itself. Evidently,

$$d_{p+1}t_p + t_{p-1}d_p = I_{\Lambda_p(\Sigma_{bd}, S_{bd})} - sd_po\Lambda_p(\chi).$$

This completes the construction of the chain homotopy t.   ♯

Corollary 3.1.16 *For each q, $H_q(sd)$ and $H_q(\Lambda(\chi))$ are inverses to each other. In particular, $H_q(\Sigma_{bd}^n, S_{bd}^n) \approx H_q(\Sigma, S)$ for all $n \geq 0$ and $q \geq 0$.*   ♯

Let X be a topological space. A **triangulation** of X is a pair $((\Sigma, S), f)$, where (Σ, S) is a simplicial complex and f is a homeomorphism from $\mid (\Sigma, S) \mid$ to X. A topological space may not admit any triangulation, and it may admit more than one nonisomorphic triangulations. A topological space X is called a **polyhedron** if it admits a triangulation.

Example 3.1.17 Consider the simplicial complex (Σ, S), where $\Sigma = \{v_0, v_1, \ldots, v_q\}$ and S is the set of nonempty subsets of Σ. Then $\mid (\Sigma, S) \mid$ is naturally identified with the subspace

$$\Delta^q = \left\{(a_0, a_1, \ldots, a_q) \in \mathbb{R}^{q+1} \mid a_i \geq 0 \text{ and } \sum_{i=0}^q a_i = 1\right\} of \mathbb{R}^{q+1}.$$

Evidently, Δ^q is homeomorphic to the disk D^q of dimension q. Thus, D^q is a polyhedron for all q. Also if we take S to be the set of proper subsets of $\Sigma = \{v_0, v_1, \ldots, v_q\}$, the $\mid (\Sigma, S) \mid$ is naturally identified with the boundary of Δ^q, and it is homeomorphic to the sphere S^{n-1}. Thus, S^{n-1} is also a polyhedron.

Example 3.1.18 Let $\Sigma = \mathbb{Z}^n$, $n \geq 1$. Let us denote the vector $(x_1, x_2, \ldots, x_n)$ of $\mathbb{Z}^n$ by $\bar{x}$. There is a partial order $\leq$ on Σ defined by putting $\bar{x} \leq \bar{y}$ if $x_i \leq y_i$ for all i. For each $m \in \mathbb{Z}$, let $\bar{m}$ denote the vector $(m, m, \ldots, m) \in \mathbb{Z}^n$, and $T_m = \{\bar{m} + \bar{\epsilon} \mid \bar{\epsilon} \in \{0, 1\}^n\}$. Note that $T_m \bigcap T_{m+1} = \{\overline{m+1}\}$ and $T_m \bigcap T_n = \emptyset$ whenever $\mid m - n \mid \geq 2$. Let S_m denote the set of all totally ordered subsets of T_m. Let $S = \bigcup_{m \in \mathbb{Z}} \wp(S_m)$. Evidently, S is a set of finite subsets of Σ, and also all subsets of members of S are in S. Thus, (Σ, S) is a simplicial complex. Observe that (Σ, S) is a locally finite simplicial complex. It is easily observed that a map α from Σ to $[0, 1]$ is a member of $\mid (\Sigma, S) \mid$ if and only if there is a $m \in \mathbb{Z}$ such that α is 0 on $\Sigma - S_m$ and $\sum_{\bar{a} \in S_m} \alpha(\bar{a}) = 1$. The map ϕ from $\mid (\Sigma, S) \mid$ to $\mathbb{R}^n$ defined by $\phi(\alpha) = \sum_{\bar{x} \in \mathbb{Z}^n} \alpha(\bar{x})\bar{x}$ can be easily seen to be a homeomorphism. Thus, the pair $((\Sigma, S), \phi)$ is a triangulation of $\mathbb{R}^n$. In particular, $\mathbb{R}^n$ is a polyhedron.

Example 3.1.19 A cylinder $S^1 \times [0, 1]$ is obtained by identifying one pair of parallel sides of a rectangle. It is homeomorphic to an open prism

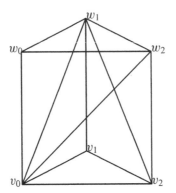

The above picture gives the simplicial complex (Σ, S), where
$\Sigma = \{v_0, v_1, v_2, w_0, w_1, w_2\}$ and
$S = \{\{v_0\}, \{v_1\}, \{v_2\}, \{w_0\}, \{w_1\}, \{w_2\},$
$\{v_0, v_1\}, \{v_0, v_2\}, \{v_1, v_2\}, \{w_0, w_1\}, \{w_0, w_2\}, \{w_1, w_2\},$
$\{v_0, w_0\}, \{v_0, w_1\}, \{v_0, w_0, w_1\}, \{v_1, w_1\},$
$\{v_0, v_1, w_1\}, \{v_0, w_2\}, \{v_0, w_0, w_2\}, \{v_2, w_2\},$
$\{v_0, v_2, w_2\}, \{v_1, v_2, w_1\}, \{v_2, w_1\}, \{v_2, w_1, w_2\}\},$
and which is a triangulation of an open and hollow prism. In turn, it gives a triangulation of a cylinder. Thus, a cylinder is a polyhedron. The scheme of triangulation may easily be demonstrated by the following figure.

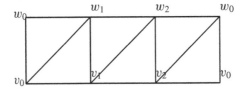

Example 3.1.20 A torus T of genus 1 is the surface obtained by identifying parallel sides of a rectangle. A scheme for triangulation of a torus is demonstrated by the following figure.

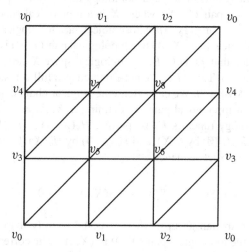

The reader is asked to write the corresponding simplicial complex. Thus, a torus is also a polyhedron.

A polyhedron may have several nonisomorphic triangulations. For example, (Σ, S) and (Σ', S') are triangulations of the circle S^1, where $\Sigma = \{v_0, v_1, v_2\}$, $S = \{\{v_0\}, \{v_1\}, \{v_2\}, \{v_0, v_1\}, \{v_0, v_2\}, \{v_1, v_2\}\}$ and $\Sigma' = \{v_0, v_1, v_2, v_3\}$, $S' = \{\{v_0\}, \{v_1\}, \{v_2\}, \{v_3\}, \{v_0, v_2\}, \{v_1, v_2\}, \{v_0, v_3\}, \{v_3, v_1\}\}$.

We state the following theorem without proof. For the proof, the reader may refer to the "Algebraic topology" by Spanier.

Theorem 3.1.21 *Let (Σ, S) be a simplicial complex. Then $\phi = \{\phi_q \mid q \geq 0\}$ is a chain equivalence from $\Omega(\Sigma, S)$ to the singular chain complex $S(\mid (\Sigma, S) \mid)$ of $\mid (\Sigma, S) \mid$, where ϕ_q is a homomorphism from $\Omega_q(\Sigma, S)$ to $S_q(\mid (\Sigma, S) \mid)$ defined by $\phi_q(, \alpha))(\overline{a})(\alpha(i)) = a_i$, $0 \leq i \leq q$ and $\phi_q((\sigma, \alpha))(\overline{a})(v) = 0$ if $v \notin \sigma$, where $\overline{a} = (a_0, a_1, \ldots, a_q) \in \Delta^q$.* ♮

Corollary 3.1.22 *Let X be a polyhedron with triangulation $((\Sigma, S), f)$. Then $H_q(X) \approx H_q(\Sigma, S)$ for all q. In particular, if $((\Sigma', S'), f')$ is another triangulations X, then $H_q(\Sigma, S) \approx H_q(\Sigma', S')$ for all q.*

Proof Since f is a homeomorphism from $\mid (\Sigma, S) \mid$ to X, $S(f)$ is a chain isomorphism from $S(\mid (\Sigma, S) \mid)$ to $S(X)$. From the above theorem, we have a chain equivalence ϕ from $\Omega(\Sigma, S)$ to $S(\mid (\Sigma, S) \mid)$. In turn, we have a chain equivalence $S(f) o \phi$ from $\Omega(\Sigma, S)$ to $S(X)$. It follows that $H_q(S(f) o \phi)$ is an isomorphism from $H_q(\Sigma, S)$ to $H_q(X)$ for all q. The rest is immediate consequence. ♮

Let X be a polyhedron with a triangulation $((\Sigma, S), f)$. Without any loss, we can term $H_q(\Sigma, S)$ as the qth simplicial homology of X. Thus, simplicial homologies and the singular homologies of a polyhedron are the same. To compute the homologies of a polyhedron X, either we triangulate it suitably and then compute the simplicial homologies or use some results of singular homology theory to compute it.

A topological pair is a pair (X, A), where X is a topological space and A is a subspace of X. We have a category $\Im p$ whose objects are topological pairs, and a morphism from a topological pair (X, A) to a topological pair (Y, B) is a continuous map f from X to Y such that $f(A) \subseteq B$. A topological space X can also be treated as a topological pair $(X, \emptyset)$. Thus, TOP can be treated as a full subcategory of $\Im p$. By convention we put $S(\emptyset) = 0$. We can extend the singular chain complex functor S to the category $\Im p$ of topological pairs by defining $S(X, A) = S(X)/S(A)$, and also the singular homology functors H_n by putting $H_n(X, A) = H_n(S(X, A))$. We identify $S(X, \emptyset) = S(X)/S(\emptyset)$ by $S(X)$, and $H_n(X, \emptyset)$ by $H_n(X)$. Every topological pair (X, A) gives a short exact sequence

$$0 \longrightarrow S(A) \xrightarrow{S(i)} S(X) \xrightarrow{S(j)} S(X, A) \longrightarrow 0$$

of chain complexes of abelian groups, where j is the identity map from X to X treated as a map from the topological pair $(X, \emptyset)$ to (X, A). The corresponding long exact homology sequence is given by

$$\cdots \xrightarrow{\partial_{n+1}} H_n(A) \xrightarrow{H_n(i)} H_n(X) \xrightarrow{H_n(j)} H_n(X, A) \xrightarrow{\partial_n} \cdots .$$

This exact sequence will be termed as long exact sequence associated with the topological pair (X, A).

Let (X, A) be a pair, and (Y, B) be a subpair in the sense that Y is a subspace of X and $B \subset A$. The inclusion map i from (Y, B) to (X, A) is said to be an excision map if $Y - B = X - A$. Thus, the inclusion map i from $(X_1, X_1 \cap X_2)$ to $(X_1 \cup X_2, X_2)$ is an excision map. We say that a pair $\{X_1, X_2\}$ of subspaces of a space X is an **excisive couple** if the inclusion chain transformation i from $S(X_1) + S(X_2)$ to $S(X_1 \cup X_2)$ induces isomorphisms on their homology groups.

Proposition 3.1.23 $\{X_1, X_2\}$ *is an excisive couple if and only if inclusion map i from $(X_1, X_1 \cap X_2)$ to $(X_1 \cup X_2, X_2)$ induces isomorphism on the corresponding singular homologies.*

Proof The excision map i induces chain transformation $S(i)$ from $S(X_1)/S(X_1 \cap X_2)$ to $S(X_1 \cup X_2)/S(X_2)$. Evidently, $S(X_1 \cap X_2) = S(X_1) \cap S(X_2)$. By the second isomorphism theorem (for chain complexes), we have a chain isomorphism ϕ from $S(X_1)/S(X_1 \cap X_2)$ to $S(X_1) + S(X_2)/S(X_2)$. Clearly, $\tilde{i} o \phi = S(i)$, where $\tilde{i}$ is the chain transformation induced by the inclusion chain transformation i from $S(X_1) + S(X_2)$ to $S(X_1 \cup X_2)$. Further, we have the following commutative diagram

$$0 \longrightarrow S(X_2) \longrightarrow S(X_1) + S(X_2) \longrightarrow (S(X_1 + S(X_2))/S(X_2) \longrightarrow 0$$

$$\downarrow I \qquad\qquad \downarrow i \qquad\qquad \downarrow \tilde{i}$$

$$0 \longrightarrow S(X_2) \longrightarrow S(X_1 \bigcup X_2) \longrightarrow S(X_1 \bigcup X_2)/S(X_2) \longrightarrow 0$$

where the rows are short exact sequences of chain complexes. Using the long exact homology sequences and the five lemmas, it follows that i induces isomorphisms on the homology groups if and only if $\tilde{i}$ induces isomorphism on the homology groups. Since $\tilde{i} \circ \phi = S(i)$, it follows that $\{X_1, X_2\}$ is an excisive couple if and only if $S(i)$ induces isomorphisms on the singular homology groups. ♯

Proposition 3.1.24 (Mayer–Vietoris sequence) *Let $\{X_1, X_2\}$ be an excisive couple with $X_1 \bigcup X_2 = X$. Then we have the long exact sequence*

$$\cdots \overset{\partial_{n+1}}{\to} H_n(X_1 \bigcap X_2) \overset{(H_n(i_1), -H_n(i_2))}{\to} H_n(X_1) \oplus H_n(X_2) \overset{H_n(j_1) \oplus H_n(j_2)}{\to} H_n(X) \overset{\partial_n}{\to} \cdots,$$

where $i_1, i_2, j_1,$ and j_2 are the corresponding inclusion maps.

Proof We have the short exact sequence of

$$0 \longrightarrow S(X_1 \bigcap X_2) \overset{(S(i_1), -S(i_2))}{\to} S(X_1) \oplus S(X_2) \overset{S(j_1) + S(j_2)}{\to} S(X_1) + S(X_2) \longrightarrow 0$$

of chain complexes, where $i_1, i_2, j_1,$ and j_2 are corresponding inclusion maps. In turn, we get the long exact homology sequence

$$\cdots \overset{\partial_{n+1}}{\to} H_n(X_1 \bigcap X_2) \overset{(H_n(i_1), -H_n(i_2))}{\to} H_n(X_1) \oplus H_n(X_2) \overset{H_n(j_1) + H_n(j_2)}{\to}$$
$$H_n(S(X_1) + S(X_2)) \overset{\partial_n}{\to} \cdots.$$

Since $\{X_1, X_2\}$ is an excisive couple, the inclusion map i from $S(X_1) + S(X_2)$ to $S(X_1 \bigcup S(X_2)) = S(X)$ induces isomorphism on homology groups. Substituting $H_n(X)$ for $H_n(S(X_1) + S(X_2))$ with suitable maps, we get the desired exact sequence. ♯

We state the following theorem without proof, and the proof can be found in the "Algebraic topology" by Spanier.

Theorem 3.1.25 *Let X_1 and X_2 be subspaces of X such that $X = X_1^0 \bigcup X_2^0$. Then $\{X_1, X_2\}$ is an excisive couple.* ♯

Corollary 3.1.26 *Let U, and A be subspaces of X such that $\bar{U} \subseteq A^0$. Then $\{X - U, A - U\}$ is an excisive couple, and the inclusion map i from $(X - U, A - U)$ to (X, A) induce isomorphism on their homologies.*

Proof Since $\bar{U} \subseteq A^0$, $X - A^0 \subseteq X - \bar{U} \subseteq X - U$. Since $X - \bar{U}$ is open, $(X - U)^0 \supset X - A^0$. This means that $(X - U)^0 \bigcup A^0 = X$. From the above theorem, it follows that $\{X - U, A\}$ is an excisive couple. Further, from Proposition 3.1.23, it follows that the inclusion map i from $(X - U, A - U)$ to (X, A) induces isomorphism on their homologies. ♮

Let f and g be maps from a topological pair (X, A) to a topological pair (Y, B). Recall that f is said to be homotopic to g if there is a continuous map H from $X \times [0, 1]$ to Y such that (i) $H(A \times [0, 1]) \subseteq B$, (ii) $H(x, 0) = f(x)$, and (iii) $H(x, 1) = g(x)$ for all x. We use the notation $f \sim g$ to say that f is homotopic to g. For example, the map H from $D^2 \times [0, 1]$ to D^2 defined by $H(z, t) = ze^{it\theta}$ is a homotopy from the identity map on (D^2, S^1) to the rotation map through an angle θ, where D^2 represents the 2-disk $\{z \in \mathbb{C} \mid |z| \leq 1\}$.

Proposition 3.1.27 *The relation $\sim$ is an equivalence relation on the set $Map((X, A), (Y, B))$ of all maps from the topological pair (X, A) to (Y, B).*

Proof Let f be a member of $Map((X, A), (Y, B))$. The map H from $(X, A) \times [0, 1]$ to (Y, B) given by $H(x, t) = f(x)$ is a continuous map, and it is a homotopy from f to f. Hence $f \sim f$. Suppose that $f \sim g$. Then there is a homotopy H from f to g. Then the map H' from $(X, A) \times [0, 1]$ to (Y, B) given by $H'(x, t) = H(x, 1 - t)$ is a homotopy from g to f. This shows that $g \sim f$. Suppose that $f \sim g$ and $g \sim h$. Let H be a homotopy from f to g and K is a homotopy from g to h. Define a map $\tilde{H}$ from $(X, A) \times [0, 1]$ to (Y, B) by
$\tilde{H}(x, t) = H(x, 2t)$ for $0 \leq t \leq \frac{1}{2}$, and
$\tilde{H}(x, t) = H'(x, 2t - 1)$ for $\frac{1}{2} \leq t \leq 1$.
Since the restrictions of $\tilde{H}$ to the closed subsets $X \times [0, \frac{1}{2}]$ and $X \times [\frac{1}{2}, 1]$ of $X \times [0, 1]$ are continuous, it follows that $\tilde{H}$ is continuous. Evidently, $\tilde{H}$ is a homotopy between f and h. This shows that $\sim$ is transitive. ♮

Proposition 3.1.28 *Let f and g be homotopic maps from (X, A) to (Y, B), and let f' and g' be homotopic maps from (Y, B) to (Z, C). The $f'of$ and $g'og$ are homotopic maps from (X, A) to (Z, C).*

Proof Let H be a homotopy from f to g, and H' be a homotopy from f' to g'. Evidently, $f'oH$ is a homotopy from $f'of$ to $f'og$. Also the map $\tilde{H}$ from $X \times [0, 1]$ to Z defined by $\tilde{H}(x, t) = H'(g(x), t)$ is a homotopy from $f'og$ to $g'og$. From the above proposition, it follows that $f'of$ is homotopic $g'og$. ♮

The above proposition allows us to have a category $[\Im p]$ whose objects are topological pairs, and the morphisms are homotopy classes of maps between topological pairs. This category is called the **homotopy category** of topological pairs. Two topological pairs (X, A) and (Y, B) are said to be of the same homotopy type if they are isomorphic objects in $[\Im p]$. More explicitly, the topological pairs (X, A) and (Y, B) are of the same homotopy type if there is a map f from (X, A) to (Y, B), and a map g from (Y, B) to (X, A) such that gof and fog are homotopic to the corresponding identity maps.

Again, we state an important theorem without proof, and the proof can be found in "Algebraic topology" by Spanier.

Theorem 3.1.29 *If f and g are homotopic maps from a topological pair (X, A) to a topological pair (Y, B), then $S(f)$ and $S(g)$ are homotopic chain transformations. In particular, $H_n(f) = H_n(g)$ for all n.* ♯

Corollary 3.1.30 *If (X, A) and (Y, B) are of the same homotopy type, then $H_n(X, A)$ is isomorphic to $H_n(Y, B)$ for all n.*

Proof Let f be a map from (X, A) to (Y, B), and g be a map from (Y, B) to (X, A) such that gof and fog are homotopic to the corresponding identity maps. Also $I_{H_n(X,A)} = H_n(I_{(X,A)}) = H_n(gof) = H_n(g)oH_n(f)$, and $I_{H_n(Y,B)} = H_n(I_Y) = H_n(fog) = H_n(f)oH_n(g)$. This shows that $H_n(f)$ is an isomorphism. ♯

In the light of the above theorem, the singular chain complex functor S and the singular homology functors H_n may be treated as functors from the homotopy category $[\Im p]$ of topological pairs.

Proposition 3.1.31 *Let f be a continuous map from $\mid (\Sigma_1, S_1) \mid$ to $\mid (\Sigma_2, S_2) \mid$, where (Σ_1, S_1) and (Σ_2, S_2) are simplicial complexes. Let ϕ be a simplicial approximation of f. Then f is homotopic to $\mid \phi \mid$.*

Proof Let α be a member of $\mid (\Sigma_1, S_1) \mid$. Then there is a unique simplex $\sigma_2 \in S_2$ such that $f(\alpha) \in < \sigma_2 >$. Since ϕ is a simplicial approximation of f, $\mid \phi \mid (\alpha) \in \mid \sigma_2 \mid$. Since $\mid \sigma_2 \mid$ is convex, $tf(\alpha) + (1 - t) \mid \phi \mid (\alpha)$ belongs to $\mid \sigma_2 \mid$. In turn, we get a continuous map H from $\mid (\Sigma_1, S_1) \mid \times [0, 1]$ to $\mid (\Sigma_2, S_2) \mid$ given by

$$H(\alpha, t) = tf(\alpha) + (1 - t) \mid \phi \mid (\alpha).$$

Evidently, H is a homotopy between f and $\mid \phi \mid$. ♯

A topological space X is said to be a **contractible space** if it is of the same homotopy type as the singleton space.

Example 3.1.32 Consider the Euclidean space $\mathbb{R}^n$. Let c denote the constant map from $\mathbb{R}^n$ to $\{\bar{o}\}$ and i denote the inclusion map from $\{\bar{o}\}$ to $\mathbb{R}^n$. Then coi is the identity map on $\{\bar{0}\}$, and the map H from $\mathbb{R}^n \times [0, 1]$ to $\mathbb{R}^n$ defined by $H(\bar{x}, t) = (1 - t)\bar{x}$ is a homotopy from $I_{\mathbb{R}^n}$ to ioc. This shows that $\mathbb{R}^n$ is contractible. Similarly, the disk D^n is contractible.

Proposition 3.1.33 *If $X = \{x_0\}$ is a singleton space, then $H_n(X) = 0$ for all $n \geq 1$ and $H_0(X) = \mathbb{Z}$.*

Proof If X is a singleton space, then there is only one singular q-simplex for all $q \geq 0$. Thus, $S_q(X) = \mathbb{Z}$ for all $q \geq 0$. It is also evident from the definition that d_q is zero map if q is odd, and d_q is isomorphism for even q. This shows that $H_0(X) = \mathbb{Z}$, and $H_q(X) = 0$ for all $q \geq 1$. ♯

The following corollary is immediate from Corollary 3.1.30, and the above proposition.

Corollary 3.1.34 *If X is a contractible space, then $H_n(X) = 0$ for all $n \geq 1$ and $H_0(X) = \mathbb{Z}$.* ♯

Thus, $H_n(\mathbb{R}^m) = H_n(D^m) = 0$ for all $n \geq 1$ and $H_0(\mathbb{R}^m) = H_0(D^m) = \mathbb{Z}$.

Proposition 3.1.35 *Let p_n denote the north pole $(0, 0, \ldots, 1)$ of S^n, and p_s denote the south pole $(0, 0, \ldots, -1)$ of S^n. Then $S^n - \{p_n, p_s\}$ and S^{n-1} are of the same homotopy type.*

Proof We have the injective map i from S^{n-1} to $S^n - \{p_n, p_s\}$ given by $i((x_1, x_2, \ldots, x_n)) = (x_1, x_2, \ldots, x_n, 0)$. We also have the retraction map r from $S^n - \{p_n, p_s\}$ to S^{n-1} given by $r((x_1, x_2, \ldots, x_n, x_{n+1})) = (\frac{x_1}{\sqrt{1-x_{n+1}^2}}, \frac{x_2}{\sqrt{1-x_{n+1}^2}}, \ldots, \frac{x_n}{\sqrt{1-x_{n+1}^2}})$. Clearly, $r \circ i = I_{S^{n-1}}$. Further, the map H from $(S^n - \{p_n, p_s\}) \times [0, 1]$ to $S^n - \{p_n, p_s\}$ defined by

$$H((x_1, x_2, \ldots, x_n, x_{n+1}), t) = \frac{(x_1, x_2, \ldots, x_n, (1-t)x_{n+1})}{\sqrt{1 - x_{n+1}^2 + (1-t)^2 x_{n+1}^2}}$$

is a homotopy from the identity map on $S^n - \{p_n, p_s\}$ to the map $i \circ r$. This shows that $S^n - \{p_n, p_s\}$ and S^{n-1} are of the same homotopy type. ♯

Proposition 3.1.36 *S^{n-1} and $\mathbb{R}^n - \{\bar{0}\}$ are of the same homotopy type.*

Proof We have inclusion map i from S^{n-1} to $\mathbb{R}^n - \{\bar{0}\}$, and the retraction map r from $\mathbb{R}^n - \{\bar{0}\}$ to S^{n-1} defined by $r(\bar{x}) = \frac{\bar{x}}{|\bar{x}|}$. The map H from $(\mathbb{R}^n - \{\bar{0}\}) \times [0, 1]$ to $(\mathbb{R}^n - \{\bar{0}\})$ defined by

$$H(\bar{x}, t) = t\bar{x} + (1-t)\frac{\bar{x}}{|\bar{x}|}$$

is a homotopy from the identity map to $i \circ r$. ♯

Exercises

3.1.1 Show that one point join of two polyhedrons is a polyhedron.

3.1.2 Let X and H be compact polyhedrons. Show that $X \times Y$ is also a compact polyhedron.

3.1.3 Is every compact metric space a polyhedron? Support.

3.1.4 Get a triangulation of the real projective space $\mathbb{R}P^2$.

3.1.5 Let (Σ, S) be the simplicial complex of Example 1.3.11. Let Σ' denote $\{w_0, w_1, w_2\}$, where $w_0 = \frac{\hat{v}_0 + \hat{v}_1}{2}$, $w_1 = \frac{\hat{v}_0 + \hat{v}_2}{2}$, $w_2 = \frac{\hat{v}_1 + \hat{v}_2}{2}$, and S' denote the set of nonempty subsets of Σ'. Is (Σ', S') a subdivision of (Σ, S)? Support. Is $|(\Sigma, S)|$ homeomorphic to $|(\Sigma', S')|$?

3.1.6 Which of the following spaces are polyhedrons?

  (i) The subspace $(0, 1)$ of $\mathbb{R}$.
 (ii) The subspace $[0, 1)$ of $\mathbb{R}$.
(iii) The subspace $[0, 1]$ of $\mathbb{R}$.
(iv) The subspace $\{\frac{1}{n} \mid n \in \mathbb{N}\} \cup \{0\}$ of $\mathbb{R}$.

3.2 Applications

Theorem 3.2.1 *For* $n \geq 1$,

$$H_q(S^n) \approx \begin{cases} \mathbb{Z} \ if \ q \in \{0, n\} \\ 0 \ \ otherwise. \end{cases}$$

Also

$$H_q(S^0) \approx \begin{cases} \mathbb{Z} \oplus \mathbb{Z} \ \ if \ q = 0 \\ \ \ 0 \ \ \ \ \ otherwise. \end{cases}$$

Proof Since $S^0 = \{1, -1\}$ is a discrete space containing two points, from Propositions 1.3.12 and 3.1.33, it follows that $H_0(S^0) \approx \mathbb{Z} \oplus \mathbb{Z}$ and $H_q(S^0) = 0$ for $q \neq 0$.

Now, suppose that $n = 1$. The simplicial complex (Σ, S) in Example 1.3.11 gives a triangulation of S^1. From Corollary 3.1.22, $H_q(\Sigma, S) \approx H_q(S^1)$ for each q. Thus, from Example 1.3.11, it follows that $H_0(S^1) \approx \mathbb{Z} \approx H_1(S^1)$, and $H_q(S^1) = 0$ for $q \notin \{0, 1\}$.

Suppose that $n \geq 2$. Since $S^n - \{p_n\}$ and $S^n - \{p_s\}$ are open sets such that $S^n = (S^n - \{p_n\}) \cup (S^n - \{p_s\})$, it follows that $\{S^n - \{p_n\}, S^n - \{p_s\}\}$ is an excisive couple. We have the corresponding Mayer–Vietoris long exact sequence

$$\cdots \overset{(H_q(j_1), H_q(j_2))}{\to} H_q(S^n - \{p_n\}) \oplus H_q(S^n - \{p_s\}) \overset{H_q(j_1) + H_q(j_2)}{\to} H_q(S^n) \overset{\partial_q}{\to}$$
$$H_{q-1}(S^n - \{p_n, p_s\}) \overset{(H_{q-1}(i_1), -H_{q-1}(i_2))}{\to}$$
$$H_{q-1}(S^n - \{p_n\}) \oplus H_{q-1}(S^n - \{p_s\}) \overset{H_{q-1}(j_1) + H_{q-1}(j_2)}{\to} \cdots .$$

Further, $S^n - \{p_n\}$ is homeomorphic to $\mathbb{R}^n$. Indeed, the map ϕ from $S^n - \{p_n\}$ to $\mathbb{R}^n$ defined by

$$\phi(x_1, x_2, \ldots, x_n, x_{n+1}) = \left(\frac{x_1}{1 - x_{n+1}}, \frac{x_2}{1 - x_{n+1}}, \ldots, \frac{x_n}{1 - x_{n+1}} \right)$$

can easily be seen to be a homeomorphism. Similarly, $S^n - \{p_s\}$ is homeomorphic to $\mathbb{R}^n$. Thus, $S^n - \{p_n\}$ and $S^n - \{p_s\}$ are contractible. In turn, $H_q(S^n - \{p_n\}) = 0 = H_q(S^n - \{p_s\})$ for each $q \geq 1$. Substituting in the Mayer–Vietoris sequence, we find that

$$H_q(S^n) \approx H_{q-1}(S^n - \{p_n, p_s\})$$

for all $q \geq 2$. Further, $S^n - \{p_n, p_s\}$ is homeomorphic to $\mathbb{R}^n - \{0\}$. From Proposition 3.1.36, $\mathbb{R}^n - \{0\}$ is of the same homotopy type as S^{n-1}. Hence, $H_q(S^n)$ is isomorphic to $H_{q-1}S^{n-1}$ for all $q \geq 2$. Consider $H_1(S^n)$, and the part

$$\cdots \longrightarrow H_1(S^n - \{p_n\}) \oplus H_1(S^n - \{p_s\}) \stackrel{H_1(j_1)+H_1(j_2)}{\to} H_1(S^n) \stackrel{\partial_1}{\to}$$
$$H_0(S^{n-1}) \stackrel{(H_0(i_1), -H_0(i_2))}{\to} H_0(S^{n-1}) \oplus H_0(S^{n-1}) \longrightarrow \cdots$$

of the Mayer–Vietoris sequence. Evidently, $(H_0(i_1), -H_0(i_2))$ is the map which takes a to $(a, -a)$, and so it is injective. Already, $H_1(S^n - \{p_n\}) \oplus H_1(S^n - \{p_s\}) = 0$. Hence $H_1(S^n) = 0$. We have established the result for $q = 1$ and for all n. Since $H_q(S^n)$ is isomorphic to $H_{q-1}S^{n-1}$ for all $q \geq 2$, the result follows by the induction on q.   ♯

Corollary 3.2.2 S^n and S^m are of the same homotopy type if and only if $n = m$.

Proof Suppose that $n \neq m$. Then $H_n(S^n) \approx \mathbb{Z}$ whereas $H_n(S^m) = 0$. The result follows from Corollary 3.1.30.   ♯

Corollary 3.2.3 (Invariance of dimension) $\mathbb{R}^n$ is homeomorphic to $\mathbb{R}^m$ if and only if $n = m$.

Proof Suppose that ϕ is a homeomorphism from $\mathbb{R}^n$ to $\mathbb{R}^m$. Suppose that $\phi(\bar{0}) = \bar{b} \in \mathbb{R}^m$. Then we have the homeomorphism ψ from $\mathbb{R}^n$ to $\mathbb{R}^m$ given by $\psi(\bar{x}) = \phi(\bar{x}) - \bar{b}$ such that $\psi(\bar{0}) = \bar{0}$. In turn, ψ induces a homeomorphism from $\mathbb{R}^n - \{\bar{0}\}$ to $\mathbb{R}^m - \{\bar{0}\}$. Further, by Proposition 3.1.36, $\mathbb{R}^n - \{\bar{0}\}$ is of the same homotopy type as S^{n-1}. Thus, S^{n-1} is of the same homotopy type as as S^{m-1}. From the above proposition, $n - 1 = m - 1$.   ♯

Corollary 3.2.4 S^{n-1} is not a retract of D^n.

Proof Suppose that $n = 1$. Since S^0 is disconnected discrete space on two elements and D^1 is connected, there is no retraction from D^1 to S^0. Next, suppose that $n \geq 2$ and there is a continuous map r from D^n to S^{n-1} such that $r \circ i = I_{S^{n-1}}$. Then

$$H_{n-1}(S^{n-1}) \stackrel{I_{H_{n-1}(S^{n-1})}}{\to} H^{n-1}(S^{n-1}) = H_{n-1}(S^{n-1}) \stackrel{H_{n-1}(i)}{\to} H_{n-1}(D^n) \stackrel{H_{n-1}(r)}{\to}$$
$$H_{n-1}(S^{n-1}).$$

Since $H_{n-1}(S^{n-1}) \approx \mathbb{Z}$ and $H_{n-1}(D^n) = \{0\}$, we arrive at a contradiction.   ♯

Corollary 3.2.5 *For $n \geq 1$,*

$$H_q(D^n, S^{n-1}) \approx \begin{cases} \mathbb{Z} & \text{if } q = n \\ 0 & \text{otherwise.} \end{cases}$$

Proof For the pair (D^n, S^{n-1}), we have the long exact homology sequence

$$\cdots \overset{\partial_{q+1}}{\to} H_q(S^{n-1}) \overset{H_q(i)}{\to} H_q(D^n) \overset{H_q(j)}{\to} H_q((D^n, S^{n-1})) \overset{\partial_q}{\to} H_{q-1}(S^{n-1}) \overset{H_{q-1}(i)}{\to}$$
$$H_{q-1}(D^n) \overset{H_{q-1}(j)}{\to} \cdots .$$

Suppose that $q \geq 2$. Since D^n is contractible, the above long exact sequence gives the exact sequence

$$\cdots 0 \longrightarrow H_q(D^n, S^{n-1}) \overset{\partial_q}{\to} H_{q-1}(S^{n-1}) \longrightarrow 0 \cdots .$$

Hence $H_q(D^n, S^{n-1}) \approx H_{q-1}(S^{n-1})$. From the Theorem 3.2.1, it follows that $H_q(D^n, S^{n-1}) = \mathbb{Z}$ if $q = n$, and it is 0 otherwise. Suppose that $q = 1$. Then the above exact sequence gives us the exact sequence

$$\cdots 0 \longrightarrow H_1(D^n, S^{n-1}) \overset{\partial_1}{\to} H_0(S^{n-1}) \overset{H_0(i)}{\to} H_0(D^n).$$

If $n > 1$, then $H_0(S^{n-1}) \approx \mathbb{Z} \approx H_0(D^n)$ and $H_0(i)$ is an isomorphism. It follows that $H_1(D^n, S^{n-1}) = 0$. Suppose that $n = 1$. Observe that $H_0(S^0) \approx \mathbb{Z} \oplus \mathbb{Z}$, $H_0(D^1) \approx \mathbb{Z}$, and the homomorphism $H_0(i)(a, b) = a + b$. Hence $Ker\, H_0(i) \approx \mathbb{Z}$. From the above exact sequence, it follows that $H_1(D^1, S^0) \approx \mathbb{Z}$. The proof is complete. ♯

Proposition 3.2.6 *Let Δ_{D^n} denote the diagonal $\{(\overline{x}, \overline{x}) \in D^n \times D^n\}$. We have a continuous map ϕ from $D^n \times D^n - \Delta_{D^n}$ to S^{n-1} such that $\phi(\overline{x}, \overline{y}) = \overline{y}$ for all $\overline{y} \in S^{n-1}$.*

Proof Let $(\overline{x}, \overline{y})$ be a member of $D^n \times D^n - \Delta_{D^n}$. Consider the line $\overline{r}(t) = (1 - t)\overline{x} + t\overline{y}$ joining $\overline{x}$ to $\overline{y}$. There is a unique nonnegative real number $t_{\overline{x}, \overline{y}}$ such that $\overline{r}(t_{\overline{x}, \overline{y}})$ belongs to S^{n-1}. This gives us a continuous map λ from $D^n \times D^n - \Delta_{D^n}$ to $[0, \infty]$ given by $\lambda(\overline{x}, \overline{y}) = t_{\overline{x}, \overline{y}}$ (obtain the explicit formula for $t_{\overline{x}, \overline{y}}$ in terms of $\overline{x}$ and $\overline{y}$). Evidently, $t_{\overline{x}, \overline{y}} = 1$ whenever $\overline{y} \in S^{n-1}$. In turn, the map ϕ from $D^n \times D^n - \Delta_{D^n}$ to S^{n-1} defined by

$$\phi((\overline{x}, \overline{y})) = (1 - t_{\overline{x}, \overline{y}})\overline{x} + t_{\overline{x}, \overline{y}}\overline{y}$$

is a continuous map with the desired property. ♯

Corollary 3.2.7 (Brouwer fixed-point theorem) *Let f be a continuous map from D^n to D^n. Then there is a fixed point of f. More explicitly, there is an element $\overline{x} \in D^n$ such that $f(\overline{x}) = \overline{x}$.*

Proof Suppose that $f(\overline{x}) \neq \overline{x}$ for each $\overline{x} \in D^n$. Then, we have a continuous map ψ from D^n to $D^n \times D^n - \Delta_{D^n}$ defined by $\psi(\overline{x}) = (f(\overline{x}), \overline{x})$. Clearly, the continuous map $r = \phi o \psi$ from D^n to S^{n-1} is a retraction. This is a contradiction to Corollary 3.2.4.   ♯

Let A be a finitely generated abelian group. From the structure theorem for finitely generated abelian groups,

$$A \approx \underbrace{\mathbb{Z} \oplus \mathbb{Z} \oplus \cdots \oplus \mathbb{Z}}_{m} \oplus \mathbb{Z}_{n_1} \oplus \mathbb{Z}_{n_2} \oplus \cdots \oplus \mathbb{Z}_{n_r},$$

where n_i / n_{i+1} for all i. The number m of infinite cyclic summands is called the rank A and $n_1, n_2, \ldots, n_r$ are called the torsion numbers of A. The rank of A is denoted by $r(A)$. Evidently, $r(A) = Dim_{\mathbb{Q}}(A \otimes_{\mathbb{Z}} \mathbb{Q})$.

Proposition 3.2.8 *Let A be a finitely generated abelian group and B be a subgroup of A. Then B is also finitely generated and $r(A) = r(B) + r(A/B)$.*

Proof We have the short exact sequence

$$0 \longrightarrow B \overset{i}{\to} A \overset{\nu}{\to} A/B \longrightarrow 0.$$

Since $\mathbb{Q}$ is a torsion-free abelian group, the sequence

$$0 \longrightarrow B \otimes_{\mathbb{Z}} \mathbb{Q} \overset{i \otimes I_{\mathbb{Q}}}{\to} A \otimes_{\mathbb{Z}} \mathbb{Q} \overset{\nu \otimes I_{\mathbb{Q}}}{\to} (A/B) \otimes_{\mathbb{Z}} \mathbb{Q} \longrightarrow 0$$

is the exact sequence of $\mathbb{Q}$-vector spaces. Hence

$$r(A) = Dim_{\mathbb{Q}} A \otimes_{\mathbb{Z}} \mathbb{Q} = Dim_{\mathbb{Q}} B \otimes_{\mathbb{Z}} \mathbb{Q} + Dim_{\mathbb{Q}}(A/B) \otimes_{\mathbb{Z}} Q =$$
$$r(B) + r(A/B).$$
                                                                                  ♯

Let $A = \{A_q \mid q \in \mathbb{Z}\}$ be a finitely generated graded abelian group in the sense that A_q is finitely generated abelian group for each q and $A_q = 0$ for all but finitely many q. Then the number $\sum_{q=0}^{\infty}(-1)^q r(A_q)$ is called the **Euler–Poincare characteristic** of A and it is denoted by $\chi(A)$.

Theorem 3.2.9 (Euler–Poincare) *Let*

$$\Omega \equiv \cdots \overset{d_{q+1}}{\to} \Omega_q \overset{d_q}{\to} \Omega_{q-1} \overset{d_{q-1}}{\to} \cdots$$

be a finitely generated chain complex of abelian groups. Then $H(\Omega) = \{H_q(\Omega) \mid q \in \mathbb{Z}\}$ is a finitely generated graded abelian group and $\chi(\Omega) = \chi(H(\Omega))$.

Proof Since Ω_q is a finitely generated abelian group, $C_q(\Omega)$, $B_q(\Omega)$, and $H_q(\Omega) = C_q(\Omega)/B_q(\Omega)$ are finitely generated. From the above proposition,

$$r(C_q(\Omega)) = r(H_q(\Omega)) + r(B_q(\Omega)).$$

Again, since $\Omega_q / C_q(\Omega) \approx B_{q-1}(\Omega)$,

$$r(\Omega_q) = r(C_q(\Omega)) + r(B_{q-1}(\Omega)).$$

From the above two equations, it follows that

$$r(\Omega_q) = r(H_q(\Omega)) + r(B_q(\Omega)) + r(B_{q-1}(\Omega)).$$

Multiplying by $(-1)^q$ and summing over q, we find that $\chi(\Omega) = \chi(H(\Omega))$.   ♯

Let X be a topological space such that the graded singular homology group $H(X) = \{H_q(X) \mid q \geq 0\}$ is a finitely generated graded abelian group. The rank of $H_q(X)$ is called the qth **Betti number**, and it is denoted by $b_q(X)$. The torsion numbers of $H_q(X)$ are called the qth torsion numbers of X. The Euler–Poincare characteristic $\chi(H(X)) = \sum_{q=0}^{n}(-1)^q b_q(X)$ of $H(X)$ is called the **Euler–Poincare characteristic** of X, and it is denoted by $\chi(X)$. These are all invariants of the space up to homotopy.

Thus, the Euler–Poincare characteristic $\chi(X)$ of any contractible space is 1. From Theorem 3.2.1, it follows that $\chi(S^{2n}) = 2$ and $\chi(S^{2n+1}) = 0$. Since the cylinder $S^1 \times [0, 1]$ is of the same homotopy type as S^1, the Euler characteristic of a cylinder is 0.

Proposition 3.2.10 *Suppose that $\{X_1, X_2\}$ is an excisive couple. Then*

$$\chi(X_1 \bigcup X_2) = \chi(X_1) + \chi(X_2) - \chi(X_1 \bigcap X_2).$$

Proof From Theorem 3.1.24, we have the following exact sequence:

$$\cdots \xrightarrow{H_{q+1}(j_1) \oplus H_{q+1}(j_2} H_{q+1}(X_1 \bigcup X_2) \xrightarrow{\partial_{q+1}} H_q(X_1 \bigcap X_2) \xrightarrow{(H_q(i_1), -H_q(i_2))}$$
$$H_q(X_1) \oplus H_q(X_2) \xrightarrow{H_q(j_1) + H_q(j_2)} H_q(X_1 \bigcup X_2) \xrightarrow{\partial_q} \cdots ,$$

where i_1, i_2, j_1, and j_2 are the corresponding inclusion maps. Denote the above exact sequence by E. From Theorem 3.2.9, the Euler–Poincare characteristic of an exact sequence is 0. Hence

$$0 = \chi(E) =$$
$$\sum_q (-1)^q [r(H_q(X_1 \bigcap X_2)) - (r(H_q(X_1)) + r(H_q(X_2))) + r(H_q(X_1 \bigcup X_2))].$$

This shows that

$$\chi(X_1 \bigcup X_2) = \chi(X_1) + \chi(X_2) - \chi(X_1 \bigcap X_2).$$

♯

Consider the torus $T = S^1 \times S^1$. Clearly, $\{X_1, X_2\}$ is an excisive couple, where $X_1 = S^1 \times \{e^{i\theta} \mid 0 \leq \theta \leq \frac{3\pi}{2}\}$ and $X_2 = S^1 \times \{e^{i\theta} \mid \frac{\pi}{2} \leq \theta \leq 2\pi\}$. Also $X_1 \bigcap X_2$ is of the same homotopy type as S^1 and $X_1 \bigcup X_2 = T$. From the above identity, it follows that $\chi(T) = 0$. Similarly, it can be shown that the Euler characteristic of the double torus (the surface obtained by removing the interior of disks on the surfaces of the two different tori and then gluing them along the boundary of the disks) is -2.

We state the following theorem of Eilenberg–Zilber without proof. This theorem together with the Kunneth formula can be used very effectively to compute the singular homologies of the product of two spaces.

Theorem 3.2.11 (Eilenberg–Zilber) *Let X and Y be topological spaces. Then $S(X \times Y)$ is chain equivalent to $S(X) \otimes S(Y)$.*   ♯

Thus, if $H_q(X)$ or $H_q(Y)$ is torsion free, then by the Kunneth formula (Theorem 2.4.3), $H_n(X \times Y) = \sum_{p+q=n} H_p(X) \otimes H_q(Y)$, and $\chi(X \times Y) = \chi(X)\chi(Y)$. In particular, $\chi(S^2 \times S^2) = 4$.

Proposition 3.2.12 *Let X be a compact polyhedron having a triangulation $((\Sigma, S), f)$, where $\mid \Sigma \mid = n$. Then*

$$\chi(X) = \sum_{q=0}^{n}(-1)^q \lambda_q,$$

where λ_q is the number of q-simplexes in (Σ, S).

Proof By the Euler–Poincare theorem, $\chi(H(\Lambda(\Sigma, S))) = \chi(\Lambda(\Sigma, S))$. Since $H_q(X) = H_q(\Lambda(\Sigma, S))$ for each q, $\chi(X) = \chi(H(\Lambda(\Sigma, S))) = \chi(\Lambda(\Sigma, S))$. Further, by the definition, $\Lambda_q(\Sigma, S)$ is the abelian group generated by the set of oriented q-simplexes $[\sigma, \alpha]$ subject to the relation $[\sigma, \alpha] + [\sigma, \beta] = 0$ whenever $[\sigma, \alpha] \neq [\sigma, \beta]$. Hence $r(\Lambda_q(\Sigma, S))$ is the number λ_q of q-simplexes in (Σ, S). It follows that $\chi(X) = \sum_{q=0}^{n}(-1)^q \lambda_q$.   ♯

Corollary 3.2.13 *If (Σ, S) is a simplicial complex giving a triangulation of S^2, then $V - E + F = 2$, where V is the number of vertices, E is the number of edges, and F is the number of faces.*   ♯

A **simple polyhedron** is a solid object in $\mathbb{R}^3$ whose surface is homeomorphic to S^2, and which is built by several two-dimensional convex polygonal faces in such a manner that the intersection of two distinct faces is an edge and the intersection of two edges is a vertex of a polygon. It is said to be a regular simple polyhedron if all the polygonal faces are congruent and the polygonal angles at the vertices are the same. A regular simple polyhedron is also called a **platonic solid**. For example, the solid regular tetrahedron

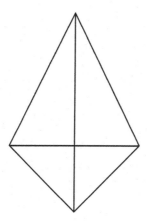

and the solid cube

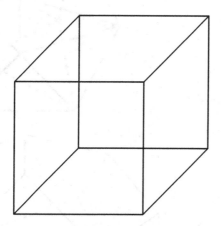

are platonic solids. It is known from ancient Greek times there are three more platonic solids, viz., regular octahedron, dodecahedron, and icosahedron. We shall see soon that these constitute all platonic solids (Figs. 3.1, 3.2, and 3.3).

Theorem 3.2.14 (Euler) *Suppose that V is the number of vertices, E is the number of edges, and F is the number of the polygonal faces describing the surface S of a simple solid polyhedron. Then*

$$V - E + F = 2.$$

Proof Suppose that there is a polygonal face A on the surface which is not a triangle, and which has n vertices, m edges, and of course 1 face. We subdivide A into triangles by taking a vertex v_0 at the centroid of the face and joining the vertex v_0 with other

Fig. 3.1 Octahedron

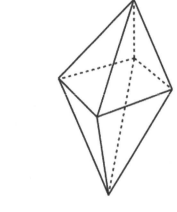

Fig. 3.2 Icosahedron

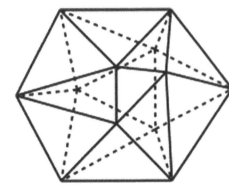

Fig. 3.3 Dodecahedron

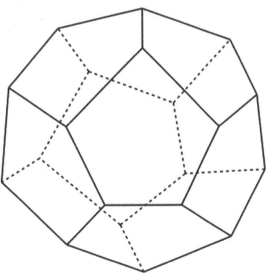

vertices of the polygon through edges. In this subdivision of A into triangles, the number of vertices is $n + 1$, the number of edges is $2m$, and the number of faces is m. Since $n + 1 - 2m + m = n - m + 1$, it follows that the number $V - E + F$ is invariant under this subdivision. Proceeding inductively, we may assume that all polygonal faces are triangles which give a triangulation of S^2. The result follows from Corollary 3.2.13. ♯

Let S be a regular simple polyhedron. Suppose that every vertex is common to m edges, and every face has n edges, $n \geq 3$. Since any two faces have exactly one common edge, and an edge has exactly two vertices,

$$mV = 2E = nF,$$

where V is the number of vertices, E is the number of edges, and F is the number of faces. Thus, $V = \frac{nF}{m}$, and $E = \frac{nF}{2}$. By the Euler theorem

$$\frac{nF}{m} - \frac{nF}{2} + F = 2.$$

Hence $F(2n - mn + 2m) = 4m$. Since $m > 0$, $(2n - mn + 2m) > 0$. Since $m \geq 2$ and $n \geq 3$, $2m > n(m - 2) \geq 3m - 6$. In turn, $m \leq 5$. Suppose that $m = 5$. Then $F(2n - 5n + 10) = 20$. The only possible solution is $(m, n, F) = (5, 3, 20)$. Further, for $m = 4$, the only possible solution is $(m, n, F) = (4, 3, 8)$. For $m = 3$, the possible solutions are $(3, 4, 6)$ and $(3, 5, 12)$. Thus, there are at the most 5 platonic solids corresponding to $(3, 3, 4)$, $(3, 4, 6)$, $(4, 3, 8)$, $(3, 5, 12)$, and $(5, 3, 20)$. Indeed, in each case, we have a platonic solid which is unique up to similarity. Associated with $(3, 3, 4)$, we have a regular tetrahedron. For $(3, 4, 6)$, we have a cube. We have a regular octahedron associated with $(4, 3, 8)$ which is obtained by joining the centroids of the faces of a cube. For $(3, 5, 12)$, we have a regular dodecahedron with 12 congruent regular pentagonal faces such that each vertex is common to 3 pentagons. Finally, we have a regular icosahedron associated with $(5, 3, 20)$ having 20 triangular faces with each vertex common to 5 triangles. The construction of the last two platonic solids can be found in "Geometry" by Michele Audin.

The above discussion establishes the following interesting and important theorem.

Theorem 3.2.15 *There are exactly five platonic solids, viz., regular tetrahedron, cube, regular octahedron, dodecahedron, and icosahedron.* ♯

Lefschetz Fixed-Point Theorem

Proposition 3.2.16 *Let W be a subspace of a vector space V over a field F. Let f be an endomorphism of V such that $f(W) \subseteq W$. Then*

$$Trace(f) = Trace(f/W) + Trace(\bar{f}),$$

where $\bar{f}$ is the endomorphism of V/W given by $\bar{f}(x + W) = f(x) + W$.

Proof Let $\{x_1, x_2, \ldots, x_r\}$ be a basis of W. Extend it to a basis $\{x_1, x_2, \ldots, x_r, x_{r+1}, \ldots, x_n\}$ of V. Suppose that $f(x_j) = \sum_{i=1}^{n} \alpha_{ij} x_i$. By definition, $Trace(f) = \sum_{i=1}^{n} \alpha_{ii}$. Evidently, $Trace(f/W) = \sum_{i=1}^{r} \alpha_{ii}$. Again, since $\{x_{r+1} + W, x_{r+2} + W, \ldots, x_n + W\}$ is a basis of V/W and $\bar{f}(x_{r+l} + W) = \sum_{t=1}^{n-r} \alpha_{(r+t)(r+l)} (x_{r+t} + W)$, $Trace(\bar{f}) = \sum_{i=r+1}^{n} \alpha_{ii}$. This shows that $Trace(f) = Trace(f/W) + Trace(\bar{f})$. ♯

Let M be a finitely generated abelian group, and f be an endomorphism of M. Then f induces an endomorphism $\tilde{f}$ of $M/T(M)$ given by $\tilde{f}(a + T(M)) = f(a) + T(M)$. Since $M/T(M)$ is a finitely generated free abelian group, we can talk of the **trace** of $\tilde{f}$. The trace of $\tilde{f}$ is also called the **trace** of f, and it is denoted by $tr(f)$. We have the split short exact sequence

$$0 \longrightarrow T(M) \overset{i}{\to} M \overset{\nu}{\to} M/T(M) \longrightarrow 0.$$

Since $\mathbb{Q}$ is a divisible abelian group, $T(M) \otimes_{\mathbb{Z}} \mathbb{Q} = 0$. In turn, $\nu \otimes I_{\mathbb{Q}}$ is an isomorphism from the $\mathbb{Q}$-vector space $M \otimes_{\mathbb{Z}} \mathbb{Q}$ to $M/T(M) \otimes_{\mathbb{Z}} \mathbb{Q}$. It is also evident that $tr(f \otimes I_{\mathbb{Q}}) = tr(\tilde{f} \otimes I_{\mathbb{Q}}) = tr(f)$.

Proposition 3.2.17 *Let f be an endomorphism of a finitely generated abelian group M, and let N be a submodule of M such that $f(N) \subseteq N$. Let $\bar{f}$ denote the endomorphism of M/N given by $\bar{f}(x + N) = f(x) + N$. Then*

$$tr(f) = tr(f/N) + tr(\bar{f}).$$

Proof Since $\mathbb{Q}$ is a torsion-free $\mathbb{Z}$-module, we have the exact sequence

$$0 \longrightarrow N \otimes_{\mathbb{Z}} \mathbb{Q} \overset{i \otimes I_{\mathbb{Q}}}{\to} M \otimes_{\mathbb{Z}} \mathbb{Q} \overset{\nu \otimes I_{\mathbb{Q}}}{\to} M/N \otimes_{\mathbb{Z}} \mathbb{Q} \longrightarrow 0.$$

Further, $f \otimes I_{\mathbb{Q}}$ is an endomorphism of the $\mathbb{Q}$-vector space $M \otimes_{\mathbb{Z}} \mathbb{Q}$ such that $(f \otimes I_{\mathbb{Q}})(N \otimes_{\mathbb{Z}} \mathbb{Q}) \subseteq N \otimes_{\mathbb{Z}} \mathbb{Q}$. From the above proposition,

$$tr(f) = Trace(f \otimes I_{\mathbb{Q}}) = Trace(f/N \otimes I_{\mathbb{Q}}) + Trace(\bar{f} \otimes I_{\mathbb{Q}}) = $$
$$tr(f/N) + tr(\bar{f}).$$
♯

Let $A = \{A_q \mid q \in \mathbb{Z}\}$ be a finitely generated $\mathbb{Z}$-graded abelian group in the sense that all A_q are finitely generated and $A_q = 0$ for all but finitely many $q \in \mathbb{Z}$. Let $f = \{f_q \mid q \in \mathbb{Z}\}$ be a graded endomorphism of A of degree 0. Then the number $L(f, A) = \sum_{q \in \mathbb{Z}} (-1)^q tr(f_q)$ is called the **Lefschetz number** of f. In particular, $L(I_A, A)$ is the Euler–Poincare characteristic $\chi(A)$ of A.

Theorem 3.2.18 (Hopf Trace Formula) *Let*

$$A \equiv 0 \longrightarrow A_n \overset{d_n}{\to} A_{n-1} \overset{d_{n-1}}{\to} \cdots \overset{d_2}{\to} A_1 \overset{d_1}{\to} A_0 \longrightarrow 0$$

be a finite chain complex of finitely generated abelian groups. Let $f = \{f_q \mid 0 \leq q \leq n\}$ be a chain transformation from A to itself. Then $L(f, A) = L(H(f), H(A))$. More explicitly, $\sum_{q=0}^{n}(-1)^q tr(f_q) = \sum_{q=0}^{n}(-1)^q tr(H_q(f))$.

Proof Let Z_q denote the group of q-cycles, and B_q denote the group of q-boundaries. Then $f_q(Z_q) \subseteq Z_q$, and $f_q(B_q) \subseteq B_q$. Also $A_q/Z_q \approx B_{q-1}$, and $Z_q/B_q \approx H_q(A)$. Thus, from the Proposition 3.2.16, we have the identities:

$$tr(f_q) = tr(f_q/Z_q) + tr(f_{q-1}/B_{q-1}),$$

and

$$tr(f_q/Z_q) = tr(H_q(f)) + tr(f_q/B_q).$$

Substituting the value of $tr(f_q/Z_q)$ from the second equation to the first equation, we obtain

$$tr(f_q) = tr(H_q(f)) + tr(f_q/B_q) + tr(f_{q-1}/B_{q-1}).$$

Multiplying the last equation by $(-1)^q$ and summing over q from 0 to n, we obtain the desired result.   ♯

Let X be a topological space such that the graded homology group $H(X) = \{H_q(X) \mid q \geq 0\}$ is finitely generated. Let f be a continuous map from X to itself. Then the Lefschetz number $L(H(f), H(X))$ is called the **Lefschetz number** of f, and it is denoted by $L(f, X)$. In particular, if X is a compact polyhedron of dimension n and f is a continuous map from X to itself, then the Lefschetz number $L(f, X) = \sum_{q=0}^{n}(-1)^q tr(H_q(f))$ is defined. The following fixed-point theorem due to Solomon Lefschetz (1884–1972) which he established in 1926 is one of the most important fixed-point theorems.

Theorem 3.2.19 (Lefschetz fixed-point theorem) *Let f be a continuous map from a compact polyhedron X to itself such that $L(f, X) \neq 0$. Then f has a fixed point.*

Proof Since X is a compact polyhedron, there is a finite simplicial complex (Σ, S) of dimension n, together with a homeomorphism ϕ from $\mid (\Sigma, S) \mid$ to X. Consider the continuous map $\psi = \phi^{-1} \circ f \circ \phi$ from $\mid (\Sigma, S) \mid$ to itself. The map f has a fixed point $x \in X$ if and only if $(\phi^{-1} \circ f \circ \phi)(\phi^{-1}(x)) = \phi^{-1}(x)$. Thus, f has a fixed point if and only if ψ has a fixed point. Further, $H_q(\psi) = H_q(\phi^{-1})H_q(f)H_q(\phi) = (H_q(\phi))^{-1}H_q(f)H_q(\phi)$. Hence $tr(H_q(\psi)) = tr(H_q(f))$. In turn, $L(f, X) = L(\psi, \mid (\Sigma, S) \mid)$. It is sufficient, therefore, to assume that $X = \mid (\Sigma, S) \mid$ and $f = \psi$. Suppose that ψ has no fixed point. We need to show that $L(\psi, \mid (\Sigma, S) \mid) = 0$. Let d be the usual metric on $\mid (\Sigma, S) \mid$. Since ψ is continuous, and the metric d is also continuous on the product, the map $x \mapsto d(x, \psi(x))$ is a continuous map from $\mid (\Sigma, S) \mid$ to the space $\{x \in \mathbb{R} \mid x \geq 0\}$ of nonnegative real numbers. Since $\mid (\Sigma, S) \mid$ is compact and $d(x, \psi(x)) > 0$ for all x, there is a positive real number ϵ such that $d(x, \psi(x)) \geq \epsilon$ for all $x \in \mid (\Sigma, S) \mid$. Let N be sufficiently large so that the *mesh* of $(\Sigma_{bd}^N, S_{bd}^N)$ with respect to the metric d_λ on $\mid (\Sigma_{bd}^N, S_{bd}^N) \mid$ induced by the usual metric d on

$| (\Sigma, S) |$ through the linear homeomorphism λ from $| (\Sigma_{bd}^N, S_{bd}^N) |$ to $| (\Sigma, S) |$ is less than $\frac{\epsilon}{3}$. Again consider $\xi = \lambda^{-1} o \psi o \lambda$. Since ψ has no fixed point, ξ also has no fixed point. Also $H_q(\xi) = (H_q(\lambda))^{-1} o H_q(\xi) o H_q(\lambda)$. Thus, it is sufficient to show that the Lefschetz number $L(\xi, | (\Sigma_{bd}^N, S_{bd}^N) |)$ is 0. Let η be a simplicial map from $((\Sigma_{bd}^N)_{bd}^m, (S_{bd}^N)_{bd}^m)$ to $(\Sigma_{bd}^N, S_{bd}^N)$ which is a simplicial approximation of ξ (Corollary 3.1.14). We have the following commutative diagram:

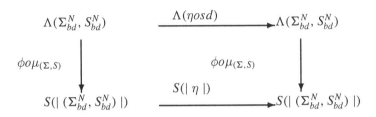

where $\mu_{(\Sigma, S)}$ is the chain equivalence described in Theorem 1.3.10, and ϕ is the chain equivalence described in Theorem 3.1.21. Further, by Proposition 3.1.31, and Theorem 3.1.29, $S(| \eta |)$ and $S(\xi)$ are chain homotopic. In turn, $H_q(| \eta |) = H_q(\xi)$ for all q. From the above commutative diagram, $tr(H_q(\xi)) = tr(H_q(\eta o sd))$. Thus, $L(\xi, | (\Sigma_{bd}^N, S_{bd}^N) |) = \sum_{q=0}^n (-1)^q tr(H_q(\eta o sd))$. From the Hopf trace formula, $\sum_{q=0}^n (-1)^q tr(H_q(\eta o sd)) = \sum_{q=0}^n (-1)^q tr(\eta_q o sd_q))$. It is sufficient, therefore, to show that $tr(\eta_q o sd_q) = 0$ for all q. Let $[\sigma, \alpha]$ be an oriented q-simplex in $(\Sigma_{bd}^N, S_{bd}^N)$. Since η is a simplicial approximation of ξ, for each $\alpha \in | (\Sigma, S) |$, there is a simplex $\sigma \in S_{bd}^N$ such that $| \eta | (\alpha)$ and $\xi(\alpha)$ are members of $| \sigma |$. Hence $d_\lambda(| \eta | (\alpha), \xi(\alpha)) < \frac{\epsilon}{3}$ for each $\alpha \in S_{bd}^N$. Let σ be a simplex in $(\Sigma_{bd}^N, S_{bd}^N)$. Suppose that $\alpha \in | \eta | (| \sigma |) \bigcap | \sigma |$. Then $\alpha = | \eta | (\beta)$ for some $\beta \in | \sigma |$. In turn,

$$d_\lambda(\beta, \xi(\beta)) \le d_\lambda(\beta, | \eta | (\beta)) + d_\lambda(| \eta | (\beta), \xi(\beta)) = d_\lambda(\beta, \alpha) + d_\lambda(| \eta | (\beta), \xi(\beta)) \le 2\frac{\epsilon}{3} < \epsilon.$$

This is a contradiction, and hence $| \eta | (| \sigma |) \bigcap | \sigma | = \emptyset$ for all simplexes in S_{bd}^N.

Let $[\sigma, \mu]$ be an oriented q-simplex in $(\Sigma_{bd}^N, S_{bd}^N)$. Consider $sd_q([\sigma, \mu])$. Let $[\tau, \nu]$ be an oriented q-simplex in $((\Sigma_{bd}^N)^m, (S_{bd}^N)_{bd}^m)$ which appears in the chain $sd_q([\sigma, \mu])$ with nonzero coefficient. Then $\tau \subseteq | \sigma |$. It follows, from the previous observation, that $| \eta | (\tau)$ is disjoint from $| \sigma |$. Thus, in the chain $sd_q([\sigma, \mu])$, the coefficient of $[\sigma, \mu]$ is zero. This means that $tr(\eta_q o sd_q) = 0$. This completes the proof of the Lefschetz fixed-point theorem.   ♯

Corollary 3.2.20 *Let X be a compact contractible polyhedron. then any continuous map f from X to itself has a fixed point.*

Proof If X is a contractible polyhedron, then $H_0(X) \approx \mathbb{Z}$ and $H_q(X) = 0$ for all $q \ge 1$. Further, given any continuous map f from X to X, $H_0(f)$ is an isomorphism. Hence $L(f, X) = \pm 1$. From the Lefschetz fixed-point theorem, f has a fixed point.
♯

The Brouwer fixed-point theorem (Corollary 3.2.7) follows as a particular case of the above corollary.

Let f be a continuous map from S^n to S^n, $n \geq 1$. Then $H_n(f)$ is a homomorphism from $H_n(S^n)$ to itself. Since $H_n(S^n)$ is an infinite cyclic group, there is a unique integer m such that $H_n(f)(a) = a^m$ for all $a \in H_n(S^n)$. This number m is called the **degree** of f, and it is denoted by $deg(f)$. Thus, $H_n(f)(a) = a^{deg(f)}$ for all $a \in H_n(S^n)$. The following proposition relates the Lefschetz number and the degree of a continuous self map on S^n.

Proposition 3.2.21 *Let f be a continuous map from S^n to itself. Then $L(f, S^n) = 1 + (-1)^n deg(f)$.*

Proof Since $tr(H_n(f)) = deg(f)$, the result is immediate from Theorem 3.2.1. ♯

Corollary 3.2.22 *If a continuous map f from S^n to itself has no fixed point, then $deg(f) = (-1)^{n+1}$. In particular, the degree of the antipodal map A from S^n to itself ($A(\overline{x}) = -\overline{x}$) is $(-1)^{n+1}$.* ♯

Theorem 3.2.23 *There is a continuous map f from S^n to itself such that $\overline{x}$ is orthogonal to $f(\overline{x})$ for all $\overline{x} \in S^n$ if and only if n is odd.*

Proof Suppose that there is a continuous map f from S^n to itself such that $\overline{x}$ and $f(\overline{x})$ are orthogonal to each other for all $\overline{x} \in S^n$. Then

$$((\| t\overline{x} + (1-t)f(\overline{x}) \|)^2 = t^2 + (1-t)^2 \neq 0$$

for all $t \in [0, 1]$. Thus, we have a continuous map H from $S^n \times [0, 1]$ to S^n given by

$$H(\overline{x}, t) = \frac{t\overline{x} + (1-t)f(\overline{x})}{\| t\overline{x} + (1-t)f(\overline{x}) \|}.$$

Clearly, H is a homotopy from I_{S^n} to f. Hence $L(f, S^n) = L(I_{S^n}, S^n) = 1 + (-1)^n$. If n is even, then $L(f, S^n) = 2 \neq 0$. By the Lefschetz fixed-point theorem, f has a fixed point $\overline{x} \in S^n$. This is a contradiction, for $< \overline{x}, f(\overline{x}) > = 0$. Hence n is odd. Conversely, suppose that $n = 2m + 1$ is odd. Then the map f from S^{2m+1} to itself defined by

$$f((x_1, x_2, \ldots, x_{2m+1}, x_{2m+2})) = (x_2, -x_1, \ldots, x_{2m+2}, -x_{2m+1})$$

is a continuous map with the required property. ♯

Consider the subspace $T(S^n) = \{(\overline{v}, \overline{w}) \in S^n \times \mathbb{R}^{n+1} \mid < \overline{v}, \overline{w} > = 0\}$ of $S^n \times \mathbb{R}^{n+1}$. $T(S^n)$ together with the projection map p from $T(S^n)$ to S^n is called the tangent bundle of S^n. A continuous section s to the projection map p is called the tangent vector field over S^n. Thus, a tangent vector field to S^n determines and is uniquely determined by a continuous map ϕ from S^n to $\mathbb{R}^{n+1}$ such that $< \overline{v}, \phi(\overline{v}) > = 0$ for all $\overline{v} \in S^n$.

Corollary 3.2.24 *Everywhere nonvanishing tangent vector field on S^n exists if and only if n is odd.*

Proof If there is a nowhere-vanishing tangent vector field on S^n, then there is a continuous map ϕ from S^n to $\mathbb{R}^{n+1}$ such that $< \overline{v}, \phi(\overline{v}) > = 0$ and $\phi(\overline{v}) \neq \overline{0}$ for all $\overline{v} \in S^n$. In turn, we have a continuous map ψ from S^n to itself given by $\frac{\phi(\overline{v})}{||\overline{v}||}$ such that $< \overline{v}, \psi(\overline{v}) > = 0$ for all $\overline{v} \in S^n$. From Theorem 3.2.23, n is odd. Conversely, if n is odd, we have a nowhere vanishing tangent vector field ϕ given by

$$\phi((x_1, x_2, \ldots, x_{2m+1}, x_{2m+2})) = (x_2, -x_1, \ldots, x_{2m+2}, -x_{2m+1}).$$
♯

We say that a set $\{\phi_1, \phi_2, \ldots, \phi_r\}$ of tangent vector fields on S^n is an orthonormal set of tangent vector fields if $< \phi_i(\overline{v}), \phi_j(\overline{v}) > = \delta_{ij}$ for all $\overline{v} \in S^n$, where δ_{ij} is the Kronecker delta. Thus, for all odd n, we have at least a singleton set $\{\phi\}$ of orthonormal tangent vector fields. It is natural to ask the following question:

What are odd n for which there is an orthonormal set $\{\phi_1, \phi_2, \ldots, \phi_n\}$ of tangent vector fields on S^n?

Note that in this case, $\{\phi_1(\overline{v}), \phi_2(\overline{v}), \ldots, \phi_n(\overline{v})\}$ will form an orthonormal basis of the tangent space $T_{\overline{v}}(S^n)$ to S^n at $\overline{v}$ for all $\overline{v} \in S^n$. This problem is faithfully related to the existence of an orthogonal multiplication on $\mathbb{R}^{n+1}$ (equivalently, the existence of a nondegenerate multiplication on S^n). We demonstrate it as follows.

An **orthogonal multiplication** on $\mathbb{R}^{n+1}$ is a bilinear map μ from $\mathbb{R}^{n+1} \times \mathbb{R}^{n+1}$ to $\mathbb{R}^{n+1}$ such that

$$|| \mu(\overline{x}, \overline{y}) || = || \overline{x} |||| \overline{y} ||$$

for all $\overline{x}, \overline{y}$ in $\mathbb{R}^{n+1}$. Evidently, $\overline{x} \neq \overline{0} \neq \overline{y}$ implies that $\mu(\overline{x}, \overline{y}) \neq \overline{0}$. For example, the usual multiplication on $\mathbb{R}$, the complex multiplication on $\mathbb{R}^2$, the quaternionic multiplication on $\mathbb{R}^4$, and the octonion multiplication on $\mathbb{R}^8$ are all orthogonal multiplications. Observe that an orthogonal multiplication is always continuous.

Proposition 3.2.25 *Let μ be an orthogonal multiplication on $\mathbb{R}^{n+1}$. For each $\overline{x} \neq \overline{0}$, the map $\mu_{\overline{x}}$ from $\mathbb{R}^{n+1}$ to itself defined by $\mu_{\overline{x}}(\overline{y}) = \mu(\overline{x}, \overline{y})$ is a linear automorphism of $\mathbb{R}^n$. Further, if $\overline{x}$ is a unit vector, then $\mu_{\overline{x}}$ is an orthogonal transformation.*

Proof Evidently, $\mu_{\overline{x}}$ is a linear transformation. Suppose that $\mu(\overline{x}, \overline{y}) = \mu(\overline{x}, \overline{z})$. Then $\mu(\overline{x}, \overline{y} - \overline{z}) = \overline{0}$. Hence $|| \overline{x} |||| \overline{y} - \overline{z} || = 0$. Since $\overline{x} \neq \overline{0}$, $\overline{y} = \overline{z}$. This shows that $\mu_{\overline{x}}$ is an injective linear transformation. Since $\mathbb{R}^{n+1}$ is finite dimensional, $\mu_{\overline{x}}$ is a linear automorphism. Further, if $\overline{x}$ is a unit vector, then $|| \mu_{\overline{x}}(\overline{y}) || = || \overline{x} |||| \overline{y} || = || \overline{y} ||$ for each $\overline{y}$. This shows that $\mu_{\overline{x}}$ is an orthogonal transformation. ♯

Proposition 3.2.26 *Let μ be an orthogonal multiplication on $\mathbb{R}^{n+1}$. Let $\overline{u}$ be a unit vector of $\mathbb{R}^{n+1}$. We can deform μ to another orthogonal multiplication $\tilde{\mu}$ on $\mathbb{R}^{n+1}$ such that $\overline{u}$ is the identity of the multiplication $\tilde{\mu}$.*

Proof Define $\tilde{\mu}(\overline{x}, \overline{y}) = \mu(\overline{x}, \mu_{\overline{u}}^{-1}(\overline{y}))$. Since $\mu_{\overline{u}}$ is a linear isomorphism, $\tilde{\mu}$ is a bilinear map. Again, since $\mu_{\overline{u}}$ is an orthogonal transformation, $\| \tilde{\mu}(\overline{x}, \overline{y}) \| = \| \overline{x} \| \| \overline{y} \|$ for all $\overline{x}, \overline{y}$ in $\mathbb{R}^{n+1}$. Finally, $\tilde{\mu}(\overline{u}, \overline{y}) = \mu(\overline{u}, \mu_{\overline{u}}^{-1}(\overline{y})) = \mu_{\overline{u}}(\mu_{\overline{u}}^{-1}(\overline{y})) = \overline{y}$ for all $\overline{y}$ in $\mathbb{R}^{n+1}$. This shows that $\overline{u}$ is the identity for $\tilde{\mu}$.   ♯

Proposition 3.2.27 *Suppose that there is an orthogonal multiplication μ on $\mathbb{R}^{n+1}$. Then there exists an orthonormal set of tangent vector fields of S^n containing n elements.*

Proof In the light of the above proposition, we may assume that $\overline{e_{n+1}}$ is the identity of μ. For each i, $1 \leq i \leq n$, consider the map ϕ from S^n to itself given by $\phi_i(\overline{v}) = \mu(\overline{e_i}, \overline{v})$. Clearly, ϕ_i is continuous for each i. Further, since $\overline{e_{n+1}}$ is the identity of the multiplication,

$$\| (\mu(\overline{e_i}, \overline{v}) - \overline{v})) \|^2 = \| (\mu(\overline{e_i}, \overline{v}) - \mu(\overline{e_{n+1}}, \overline{v})) \|^2 = \| \mu(\overline{e_i} - \overline{e_{n+1}}, \overline{v}) \|^2 = 2.$$

Again, since $\mu(\overline{e_i}, \overline{v})$ and $\overline{v}$ are unit vectors, it follows that $< \phi_i(\overline{v}), \overline{v} > = 0$ for all $\overline{v} \in S^n$. This means that each ϕ_i is a unit tangent vector field on S^n. Next,

$$\| \phi_i(\overline{v}) - \phi_j(\overline{v}) \|^2 = \| (\mu(\overline{e_i}, \overline{v}) - \mu(\overline{e_j}, \overline{v})) \|^2 = \| \mu(\overline{e_i} - \overline{e_j}, \overline{v}) \|^2 = 2$$

for all $i \neq j$. Again, since $\phi_i(\overline{v})$ and $\phi_j(\overline{v})$ are unit vectors, it follows that $< \phi_i(\overline{v}), \phi_j(\overline{v}) > = 0$. This shows that $\{\phi_i \mid 1 \leq i \leq n\}$ is an orthonormal set of vector fields on S^n.   ♯

$\mathbb{R}^n$ together with the orthogonal multiplication with identity is also called a real normed algebra. It follows from the above result and Corollary 3.2.22 that if an orthogonal multiplication on $\mathbb{R}^n$ exists, then $n = 1$ or $n = 2m$ is even. We already have normed algebra structure on $\mathbb{R}$, $\mathbb{R}^2$, $\mathbb{R}^4$, and $\mathbb{R}^8$. They are the algebra of real numbers, the algebra of complex numbers, the division algebra of quaternions, and the algebra of octonions. Indeed, Adams in 1960 showed that the above mentioned normed algebras are all normed division algebras over the field of real numbers. Thus, on S^1 all maximal orthonormal sets of tangent vector fields are singletons.

Next, consider S^n. Suppose that $n + 1 = (2a + 1)2^b$, where $b = c + 4d$, a, b, c and d are integers with $0 \leq c \leq 3$. Let $\rho(n) = 2^c + 8d$. Again, Adams in 1962 (Vector fields on spheres, Ann of math 1962) established that a maximal orthonormal set of vector fields on S^n contains exactly $\rho(n) - 1$ elements. For example, if $n = 1$, then $c = 1$ and $d = 0$, and a maximal orthonormal set of vector fields contains exactly 1 element. Indeed, $z \mapsto \overline{z}$ and $z \mapsto -\overline{z}$ are the only tangent vector fields on S^1. If $n = 3$, then $\rho(4) = 4$, and hence a maximal orthonormal set of vector fields on S^3 contains 3 elements. Determine one such orthonormal set. For $n = 7$, again $\rho(8) = 8$, and so a maximal orthonormal set of vector fields on S^7 contains 7 elements.

A flow on a space X is a continuous action of the additive group $\mathbb{R}$ of real numbers. More explicitly, a flow on X is a continuous map ϕ from $X \times \mathbb{R}$ to X such that (i) $\phi(x, s + t) = \phi(\phi(x, s), t)$, and (ii) $\phi(x, 0) = x$ for all $x \in X$ and s, t in $\mathbb{R}$. Thus, on a locally compact Hausdorff space X, a flow is simply a continuous homomorphism from $\mathbb{R}$ to the group $Homeo(X)$ of homeomorphisms of X, where $Homeo(X)$ is equipped with the compact open topology. A point $x \in X$ is called a fixed point of the flow if $\phi(x, t) = x$ for all $t \in \mathbb{R}$.

Theorem 3.2.28 *Let X be a compact polyhedron whose Euler characteristic $\chi(X)$ is nonzero. Then any flow on X has a fixed point.*

Proof Let ϕ be a flow on X. Fix $t \in \mathbb{R}$. Then the map ϕ_t from X to X given by $\phi_t(x) = \phi(x, t)$ is a continuous map, and the map H from $X \times I$ to X given by $H(x, s) = \phi(x, st)$ is a homotopy from I_X to ϕ_t. In turn, $L(\phi_t, X) = L(I_X, X) = \chi(X) \neq 0$. By Lefschetz fixed-point theorem, ϕ_t has a fixed point for all $t \in \mathbb{R}$. Let F_n denote the fixed point set of $\phi_{\frac{1}{2^n}}$. Then F_n is a nonempty closed subset of X, and $F_{n+1} \subseteq F_n$ for all n. Let $F = \bigcap_n F_n$. Since X is compact, F is a nonempty closed set. Thus, $\phi_{\frac{1}{2^n}}$ fixes each point of F for each n. Since ϕ is an action, each point of F is fixed by $\phi_{\frac{m}{2^n}}$ for each m, n. Again, since the set of dyadic rational numbers forms a dense set, and the map $t \mapsto \phi_t$ is a continuous map, F is a fixed point set of the flow ϕ.   ♯

Jordan–Brouwer Separation Theorem

It appears evident that if we take a homeomorphic copy A of S^1 embedded in S^2, then $S^2 - A$ is decomposed into two connected components B and C whose common boundary is A. However, a rigorous mathematical proof of this fact is not so easy. It requires some amount of mathematical work. More generally, we have the following theorem.

Theorem 3.2.29 (Jordan–Brouwer separation theorem) *Let A be a copy of S^{n-1} embedded as a homeomorphic subspace of S^n. Then $S^n - A = B \bigcup C$, where B and C are connected components of $S^n - A$ such that A is the common boundary of B and C.*

We need some more results to establish this theorem.

Lemma 3.2.30 *Let A be a subspace of S^n which is homeomorphic to I^k, $0 \leq k \leq n$, where $I = [0, 1]$. Then the reduced homology $\tilde{H}_q(S^n - A) = 0$ for all q.*

Proof The proof is by the induction on k. If $k = 0$, then I^0 is a point, and so $S^n - A$ is homeomorphic to $\mathbb{R}^n$. Since $\mathbb{R}^n$ is contractible, $\tilde{H}^q(S^n - A) = 0$ for all q. Assume that the result is true for all $k, 0 \leq k < m, 1 \leq m \leq n$. We prove it for $m + 1$. Let ϕ be a homeomorphism from I^{m+1} to A. Suppose that $\tilde{H}_q(S^n - A) \neq 0$. Let a be a nonzero member of $\tilde{H}_q(S^n - A)$. We construct a sequence $\{[\alpha_r, \beta_r]\}$ of closed subintervals of $I = [0, 1]$ such that (i) $\alpha_r \leq \alpha_{r+1} \leq \beta_{r+1} \leq \beta_r$ for each r, (ii) $\beta_r - \alpha_r = \frac{1}{2^r}$, and (iii) $\tilde{H}_q(i_r)(a) \neq 0$, where i_r is the inclusion map from $S^n - A$ to $S^n - \phi(I^m \times [\alpha_r, \beta_r])$. We do it by the induction on r. Take $\alpha_0 = 0$ and $\beta_0 = 1$. Let $B = \phi([I^m \times [0, \frac{1}{2}])$ and $C = \phi([I^m \times [\frac{1}{2}, 1])$. Clearly, A, B, and C are closed subsets of S^n and $A = B \bigcup C$. Evidently, $S^n - A = (S^n - B) \bigcap (S^n - C)$, and $(S^n - B) \bigcup (S^n - C) = S^n - (B \bigcap C) = S^n - \phi(I^m \times \{\frac{1}{2}\})$. By the induction hypothesis, $\tilde{H}^q(S^n - (B \cap C)) = 0$ for all q. Since $S^n - B$ and $S^n - C$ are open, $\{S^n - B, S^n - C\}$ is an excisive couple, we have the corresponding Mayer–Vietoris exact sequence

$$\longrightarrow \tilde{H}_{q+1}(S^n - (B \cap C)) \longrightarrow \tilde{H}_q(S^n - A) \overset{(\tilde{H}_q(i_1), -\tilde{H}_q(i_1'))}{\to}$$
$$\tilde{H}_q(S^n - B) \oplus \tilde{H}_q(S^n - C) \overset{j_1 + j_1'}{\to} \tilde{H}_q(S^n - (B \cap C)) \longrightarrow \cdots,$$

where i_1 is the inclusion map from $S^n - A$ to $S^n - B$ and i_1' is the inclusion map from $S^n - A$ to $S^n - C$. Since the extreme terms are 0, $(\tilde{H}_q(i_1), -\tilde{H}_q(i_1'))$ is an isomorphism. Again, since $a \neq 0$, $\tilde{H}_q(i_1)(a) \neq 0$ or $\tilde{H}_q(i_1')(a) \neq 0$. If $\tilde{H}_q(i_1)(a) \neq 0$, take $\alpha_1 = 0$ and $\beta_1 = \frac{1}{2}$, and if not, take $\alpha_1 = \frac{1}{2}$ and $\beta_1 = 1$. Evidently, the required conditions are satisfied. Assume that we have already constructed the closed intervals $[\alpha_t, \beta_t]$ for all $t \leq r$ with the required conditions. We construct the closed interval $[\alpha_{r+1}, \beta_{r+1}]$ so that the required conditions are satisfied. Let A_t denote $\phi(I^m \times [\alpha_t, \beta_t])$, $t \leq r$. Let $D = \phi(I^m \times [\alpha_r, \alpha_r + \frac{1}{2^{r+1}}])$ and $E = [\alpha_r + \frac{1}{2^{r+1}}, \beta_r]$. Again, using the Mayer–Vietoris sequence for the excisive couple $\{S^n - D, S^n - E\}$, we obtain that $\tilde{H}_q(i_{r+1})(a) \neq 0$ or $\tilde{H}_q(i_{r+1}')(a) \neq 0$. If $\tilde{H}_q(i_{r+1})(a) \neq 0$, take $\alpha_{r+1} = \alpha_r$ and $\beta_{r+1} = \alpha_r + \frac{1}{2^{r+1}}$, and if not, take $\alpha_{r+1} = \alpha_r + \frac{1}{2^{r+1}}$ and $\beta_{r+1} = \beta_r$. It is clear that $\{[\alpha_t, \beta_t] \mid t \leq r + 1\}$ satisfies the required condition. By the induction, the construction of the sequence $\{[\alpha_m, \beta_m]\}$ with the required condition is complete. Since $(\beta_m - \alpha_m) = \frac{1}{2^m}$, $\overset{Lim}{\to} \alpha_m = \overset{Lim}{\to} \beta_m = \alpha(\text{say})$. Put $A_r = \phi([I^m \times [\alpha_r, \beta_r])$. Then we have the chain $A \supseteq A_1 \supseteq A_2 \supseteq \cdots \supseteq A_r \supseteq A_{r+1} \supseteq \cdots$ of the closed subsets of S^n such that $\bigcap_{r=1}^{\infty} A_r = \phi(I^m \times \{\alpha\})$ is isomorphic to I^m. Further, $S^n - A \supseteq S^n - A_1 \supseteq S^n - A_2 \supseteq \cdots \supseteq S^n - A_r \supseteq S^n - A_{r+1} \supseteq \cdots$ is a chain of open subsets of S^n such that $\bigcup_{r=1}^{\infty}(S^n - A_r) = S^n - \phi(I^m \times \{\alpha\})$. Since Δ^q is compact for each q, any singular q-chain σ in $S^n - \phi(I^m \times \{\alpha\})$ is a singular q-chain in $S^n - A_r$ for some r. It follows that $\tilde{H}_q(S^n - \phi(I^m \times \{\alpha\})) = Lim_r \tilde{H}_q(S^n - A_r)$ and the sequence $\tilde{H}_q(i_r)(a)$ represents a nonzero element of $\tilde{H}_q(S^n - \phi(I^m \times \{\alpha\}))$. This is a contradiction, since $\tilde{H}_q(S^n - \phi(I^m \times \{\alpha\})) = 0$. This shows that $\tilde{H}_q(S^n - A) = 0$.

♯

Corollary 3.2.31 *Let A be a subspace of S^n which is homeomorphic to S^k for some k, $0 \leq k \leq n - 1$. Then $\tilde{H}_q(S^n - A) = 0$ for all $q \neq n - k - 1$ and $\tilde{H}_{n-k-1}(S^n - A) \approx \mathbb{Z}$.*

Proof The proof is by the induction on k. Suppose $k = 0$. Then $A \approx S^0$ consists of two points. Then $S^n - A$ is of the same homotopy type as S^{n-1}. Since $\tilde{H}_q(S^{n-1}) = 0$ for all $q \neq n - 1$ and $\tilde{H}_{n-1}(S^{n-1}) \approx \mathbb{Z}$, the result follows for $k = 0$. Assume that the result holds for all $k < m \leq n - 1$. We prove it for m. Let ϕ be a homeomorphism from S^m to A. Let $A_+ = \phi(S_+^m)$ and $A_- = \phi(S_-^m)$, where S_+^m denotes the upper closed m-hemisphere and S_-^m denotes the lower closed m-hemisphere. Evidently, A_+, A_-, and A are closed subsets of S^n such that $A_+ \bigcup A_- = A$, $A_+ \bigcap A_- = \phi(S_+^m \bigcap S_-^m) \approx S^{m-1}$. Applying the Mayer–Vietoris sequence for the excisive couple $\{S^n - A_+, S^n - A_-\}$, we obtain the following exact sequence:

$$\longrightarrow \tilde{H}_{q+1}(S^n - A_+) \oplus \tilde{H}_{q+1}(S^n - A_-) \overset{j_1 + j_1'}{\to} \tilde{H}_{q+1}(S^n - (A_+ \bigcap A_-)) \overset{\partial_{q+1}}{\to}$$
$$\tilde{H}_q(S^n - A) \overset{(\tilde{H}_q(i_1), -\tilde{H}_q(i_1'))}{\to} \tilde{H}_q(S^n - A_+) \oplus \tilde{H}_q(S^n - A_-) \longrightarrow \cdots,$$

where i_1 is the inclusion map from $S^n - A$ to $S^n - A_+$ and i_1' is the inclusion map from $S^n - A$ to $S^n - A_-$. Since A_+ and A_- are isomorphic to I^m, it follows (Lemma 3.2.30) that extreme two terms are 0. Hence $\tilde{H}_{q+1}(S^n - (A_+ \bigcap A_-))$ is isomorphic to $\tilde{H}_q(S^n - A)$. Since $(A_+ \bigcap A_-))$ is isomorphic to S^{m-1}, by the induction assumption, it follows that $\tilde{H}_q(S^n - A) = 0$ for all $q \neq n - m - 1$ and $\tilde{H}_{n-m-1}(S^n - A) \approx \mathbb{Z}$.   ♯

Proof of the Jordan–Brouwer Separation theorem. Let A be a subspace of S^n which is homeomorphic to S^{n-1}. Then A is a closed subset of S^n, and from the above Corollary, $\tilde{H}_0(S^n - A) \approx \mathbb{Z}$ and $\tilde{H}_q(S^n - A) = 0$ for $q \neq 0$. Thus, $S^n - A = B \bigcup C$, where B and C are path components of $S^n - A$. Further, since $S^n - A$ is an open subset of S^n, it is locally path connected. Hence B and C are connected components of $S^n - A$. Finally, we need to show that $A = bd(B) = bd(C)$. Now, since B is open, $bd(B) = cl(B) \bigcap cl(S^n - B) \subseteq cl(B) - B$. Again, since C is open and $B \bigcap C = \emptyset$, $bd(B) \subseteq cl(B) - C$. Thus, $bd(B) \subseteq cl(B) - (B \bigcup C) \subseteq A$. Similarly, $bd(C) \subseteq A$. Finally, we need to show that $A \subseteq (cl(B) \bigcap cl(C))$. Let $x \in A$, and let U be an open neighborhood of x in S^n. We need to show that $U \bigcap cl(B) \neq \emptyset \neq U \bigcap cl(C)$. Since A is homeomorphic to S^{n-1} and $U \bigcap A$ is an open subset of A, there is a subset D of $U \bigcap A$ such that $A - D$ is homeomorphic to I^{n-1}. By Lemma 3.2.30, $\tilde{H}_q(S^n - (A - D)) = 0$ for all q. Hence $S^n - (A - D)$ is path connected. Let $b \in B$ and $c \in C$. Clearly, b, c both are in $S^n - (A - D)$. Let σ be a path in $S^n - (A - D)$ from b to c. Since b and c are in different path components B and C of $S^n - A$, $\sigma(I) \bigcap D \neq \emptyset$. This means that D contains a point of $cl(B)$ and also a point of $cl(C)$. In turn, $U \bigcap cl(B) \neq \emptyset \neq U \bigcap cl(C)$. Thus, $x \in cl(B) \bigcap cl(C)$.   ♯

Theorem 3.2.32 (Invariance of domain) *Let U be an open subset of S^n, and let V be a subspace of S^n which is homeomorphic to U. Then V is also an open subset of S^n.*

Proof Let ϕ be a homeomorphism from U to V. Let $y \in V$, and x be the unique element of U such that $\phi(x) = y$. Then there is a neighborhood A of x contained in U which is homeomorphic to I^n whose boundary B is homeomorphic to S^{n-1}. Evidently, $\phi(A)$ is homeomorphic to I^n and $\phi(B)$ is homeomorphic to S^{n-1}. From the above Jordan–Brouwer separation theorem, $S^n - \phi(B)$ has two path components. By Lemma 3.2.30, $S^n - \phi(A)$ is path connected. Evidently, $\phi(A) - \phi(B)$, being homeomorphic to the interior of I^n, is path connected. Since $S^n - \phi(B) = (S^n - \phi(A)) \bigcup (\phi(A) - \phi(B))$, it follows that $S^n - \phi(A)$ and $\phi(A) - \phi(B)$ are components of $S^n - \phi(B)$. Since $S^n - \phi(B)$ is open, $\phi(A) - \phi(B)$ is open. Clearly, $y \in \phi(A) - \phi(B) \subseteq V$. This shows that V is an open subset of S^n.   ♯

Corollary 3.2.33 *Let U be an open subset of $\mathbb{R}^n$ which is homeomorphic to a subspace V of $\mathbb{R}^n$. Then V is an open subset of $\mathbb{R}^n$.*

Proof $\mathbb{R}^n$ can be treated as an open subspace of its one point compactification S^n. Hence U is an open subspace of S^n and V is a subspace of S^n which is contained in $\mathbb{R}^n$ and which is homeomorphic to U. From the above theorem, V is an open subset of S^n which is contained in $\mathbb{R}^n$. This means that V is an open subset of $\mathbb{R}^n$. ♯

Corollary 3.2.34 *Let ϕ be an injective continuous map from S^n to itself. Then ϕ is a homeomorphism.*

Proof Since S^n is compact Hausdorff, $\phi(S^n)$ is a closed subset of S^n which is homeomorphic to S^n. From the earlier theorem, $\phi(S^n)$ is also open. Since S^n is connected, ϕ is surjective. ♯

Observe that the above corollary is not true for $\mathbb{R}^n$. However, it follows that a bijective continuous map from $\mathbb{R}^n$ to itself is a homeomorphism.

Borsuk–Ulam Theorem

Consider the Euclidean space $\mathbb{R}^{n+1}$. The map A from $\mathbb{R}^{n+1}$ to itself given by $A(\bar{x}) = -\bar{x}$ is called the antipodal map. Evidently, A is an orthogonal transformation on $\mathbb{R}^{n+1}$ of determinant $(-1)^{n+1}$. A subset X of $\mathbb{R}^{n+1}$ is said to be invariant under antipodal map A if $A(X) = X$. For example, S^n, D^{n+1}, and the cube $\dot{I}^n = \{\bar{x} \in \mathbb{R}^{n+1} \mid max\{| x_i |\} = 1\}$ are A-invariant subspaces of $\mathbb{R}^{n+1}$. Let X and Y be A-invariant subspaces of $\mathbb{R}^{n+1}$. A continuous map f from X to Y is called an **antipodes** preserving map (also called an odd map) if $f(A\bar{x}) = A(f(\bar{x}))$ for each $\bar{x} \in X$. Thus, the antipodal map A is an antipodes preserving map. The map f from $\dot{I}^n$ to S^n given by $f(\bar{x}) = \frac{\bar{x}}{|\bar{x}|}$ is an antipodes preserving map. If $m \le n$, then the inclusion map i from S^m to S^n given by $i(x_0, x_1, \ldots, x_m) = (x_0, x_1, \ldots, x_m, 0, 0, \ldots, 0)$ is antipodes preserving continuous map. However, we shall establish the theorem of Borsuk–Ulam which asserts that such a map from S^m to S^n for $m > n$ does not exist.

There are several equivalent formulations of the theorem of Borsuk–Ulam.

Theorem 3.2.35 *The following statements are equivalent:*

1. *If $m > n$, then there is no antipodes preserving continuous map from S^m to S^n.*
2. *There is no antipodes preserving continuous map from S^n to S^{n-1} for any $n \in \mathbb{N}$.*
3. *Given a natural number n and a continuous map f from S^n to $\mathbb{R}^n$, there is an element $\bar{x} \in S^n$ such that $f(-\bar{x}) = f(\bar{x})$.*
4. *Given a natural number n and a continuous antipodes preserving map f from S^n to $\mathbb{R}^n$, there is an element $\bar{x} \in S^n$ such that $f(\bar{x}) = \bar{0}$.*

Proof $(1 \Rightarrow 2)$. Evident.

$(2 \Rightarrow 1)$. Assume 2. Suppose that $m > n$ and f is a continuous antipodes preserving map from S^m to S^n. Let i be the inclusion map from S^{n+1} to S^m given by $i((x_0, x_1, \ldots, x_{n+1})) = (x_0, x_1, \ldots, x_{n+1}, 0, 0, \ldots, 0)$. Then i is an antipodes preserving continuous map, and so $f \circ i$ is an antipodes preserving continuous map from S^{n+1} to S^n. This is a contradiction to 2.

$(2 \Rightarrow 3)$. Assume 2. Let f be a continuous map from S^n to $\mathbb{R}^n$. Suppose that $f(-\bar{x}) \ne f(\bar{x})$ for all $\bar{x} \in S^n$. Then the map g from S^n to S^{n-1} given by

$$g(\bar{x}) = \frac{f(\bar{x}) - f(-\bar{x})}{\|\, f(\bar{x}) - f(-\bar{x})\,\|}$$

is such that $g(-\bar{x}) = -g(\bar{x})$ for all $\bar{x} \in S^n$. This is a contradiction to 2.

$(3 \Rightarrow 4)$. Assume 3. Let f be a continuous antipodes preserving continuous map from S^n to $\mathbb{R}^n$. From 3, there is an element $\bar{x} \in S^n$ such that $f(-\bar{x}) = f(\bar{x})$. Since f is antipodes preserving, $f(-\bar{x}) = -f(\bar{x})$. This means that $f(\bar{x}) = -f(\bar{x})$. Hence $f(\bar{x}) = \bar{0}$.

$(4 \Rightarrow 2)$. Assume 4. Let f be a continuous antipodes preserving continuous map from S^n to S^{n-1}. Then f can also be treated as a continuous antipodes preserving continuous map from S^n to $\mathbb{R}^n$. By 4, there is an element $\bar{x} \in S^n$ such that $f(\bar{x}) = 0$. This is a contradiction, since $\bar{0} \notin S^{n-1}$. ♯

We state the following theorem without proof. The proof can be found in the Algebraic topology by Spanier.

Theorem 3.2.36 *Let f be an antipodes preserving continuous map from S^n to itself. Then $deg(f)$ is an odd integer.* ♯

Theorem 3.2.37 (Borsuk–Ulam Theorem) *All the statements of Theorem 3.2.35 are true.*

Proof Since all the statements of Theorem 3.2.35 are equivalent, it is sufficient to prove 1. Let f be an antipodes preserving continuous map from S^m to S^n, where $m > n$. Let i denote the inclusion map from S^n to S^m. Then i is an antipodes preserving continuous map, and hence $i \circ f$ is an antipodes preserving continuous map from S^m to itself. Since $H_m(S^n) = 0$, $H_m(i) = 0$. This means that $H_m(f \circ i) = H_m(f) \circ H_m(i) = 0$. This is a contradiction, since $deg(f \circ i)$ is odd. ♯

Corollary 3.2.38 *S^n cannot be embedded in $\mathbb{R}^n$.*

Proof Let f be a continuous map from S^n to $\mathbb{R}^n$. From the Borsuk–Ulam theorem, and Theorem 3.2.35 (3), there exists $\bar{x} \in S^n$ such that $f(-\bar{x}) = f(\bar{x})$. Hence f cannot be injective. ♯

Following are some interesting and important consequences of the Borsuk–Ulam theorem.

1. Treat the Earth as a sphere S^2. Let $T(\bar{x})$ denote the temperature, and $P(\bar{x})$ denote the pressure at $\bar{x}$. Then the function f from S^2 to $\mathbb{R}^2$ defined by $f(\bar{x}) = (T(\bar{x}), P(\bar{x}))$ is a continuous function. From the theorem of Borsuk–Ulam, there are diametrically opposite two places $\bar{x}$ and $-\bar{x}$ where temperature and pressure are the same.

2. (Lusternik–Schnirelmann Theorem) Let $\{X_1, X_2, \ldots, X_n, X_{n+1}\}$ be a covering of S^n consisting of closed subsets of S^n. Then there is an element $\bar{x} \in S^n$ and an X_i such that both the antipodal points $\bar{x}$ and $-\bar{x}$ belong to X_i.

Proof Observe that for any subset A of S^n, the distance function $\bar{x} \mapsto d(\bar{x}, A)$ from S^n to $\mathbb{R}$ is a continuous function. Also A is closed if and only if $d(\bar{x}, A) = 0$ implies that $\bar{x} \in A$. Thus, the function f from S^n to $\mathbb{R}^n$ defined by $f(\bar{x}) = (d(\bar{x}, X_1), d(\bar{x}, X_2), \ldots, d(\bar{x}, X_n))$ is a continuous function. By the Borsuk–Ulam theorem, there is an element $\bar{x} \in S^n$ such that $f(-\bar{x}) = f(\bar{x})$. This means that $d(\bar{x}, X_i) = d(-\bar{x}, X_i)$ for all $i \leq n$. If $d(\bar{x}, X_i) = 0 = d(-\bar{x}, X_i)$ for some i, then $\bar{x}$ and $-\bar{x}$ belong to X_i. Suppose that $d(\bar{x}, X_i) > 0$ for all $i \leq n$, then $d(-\bar{x}, X_i) > 0$ for all $i \leq n$. This means that $\bar{x} \notin X_i$ and $-\bar{x} \notin X_i$ for each $i \leq n$. Since $\{X_1, X_2, \ldots, X_n, X_{n+1}\}$ is a covering of S^n, $\bar{x} \in X_{n+1}$ and $-\bar{x} \in X_{n+1}$.   ♯

Hurewicz Theorem, An Application of Spectral Sequence

Our aim is to use spectral sequence (Sect. 2.5) arguments to establish an important theorem due to Hurewicz which relates the fundamental groups and homology groups of a space.

Let X be a path connected space with a base point $x_0 \in X$. Recall (see Example 1.1.35) the loop space $\Omega(X, x_0)$ of all continuous loops in X around x_0. A path in $\Omega(X, x_0)$ from a loop σ to a loop τ is, in fact, a homotopy H from σ to τ. Let $\pi_1(X, x_0) = \pi_0(\Omega(X, x_0))$ denote the set of all path components of $\Omega(X, x_0)$. Thus, $\pi_1(X, x_0)$ is the set of homotopy classes of loops in X around x_0. A homotopy class of loops determined by σ will be denoted by $[\sigma]$. Let σ and τ be members of $\Omega(X, x_0)$. Define a map $\sigma \star \tau$ from I to X by putting $(\sigma \star \tau)(t) = 2t$ for $t \in [0, \frac{1}{2}]$ and $(\sigma \star \tau)(t) = 2t - 1$ for $t \in [\frac{1}{2}, 1]$. Clearly, $\sigma \star \tau \in \Omega(X, x_0)$. The notation $\sigma \approx \tau$ will mean that σ is homotopic to τ. If $\sigma \approx \sigma'$ and $\tau \approx \tau'$, then it can be seen easily that $\sigma \star \tau \approx \sigma' \star \tau'$. Thus, we have a product $\cdot$ in $\pi_1(X, x_0)$ given by $[\sigma] \cdot [\tau] = [\sigma \star \tau]$. It can be verified that $\pi_1(X, x_0)$ is a group with respect to this operation. The homotopy class $[e_0]$ is the identity, where e_0 is the constant loop given by $e_0(t) = x_0$ for all t. The inverse of $[\sigma]$ is $[\sigma^{-1}]$, where $\sigma^{-1}(t) = \sigma(1 - t)$. The group $\pi_1(X, x_0)$ is called the **first fundamental group** or the **homotopy group** of X based at x_0. Further, $\pi_1(\Omega(X, x_0), e_0)$ is called the second fundamental group of X based at x_0 and it is denoted by $\pi_2(X, x_0)$. Inductively, we can define all higher fundamental groups $\pi_n(X, x_0)$. It can be seen that $\pi_n(X, x_0)$ is abelian for all $n \geq 2$. Let $\sigma \in \Omega(X, x_0)$. Then σ is a 1-singular simplex in $S^1(X)$. Indeed, σ represents a 1-cycle and determines an element of $H_1(X, \mathbb{Z})$ which we denote by $\hat{\sigma}$. If $\sigma \approx \tau$, then it can be easily observed that $\hat{\sigma} = \hat{\tau}$. This defines a map χ from $\pi_1(X, x_0)$ to $H_1(X, \mathbb{Z})$. Since X is path connected, χ is surjective. It can also be shown that χ is a homomorphism whose kernel is the commutator $[\pi_1(X, x_0), \pi_1(X, x_0)]$ of $\pi_1(X, x_0)$ (see Algebraic Topology by Greenberg or Algebraic Topology by Spanier for details). Thus, $H_1(X, \mathbb{Z})$ is naturally isomorphic to the abelianizer of the fundamental group $\pi_1(X, x_0)$. In particular, if $\pi_1(X, x_0)$ is abelian, then $\pi_1(X, x_0) \approx H_1(X, x_0)$. If X is path connected and $\pi_1(X, x_0) = \{0\} = \pi_0(\Omega(X, x_0))$, then $\Omega(X, x_0)$ is path connected. In turn, $\pi_2(X, x_0) = \pi_1(\Omega(X, x_0), e_0) \approx H_1(\Omega(X, x_0), \mathbb{Z})$.

A continuous map $E \xrightarrow{p} B$ is called a **Hurewicz Fibration** if it has the homotopy lifting property with respect to any space X in the following sense: "Given a homotopy H from $X \times I$ to B and a continuous map f' from X to E such that $H(x, 0) = pof'(x)$ for each $x \in X$, there is a continuous map $\hat{H}$ from $X \times I$ to E such that $\hat{H}(x, 0) = f'(x)$ for each x and $po\hat{H} = H$".

A continuous map $E \xrightarrow{p} B$ is called a **Weak Fibration** or a **Serre fibration** if it has the homotopy lifting property with respect to any cube I^n, $n \geq 0$. Thus, a Hurewicz fibration is a Serre fibration. For an element, $b \in B$, $p^{-1}(b)$ is called the fiber over b.

Example 3.2.39 1. The exponential map exp from $\mathbb{R}$ to S^1 given by $exp(x) = e^{2\pi i x}$ is a fibration. Indeed, it is a covering projection.

 2. $SO(n + 1)$ acts transitively on S^n in an obvious manner. The map p from $SO(n + 1)$ to S^n given by $p(A) = e_{n+1}A$ is a continuous map which is a weak fibration. The fiber $p^{-1}(e_{n+1}) \approx SO(n)$.

 3. Let X be a path connected space and $x_0 \in X$. Let $P(X, x_0)$ denote the space of all paths with initial point x_0. The map p from $P(X, x_0)$ to X defined by $p(\sigma) = \sigma(1)$ is a fibration with fiber $p^{-1}(x_0) = \Omega(X, x_0)$. This fibration is called the **path space fibration**.

We state (without proof) the following theorem due to Serre. The proof can be found in the "Algebraic Topology" by Spanier.

Theorem 3.2.40 (Serre) *Let $E \xrightarrow{p} B$ be a Serre fibration with fiber $F = p^{-1}(b)$. Suppose that $\pi_1(B) = \{0\} = \pi_0(F)$. Then there is a first quadrant spectral sequence $\{(E^r, d^r) \mid r \geq 2\}$ ($E^r_{p,q} = 0$ whenever $p < 0$ or $q < 0$) converging to E^∞ such that $E^2_{p,q} = H_p(B, H_q(F))$ for all p, q and for each n, there is a descending chain*

$$H_n(E) = A_{n,o} \supseteq A_{n-1,1} \supseteq \cdots \supseteq A_{0,n} \supseteq A_{-1,n+1} = \{0\}$$

of abelian groups with $E^\infty_{p,q} = A_{p,q}/A_{p-1,q+1}$ for each p, q. In particular, $H_n(E) = \{0\}$ if and only if $E^\infty_{p,q} = 0$ for all p, q with $p + q = n$. Further, if we take the coefficients of the homologies in a field, then the above chain is a chain of vector spaces and so $H_n(E) = \oplus \sum_{p+q=n} E^\infty_{p,q}$.

A path connected space X is said to be an n-connected space if $\pi_r(X, x_0) = 0$ for all $r \leq n$. Using the above theorem of Serre, we shall establish the following important theorem due to Hurewicz.

Theorem 3.2.41 (Hurewicz) *Let X be an n-connected space, $n \geq 1$. Then the reduced homology group $\hat{H}_r(X) = 0$ for all $r \leq n$ and $\pi_{n+1}(X, x_0) \approx H_{n+1}(X) = H_{n+1}(X, x_0)$.*

Proof The proof is by the induction on n. Assume that $n = 1$. By the hypothesis, $0 = \pi_0(X, x_0)$ and also $0 = \pi_1(X, x_0) = \pi_0(\Omega(X, x_0), e_0)$. This means that $\hat{H}_0(X) = 0$ and also $H_1(X, x_0) = \pi_1(X, x_0)/[\pi_1(X, x_0), \pi_1(X, x_0)] = 0$. Also $\Omega(X, x_0)$ is path connected. Now, since $\pi_2(X, x_0)$ is abelian,

$$\pi_2(X, x_0) = \pi_1(\Omega(X, x_0), e_0) \approx H_1(\Omega(X)).$$

Thus, it remains to show that $H_2(X) \approx H_1(\Omega(X))$. Consider the path space fibration $(P(X, x_0), e_0) \xrightarrow{p} (X, x_0)$, where $p(\sigma) = \sigma(1)$. The fiber is $p^{-1}(x_0) = \Omega(X, x_0)$. The fibration p satisfies the hypothesis of Theorem 3.2.40. Hence, we have a first quadrant spectral sequence $\{(E^r, d^r) \mid r \geq 2\}$ ($E_{p,q}^r = 0$ whenever $p < 0$ or $q < 0$) such that $E_{p,q}^2 = H_p(X, H_q(\Omega(X, x_0)))$ for all p, q and for each n, we have a finite descending chain

$$H_n(P(X, x_0)) = A_{n,0} \supseteq A_{n-1,1} \supseteq \cdots \supseteq A_{0,n} \supseteq A_{-1,n-1} = 0,$$

where $E_{p,q}^\infty = A_{p,q}/A_{p-1,q+1}$ for all p, q. The path space $P(X, x_0)$ is contractible, since the map H from $P(X, x_0) \times I$ to $P(X, x_0)$ given by $H(\sigma, s)(t) = \sigma(st)$ is a homotopy from the constant map $\sigma \mapsto e_0$ to the identity map $I_{P(X,x_0)}$. Consequently, $H_n(P(X, x_0)) = 0$ for all $n > 0$. This means that $E_{p,q}^\infty = 0$ for all p, q with $p, q > 0$. Now,

$$H_2(X, \mathbb{Z}) = E_{2,0}^2 \xrightarrow{d_{2,0}^2} E_{0,1}^2 = H_0(X, H_1(\Omega(X, x_0))) = H_1(\Omega(X, x_0)) =$$
$$\pi_1(\Omega(X, x_0)) = \pi_2(X, x_0).$$

It is sufficient, therefore, to show that $d_{2,0}^2$ is an isomorphism. Suppose it is contrary. Since it is a first quadrant spectral sequence, $E_{0,1}^3 \neq 0$ and $E_{2,0}^3 \neq 0$. Since $d^r = 0$ for $r \geq 3$, $E_{0,1}^\infty = E_{0,1}^3 \neq 0$ and $E_{2,0}^\infty = E_{2,0}^3 \neq 0$. This is a contradiction to the observed fact that $E_{p,q}^\infty = 0$ for $p, q > 0$. This establishes the result for $n = 1$.

Assume the result for $n \geq 1$. Let X be an $n + 1$-connected space. Then $\Omega(X, x_0)$ is n-connected. By the induction assumption, $\hat{H}_r(\Omega(X, c_0)) = 0$ for all $r \leq n$ and $\pi_{n+1}(\Omega(X, x_0), e_0) \approx H_{n+1}(\Omega(X, x_0))$. Now, $\pi_{n+2}(X, x_0) \approx \pi_{n+1}(\Omega(X, x_0), e_0) \approx H_{n+1}(\Omega(X, x_0))$. It is sufficient, therefore, to show that $H_{n+1}(\Omega(X, x_0)) \approx H_{n+2}(X)$. By the induction hypothesis, $H_q(\Omega(X, x_0)) = 0$ for all $0 < q < n + 1$. Consequently, by the universal coefficient theorem for homology (Corollary 2.4.6),

$$E_{p,q}^2 = H_p(X, H_q(\Omega(X, x_0))) \approx$$
$$H_p(X) \oplus H_q(\Omega(X, x_0)) \oplus Tor(H_{p-1}(X), H_q(\Omega(X, x_0))) = 0$$

for all $0 < q < n + 1$. Thus, the differentials $d^r, 2 \leq r \leq n$, have no impact on p axis for all $p \leq n + 1$. Further, d^{n+1} also does not disturb the entries. Since $P(X, x_0)$ is contractible, $E_{p,q}^\infty = 0$ except when $p = 0 = q$. This means that $H_r(X) = 0$ for all r, $1 \leq r \leq n + 1$ and consequently $H_{n+2}(X) \xrightarrow{d^{n+2}} H_{n+1}(\Omega(X, x_0))$ is an isomorphism. This completes the proof. ♯

Exercises

3.2.1 Show that S^2 is not of the same homotopy type as $S^1 \times S^1$.

3.2.2 Find $\chi(S^1 \times S^1 \times S^1)$.

3.2.3 Find $\chi(S^1 \vee_p S^1)$.

3.2.4 Let X be a compact polyhedron, and f be a map from X to itself which is homotopic to a constant map. Show that f has a fixed point.

3.2.5 Let G be a path connected Hausdorff topological group. Let $g \in G$. Show that the left multiplication L_g by g is homotopic to the identity map. Hint: Consider the map H from $G \times [0, 1]$ to G defined by $H(x, t) = \sigma_g(t)x$, where σ_g is a path from e to g.

3.2.6 Let G be a Compact group which is a polyhedron. Show that $\chi(G) = 0$. Deduce that if S^n has a topological group structure, then n is odd. In particular, S^2 cannot be given a topological group structure.

3.2.7 Let A be an $n \times n$ orthogonal matrix. Treating A as a map from S^{n-1} to itself, find its degree, $n \geq 2$.

3.3 Co-homology of Groups

The origin of the co-homology theory of groups lies in the work of Hurewicz in Topology. Around the mid-1930s, Hurewicz introduced the concept of higher homotopy groups $\pi_n(X)$, $n \geq 2$ of spaces, and studied path connected CW complexes all of whose higher homotopy groups are trivial. Such spaces are called **aspherical spaces**. He established that two aspherical spaces X and Y are of the same homotopy type if and only if $\pi_1(X) \approx \pi_1(Y)$. With the work of Eilenberg–MacLane, it also became clear that for any group π, there is a path connected CW complex $K(\pi, 1)$ (unique up to homotopy) called an Eilenberg–MacLane space which is aspherical with a fundamental group π. Indeed, K defines a faithful functor from the category of groups to the category of homotopy classes of aspherical spaces. Thus, the homology and co-homology theory of spherical spaces is basically the homology and co-homology theory of groups. We shall denote $H_q(K(\pi, 1), A)$ by $H_q(\pi, A)$ and $H^q(K(\pi, 1), A)$ by $H^q(\pi, A)$. Evidently, $H_0(\pi) = \pi$ and $H_1(\pi) = \pi_{ab}$. For $q \geq 2$, the algebraic description of $H_q(\pi, A)$ and $H^q(\pi, A)$ is by no means evident. Hopf was the first to give an algebraic description of $H_2(\pi)$ as $H_2(\pi) = \frac{R \cap [F,F]}{[R,F]}$, where

$$1 \longrightarrow R \longrightarrow F \longrightarrow \pi \longrightarrow 1$$

is a presentation of π (see also Chap. 10, Algebra 2). Finally, Eilenberg and MacLane gave an algebraic description of homologies and co-homologies of groups during the mid-1940s. Interestingly, the low-dimensional homologies and co-homologies agreed with those given by Schur, Schreier, and Brauer in terms of crossed homomorphisms and factor systems (see Chap. 10, Algebra 2). With that, co-homology theory of groups became an important object of study having its application in algebra, number theory, and topology. This section is devoted to a very brief description of co-homology theory of groups, and we shall use the algebraic definition of co-homology groups. For group theoretic interpretations of low-dimensional homology and co-homology groups, we refer to Chap. 10 of Algebra 2.

Let G be a group. Consider the integral group ring $\mathbb{Z}(G)$. Let A be a $\mathbb{Z}(G)$-module. By the definition, the nth homology group of G with coefficient in A is given by $H_n(G, A) = Tor_n^{\mathbb{Z}(G)}(\mathbb{Z}, A)$, and the nth co-homology with coefficient in A is given by $H^n(G, A) = EXT_{\mathbb{Z}(G)}^n(\mathbb{Z}, A)$, where $\mathbb{Z}$ is treated as a trivial G-module.

Let G be a group. A left G-module is an abelian group $(A, +)$ together with an external multiplication $\cdot$ of G on A such that (i) $e \cdot a = a$, (ii) $g \cdot (a + b) = g \cdot a + g \cdot b$, and $(gh) \cdot a = g \cdot (h \cdot a)$ for all $g, h \in G$ and $a, b \in A$ (here e denotes the identity of the group G). Thus, a left G-module $(A, +)$ is essentially a homomorphism ϕ from G to $Aut(A, +)$ given by $\phi(g)(a) = g \cdot a$, $g \in G$ and $a \in A$. Similarly, a right G-module can be defined. Further, a left G-module structure on an abelian group $(A, +)$ is essentially a left $\mathbb{Z}(G)$-module structure on $(A, +)$ given by the relation $\left(\sum_{g \in G} \alpha_g g \right) \cdot a = \sum_{g \in G} (\alpha_g (g \cdot a))$ (observe that all the sums are finite). In turn, a $\mathbb{Z}(G)$-module $(A, +)$ is essentially a ring homomorphism ϕ from $\mathbb{Z}(G)$ to the ring $End(A, +)$ such that $\phi(G) \subseteq Aut(A, +)$. Every abelian group $(A, +)$ is a $\mathbb{Z}(G)$-module given by $\left(\sum_{g \in G} \alpha_g g \right) \cdot a = \sum_{g \in G} (\alpha_g) a$. This $\mathbb{Z}(G)$-module A is called the trivial $\mathbb{Z}(G)$-module structure on A. In particular, the trivial $\mathbb{Z}(G)$-module structure on $\mathbb{Z}$ is given by $\left(\sum_{g \in G} \alpha_g g \right) \cdot n = \sum_{g \in G} (\alpha_g) n$ for all $n \in \mathbb{Z}$. Consider the map ϵ from $\mathbb{Z}(G)$ to $\mathbb{Z}$ given by $\epsilon \left(\sum_{g \in G} \alpha_g g \right) = \sum_{g \in G} \alpha_g$. This map is called the augmentation map. Treating $\mathbb{Z}(G)$ as $\mathbb{Z}(G)$-module, it is easily observed that the augmentation map ϵ is $\mathbb{Z}(G)$ homomorphism from $\mathbb{Z}(G)$ to $\mathbb{Z}$, and indeed, it is also a ring homomorphism. The kernel $ker\epsilon$ of ϵ is a two-sided ideal of $\mathbb{Z}(G)$. It is denoted by $I(G)$, and it is called the **augmentation ideal** of G. Thus, $I(G) = \left\{ \sum_{g \in G} \alpha_g g \mid \sum_{g \in G} \alpha_g = 0 \right\}$.

Let us describe $H_*(-, -)$ and $H^*(-, -)$ as bi-functors. Let G and G' be groups, and α be a group homomorphism from G to G'. Let B be a G'-module. Then B can also be treated as a G-module by defining $g \cdot b = \alpha(g) \cdot b$. Let Ω denote the category whose objects are pairs (G, A), where G is a group and A is a G-module. A morphism from a pair (G, A) to (G', A') is a pair (α, f), where α is a group homomorphism from G to G' and f is a group homomorphism from A to A' satisfying $f(g \cdot a) = \alpha(g) \cdot f(a)$. More explicitly, f is a G-homomorphism, where A' is treated as a G-module through the homomorphism α. Let (α, f) be a morphism in Ω from (G, A) to (G', A'). Let F be a G-free resolution of the trivial G-module $\mathbb{Z}$, and let F' be a G' free resolution of the trivial G'-module $\mathbb{Z}$. Then F' can be treated as a chain complex of G-modules through the homomorphism α. Evidently, F' treated as a chain complex of G-modules is acyclic and so it is a resolution (it may not be projective) over the trivial G-module $\mathbb{Z}$. From Theorem 2.1.2, the identity $I_\mathbb{Z}$ can be lifted to a chain transformation $\tilde{\alpha} = \{\alpha_n \mid n \geq 0\}$ from F to F'. Also, f is a G-homomorphism when A' is also treated as a G-module. Hence $\alpha \otimes_G f = \{\alpha_n \otimes f \mid n \geq 0\}$ is a chain transformation from the chain complex $F \otimes_G A$ to the chain complex $F' \otimes_{G'} A'$ of abelian groups. In turn, for each n, we have a homomorphism $H_n(\alpha, f) = H_n(\alpha \otimes_G f)$ from $H_n(G, A)$ to $H^n(G', A')$. It can be easily observed that $H_n(I_G, I_A) = I_{H_n(G,A)}$, and $H_n(\beta o\alpha, gof) = H_n(\beta, g)oH_n(\alpha, f)$. This introduces the bi-functor $H_\star(-, -)$. Similarly, we can introduce the contra-variant bi-functor $H^\star(-, -)$.

Proposition 3.3.1 *(i)* $(I(G), +)$ *is a free abelian group with basis* $S = \{g - 1 \mid g \in G - \{1\}\}$.

(ii) If X is a set of generators of the group G, then $\{1 - x \mid x \in X\}$ generates $I(G)$ as a G-module.

Proof (i) Evidently, S is $\mathbb{Z}$-linearly independent. We need to show that S generates $I(G)$ as a group. Let $\left(\sum_{g \in G} \alpha_g g\right)$ be a member of $I(G)$. Then $\sum_{g \in G} \alpha_g = 1$. Hence $\sum_{g \in G} \alpha_g g = \sum_{g \in G} \alpha_g(g - 1)$. This shows that S generates $(I(G), +)$.

(ii) We need to show that $g - 1 \in\ <X>$, where $<X>$ denote the submodule generated by X. For each $x \in X$, $x^{-1} - 1 = -(x^{-1}(x - 1))$. Hence $x^{-1} - 1 \in\ <X>$ for each $x \in X$. Thus, we may assume that X is closed under inversion. Every element g of G is expressible as $g = x_1 x_2 \cdots x_r$. We prove the assertion by the induction on r. If $r = 1$, it is evident. Assume the result for r. Suppose that $g = x_1 x_2 \cdots x_r x_{r+1}$. Then $g - 1 = x_1 x_2 \cdots x_r x_{r+1} - 1 = x_1(x_2 x_3 \cdots x_{r+1} - 1) + (x_1 - 1)$. By the induction assumption, $x_2 x_3 \cdots x_{r+1} - 1 \in\ <X>$. Hence $g - 1 \in\ <X>$.   ♯

Proposition 3.3.2 *Let H be a subgroup of G. Then $(\mathbb{Z}(G), +)$ is a free $\mathbb{Z}(H)$-module of rank $[G, H]$.*

Proof Given $x \in G$, Hx is a left $\mathbb{Z}(H)$-submodule of $\mathbb{Z}(G)$ which is isomorphic to $\mathbb{Z}(H)$ considered as a $\mathbb{Z}(H)$-module. If S is a right transversal to H in G, then $\mathbb{Z}(G)$ considered as a $\mathbb{Z}(H)$-module is the direct sum $\oplus \sum_{x \in S} Hx$. The result follows.   ♯

Corollary 3.3.3 *Every projective $\mathbb{Z}(G)$-module is also a projective $\mathbb{Z}(H)$-module.* ♯

Low Dimensional Homology and Co-homology

Proposition 3.3.4 $H_0(G, A) \approx A/I(G)A$.

Proof By definition, $H_0(G, A) = Tor_0^{\mathbb{Z}(G)}(\mathbb{Z}, A)$. From Proposition 2.2.28, $Tor_0^{\mathbb{Z}(G)}$ $(\mathbb{Z}, A) = \mathbb{Z} \otimes_{\mathbb{Z}(G)} A$. We have the short exact sequence

$$0 \longrightarrow I(G) \longrightarrow \mathbb{Z}(G) \overset{\epsilon}{\to} \mathbb{Z} \longrightarrow 0.$$

Since tensoring is a right exact functor, we have the exact sequence

$$I(G) \otimes_{\mathbb{Z}(G)} A \overset{i \otimes I_A}{\to} \mathbb{Z}(G) \otimes_{\mathbb{Z}(G)} A \overset{\epsilon \otimes I_A}{\to} \mathbb{Z} \otimes_{\mathbb{Z}(G)} A \longrightarrow 0.$$

It follows that $\mathbb{Z} \otimes_{\mathbb{Z}(G)} A$ is isomorphic to $(\mathbb{Z}(G) \otimes_{\mathbb{Z}(G)} A)/imagei \otimes I_A$. Further, we have the isomorphism ϕ from $\mathbb{Z}(G) \otimes_{\mathbb{Z}(G)} A$ to A give by $\phi(r \otimes a) = ra$. Thus, $\mathbb{Z} \otimes_{\mathbb{Z}(G)} A$ is isomorphic to $A/I(G)A$. This shows $H_0(G, A)$ is isomorphic to $A/I(G)A$.   ♯

In particular, if A is a trivial G-module, then $H_0(G, A) \approx A$.

Proposition 3.3.5 $H^0(G, A)$ *is isomorphic to the submodule* $A^G = \{a \in A \mid ga = a \text{ for all } g \in G\}$ *of* A.

Proof By definition, $H^0(G, A) = EXT^0_{\mathbb{Z}(G)}(\mathbb{Z}, A)$. By Proposition 2.1.6, $EXT^0_{\mathbb{Z}(G)}$ $(\mathbb{Z}, A)$ is isomorphic to $Hom_{\mathbb{Z}(G)}(\mathbb{Z}, A)$. Define a map χ from $Hom_{\mathbb{Z}(G)}(\mathbb{Z}, A)$ to A by $\chi(\alpha) = \alpha(1)$. Evidently, χ is an injective homomorphism. Further, if $\alpha \in Hom_{\mathbb{Z}(G)}(\mathbb{Z}, A)$, then $g\alpha(1) = \alpha(g1) = \alpha(1)$. Thus, image $\chi \subseteq \{a \in A \mid ga = a \text{ for all } g \in G\}$. Also, if $ga = a$ for all $g \in G$, then the map α from $\mathbb{Z}$ to A given by $\alpha(n) = na$ is easily seen to be a member of $Hom_{\mathbb{Z}(G)}(\mathbb{Z}, A)$ such that $\chi(\alpha) = a$ This shows that $H^0(G, A) \approx A^G$.   ♯

In particular, if A is a trivial G-module, then $H^0(G, A) \approx A$.

Proposition 3.3.6 $H_1(G, A)$ *is isomorphic to the kernel of* $i_\star$, *where* $i_\star$ *is the homomorphism from* $I(G) \otimes_{\mathbb{Z}(G)} A$ *to* A *given by* $i_\star(r \otimes a) = ra$, $r \in I(G)$, $a \in A$.

Proof By definition, $H_1(G, A) = Tor^{\mathbb{Z}(G)}(\mathbb{Z}, A)$. Consider the short exact sequence

$$0 \longrightarrow I(G) \overset{i}{\to} \mathbb{Z}(G) \overset{\epsilon}{\to} \mathbb{Z} \longrightarrow 0.$$

Since $\mathbb{Z}(G)$ is a free $\mathbb{Z}(G)$-module, $Tor^{\mathbb{Z}(G)}(\mathbb{Z}(G), A) = 0$. Now, taking the tensor product by A from right, and using Theorem 2.2.26, we get the exact sequence

$$0 \longrightarrow Tor^{\mathbb{Z}(G)}(\mathbb{Z}, A) \overset{\partial}{\to} I(G) \otimes_{\mathbb{Z}(G)} A \overset{i \otimes I_A}{\to} \mathbb{Z}(G) \otimes_{\mathbb{Z}(G)} A \overset{\epsilon \otimes I_A}{\to} 0.$$

The result follows if we observe that the map ϕ from $\mathbb{Z}(G) \otimes_{\mathbb{Z}(G)} A$ to A given by $\phi(r \otimes a) = ra$ is an isomorphism.   ♯

Corollary 3.3.7 *If* A *is a trivial* G-module, then $H_1(G, A)$ *is isomorphic to* $I(G)/I(G)^2 \otimes_{\mathbb{Z}(G)} A$.

Proof If the action of G on A is trivial, then $r \cdot a = 0$ for all $r \in I(G)$ and $a \in A$. Further, we have the exact sequence

$$0 \longrightarrow I(G)^2 \overset{i'}{\to} I(G) \overset{\nu}{\to} I(G)/I(G)^2 \longrightarrow 0$$

of $\mathbb{Z}(G)$-modules, where i' is the inclusion map. Tensoring with A, we get the exact sequence

$$I(G)^2 \otimes_{\mathbb{Z}(G)} A \overset{i_\star}{\to} I(G) \otimes_{\mathbb{Z}(G)} A \overset{\nu_\star}{\to} I(G)/I(G)^2 \otimes_{\mathbb{Z}(G)} A \longrightarrow 0$$

of groups. Now, in $I(G) \otimes_{\mathbb{Z}(G)} A$,

$$(x - 1)(y - 1) \otimes a = (x - 1)y \otimes a - (x - 1) \otimes a = (x - 1) \otimes ya - (x - 1) \otimes a = 0.$$

This shows that $i_\star$ is zero map. The result follows from the Proposition 3.3.6.   ♯

Proposition 3.3.8 *The group $I(G)/I(G)^2$ is isomorphic to G_{ab}.*

Proof By Proposition 3.3.1(i), $I(G)$ is a free abelian group on the set $S = \{g - 1 \mid g \in G - \{1\}\}$. Hence we have a unique group homomorphism ϕ from $I(G)$ to G_{ab} given by $\phi(g - 1) = g[G, G]$. Since $(g - 1)(h - 1) = (gh - 1) - (g - 1) - (h - 1)$, it follows that $\phi(g - 1)(h - 1) = 0$ for all $g, h \in G$. Thus, ϕ induces a homomorphism $\bar{\phi}$ from $I(G)/I(G)^2$ to G_{ab}. Define a map ψ from G to $I(G)/I(G)^2$ by $\psi(g) = (g - 1) + I(G)^2$. Again, since $(g - 1)(h - 1) = (gh - 1) - (g - 1) - (h - 1)$, it follows that ψ is a homomorphism. Further, since $I(G)/I(G)^2$ is an abelian group, ψ induces a homomorphism $\bar{\psi}$ from G_{ab} to $I(G)/I(G)^2$ which is the inverse of $\bar{\phi}$.   ♯

Corollary 3.3.9 *If A is a trivial G-module, then $H_1(G, A)$ is isomorphic to $G_{ab} \otimes A$. In particular, $H_1(G, \mathbb{Z})$ is G_{ab}.*   ♯

Now, we look at $H^1(G, A)$. By definition, $H^1(G, A) = EXT_{\mathbb{Z}(G)}(\mathbb{Z}, A)$. We have the exact sequence

$$0 \longrightarrow I(G) \overset{i}{\to} \mathbb{Z}(G) \overset{\epsilon}{\to} \mathbb{Z} \longrightarrow 0$$

of $\mathbb{Z}(G)$-modules. Applying the functor $Hom_{\mathbb{Z}(G)}(-, A)$, using Theorem 1.4.12, and observing that $EXT_{\mathbb{Z}(G)}(\mathbb{Z}(G), A) = 0$, we obtain the exact sequence

$$0 \longrightarrow Hom_{\mathbb{Z}(G)}(\mathbb{Z}, A) \overset{\epsilon^*}{\to} Hom_{\mathbb{Z}(G)}(\mathbb{Z}(G), A) \overset{i^*}{\to} Hom_{\mathbb{Z}(G)}(I(G), A) \overset{\partial}{\to}$$
$$EXT_{\mathbb{Z}(G)}(\mathbb{Z}, A) \longrightarrow 0.$$

Hence $\quad EXT_{\mathbb{Z}(G)}(\mathbb{Z}, A) \approx Hom_{\mathbb{Z}(G)}(I(G), A)/Ker\partial = Hom_{\mathbb{Z}(G)}(I(G), A)/$ *image* i^*. Evidently, $Hom_{\mathbb{Z}(G)}(\mathbb{Z}(G), A) \approx A$ through the map $f \mapsto f(1)$. Now, we interpret $Hom_{\mathbb{Z}(G)}(I(G), A)$. Define a map η from $Hom_{\mathbb{Z}(G)}(I(G), A)$ to the set $Map(G, A)$ of all maps from G to A by $\eta(\phi)(g) = \phi(g - 1)$. Since $I(G)$ is generated by $S = \{g - 1 \mid g \in G - \{1\}\}$, η is an injective map. Further,

$$\eta(\phi)(gh) = \phi(gh - 1) = \phi(g(h - 1) + g - 1) = \phi(g - 1) + g\phi(h - 1) = g\eta(\phi(h)) + \eta(\phi)(g)$$

for all $g, h \in G$. Thus, $\eta(\phi)$ is a crossed homomorphism (see Algebra 2, def 8.6.17) from G to A in the sense of the following definition.

Definition 3.3.10 Let G be a group, and A be a G-module. A map χ from G to A is called a **crossed homomorphism** or a **derivation** if

$$\chi(gh) = g\chi(h) + \chi(g)$$

for all $g, h \in G$.

If A is a trivial G-module, then a crossed homomorphism is precisely a homomorphism. Further, if A is a G-module (not necessarily trivial) and $a \in A$, then the map χ_a from G to A given by $\chi_a(g) = ga - a$ can easily be seen to be a crossed homomorphism. This crossed homomorphism is called the **inner crossed homomorphism** or the **inner derivation** determined by a. The set $Cross(G, A)$ of all crossed homomorphisms from G to A is a group under the usual pointwise addition, and the subset $Icross(G, A)$ of all inner crossed homomorphisms form a subgroup of $Cross(G, A)$.

Theorem 3.3.11 $H^1(G, A)$ *is isomorphic to* $Cross(G, A)/Icross(G, A)$.

Proof It follows from the above discussion that $H^1(G, A)$ is isomorphic to $Hom_{\mathbb{Z}(G)}(I(G), A)/image\ i^*$. Thus, it is sufficient to show that $Hom_{\mathbb{Z}(G)}(I(G), A)/image\ i^* \approx Cross(G, A)/Icross(G, A)$. Further, it also follows from the above discussion that the map η from $Hom_{\mathbb{Z}(G)}(I(G), A)$ to $Cross(G, A)$ defined by $\eta(\phi)(g) = \phi(g - 1)$ is an injective homomorphism. Let χ be a member of $Cross(G, A)$. Since $I(G)$ is a free abelian group on $S = \{g - 1 \mid g \in G - \{1\}\}$, we have a group homomorphism ϕ from $I(G)$ to A given by $\phi(g - 1) = \chi(g)$. Now

$$\phi(g(h - 1)) = \phi(gh - 1 - (g - 1)) = \phi(gh - 1) - \phi(g - 1) =$$
$$\chi(gh) - \chi(g) = g\chi(h) = g\phi(h - 1).$$

This shows that $\phi \in Hom_{\mathbb{Z}(G)}(I(G), A)$ such that $\eta(\phi) = \chi$. In turn, it follows that η is an isomorphism from $Hom_{\mathbb{Z}(G)}(I(G), A)$ to $Cross(G, A)$. Let ϕ be a member of $image\ i^*$. Then ϕ is the restriction of a G-homomorphism ψ from $\mathbb{Z}(G)$ to A. The homomorphism ψ is uniquely determined by $\psi(1) = a \in A$. But then $\phi(g - 1) = \psi(g1) - \psi(1) = ga - a$. This shows that $\eta(image\ i^*) = Icross(G, A)$. The result follows from the first isomorphism theorem.   ♯

Corollary 3.3.12 *If* A *is a trivial* G-module, then $H^1(G, A)$ *is isomorphic to* $Hom(G, A) \approx Hom(G_{ab}, A) \approx Hom(H_1(G, \mathbb{Z}), A)$.

Proof If A is a trivial G-module, then every crossed homomorphism is a homomorphism and every inner crossed homomorphism is the trivial homomorphism. The result follows.   ♯

Remark 3.3.13 The above corollary is a special case of the universal coefficient Theorem 1.4.14.

Let A be a G-module. This amounts to have a homomorphism σ from G to $Aut(A)$ with $\sigma(g)(a) = g \cdot a$. In turn, we have the semi-direct product $A \succ_\sigma G$ of A with G in relation to the G-module structure σ on A. The set part of $A \succ_\sigma G$ is the Cartesian product $A \times G$ and the binary operation $\cdot$ is given by $(a, g) \cdot (b, h) = (a + g \cdot b, gh)$. We get the split short exact sequence

$$E_\sigma \equiv 0 \longrightarrow A \xrightarrow{i_1} A \succ_\sigma G \xrightarrow{p_2} 1,$$

where $i_1(a) = (a, 1)$, $p_2(a, g) = g$. A splitting t of this exact sequence is given by $t(g) = (0, g)$.

Proposition 3.3.14 *We have a bijective map η from the group $Cross(G, A)$ of all crossed homomorphisms from G to A to the set $Split\, E_\sigma$ of all splittings of E_σ.*

Proof Any section s of E_σ is given by $s(g) = (\chi(g), g)$, where χ is a map from G to A such that $\chi(1) = 0$. Further, s is a splitting of E_σ if and only if

$$(\chi(gh), gh) = s(gh) = s(g)s(h) = (\chi(g), g)(\chi(h), h) =$$
$$(\chi(g) + g \cdot \chi(h), gh).$$

Hence s is a splitting if and only if $\chi \in Cross(G, A)$. The result follows.   ♯

The first temptation is to compute $H_n(F, A)$ and $H^n(F, A)$ for free groups. Low-dimensional homology and co-homology groups have already been discussed.

Theorem 3.3.15 *Let F be a free group on X. Then the augmentation ideal $I(F)$ is a free $\mathbb{Z}(F)$-module with basis $X - 1 = \{x - 1 \mid x \in X\}$.*

Proof To show that $I(F)$ is a free $\mathbb{Z}(F)$-module with basis $X - 1$, we need to show that any map f from $X - 1$ to a $\mathbb{Z}(F)$-module A can be uniquely extended to a $\mathbb{Z}(F)$-homomorphism from $I(F)$ to A. By Proposition 3.3.1, $X - 1$ is a set of generators for the $\mathbb{Z}(F)$-module $I(F)$. Thus, if the extension exists, it is unique. We show the existence of the extension. Since F is a free group on X, we have a unique group homomorphism $\tilde{f}$ from F to $A \succ_\sigma F$ given by $\tilde{f}(x) = (f(x - 1), x)$. Clearly $\tilde{f}$ is a splitting of

$$E_\sigma \equiv 0 \longrightarrow A \overset{i_1}{\to} A \succ_\sigma F \overset{p_2}{\to} F \longrightarrow 1.$$

From Proposition 3.3.14, the map ϕ from F to A given by $\phi(x) = f(x - 1)$ is a crossed homomorphism from F to A. Again by Proposition 3.3.1(i), f can be extended to a group homomorphism $\check{f}$ from $I(F)$ to A. As in the proof of Theorem 3.3.11,

$$\check{f}(g(h - 1)) = \check{f}(gh - 1 - (g - 1)) = \check{f}(gh - 1) - \check{f}(g - 1) =$$
$$\phi(gh) - \phi(g) = g \cdot \phi(h) = g \cdot \check{f}(h - 1).$$

This shows that $\check{f}$ is $\mathbb{Z}(F)$-homomorphism from $I(F)$ to A which is an extension of f.   ♯

Corollary 3.3.16 *Let F be a free group. Then for all F-modules A and $n \geq 2$, $H^n(F, A) = 0 = H_n(F, A)$.*

Proof By definition, $H^n(F, A) = EXT^n_{\mathbb{Z}(F)}(\mathbb{Z}, A)$ and $H_n(F, A) = Tor^{\mathbb{Z}(F)}_n(\mathbb{Z}, A)$. The result follows from the observation that

$$0 \longrightarrow I(F) \overset{i}{\to} \mathbb{Z}(F) \overset{\epsilon}{\to} \mathbb{Z} \longrightarrow 0$$

is a projective resolution of the trivial F-module $\mathbb{Z}$. ♯

In particular, if G is an infinite cyclic group, then $H^n(G, A) = 0 = H_n(G, A)$ for $n \geq 2$. Our next aim is to compute the homologies and the co-homologies of cyclic groups, and then to compute the homologies and co-homologies of the direct sum of two groups in terms of homologies and the co-homologies of individual groups. This, in turn, will enable us to compute the homologies and the co-homologies of all finitely generated abelian groups.

Let $G = \langle g \rangle$ be a cyclic group of order m generated by g, and let A be a G-module. For each $n \geq 0$, put $X_n = \mathbb{Z}(G)$. Let σ denote the map from $\mathbb{Z}(G)$ to $\mathbb{Z}(G)$ given by $\sigma\left(\sum_{r=0}^{m-1} \alpha_r g^r\right) = (1 - g)\left(\sum_{r=0}^{m-1} \alpha_r g^r\right)$, and τ the map from $\mathbb{Z}(G)$ to $\mathbb{Z}(G)$ given by $\tau\left(\sum_{r=0}^{m-1} \alpha_r g^r\right) = (1 + g + g^2 + \cdots + g^{m-1})\left(\sum_{r=0}^{m-1} \alpha_r g^r\right)$. Evidently, σ and τ are $\mathbb{Z}(G)$-homomorphisms.

Proposition 3.3.17 *The chain*

$$X \equiv \cdots \xrightarrow{\tau} X_{2n+1} \xrightarrow{\sigma} X_{2n} \xrightarrow{\tau} \cdots \xrightarrow{\sigma} X_2 \xrightarrow{\tau} X_1 \xrightarrow{\sigma} X_0 \xrightarrow{\epsilon} \mathbb{Z} \longrightarrow 0$$

is a projective resolution of the trivial G-module $\mathbb{Z}$.

Proof Since $(1 - g)(1 + g + g^2 + \cdots + g^{m-1}) = (1 - g^m) = 0 = (1 + g + g^2 + \cdots + g^{m-1})(1 - g)$, it follows that $\sigma\tau = 0 = \tau\sigma$. It is also evident that $\epsilon\sigma = 0$. Thus, $image\ \tau \subseteq Ker\ \sigma$, $image\ \sigma \subseteq Ker\ \tau$, and $image\ \sigma \subseteq Ker\ \epsilon$. Finally, we need to show that $Ker\sigma \subseteq image\ \tau$, $Ker\tau \subseteq image\ \sigma$, and also $Ker\ \epsilon \subseteq image\ \sigma$. Let $\sum_{r=0}^{m-1} \alpha_r g^r$ be a member of $Ker\ \epsilon$. Then $\sum_{r=0}^{m-1} \alpha_r = 0$. Hence

$$\sum_{r=0}^{m-1} \alpha_r g^r = \sum_{r=0}^{m-1} \alpha_r g^r - \left(\sum_{r=0}^{m-1} \alpha_r\right)1 = -(1 - g)\sum_{r=1}^{m-1}(\alpha_r b_r),$$

where $b_r = g^{r-1} + g^{r-2} + \cdots + 1$. This shows that $\sum_{r=0}^{m-1} \alpha_r g^r$ belongs to the image of σ. Consequently, $Ker\ \epsilon = image\ \sigma$.

Now, let $\sum_{r=0}^{m-1} \alpha_r g^r$ be a member of $Ker\ \sigma$. Then $(1 - g)(\sum_{r=0}^{m-1} \alpha_r g^r) = 0$. This amounts to say that

$$\sum_{r=0}^{m-1}(\alpha_r - \alpha_{r-1})g^r = 0.$$

It follows that $\alpha_r = \alpha_0$ for all r. Hence $\sum_{r=0}^{m-1} \alpha_r g^r = \alpha_0 \sum_{r=0}^{m-1} g^r$, and it belongs to the image of τ. Consequently, $Ker\ \sigma = image\ \tau$. Similarly, we can show that $Ker\ \tau = image\ \sigma$. ♯

Corollary 3.3.18 *Let $G = \langle g \rangle$ be a cyclic group of order m which is generated by g, and let A be a G-module. Then*

(i) $H^0(G, A)$ is isomorphic to the subgroup $\{a \in A \mid g \cdot a = a$ for all $g \in G\}$,

(ii) $H^{2n+1}(G, A) = Ker\ \tau^*/image\ \sigma^* = H^1(G, A)$ is isomorphic to the sub-quotient group B/C of A, where $B = \{a \in A \mid \sum_{r=0}^{m-1} g^r \cdot a = 0\}$ and $C = \{g \cdot a - a \mid a \in A\}$, $n \geq 0$.

(iii) $H^{2n}(G, A) = Ker\ \sigma^\star/image\ \tau^\star = H^2(G, A)$ *is isomorphic to the subquotient group* D/E, *where* $D = \{a \in A \mid g \cdot a = a\}$ *and* $E = \left(\sum_{r=0}^{m-1} g^r\right) \cdot A, n \geq 1$.

Proof By definition, $H^n(G, A) = H^n(Hom_{\mathbb{Z}(G)}(X, A))$. Thus, $H^0(G, A) = Hom_{\mathbb{Z}(G)}(\mathbb{Z}, A)$. Evidently, $Hom_{\mathbb{Z}(G)}(\mathbb{Z}, A)$ is a subgroup of $Hom_{\mathbb{Z}}(\mathbb{Z}, A)$. In fact an element $f \in Hom_{\mathbb{Z}}(\mathbb{Z}, A)$ is a member of $Hom_{\mathbb{Z}(G)}(\mathbb{Z}, A)$ if and only if $g \cdot f(1) = 1$. Further the map ϕ from $Hom_{\mathbb{Z}}(\mathbb{Z}, A)$ to A given by $\phi(f) = f(1)$ is a group isomorphism such that $\phi(Hom_{\mathbb{Z}(G)}(\mathbb{Z}, A)) = \{a \in A \mid g \cdot a = a\}$. This proves (i). (ii) and (iii) follow immediately if we observe that $\phi(Ker\ \tau^\star) = B$, $\phi(image\ \tau^\star) = C$, $\phi(Ker\ \sigma^\star) = D$, and $\phi(image\ \tau^\star) = E$.   ♯

Corollary 3.3.19 *Let* $G = <g>$ *be a cyclic group of order m which is generated by g, and let A be a G-module. Then*

(i) $H_0(G, A)$ *is isomorphic to* $A/I(G)A$,

(ii) $H_{2n+1}(G, A) = Ker\ \sigma_\star/image\ \tau_\star = H_1(G, A)$ *is isomorphic to the subquotient group* D/E *of A, where D and E are as in the above corollary.*

(iii) $H_{2n}(G, A) = Ker\ \tau_\star/image\ \sigma_\star = H_2(G, A)$ *is isomorphic to the subquotient group* B/C, *where B and C are as in the above corollary.*

Proof (i) follows from Proposition 3.3.4. By definition, $H_n(G, A) = H_n(X \otimes_{\mathbb{Z}(G)} A)$. The map ψ from $\mathbb{Z}(G) \otimes_{\mathbb{Z}(G)} A$ to A given by $\psi\left(\sum_{g \in G} \alpha_g g \otimes a\right) = \sum_{g \in G} \alpha_g g \cdot a$ is a group isomorphism. (ii) and (iii) follow immediately if we observe that $\psi(Ker\ \sigma_\star) = D$, $\psi(image\ \tau_\star) = E$, $\psi(Ker\ \tau_\star) = B$, and $\psi(image\ \sigma_\star) = C$.   ♯

Corollary 3.3.20 *If G is a cyclic group of order m, and A is a trivial G-module, then*

(i) $H^0(G, A) \approx A$,

(ii) $H^{2n+1}(G, A) \approx \{a \in A \mid ma = 0\}, n \geq 0$,

(iii) $H^{2n}(G, A) \approx A/mA, n \geq 0$,

(iv) $H_0(G, A) \approx A$,

(v) $H_{2n+1}(G, A) \approx A/mA, n \geq 0$, *and*

(vi) $H_{2n}(G, A) \approx \{a \in A \mid ma = 0\}$.   ♯

For any group G, $H_n(G, \mathbb{Z})$ is denoted by $H_n(G)$ and $H^n(G, \mathbb{Z})$ is denoted by $H^n(G)$, where $\mathbb{Z}$ is treated as a trivial G-module.

Corollary 3.3.21 *If G is a finite cyclic group and A is a finite G-module, then* $\mid H^{2n+1}(G, A) \mid = \mid H^{2n}(G, A) \mid$.

Proof By the fundamental theorem of homomorphism, $Hom_{\mathbb{Z}(G)}(\mathbb{Z}(G), A)/Ker\sigma^\star \approx image\ \sigma^\star$ and $Hom_{\mathbb{Z}(G)}(\mathbb{Z}(G), A)/Ker\tau^\star \approx image\ \tau^\star$. Thus, $\mid Hom_{\mathbb{Z}(G)}(\mathbb{Z}(G), A)/Ker\sigma^\star \mid = \mid image\ \sigma^\star \mid$ and $\mid Hom_{\mathbb{Z}(G)}(\mathbb{Z}(G), A)/Ker\tau^\star \mid = \mid image\ \tau^\star \mid$. Eliminating $\mid Hom_{\mathbb{Z}(G)}(\mathbb{Z}(G), A) \mid$, we obtain the desired result.   ♯

Theorem 3.3.22 *Let G be a group, and A be a trivial G-module. Then for each $n \geq 1$, we have the following natural short exact sequences*

$$0 \longrightarrow H_n(G) \otimes_{\mathbb{Z}} A \xrightarrow{\lambda_n} H_n(G, A) \xrightarrow{\nu_n} Tor^{\mathbb{Z}}(H_{n-1}(G), A) \longrightarrow 0,$$

and

$$0 \longrightarrow EXT_{\mathbb{Z}}(H_{n-1}(G), A) \xrightarrow{\rho_n} H^n(G, A) \xrightarrow{\mu_n} Hom(H_n(G), A) \longrightarrow 0.$$

Further, these sequences split.

Proof First observe that $\mathbb{Z}(G) \otimes_{\mathbb{Z}(G)} \mathbb{Z}$ is isomorphic to $\mathbb{Z}$. Hence if F is a free $\mathbb{Z}(G)$-module, then $F \otimes_{\mathbb{Z}(G)} \mathbb{Z}$ is a free $\mathbb{Z}$-module. Further, since $\mathbb{Z}$ is bi-$(\mathbb{Z}(G), \mathbb{Z})$-module, $(F \otimes_{\mathbb{Z}(G)} \mathbb{Z}) \otimes_{\mathbb{Z}} A$ is naturally isomorphic to $F \otimes_{\mathbb{Z}(G)} A$ and $Hom_{\mathbb{Z}}(F \otimes_{\mathbb{Z}(G)} \mathbb{Z}, A)$ is naturally isomorphic to $Hom_{\mathbb{Z}(G)}(F, A)$. Let

$$P \equiv \cdots \xrightarrow{d_{n+1}} P_n \xrightarrow{d_n} \cdots \xrightarrow{d_2} P_1 \xrightarrow{d_1} P_0 \xrightarrow{d_0 = \epsilon} \mathbb{Z} \longrightarrow 0$$

be a free G resolution of the trivial G-module $\mathbb{Z}$. Then

$$P \otimes_{\mathbb{Z}(G)} \mathbb{Z} \equiv \cdots \xrightarrow{(d_{n+1})_*} P_n \otimes_{\mathbb{Z}(G)} \mathbb{Z} \xrightarrow{(d_n)_*} \cdots \xrightarrow{(d_2)_*} P_1 \otimes_{\mathbb{Z}(G)} \mathbb{Z} \xrightarrow{(d_1)_*}$$
$$P_0 \otimes_{\mathbb{Z}(G)} \mathbb{Z} \xrightarrow{(d_0)_* = \epsilon_*} \mathbb{Z} \otimes_{\mathbb{Z}(G)} \mathbb{Z} \longrightarrow 0$$

is a chain complex of free abelian groups. From the above observation, the chain complex $(P \otimes_{\mathbb{Z}(G)} \mathbb{Z}) \otimes_{\mathbb{Z}} A$ of abelian groups is naturally chain isomorphic to the chain complex $P \otimes_{\mathbb{Z}(G)} A$, and the co-chain complex $Hom_{\mathbb{Z}}((P \otimes_{\mathbb{Z}(G)} \mathbb{Z}), A)$ is naturally co-chain isomorphic to $Hom_{\mathbb{Z}(G)}(P, A)$. Further, from Corollary 2.4.6 (universal coefficient theorem for homology), we have a natural short exact sequence

$$0 \longrightarrow H_n((P \otimes_{\mathbb{Z}(G)} \mathbb{Z}) \otimes_{\mathbb{Z}} A) \xrightarrow{\lambda_n} H_n((P \otimes_{\mathbb{Z}(G)} \mathbb{Z}) \otimes_{\mathbb{Z}} A) \xrightarrow{\nu_n}$$
$$Tor^{\mathbb{Z}}(H_{n-1}(P \otimes_{\mathbb{Z}(G)} \mathbb{Z}), A) \longrightarrow 0$$

which splits (nonnaturally). Using the chain isomorphism from $(P \otimes_{\mathbb{Z}(G)} \mathbb{Z}) \otimes_{\mathbb{Z}} A$ to $P \otimes_{\mathbb{Z}(G)} A$, and the definition of the homology groups, we obtain the short exact sequence

$$0 \longrightarrow H_n(G) \otimes_{\mathbb{Z}} A \xrightarrow{\lambda_n} H_n(G, A) \xrightarrow{\nu_n} Tor^{\mathbb{Z}}(H_{n-1}(G), A) \longrightarrow 0,$$

which splits but not naturally. Similarly, using Theorem 1.4.14 (universal coefficient theorem for co-homology), we obtain natural short exact sequence

$$0 \longrightarrow EXT_{\mathbb{Z}}(H_{n-1}(G), A) \xrightarrow{\rho_n} H^n(G, A) \xrightarrow{\mu_n} Hom(H_n(G), A) \longrightarrow 0,$$

which also splits naturally.   ♯

Let K and L be groups. Let A be a K-module, and B be an L-module. Then $A \otimes_{\mathbb{Z}} B$ can be given a $K \times L$-module structure by defining $(k, l) \cdot (a \otimes b) = ka \otimes lb$ (verify). Again, if f is a K-homomorphism from A to A' and g an L-homomorphism from B to B', then we have a $K \times L$-homomorphism $f \otimes g$ from $A \otimes_{\mathbb{Z}} B$ to $A' \otimes_{\mathbb{Z}} B'$ given by $(f \otimes g)(a \otimes b) = f(a) \otimes g(b)$. Indeed, $\otimes_{\mathbb{Z}}$ defines a functor from the product category K-Mod $\times$ L-Mod to the category $K \times L$-Mod of $K \times L$-modules. Further, the $K \times L$-module $\mathbb{Z}(K) \otimes_{\mathbb{Z}} \mathbb{Z}(L)$ is isomorphic $\mathbb{Z}(K \times L)$ considered as $\mathbb{Z}(K \times L)$-module. Indeed, the map ϕ from $\mathbb{Z}(K \times L)$ to $\mathbb{Z}(K) \otimes_{\mathbb{Z}} \mathbb{Z}(L)$ defined by $\phi\left(\sum_{(k,l) \in K \times L} \alpha_{k,l}(k, l)\right) = \sum_{(k,l) \in K \times L} \alpha_{k,l} k \otimes l$ (observe that $\left(\sum_{k \in K} \alpha_k k\right) \otimes \left(\sum_{l \in L} \beta_l l\right) = \sum_{(k,l) \in K \times L} \alpha_k \beta_l (k \otimes l)$) is easily seen to be an isomorphism. In turn, the $\mathbb{Z}(K \times L)$-module $(\mathbb{Z}(K)^r \otimes \mathbb{Z}(L)^s)$ is isomorphic to the $\mathbb{Z}(K \times L)$-module $\mathbb{Z}(K \times L)^{rs}$. More generally, if Ω is a free K-module and Λ be a free L-module, then $\Omega \otimes_{\mathbb{Z}} \Lambda$ is a free $\mathbb{Z}(K \times L)$-module.

Using Theorem 2.4.5 (Kunneth formula), we establish the following.

Theorem 3.3.23 *Let K and L be groups. Then for each $n \geq 1$, we have the following natural short exact sequence*

$$
0 \longrightarrow \oplus \sum_{p+q=n} H_p(K) \otimes_{\mathbb{Z}} H_q(L) \overset{\lambda_n}{\to} H_n(K \times L) \overset{\mu_n}{\to}
$$
$$
\oplus \sum_{p+q=n-1} Tor^{\mathbb{Z}}(H_p(K), H_q(L)) \longrightarrow 0
$$

which splits but not naturally.

Proof Let

$$
F \equiv \cdots \overset{d_{n+1}}{\to} F_n \overset{d_n}{\to} F_{n-1} \overset{d_{n-1}}{\to} \cdots \overset{d_2}{\to} F_1 \overset{d_1}{\to} F_0 \overset{\epsilon}{\to} \mathbb{Z} \longrightarrow 0
$$

be a K-free resolution of the trivial K-module $\mathbb{Z}$, and let

$$
F' \equiv \cdots \overset{d'_{n+1}}{\to} F'_n \overset{d'_n}{\to} F'_{n-1} \overset{d'_{n-1}}{\to} \cdots \overset{d'_2}{\to} F'_1 \overset{d'_1}{\to} F'_0 \overset{\epsilon'}{\to} \mathbb{Z} \longrightarrow 0
$$

be an L-free resolution of the trivial L-module $\mathbb{Z}$. Evidently, F_n and F'_n are free $\mathbb{Z}$-modules, $H_0(F) = \mathbb{Z} = H_0(F')$ and $H_n(F) = 0 = H_n(F')$ for all $n \geq 1$. By Theorem 2.4.3 (Kunneth formula) $H_n(F \otimes_{\mathbb{Z}} F') = 0$ for all $n \geq 1$, and $H_0(F \otimes_{\mathbb{Z}} F') = \mathbb{Z} \otimes_{\mathbb{Z}} \mathbb{Z} \approx \mathbb{Z}$. Thus, from the discussion prior to the statement of the theorem, it follows that $F \otimes_{\mathbb{Z}} F'$ is a free $K \times L$ resolution of the trivial $K \times L$-module $\mathbb{Z}$. Furthermore, we have a natural isomorphism from $(F_p \otimes_{\mathbb{Z}} F'_q) \otimes_{K \times L} \mathbb{Z}$ to $(F_p \otimes_K \mathbb{Z}) \otimes_{\mathbb{Z}} (F'_q \otimes_L \mathbb{Z})$ given by $((x \otimes y) \otimes n) \longmapsto (x \otimes n) \otimes (y \otimes 1)$. This, in turn, gives us a natural chain isomorphism from the chain complex $(F \otimes_{\mathbb{Z}} F') \otimes_{K \otimes L} \mathbb{Z}$ to $(F \otimes_K \mathbb{Z}) \otimes_{\mathbb{Z}} (F' \otimes_L \mathbb{Z})$. Evidently, $F \otimes_K \mathbb{Z}$ and $F' \otimes_L \mathbb{Z}$ are chain complexes of free abelian groups. Again applying Theorem 2.4.3 (Kunneth formula) and using the definition of the homology groups, we obtain the desired result.   ♮

Remark 3.3.24 Using the above two theorems, we can compute the integral homology and co-homology groups of finitely generated abelian groups with the coefficient in a trivial module. In particular they are finitely generated abelian groups.

Example 3.3.25 We compute the integral homology $H_n(K \times L)$ and the integral co-homology group $H^n(K \times L)$, where K is a cyclic group of order k and L is a cyclic group of order l. From Proposition 3.3.4, $H_0(K \times L) = H_0(K \times L, \mathbb{Z}) \approx \mathbb{Z}$. Suppose that $n \geq 1$. From Corollary 3.3.20, for a cyclic group G of order m, $H_0(G) \approx \mathbb{Z}$, $H_{2r+1}(G) \approx \mathbb{Z}_m$, and $H_{2s}(G) = \mathbb{Z}_m$, $s \geq 1$. Suppose that $n = 2r + 1$, $r \geq 0$. Suppose that $p + q = 2r + 1$. If $p = 0$, then $H_p(K) \otimes_{\mathbb{Z}} H_q(L) \approx \mathbb{Z} \otimes_{\mathbb{Z}} \mathbb{Z}_l \approx \mathbb{Z}_l$, and if $q = 0$, then $H_p(K) \otimes_{\mathbb{Z}} H_q(L) \approx \mathbb{Z}_k \otimes_{\mathbb{Z}} \mathbb{Z} \approx \mathbb{Z}_k$. Next, if $p \neq 0 \neq q$, then either p is even or q is even. In turn, $H_p(K) \otimes_{\mathbb{Z}} H_q(L) = 0$. This means that $\oplus \sum_{p+q=n} H_p(K) \otimes_{\mathbb{Z}} H_q(L) \approx \mathbb{Z}_k \times \mathbb{Z}_l \approx K \times L$. Now, suppose that $p + q = n - 1 = 2r$. If $p = 0$, then $Tor^{\mathbb{Z}}(H_p(K), H_q(L)) \approx Tor^{\mathbb{Z}}(\mathbb{Z}, \mathbb{Z}_l) = 0$, and if $q = 0$, then $Tor^{\mathbb{Z}}(H_p(K), H_q(L)) \approx Tor^{\mathbb{Z}}(\mathbb{Z}_k, \mathbb{Z}) = 0$. Again, if $p \neq 0 \neq q$, $p + q = 2r$, then p and q both are even or both are odd. If both are even, then $Tor^{\mathbb{Z}}(H_p(K), H_q(L)) = Tor^{\mathbb{Z}}(0, 0) = 0$. Suppose that $p = 2s + 1$ and $q = 2t + 1$ both are odd. Then $Tor^{\mathbb{Z}}(H_p(K), H_q(L)) \approx Tor^{\mathbb{Z}}(\mathbb{Z}_k, \mathbb{Z}_l) \approx \mathbb{Z}_d$, where d is the g.c.d. of k and l. Further, the pairs (s, t) such that $2s + 2t + 2 = 2r$ is r in number. Hence, from Theorem 3.3.23, we get the split exact sequence

$$0 \longrightarrow \mathbb{Z}_k \times \mathbb{Z}_l \overset{\lambda_n}{\longrightarrow} H_n(K \times L) \overset{\mu_n}{\longrightarrow} \mathbb{Z}_d^r \longrightarrow 0,$$

where $n = 2r + 1$ is odd. Thus,

$$H_{2r+1}(K \times L) \approx \mathbb{Z}_k \times \mathbb{Z}_l \times \mathbb{Z}_d^r.$$

In particular, $H_1(\mathbb{Z}_k \times \mathbb{Z}_l) \approx \mathbb{Z}_k \times \mathbb{Z}_l$. Using the same argument we can show that

$$H_{2r}(K \times L) \approx \mathbb{Z}_d^r,$$

where $r \geq 0$.

Theorems 3.3.22 and 3.3.23 express the homologies and the co-homologies of the direct product of groups in terms of the homologies and the co-homologies of the individual groups. Our next aim is to express the homologies and the co-homologies of the free product of groups in terms of the homologies and the co-homologies of the individual groups.

Theorem 3.3.26 *Let $G = K \star L$ denote the free product of K and L. Let A be a left G-module. Then for $n \geq 2$,*

(i) *$H^n(G, A) \approx H^n(K, A) \oplus H^n(L, A)$, and*
(ii) *$H_n(G, A) \approx H_n(K, A) \oplus H_n(L, A)$, where A is treated as a K-module and also as an L-module through the natural inclusion homomorphism.*

Before proving this theorem, we establish some elementary facts.

Proposition 3.3.27 *Let K be a subgroup of a group G. Let A be a left K-module, and B be a left G-module. Then*

$$EXT^n_{\mathbb{Z}(G)}(\mathbb{Z}(G) \otimes_{\mathbb{Z}(K)} A, B) \approx EXT^n_{\mathbb{Z}(K)}(A, B).$$

Proof Let

$$F \equiv \cdots \overset{d_{n+1}}{\to} F_n \overset{d_n}{\to} \cdots \overset{d_2}{\to} F_1 \overset{d_1}{\to} F_0 \overset{d_0}{\to} A \longrightarrow 0$$

be a free $\mathbb{Z}(K)$ resolution of A. Since $\mathbb{Z}(G)$ is a free $\mathbb{Z}(K)$-module, $\mathbb{Z}(G) \otimes_{\mathbb{Z}(K)} F$ is a $\mathbb{Z}(G)$-free resolution of $\mathbb{Z}(G) \otimes_{\mathbb{Z}(K)} A$. Applying the functor $Hom_{\mathbb{Z}(G)}(-, B)$ on the resolution $\mathbb{Z}(G) \otimes_{\mathbb{Z}(K)} F$, and using the natural isomorphisms from $Hom_{\mathbb{Z}(G)}$ $(\mathbb{Z}(G) \otimes_{\mathbb{Z}(K)} F_n, B)$ to $Hom_{\mathbb{Z}(K)}(F_n, B)$, we find that the co-chain complex $Hom_{\mathbb{Z}(G)}(\mathbb{Z}(G) \otimes_{\mathbb{Z}(K)} F, B)$ is co-chain isomorphic to $Hom_{\mathbb{Z}(K)}(F, B)$. In turn, $EXT^n_{\mathbb{Z}(G)}(\mathbb{Z}(G) \otimes_{\mathbb{Z}(K)} A, B) = H^n(Hom_{\mathbb{Z}(G)}(\mathbb{Z}(G) \otimes_{\mathbb{Z}(K)} F, B))$ is isomorphic to $H^n(Hom_{\mathbb{Z}(K)}(F, B)) = EXT^n_{\mathbb{Z}(K)}(A, B)$.   ♯

Proposition 3.3.28 *Let G denote the free product $K \star L$ of K and L. Let A be a left G-module. Then we have a natural (see the remark below for explicit meaning of being natural) isomorphism from $Cross(G, A)$ to $Cross(K, A) \oplus Cross(L, A)$.*

Proof If $\chi \in Cross(G, A)$, then $\chi/K \in Cross(K, A)$ and $\chi/L \in Cross(L, A)$. Evidently, the map ϕ from $Cross(G, A)$ to $Cross(K, A) \oplus Cross(L, A)$ defined by $\phi(\chi) = (\chi/K, \chi/L)$ is a natural homomorphism. We construct the inverse of ϕ. Let $(\chi_1, \chi_2) \in Cross(K, A) \oplus Cross(L, A)$. The G-module structure on A determines the semi-direct product $A \succ G$. Evidently, $A \succ K$ and $A \succ L$ are subgroups of $A \succ G$. The crossed homomorphism χ_1 determines a homomorphism f_1 from K to $A \succ K \subseteq A \succ G$ such that $p_2 o f_1$ is the inclusion map i_1 from K to G. Similarly, χ_2 determines a homomorphism f_2 from L to $A \succ G$ such that $p_2 o f_2$ is the inclusion map i_2 from L to G. From the universal property of the free product, we obtain a homomorphism f from G to $A \succ G$ such that $p_2 o f$ is the identity map I_G on G. In turn, it gives an element $\chi \in Cross(G, A)$. Define $\psi((\chi_1, \chi_2)) = \chi$ thus obtained. Evidently, $\chi/K = \chi_1$ and $\chi/L = \chi_2$. This shows that ψ is the inverse of ϕ.   ♯

Remark 3.3.29 Let K and L be groups. Every $K \star L$-module structure on A determines and is determined uniquely by a K-module and an L-module structure on A. This is because every homomorphism from $K \star L$ to $Aut(A, +)$ determines and is determined uniquely by a pair (ϕ, ψ), where ϕ is a homomorphism from K to $Aut(A, +)$ and ψ is a homomorphism from L to $Aut(A, +)$. Let Λ denote the category whose objects are triples (K, L, A), where K and L are groups and A is a K-module as well as an L-module. A morphism from (K, L, A) to (K', L', A') is a triple (ϕ, ψ, f), where ϕ is a group homomorphism from K' to K, ψ is a group homomorphism from L' to L, and f is a group homomorphism from A to A' such that

$$f((\phi \star \psi)(g') \cdot a) = g' \cdot f(a), \ g' \in K \star L, \ a \in A.$$

The composition of morphisms are defined in the obvious manner. We have the functors $Cross(- \star -, \ -)$ and $Cross(-, -) \oplus Cross(-, -)$ from Λ to the category AB of abelian groups defined by $Cross(- \star -, \ -)(K, L, A) = Cross(K \star L, A)$ and $(Cross(-, -) \oplus Cross(-, -))(K, L, A) = Cross(K, A) \oplus Cross(L, A)$. A more precise way of expressing the above proposition is to say that these two functors are naturally isomorphic. It is evident from the proof of the above proposition that $\phi = \{\phi_{(K,L,A)}\}$ is a natural isomorphism, where $\phi_{(K,L,A)}(\chi) = (\chi/K, \chi/L)$.

Corollary 3.3.30 *Let K and L be groups. Then the $\mathbb{Z}(K \star L)$-module $I(K \star L)$ is naturally isomorphic to the $\mathbb{Z}(K \star L)$-module $(\mathbb{Z}(K \star L) \otimes_{\mathbb{Z}(K)} I(K)) \oplus (\mathbb{Z}(K \star L) \otimes_{\mathbb{Z}(L)} I(L))$.*

Proof For any group G and a G-module A, $Hom_{\mathbb{Z}(G)}(I(G), A)$ is naturally isomorphic to $Cross(G, A)$ (see the proof of Theorem 3.3.11). In turn, using Proposition 3.3.28, we get the following chain of natural isomorphisms:

$$Hom_{\mathbb{Z}(K \star L)}(I(K \star L), A) \approx Cross(K \star L, A) \approx$$
$$Cross(K, A) \oplus Cross(L, A) \approx Hom_{\mathbb{Z}(K)}(I(K), K) \oplus Hom_{\mathbb{Z}(L)}(I(L), A) \approx$$
$$Hom_{\mathbb{Z}(K \star L)}(Z(K \star L) \otimes_{\mathbb{Z}(K)} I(K), A) \oplus Hom_{\mathbb{Z}(K \star L)}(Z(K \star L) \otimes_{\mathbb{Z}(L)} I(L), A) \approx$$
$$Hom_{\mathbb{Z}(K \star L)}((\mathbb{Z}(K \star L) \otimes_{\mathbb{Z}(K)} I(K)) \oplus (\mathbb{Z}(K \star L) \otimes_{\mathbb{Z}(L)} I(L)), A).$$

Thus, the functors $Hom_{\mathbb{Z}(K \star L)}(I(K \star L), -)$ and $Hom_{\mathbb{Z}(K \star L)}((\mathbb{Z}(K \star L) \otimes_{\mathbb{Z}(K)} I(K)) \oplus (\mathbb{Z}(K \star L) \otimes_{\mathbb{Z}(L)} I(L)), -)$ from the category of $\mathbb{Z}(K \star L)$-modules to the category of abelian groups are naturally isomorphic. This shows that the $\mathbb{Z}(K \star L)$-module $I(K \star L)$ is naturally isomorphic to the $\mathbb{Z}(K \star L)$-module $(\mathbb{Z}(K \star L) \otimes_{\mathbb{Z}(K)} I(K)) \oplus (\mathbb{Z}(K \star L) \otimes_{\mathbb{Z}(L)} I(L))$. ♯

Proposition 3.3.31 *Let G be a group, and A be a left G-module. Then for $n \geq 2$, $H^n(G, A)$ is naturally isomorphic to $EXT_{\mathbb{Z}(G)}^{n-1}(IG, A)$, and for $n \geq 1$, $H_n(G, A)$ is naturally isomorphic to $Tor_{n-1}^{\mathbb{Z}(G)}(I(G), A)$.*

Proof By definition, $H^n(G, A) = EXT_{\mathbb{Z}(G)}^n(\mathbb{Z}, A)$. Consider the short exact sequence

$$0 \longrightarrow I(G) \overset{i}{\to} \mathbb{Z}(G) \overset{\epsilon}{\to} \mathbb{Z} \longrightarrow 0$$

of $\mathbb{Z}(G)$-modules. Using Corollary 2.1.17, we get the long exact sequence

$$\cdots EXT_{\mathbb{Z}(G)}^{n-1}(\mathbb{Z}(G), A) \overset{EXT_{\mathbb{Z}(G)}^{n-1}(-,A)(i)}{\to} EXT_{\mathbb{Z}(G)}^{n-1}(I(G), A) \overset{\partial_{n-1}}{\to}$$
$$EXT_{\mathbb{Z}(G)}^n(\mathbb{Z}, A) \overset{EXT_{\mathbb{Z}(G)}^n(-,A)(\epsilon)}{\to} EXT_{\mathbb{Z}(G)}^n(\mathbb{Z}(G), A) \cdots .$$

Since $\mathbb{Z}(G)$ is a free $\mathbb{Z}(G)$-module, $EXT_{\mathbb{Z}(G)}^r(\mathbb{Z}(G), A) = 0$ for all $r \geq 1$. Hence for $n \geq 2$, we get the exact sequence

$$0 \longrightarrow EXT_{\mathbb{Z}(G)}^{n-1}(I(G), A) \overset{\partial_{n-1}}{\to} EXT_{\mathbb{Z}(G)}^n(\mathbb{Z}, A) \longrightarrow 0, n \geq 2.$$

This shows that for $n \geq 2$, $H^n(G, A)$ is naturally isomorphic to $EXT_{\mathbb{Z}(G)}^{n-1}(IG, A)$. Similarly, we can establish the second part.   ♮

Proof of Theorem 3.3.26 Using Propositions 3.3.31, 3.3.27, Corollary 3.3.30, and finally again Proposition 3.3.31 in this order, we obtain the following chain of natural isomorphisms:

$$H^n(K \star L, A) \approx EXT_{\mathbb{Z}(K \star L)}^{n-1}(I(K \star L), A) \approx EXT_{\mathbb{Z}(K \star L)}^{n-1}((\mathbb{Z}(K \star L) \otimes_{\mathbb{Z}(K)}$$
$$I(K)) \oplus (\mathbb{Z}(K \star L) \otimes_{\mathbb{Z}(L)} I(L)), A) \approx EXT_{\mathbb{Z}(K \star L)}^{n-1}((\mathbb{Z}(K \star L) \otimes_{\mathbb{Z}(K)}$$
$$I(K)), A) \oplus EXT_{\mathbb{Z}(K \star L)}^{n-1}((\mathbb{Z}(K \star L) \otimes_{\mathbb{Z}(L)} I(L)), A) \approx$$
$$EXT_{\mathbb{Z}(K)}^{n-1}(I(K), A) \oplus EXT_{\mathbb{Z}(L)}^{n-1}(I(L), A) \approx H^n(K, A) \oplus H^n(L, A).$$

This proves the statement (i). The statement (ii) can be established in the same way.
                                                                                      ♮

Bar Resolutions

To compute $H^n(G, A) = EXT_{\mathbb{Z}(G)}^n(\mathbb{Z}, A)$ and $H_n(G, A) = Tor_n^{\mathbb{Z}(G)}(\mathbb{Z}, A)$, it is convenient to have some suitable and useful projective $\mathbb{Z}(G)$ resolutions of the trivial G-module $\mathbb{Z}$. Here, we describe an important projective $\mathbb{Z}(G)$ resolution of the trivial G-module $\mathbb{Z}$ called the **Bar resolution**.

Let G be a group. For each $n \geq 1$, let us denote $(G - \{e\})^n$ by X_n. For convenience, an ordered n-tuple $(a_1, a_2, \ldots, a_n)$, $a_i \in G$, will be denoted by $[a_1, a_2, \ldots, a_n]$. Thus, $X_n = \{[a_1, a_2, \ldots, a_n] \mid a_i \in G - \{e\}\}$. We denote the singleton $\{[]\}$ by X_0. Let $B(G)_n$ denote the free $\mathbb{Z}(G)$-module with X_n as a basis. Thus, $B(G)_0$ is the free $\mathbb{Z}(G)$-module on $\{[]\}$, and so $B(G)_0 = \mathbb{Z}(G)[] \approx \mathbb{Z}(G)$. For convenience, we give meaning to all ordered n-tuples $[a_1, a_2, \ldots, a_n]$ by identifying it to the identity of $B(G)_n$ whenever $a_i = e$ for some i, $1 \leq i \leq n$. Observe that if $a_i = e$ for some i, then

$$a_1 \cdot [a_2, a_3, \ldots, a_n] + \sum_{i=1}^{n-1}(-1)^i[a_1, a_2, \ldots, a_i a_{i+1}, a_{i+2}, \ldots, a_n] +$$
$$(-1)^n[a_1, a_2, \ldots, a_{n-1}] = 0.$$

Thus, since $B(G)_n$ is a free $\mathbb{Z}(G)$-module with basis X_n, we have a $\mathbb{Z}(G)$-homomorphism ∂_n from $B(G)_n$ to $B(G)_{n-1}$, $n \geq 1$ by

$$\partial_n([a_1, a_2, \ldots, a_n]) = a_1 \cdot [a_2, a_3, \ldots, a_n] +$$
$$\sum_{i=1}^{n-1}(-1)^i[a_1, a_2, \ldots, a_i a_{i+1}, a_{i+2}, \ldots, a_n] + (-1)^n[a_1, a_2, \ldots, a_{n-1}].$$

For example, $\partial_1([a_1]) = a_1[]$. We also have the augmentation $\mathbb{Z}(G)$-homomorphism ∂_0 from $B(G)_0 = \mathbb{Z}(G)[]$ to $\mathbb{Z}$ given by $\partial_0((\sum_{g \in G} \alpha_g g)[]) = \sum_{g \in G} \alpha_g$.

Theorem 3.3.32 *The chain*

$$B(G) \equiv \cdots \stackrel{\partial_{n+2}}{\to} B(G)_{n+1} \stackrel{\partial_{n+1}}{\to} B(G)_n \stackrel{\partial_n}{\to} \cdots \stackrel{\partial_2}{\to} B(G)_1 \stackrel{\partial_1}{\to} B(G)_0 \stackrel{\partial_0}{\to} \mathbb{Z} \longrightarrow 0$$

is a projective $\mathbb{Z}(G)$-resolution of the trivial $\mathbb{Z}(G)$-module $\mathbb{Z}$.

Proof For each n, $B(G)_n$ is a free $\mathbb{Z}(G)$-module and ∂_n is a module homomorphism. It is sufficient to establish the exactness of the above sequence treating it as a sequence of abelian groups. For this purpose, we define a contracting chain homotopy $\{s_n \mid n \geq 0\}$ on $B(G)$ as follows: We have a group homomorphism s_{-1} from $\mathbb{Z}$ to $B(G)_0 = \mathbb{Z}(G)[]$ given by $s_{-1}(n) = n[]$, and also a group homomorphism s_0 from $B(G)_0$ to $B(G)_1$ given by $s_0\left(\sum_{g \in G} \alpha_g g[]\right) = \sum_{g \in G} \alpha_g [g]$. Suppose that $n \geq 1$. Then $B(G)_n$ is a free abelian group with basis $\{g \cdot [g_1, g_2, \ldots, g_n] \mid g, g_i \in G\}$. Thus, we have a group homomorphism s_n from $B(G)_n$ to $B(G)_{n+1}$ given by $s_n(g \cdot [g_1, g_2, \ldots, g_n]) = [g, g_1, g_2, \ldots, g_n]$. In particular, $s_n([g_1, g_2, \ldots, g_n]) = s_n(e \cdot [g_1, g_2, \ldots, g_n] = [e, g_1, g_2, \ldots, g_n] = 0$. Evidently,

$$\partial_0 s_{-1} = I_\mathbb{Z} \tag{3.1}$$

and

$$\partial_{n+1} s_n + s_{n-1}\partial_n = I_{B(G)_n} \tag{3.2}$$

for all $n \geq 0$. In other words, $s = \{s_n \mid n \geq -1\}$ is a contracting homotopy on $B(G)$. From (3.2),

$$x = \partial_{n+1}(s_n(x)) + s_{n-1}(\partial_n(x))$$

for all $x \in B(G)_n$. This shows that

$$Ker\,\partial_n \subseteq Image\,\partial_{n+1}$$

for all n. Next, we show that $Image\,\partial_{n+1} \subseteq Ker\,\partial_n$. We need to show that $\partial_n\partial_{n+1} = 0$. We prove it by the induction on n. Composing ∂_n from left to Eq. (3.2), we obtain

$$\partial_n\partial_{n+1}s_n + \partial_n s_{n-1}\partial_n = \partial_n. \tag{3.3}$$

Putting $n = 0$ in the above equation, we get

$$\partial_0\partial_1 s_0 + \partial_0 s_{-1}\partial_0 = \partial_0.$$

Using (3.1), we get that $\partial_0\partial_1 s_0 + \partial_0 = \partial_0$. This shows that $\partial_0\partial_1 s_0 = 0$. Since s_0 is surjective, $\partial_0\partial_1 = 0$. Assume that $\partial_{n-1}\partial_n = 0$. By (3.2),

$$\partial_n s_{n-1} + s_{n-2}\partial_{n-1} = I_{B(G)_{n-1}}.$$

Composing from right with ∂_n, we get

$$\partial_n s_{n-1}\partial_n + s_{n-2}\partial_{n-1}\partial_n = \partial_n. \tag{3.4}$$

Equating (3.3) and (3.4), we get

$$s_{n-2}\partial_{n-1}\partial_n = \partial_n\partial_{n+1}s_n. \tag{3.5}$$

Since $\partial_{n-1}\partial_n = 0$, by (3.5), $\partial_n\partial_{n+1}s_n = 0$. Again, since s_n is surjective, it follows that $\partial_n\partial_{n+1} = 0$.   ♮

The projective resolution $B(G)$ is called the **standard bar resolution** of trivial $\mathbb{Z}(G)$-module $\mathbb{Z}$. Evidently, B is a contra-variant functor from the category of groups to the category of acyclic chain complexes over $\mathbb{Z}$.

Corollary 3.3.33 *Let A be a G-module. Then $H^n(G, A) = H^n(Hom_{\mathbb{Z}(G)}(B(G), A))$ and $H_n(G, A) = H_n(B(G) \otimes_{\mathbb{Z}(G)} A)$.*   ♮

The co-chain complex $Hom_{\mathbb{Z}(G)}(B(G), A)$ is given by

$$Hom_{\mathbb{Z}(G)}(B(G), A) \equiv 0 \longrightarrow Hom_{\mathbb{Z}(G)}(B(G)_0, A) \overset{\delta^0}{\to} Hom_{\mathbb{Z}(G)}(B(G)_1, A) \overset{\delta^1}{\to}$$
$$\cdots \overset{\delta^{n-1}}{\to} Hom_{\mathbb{Z}(G)}(B(G)_n, A) \overset{\delta^n}{\to} Hom_{\mathbb{Z}(G)}(B(G)_{n+1}, A) \overset{\delta^{n+1}}{\to} \cdots ,$$

where $\delta^n(f) = f \circ \partial_n$.

Next, since $B(G)_n$ is the free $\mathbb{Z}(G)$-module with basis X_n, $Hom_{\mathbb{Z}(G)}(B(G)_n, A)$ is naturally isomorphic to the group $Map(X_n, A)$ of maps from X_n to A, and $Hom_{\mathbb{Z}(G)}(B(G), A)$ is naturally chain isomorphic to the co-chain complex $Map(X, A)$ given by

$$Map(X, A) \equiv 0 \longrightarrow Map(\{[]\}, A) \overset{\delta^0}{\to} Map(X_1, A) \overset{\delta^1}{\to} \cdots \overset{\delta^{n-1}}{\to}$$
$$Map(X_n, A) \overset{\delta^n}{\to} \cdots ,$$

where
$$\delta^n(f)([x_1, x_2, \ldots, x_{n+1}]) = x_1 \cdot f([x_2, x_3, \ldots, x_{n+1}]) +$$
$$\sum_{i=1}^{n}(-1)^n f([x_1, x_2, \ldots x_{i-1}, x_i x_{i+1}, x_{i+2}, \ldots, x_{n+1}]) +$$
$$(-1)^{n+1} f([x_1, x_2, \ldots, x_n]).$$

Thus, $H^n(G, A) = Ker\delta^n / image\delta^{n-1}$. The members of $Ker\delta^n$ are called the n-co-cycles and the members $image\delta^{n-1}$ are called the n-co-boundaries. The group $Ker\delta^n$ of n-co-cycles is denoted by $Z^n(G, A)$, and the group $image\delta^{n-1}$ of co-boundaries is denoted by $B^n(G, A)$. Thus, $H^n(G, A) = Z^n(G, A)/B^n(G, A)$. Evidently, $Z^n(-, -)$ and $B^n(-, -)$ also define bi-functors.

Let us again consider $H^0(G, A)$, $H^1(G, A)$, and $H^2(G, A)$. Any element f of $Map(\{[]\}, A)$ is determined by an element $a = f([])$, then $\delta^0(f)$ is precisely the inner crossed homomorphism determined by $a = f([])$. More explicitly, $\delta^0(f)([g]) = g \cdot a - a$. Thus, $Z^0(G, A) = Ker\delta^0 \approx \{a \in A \mid g \cdot a = a \forall g \in G\}$ and $B^0(G, A) = 0$. This means that $H^0(G, A)$ is isomorphic to the subgroup $\{a \in A \mid g \cdot a = a \forall g \in G\}$ as already observed. Evidently, $B^1(G, A) = image\delta^0 = Icross(G, A)$. Next, $\delta^1(f)([x, y]) = x \cdot f([y]) - f([xy]) + f([x])$. Hence $Z^1(G, A) = Ker\delta^1 = \{f \in Map(X_1, A) \mid x \cdot f([y]) - f([xy]) + f([x]) = 0\} \approx Cross(G, A)$.   This means that $H^1(G, A) \approx Cross(G, A)/Icross(G, A)$ as already observed.

Now, we describe $H^2(G, A)$. Evidently, $B^2(G, A) = image\delta^1$ can be identified with the group of all maps f from $G \times G$ to A for which there is a map h from G to

A such that $h(e) = 0$ and $f((x, y)) = x \cdot h(y) - h(xy) + h(x)$ for all $x, y \in G$. The map δ^2 from $Map(X_2, A)$ to $Map(X_3, A)$ is given by

$$\delta^2(f)([x, y, z]) = x \cdot f([y, z]) - f([xy, z]) + f([x, yz]) - f([x, y]).$$

Thus, $Z^2(G, A) = Ker \delta^2$ can be identified with the group of all maps f from $G \times G$ to A such that $f(1, x) = f(x, 1) = 0$ and

$$f(x, y) + f(xy, z) = \sigma(x)(f(y, z)) + f(x, yz),$$

where σ is the homomorphism from G to $Aut(A, +)$ associated with the G-module structure on A. This means that (f, σ) is a factor system (see Chap. 10, Algebra 2). In Chap. 10 of Algebra 2, we used the notation $H_\sigma^2(G, A)$ for $H^2(G, A)$, where σ is the homomorphism from G to $Aut(A, +)$ associated with the G-module structure A. Further (see p 389, Algebra 2), $H^2(G, A) = H_\sigma^2(G, A)$ has another group theoretic interpretation as the group $EXT_\sigma(G, A)$ of equivalence classes of extensions of A by G under the Baer sum $\boxplus$. For the detailed study of $H^2(G, A) = H_\sigma^2(G, A)$ including its group theoretic interpretation, the Hopf formula, and also the study of Schur multipliers, the reader is referred to and advised to supplement this section with Chap. 10 of Algebra 2.

For an interpretation of $H^n(G, A), n \geq 3$ as n-fold extensions of the group A by G, the reader is referred to "An Interpretation of the Cohomology groups $H^n(G, M)$," D.F. Holt, Journal of Algebra, 60 (1979) (see also "Cohomology theory in abstract groups II", Eilenberg and Mac Lane, Annals of Math, 48 (1947)).

The computation of co-homology of general linear groups over finite fields by D. Quillen (see Ann Math. 96 (1972) 552–586) prompted Quillen to define higher algebraic K-groups in a proper setting. This was indeed, the beginning of higher algebraic K-theory.

Exercises

3.3.1 Determine $Hom_{\mathbb{Z}(Q_8)}(I(Q_8), V_4)$, where V_4 is the trivial Q_8-module.

3.3.2 Determine $Hom_{\mathbb{Z}(S_3)}(I(S_3), V_4)$, where V_4 is the nontrivial S_3-module in obvious manner.

3.3.3 Determine $H^n(U_p, \mathbb{Z}_p)$, where p is prime and the cyclic group $\mathbb{Z}_p$ is U_p-module in obvious manner.

3.3.4 Compute $H^n(\mathbb{Z}_2 \star \mathbb{Z}_3, A)$, where A is the trivial $\mathbb{Z}_2 \star \mathbb{Z}_3$-module.

3.3.5 Let G be a finite group of order m. Define a map χ from $Hom_{\mathbb{Z}(G)}(B(G)_n, A)$ to $Hom_{\mathbb{Z}(G)}(B(G)_{n-1}, A), n > 0$ by

$$\chi(\phi)([x_1, x_2, \ldots, x_{n-1}]) = \sum\nolimits_{g \in G} \phi[x_1, x_2, \ldots, x_{n-1}, g].$$

Show that

$$\sum_{g \in G} \delta^{n+1}(\phi)([x_1, x_2, \ldots, x_n, g]) = x_1 \cdots \chi(\phi)([x_2, x_3, \ldots, x_n] + \\ \sum_{i=1}^{n-1}(-1)^i \chi(\phi)([x_1, x_2, \ldots, x_i x_{i+1}, x_{i+2}, \ldots, x_n]) + \\ (-1)^n \chi(\phi)([x_1, x_2, \ldots, x_{n-1}]) + (-1)^{n+1} m\phi([x_1, x_2, \ldots, x_n]).$$

Deduce that $mC^n(G, A) \subseteq B^n(G, A)$ and $mH^n(G, A) = 0$.

3.3.6 Let G be a group of order m, and let A be a G-module such that for each $a \in A$ there is a unique $b \in A$ such that $mb = a$. Show that $mH^n(G, A) = 0$ for all $n > 0$.

3.4 Calculus and Co-homology

The germs of the co-homological concepts were present in the fundamental theorems of calculus and differential equations. This section is devoted to explore this relationship.

Consider the differential equation

$$F'(x) = f(x)$$

in one variable defined on an open subset U of R, where f is a continuous function from U to R, and F is an unknown differentiable function. Evidently, from the fundamental theorem of calculus, we have a solution of the above differential equation.

Now, consider the first-order differential equation in two variables. All functions from now onward are taken to be C^∞-functions (functions all of whose repeated partial derivatives of any order are continuous functions). Let U be an open subset of $\mathbb{R}^2$. Let $f = (f^1, f^2)$ be a C^∞-function from U to $\mathbb{R}^2$. Consider the differential equation

$$F'(x, y) = f(x, y) = (f^1(x, y), f^2(x, y)), \tag{3.6}$$

where $F(x, y)$ is an unknown C^∞-function from U to $\mathbb{R}$ and $F'(x, y)$ denote the derivative of F at (x, y). This amounts to say that

$$\frac{\partial F(x, y)}{\partial x} = f^1(x, y), \tag{3.7}$$

and

$$\frac{\partial F(x, y)}{\partial y} = f^2(x, y) \tag{3.8}$$

on U. In general, the solution to the above equation does not exist. Suppose a solution exits. Since F is a C^∞-function,

$$\frac{\partial^2 F(x, y)}{\partial y \partial x} = \frac{\partial^2 F(x, y)}{\partial x \partial y}.$$

Thus, a necessary condition for the existence of a solution of (3.6) is

$$\frac{\partial f^1(x, y)}{\partial y} = \frac{\partial f^2(x, y)}{\partial x}. \tag{3.9}$$

However, the condition (3.9) is not a sufficient condition. Consider, for example, the function $f = (f^1, f^2)$ from $U = \mathbb{R}^2 - \{(0, 0)\}$ to $\mathbb{R}^2$ given by $f^1(x, y) = \frac{-y}{x^2+y^2}$ and $f^2(x, y) = \frac{x}{x^2+y^2}$. Evidently, f is a C^∞ function such that

$$\frac{\partial f^1}{\partial y} = \frac{y^2 - x^2}{(x^2 + y^2)^2} = \frac{\partial f^2}{\partial x}.$$

Suppose that there is a C^∞- function F from $U = \mathbb{R}^2 - \{(0, 0)\}$ to $\mathbb{R}$ such that

$$F'(x, y) = (f^1(x, y), f^2(x, y)).$$

Consider the map ϕ from $\mathbb{R}$ to $\mathbb{R}$ given by $\phi(\theta) = F(cos\theta, sin\theta)$. Since the maps $\theta \mapsto (cos\theta, sin\theta)$ and F are C^∞ maps, ϕ is a C^∞ map. By the fundamental theorem of integral calculus,

$$\int_0^{2\pi} \phi'(\theta)d\theta = \phi(2\pi) - \phi(0) = F(1, 0) - F(1, 0) = 0.$$

On the other hand, using the chain rule of derivatives,

$$\phi'(\theta) = \frac{\partial F}{\partial x}\frac{dx}{d\theta} + \frac{\partial F}{\partial y}\frac{dy}{d\theta} = Sin^2\theta + Cos^2\theta = 1.$$

But then

$$\int_0^{2\pi} \phi'(\theta)d\theta = 2\pi \neq 0.$$

This is a contradiction. Hence, the condition (3.9) is not a sufficient condition for the existence of a solution (observe that $f^1dx + f^2dy = 0$ is an exact differential equation on $\mathbb{R}^2 - \{(0, 0)\}$, and in usual differential equation books, one asserts that $Tan^{-1}(\frac{x}{y})$ is a solution. Where is the fallacy?

Let U be an open subset of $\mathbb{R}^2$. Let $C^\infty(U, \mathbb{R}^2)$ denote the $\mathbb{R}$- vector space of C^∞ functions from U to $\mathbb{R}^2$, and $C^\infty(U, \mathbb{R})$ denote the $\mathbb{R}$- vector space of C^∞ functions from U to $\mathbb{R}$. Define a map D^0 from $C^\infty(U, \mathbb{R})$ to $C^\infty(U, \mathbb{R}^2)$ by

$$D^0(\phi)(x, y) = \left(\frac{\partial \phi(x, y)}{\partial x}, \frac{\partial \phi(x, y)}{\partial y}\right) = grad\phi(x, y),$$

and also a map D^1 from $C^\infty(U, \mathbb{R}^2)$ to $C^\infty(U, \mathbb{R})$ by

$$D^1(f)(x, y) = \left(\frac{\partial f(x, y)}{\partial y} - \frac{\partial f(x, y)}{\partial x} \right).$$

Evidently, D^0 and D^1 are vector space homomorphisms such that $D^1 D^0 = 0$, and we have the finite chain complex

$$\Omega(U) \equiv 0 \longrightarrow C^\infty(U, \mathbb{R}) \xrightarrow{D^0} C^\infty(U, \mathbb{R}^2) \xrightarrow{D^1} C^\infty(U, \mathbb{R}) \longrightarrow 0$$

of $\mathbb{R}$-vector spaces. The above chain complex $\Omega(U)$ is called the de Rham complex of U, and $H^0(\Omega(U)) = H^0_{dR}(U)$ and $H^1(\Omega(U)) = H^1_{dR}(U)$ are called the de Rham cohomologies of U. Observe that $H^n_{dR}(U) = 0$ for all $n \geq 2$. The following proposition is immediate.

Proposition 3.4.1 *Let U be an open subset of $\mathbb{R}^2$. Then $H^1_{dR}(U) = 0$ is necessary as well as sufficient condition for the condition (3.9) to be the sufficient condition for Eq. (3.6) to have a solution on U.* ♮

We have already observed that $H^1_{dR}(\mathbb{R}^2 - \{(0, 0)\} \neq 0$. We shall see that $H^1_{dR}(\mathbb{R}^2 - \{(0, 0)\} = \mathbb{R} = H^0_{dR}(\mathbb{R}^2 - \{(0, 0)\}$.

Further, let U be an open subset of $\mathbb{R}^3$. Let $f = (f^1, f^2, f^3)$ be a C^∞ function from U to $\mathbb{R}^3$. Consider the differential equation

$$\begin{align} \nabla F(x, y, z) = F'(x, y, z) = f(x, y, z) = \\ (f^1(x, y, z), f^2(x, y, z), f^3(x, y, z)) \end{align} \tag{3.10}$$

on the open subset U of $\mathbb{R}^3$, where F is an unknown C^∞ function from U to $\mathbb{R}$. If it has a solution F, then

$$\frac{\partial F}{\partial x} = f^1, \quad \frac{\partial F}{\partial y} = f^2, \quad \frac{\partial F}{\partial z} = f^3.$$

In turn,

$$Curl f = \left(\frac{\partial f^3}{\partial y} - \frac{\partial f^2}{\partial z}, \frac{\partial f^1}{\partial z} - \frac{\partial f^3}{\partial x}, \frac{\partial f^2}{\partial x} - \frac{\partial f^1}{\partial y} \right) = (0, 0, 0). \tag{3.11}$$

Thus, (3.11) is a necessary condition for the existence of a solution of (3.10). Again, it is not a sufficient condition for the existence of a solution. For example, consider the open set $U = \mathbb{R}^3 - S^1$, where $S^1 = \{(x, y, z) \mid z = 0 \text{ and } x^2 + y^2 = 1\}$ is the copy of the unit circle embedded in $\mathbb{R}^3$. Let $f = (f^1, f^2, f^3)$ be a C^∞ map from U to $\mathbb{R}^3$ given by

$$f^1(x, y, z) = \frac{2xz}{(x^2 + y^2 - 1)^2 + z^2},$$

$$f^2(x, y, z) = \frac{2yz}{(x^2 + y^2 - 1)^2 + z^2},$$

and

$$f^3(x, y, z) = \frac{1 - x^2 - y^2}{(x^2 + y^2 - 1)^2 + z^2}.$$

It is straightforward to observe that $Curl\, f = (0, 0, 0)$. Suppose that the C^∞ map F from U to $\mathbb{R}$ is a solution to the differential equation (3.10), where f is as above. Consider the C^∞ map σ from the open interval $(-\pi, \pi)$ to U given by

$$\sigma(t) = (\sqrt{1 + cost}, 0, sint).$$

In turn, we have a C^∞ map $\phi = Fo\sigma$ from the open interval $(-\pi, \pi)$ to $\mathbb{R}$. By the fundamental theorem of integral calculus, the indefinite integral

$$\int_{-\pi}^{\pi} \phi'(t)dt = Lim_{\epsilon \to 0} \int_{-\pi+\epsilon}^{\pi-\epsilon} \phi'(t)dt = Lim_{\epsilon \to 0}(\phi(-\pi + \epsilon) - \phi(\pi - \epsilon)) = 0.$$

On the other hand, by the chain rule of derivative, we find the $\phi'(t) = 1$ for all $t \in (-\pi, \pi)$, and hence

$$\int_{-\pi}^{\pi} \phi'(t)dt = Lim_{\epsilon \to 0} \int_{-\pi+\epsilon}^{\pi-\epsilon} \phi'(t)dt = 2\pi.$$

This is a contradiction.

Consider another equation

$$Curl\, \tilde{F} = f = (f^1, f^2, f^3), \tag{3.12}$$

where f is a given C^∞ function from an open subset U of $\mathbb{R}^3$ to $\mathbb{R}^3$, and $\tilde{F} = (\phi_1, \phi_2, \phi_3)$ is an unknown C^∞ function from U to $\mathbb{R}^3$. If such a solution exists, then

$$Curl\, \tilde{F} = \left(\frac{\partial \phi^3}{\partial y} - \frac{\partial \phi^2}{\partial z}, \frac{\partial \phi^1}{\partial z} - \frac{\partial \phi^3}{\partial x}, \frac{\partial \phi^2}{\partial x} - \frac{\partial \phi^1}{\partial y}\right) = (f^1, f^2, f^3).$$

In turn, we get a necessary condition

$$Div\, f = \frac{\partial f^1}{\partial x} + \frac{\partial f^2}{\partial y} + \frac{\partial f^3}{\partial z} = 0 \tag{3.13}$$

to have a solution. One may again see that this is not a sufficient condition.

Let U be an open set of $\mathbb{R}^3$. As in the case of $\mathbb{R}^2$, we have the chain complex

$$\Omega(U) \equiv 0 \longrightarrow C^\infty(U, \mathbb{R}) \overset{D^0}{\to} C^\infty(U, \mathbb{R}^3) \overset{D^1}{\to} C^\infty(U, \mathbb{R}^3) \overset{D^2}{\to} C^\infty(U, \mathbb{R}) \longrightarrow 0$$

of $\mathbb{R}$-vector spaces, where $D^0 = Grad$, $D^1 = Curl$, $D^2 = Div$. This chain complex $\Omega(U)$ is again called the de Rham complex of U, $H^0(\Omega(U)) = H^0_{dR}(U) = Ker D^0$, $H^1(\Omega(U)) = H^1_{dR}(U) = Ker D^1/Image D^0$, $H^2(\Omega(U)) = H^2_{dR}(U) = Ker D^2/Image D^1$, and $H^3(\Omega(U)) = H^3_{dR}(U) = C^\infty(U, \mathbb{R})/Image D^2$ are called the de Rham co-homologies of U. Note that $H^n_{dR}(U) = 0$ for all $n \geq 4$.

The following proposition is immediate.

Proposition 3.4.2 *The condition (3.11) is a sufficient condition for the existence of a solution of (3.10) if and only if $H^1_{dR}(U) = 0$. The condition (3.13) is sufficient condition for the existence of a solution of (3.12) if and only if $H^2_{dR}(U) = 0$.* ♯

Observe that $H^1_{dR}(\mathbb{R}^3 - S^1) \neq 0$. Now, our aim is to formalize and generalize the above discussion. Consider the vector space $\mathbb{R}^n$ over $\mathbb{R}$. For convenience and the compatibility with notations of calculus, we denote the standard basis of $\mathbb{R}^n$ by $\{dx^1, dx^2, \ldots, dx^n\}$. We recall (see Sect. 7.3 of Algebra 2) the concept of exterior powers and also the exterior algebra $E(V)$ of a vector space V over $\mathbb{R}$. Let V be a finite-dimensional vector space over $\mathbb{R}$, and W be another vector space over $\mathbb{R}$. A map f from V^r to W is called an r-alternating map if (i) f is linear in each coordinate, and (ii) $f(v_1, v_2, \ldots, v_n) = 0$ whenever $v_i = v_j$ for some $i \neq j$. This, of course, is equivalent to say that

$$f(v_1, v_2, \ldots, v_i, v_{i+1}, \ldots, v_j, v_{j+1}, \ldots, v_n) =$$
$$-f(v_1, v_2, \ldots, \overset{i}{\overbrace{v_j}}, v_{i+1}, \ldots, \overset{j}{\overbrace{v_i}}, v_{j+1}, \ldots, v_n).$$

The rth exterior power $\bigwedge^r V$ of the vector space V is a vector space together with an r-alternating map f from V^r to $\bigwedge^r V$ such that given any r-alternating map g from V^r to W, there is a unique linear transformation η from $\bigwedge^r V$ to W such that $\eta \circ f = g$. It follows as usual that the exterior power $\bigwedge^r V$ is unique up to isomorphism. The image $f((v_1, v_2, \ldots, v_r))$ is denoted by $v_1 \wedge v_2 \wedge \cdots \wedge v_r$. By convention, we take $\bigwedge^0 V$ to be $\mathbb{R}$. Thus,

(i) $v_1 \wedge v_2 \wedge \cdots v_{i-1} \wedge \alpha v_i + \beta v_i' \wedge v_{i+1} \cdots \wedge v_r = \alpha(v_1 \wedge v_2 \wedge \cdots v_i \wedge \cdots \wedge v_r) + \beta(v_1 \wedge v_2 \wedge \cdots v_i' \wedge \cdots \wedge v_r)$,

(ii) $v_1 \wedge v_2 \wedge \cdots \wedge v_r = 0$ whenever $v_i = v_j$ for some $i \neq j$, and

(iii) $v_1 \wedge v_2 \wedge \cdots \wedge v_i \cdots \wedge v_j \cdots \wedge v_r = -v_1 \wedge v_2 \wedge \cdots \wedge \overset{i}{\overbrace{v_j}} \wedge \cdots \wedge \overset{j}{\overbrace{v_i}} \wedge \cdots \wedge v_r$.

Evidently, $\bigwedge^r V$ is generated by $\{v_1 \wedge v_2 \wedge \cdots \wedge v_r \mid v_i \in V\}$, and $v_1 \wedge v_2 \wedge \cdots \wedge v_r = 0$ if and only if $\{v_1, v_2, \cdots, v_r\}$ is linearly dependent. Consequently, $\bigwedge^r V = 0$ for $r > Dim V$.

Further, if $\{e_1, e_2, \ldots, e_n\}$ is a basis of V, then it can be easily observed (see Sect. 7.3 of algebra 2) that $\{e_{i_1} \wedge e_{i_2} \wedge \cdots \wedge e_{i_r} \mid 1 \leq i_1 < i_2 < \cdots < i_r \leq n\}$ forms a basis of $\bigwedge^r V$. In particular, $Dim \bigwedge^r V = {}^n C_r$.

Let $E(V)$ denote the direct sum $\oplus \sum_{r=0}^n \bigwedge^r V$. The external multiplication $\cdot$ from $\bigwedge^r V \times \bigwedge^s V$ to $\bigwedge^{r+s} V$ given by

$$(v_1 \wedge v_2 \wedge \cdots \wedge v_r) \cdot (w_1 \wedge w_2 \wedge \cdots \wedge w_s) = v_1 \wedge v_2 \wedge \cdots \wedge v_r \wedge w_1 \wedge w_2 \wedge \cdots \wedge w_s$$

can be extended to a multiplication on $E(V)$ with respect to which $E(V)$ is an alternating associative algebra over $\mathbb{R}$ in the sense $ab = -ba$ for all $a, b \in E(V)$. Clearly, $Dim E(V) = 2^{Dim V}$.

Now, let us take V to be $\mathbb{R}^n$. As already mentioned, the standard basis will be denoted by $\{dx^1, dx^2, \ldots, dx^n\}$. For each p, $1 \le p \le n$. Let us denote the set $\{(i_1, i_2, \ldots, i_p) \mid 1 \le i_1 < i_2 < \cdots < i_p \le n\}$ by A_p^n. For each $I = (i_1, i_2, \ldots, i_p)$ in A_p^n, we denote $dx^{i_1} \wedge dx^{i_2} \wedge \cdots \wedge dx^{i_p}$ by dx^I. Thus, $\{dx^I \mid I \in A_p^n\}$ is a basis of $\bigwedge^p \mathbb{R}^n$. Let U be an open subset of $\mathbb{R}^n$. Let us denote the tensor product $C^\infty(U, \mathbb{R}) \otimes_{\mathbb{R}} \bigwedge^p \mathbb{R}^n$ by $\Omega^p(U)$. Thus,

$$\Omega^p(U) = \left\{ \sum\nolimits_{I \in A_p^n} \alpha_I dx^I \mid \alpha_I \in C^\infty(U, \mathbb{R}) \right\},$$

and the members of $\Omega^p(U)$ are called **p-forms** on U. In particular, $\Omega^0(U) = C^\infty(U, \mathbb{R})$, $\Omega^1(U) = C^\infty(U, \mathbb{R}^n)$ and so on. Define a map D^p from $\Omega^p(U)$ to $\Omega^{p+1}(U)$ by

$$D^p\left(\sum\nolimits_{I \in A_p^n} \alpha_I dx^I \right) = \sum\nolimits_{I \in A_p^n} \left(\sum\nolimits_{i=1}^n \frac{\partial \alpha_I}{\partial x_i} dx^i \right) \wedge dx^I.$$

Evidently, D^p is a linear transformation.

Proposition 3.4.3 $D^{p+1} \circ D^p = 0$ for all p, and

$$\Omega(U) \equiv 0 \longrightarrow \Omega^0(U) \xrightarrow{D^0} \Omega^1(U) \xrightarrow{D^1} \cdots \xrightarrow{D^{p-1}} \Omega^p(U) \xrightarrow{D^p} \cdots \xrightarrow{D^{n-1}} \Omega^n(U) \longrightarrow 0$$

is a finite chain complex of $\mathbb{R}$-vector spaces.

Proof It is sufficient to show that $(D^{p+1} \circ D^p)(\alpha_I dx^I) = 0$. Since $\frac{\partial^2 \alpha_I}{\partial x_i \partial x_j} = \frac{\partial^2 \alpha_I}{\partial x_j \partial x_i}$ and $dx^i \wedge dx^j = -dx^j \wedge dx^i$ for all i, j,

$$D^{p+1}(D^p(\alpha_I dx^I)) = D^{p+1}\left(\sum\nolimits_{i=1}^n \frac{\partial \alpha_I}{\partial x_i} dx^i \wedge dx^I \right) =$$
$$\sum\nolimits_{i=1}^n \left(\sum\nolimits_{j=1}^n \frac{\partial^2 \alpha_I}{\partial x_j \partial x_i} dx^j \right) \wedge dx^i \wedge dx^I = -\sum\nolimits_{i=1}^n \left(\sum\nolimits_{j=1}^n \frac{\partial^2 \alpha_I}{\partial x_j \partial x_i} dx^j \right) \wedge dx^i \wedge dx^I.$$

This shows that $D^{p+1}(D^p(\alpha_I dx^I)) = 0$.   ♯

The co-chain complex $\Omega(U)$ is called the **de Rham complex**, and $H^p(\Omega(U)) = H^p_{dR}(U)$ is called the pth de Rham co-homology of U. The members of $Ker D^p$ are called the **closed p-forms**, and the members of $image D^{p-1}$ are called **exact forms**.

Example 3.4.4 (i). Consider the case when $n = 0$. By convention, $\mathbb{R}^0$ is a singleton $\{0\}$, and an open subset U of $\mathbb{R}^0$ is also $\{0\}$. Evidently, $D^0(\{0\}) \approx \mathbb{R}$ and de Rham complex of a singleton is given by

$$0 \longrightarrow \mathbb{R} \longrightarrow 0 \longrightarrow 0 \cdots .$$

Hence $H_{dR}^0(\{\star\}) = \mathbb{R}$ and $H_{dR}^p(\{\star\}) = 0$ for $p \geq 1$, where $\{\star\}$ is a singleton set.

(ii). Consider the case when $n = 1$. Let $U = (a, b)$. The de Rham complex $\Omega(U)$ of U is given by

$$\Omega(U) \equiv 0 \longrightarrow C^{\infty}(U, \mathbb{R}) \overset{D^0}{\to} C^{\infty}(U, \mathbb{R}) \longrightarrow 0 \longrightarrow \cdots ,$$

where $D^0(f)(x) = f'(x)$, $x \in (a, b)$. Evidently, $Ker D^0$ is the space of constant functions on $U = (a, b)$. Hence $H_{dR}^0((a, b)) \approx \mathbb{R}$. Further, by the fundamental theorem of integral calculus, every C^{∞} function on (a, b) is the derivative of its integral. Hence $H_{dR}^p((a, b)) = 0$ for all $p \geq 1$. Similarly, if U is the disjoint union of r open intervals, then $H_{dR}^0(U) \approx \mathbb{R}^r$, and $H_{dR}^p(U) = 0$ for all $p \geq 1$. Using the induction, we shall show that $H_{dR}^0(\mathbb{R}^n) = \mathbb{R}$ for $p = 0$ and $H_{dR}^p(\mathbb{R}^n) = 0$ when $p \geq 1$.

Consider the category Σ whose objects are open subsets of finite-dimensional Euclidean spaces, and morphisms are C^{∞} maps. Let f be a C^{∞} map from an open subset U of $\mathbb{R}^n$ to an open subset V of $\mathbb{R}^m$. Let $y^1, y^2, \ldots, y^m$ denote coordinate functions on V, and $x^1, x^2, \ldots, x^n$ denote the coordinate function on U. Then $y^j \circ f$ is a C^{∞} function from U to $\mathbb{R}$ for each j. Let $J = (j_1, j_2, \ldots, j_p) \in A_p^m$. Denote

$$\left(\sum_{i=1}^n \frac{\partial(y^{j_1} \circ f)}{\partial x_i} dx^i \right) \wedge \left(\sum_{i=1}^n \frac{\partial(y^{j_2} \circ f)}{\partial x_i} dx^i \right) \wedge \cdots \wedge \left(\sum_{i=1}^n \frac{\partial(y^{j_1} \circ f)}{\partial x_i} dx^i \right)$$

by $d(y^J \circ f)$. Evidently, $d(y^J \circ f)$ is a member of $\Omega^p(U)$. In turn, we get a linear map $\Omega^p(f)$ from $\Omega^p(V)$ to $\Omega^p(U)$ defined by

$$\Omega^p(f)\left(\sum_{J \in A_p^m} \beta_J dy^J \right) = \sum_{J \in A_p^m} (\beta_J \circ f) d(y^J \circ f).$$

It can be easily seen that $\Omega(f) = \{\Omega^p(f) \mid p \geq 0\}$ is a chain transformation from $\Omega(V)$ to $\Omega(U)$. In turn, we get a contra-variant functor Ω from Σ to the category of chain complexes of vector spaces, and also de Rham co-homology functors H_{dR}^p from Σ to the category of vector spaces over $\mathbb{R}$. With a little more effort, these functors can be extended to functors from the category of differentiable manifolds to the category of vector spaces over $\mathbb{R}$. The reader is referred to "Differential forms in Algebraic topology" by Bott and Tu for details.

Proposition 3.4.5 *Let U be a connected open subset of $\mathbb{R}^n$. Then $H_{dR}^0(U) \approx \mathbb{R}$. More generally, if U is an open subset of $\mathbb{R}^n$ having m components, then $H_{dR}^0(U) \approx \mathbb{R}^m$.*

Proof Let f be a member of $Ker D^0$. Then $\frac{\partial f}{\partial x_i} = 0$ on U for all i. Since f is a C^{∞} function, all repeated partial derivatives of f are 0 on U. By Taylor's theorem, f is a locally constant function. The result follows, since U is connected.   ♯

Proposition 3.4.6 *Let U be an open subset of $\mathbb{R}^n$ and $x_0 \in U$. Let e_{x_0} denote the function from $C^\infty(U, \mathbb{R})$ to itself given by $e_{x_0}(f)(x) = f(x_0)$. Suppose that for each $p > 0$, we have a linear transformation s_p from $\Omega^p(U)$ to $\Omega^{p-1}(U)$ such that*

(i) $s_1 D^0 = I_{D^0(U)} - e_{x_0}$, and
(ii) $D^{p-1}s_p + s_{p+1}D^p = I_{D^p(U)}$ for all $p > 0$.
Then $H^0_{dR}(U) \approx \mathbb{R}$ and $H^p_{dR}(U) = 0$ for all $p > 0$.

Proof Let $f \in Ker\, D^0$. Then $0 = s_1 D^0 f = f - e_{x_0}(f)$. Hence $f(x) = f(x_0)$ for all $x \in U$. This shows that $H^0_{dR}(U)$ is the space of all constant functions on U. As such $H^0_{dR}(U) \approx \mathbb{R}$. Next, suppose that $p > 0$, and $\omega = \sum_I \alpha_I dx^I$ belongs to $Ker\, D^p$. Then from (ii), $D^{p-1}s_p(\omega) = \omega$. Hence $\omega \in image\, D^{p-1}$. This shows that $H^p_{dR}(U) = 0$ for all $p > 0$. ♯

A subset X of $\mathbb{R}^n$ is called a **star-shaped** subspace if there is a point $x_0 \in X$ such that for all $x \in X$, the line segment $\{tx_0 + (1-t)x \mid t \in [0, 1]\}$ joining x_0 with x is contained in X. For example, $\mathbb{R}^n$, upper half-plane H^n_+, lower half-plane H^n_-, and $\mathbb{R}^2 - \{(x, 0) \mid x \geq 0\}$ are all star shaped, whereas $\mathbb{R}^2 - \{(0,0)\}$ is not star shaped. Evidently, a star-shaped subspace of $\mathbb{R}^n$ is a path connected and so also a connected subspace.

Theorem 3.4.7 (Poincare Lemma) *Let U be a star-shaped open subspace of $\mathbb{R}^n$. Then $H^0_{dR}(U) \approx \mathbb{R}$ and $H^p_{dR}(U) = \{0\}$ for all $p > 0$.*

Proof Let U be a star-shaped open subspace of $\mathbb{R}^n$. Without loss of generality, we may assume that U is star shaped about $\bar{0}$. Thus, $t\bar{x} \in U$ for all $\bar{x} \in U$ and $t \in [0, 1]$. In the light of Proposition 3.4.6, it is sufficient to define a linear transformation s_p from $\Omega^p(U)$ to $\Omega^{p-1}(U)$ for all $p > 0$ such that

(i) $s_1 D^0 = I_{D^0(U)} - e_{\bar{0}}$, and
(ii) $D^{p-1}s_p + s_{p+1}D^p = I_{D^p(U)}$ for all $p > 0$.
Consider the open subset $V = U \times \mathbb{R}$ of $\mathbb{R}^{n+1}$. Let ψ be a C^∞ map from $\mathbb{R}$ to $\mathbb{R}$ satisfying the conditions (i) $\psi(t) = 0$ for $t \leq 0$, (ii) $\psi(t) = 1$ for $t \geq 1$, and (iii) $\psi(t) \in [0, 1]$ for $t \in [0, 1]$ (see "Principles of mathematical analysis" by Rudin for the existence of such a function). This gives us a C^∞ map ϕ from V to U defined by $\phi(\bar{x}, t) = \psi(t)\bar{x}$. Let $\omega = \sum_{I \in A^n_p} \alpha_I dx^I$ be a p-form in U. Then by the definition

$$\Omega^p(\phi)(\omega) = \sum_{I \in A^n_p}(\alpha_I o\phi)d(x^I o\phi).$$

Now, for fixed $I = (i_1, i_2, \ldots, i_p)$, $1 \leq i_1 < i_2 < \cdots < i_p \leq n$,

$$d(x^I o\phi) = (\psi(t)dx^{i_1} + \psi'(t)dx^{i_1} \wedge dx^{n+1}) \wedge (\psi(t)dx^{i_2} + \psi'(t)dx^{i_2} \wedge dx^{n+1})\wedge$$
$$\cdots \wedge (\psi(t)dx^{i_p} + \psi'(t)dx^{i_p} \wedge dx^{n+1}).$$

Thus,

$$\Omega^p(\phi)(\omega) = \sum_{I \in A^n_p} \alpha_I(\psi(t)\bar{x})(\psi(t))^p dx^I +$$
$$\sum_{I \in A^n_p} \alpha_I(\psi(t)\bar{x})\left(\sum_{k=1}^p (-1)^{p-k}dx^{J^k_{p-1}} \wedge dx^{n+1}\right),$$

where $J_{p-1}^k = I - \{i_k\}$. In turn,

$$\Omega^p(\omega) = \sum_{I \in A_p^n} \alpha_I (\psi(t)\bar{x})(\psi(t))^p dx^I + \sum_{J \in A_{p-1}^n} \beta_J(\bar{x}, t) dx^J \wedge dx^{n+1}.$$

Define $s_p(\omega) = \sum_{J \in A_{p-1}^n} \gamma_J(\bar{x}) dx^J$, where

$$\gamma_J(\bar{x}) = \int_0^1 \beta_J(\bar{x}, t) dt.$$

It is straightforward to verify (left as an exercise) that $\{s_p \mid p \geq 1\}$ satisfies the required conditions (i) and (ii).   ♯

Our next aim is to establish the Mayer–Vietoris exact sequence for de Rham co-homology. Let U and V be open subsets of $\mathbb{R}^n$. Let $W = U \bigcup V$. Evidently, W is an open subset of $\mathbb{R}^n$. If $U \bigcap V = \emptyset$, then $\Omega(W) = \Omega(U) \oplus \Omega(V)$, and so $H_{dR}^p(W) \approx H_{dR}^p(U) \oplus H_{dR}^p(V)$. Suppose that $U \bigcap V$ is a nonempty set. Let i_1 denote the inclusion map from $U \bigcap V$ to U, and i_2 denote the inclusion map from $U \bigcap V$ to V. Let j_1 denote the inclusion map from U to W, and j_2 denote the inclusion map from V to W. All these maps are C^∞ maps. Using the contra-variant functor Ω, we obtain a co-chain transformation $\Omega(i_1)$ from $\Omega(U)$ to $\Omega(U \bigcap V)$, a co-chain transformation $\Omega(i_2)$ from $\Omega(V)$ to $\Omega(U \bigcap V)$, a co-chain transformation $\Omega(j_1)$ from $\Omega(U \bigcup V)$ to $\Omega(U)$, and a co-chain transformation $\Omega(j_2)$ from $\Omega(U \bigcup V)$ to $\Omega(V)$.

Theorem 3.4.8 (Mayer–Vietoris sequence) *Let U and V be open subsets of $\mathbb{R}^n$. Then we have the short exact sequence*

$$0 \longrightarrow \Omega(W) \overset{(\Omega(j_1), \Omega(j_2))}{\rightarrow} \Omega(U) \oplus \Omega(V) \overset{\Omega(i_1) - \Omega(i_2)}{\rightarrow} \Omega(U \bigcap V) \longrightarrow 0$$

of de Rham co-chain complexes of vector spaces, where $W = U \bigcup V$.

Proof Evidently, $\Omega^p(j_1)$ and $\Omega^p(j_2)$ are injective for all p, and they agree on $U \bigcap V$. Further, if $\left(\sum_{I \in A_p^n} \alpha_I dx^I, \sum_{I \in A_p^n} \beta_I dx^I \right)$ belongs to the $Ker(\Omega(i_1) - \Omega(i_2))$, then $\alpha_I = \beta_I$ on $U \bigcap V$. In turn, we get a C^∞ map γ_I from W to $\mathbb{R}$ whose restriction to U is α_I and whose restriction to V is β_I. Clearly,

$$(\Omega(j_1), \Omega(j_2)) \left(\sum_{I \in A_p^n} \gamma_I dx^I \right) = \left(\sum_{I \in A_p^n} \alpha_I dx^I, \sum_{I \in A_p^n} \beta_I dx^I \right).$$

Thus, the above sequence is exact at $\Omega(W)$ and also at $\Omega(U) \oplus \Omega(V)$. We need to show that $\Omega^p(i_1) - \Omega^p(i_2)$ is surjective for all p. Let $\sum_{I \in A_p^n} \alpha_I dx^I$ be a member of $\Omega^p(U \bigcap V)$. Then α_I is a C^∞ map from $U \bigcap V$ to $\mathbb{R}$. Let $\{\chi_U, \chi_V\}$ be a C^∞ partition of unity of W subordinate to the open cover $\{U, V\}$ of W. Then χ_U and χ_V are C^∞ functions on W such that (i) $supp(\chi_U) \subseteq U$, (ii) $supp(\chi_V) \subseteq V$, and

(iii) $\chi_U + \chi_V = 1$ on W (for the proof of the existence of the partition of unity, see a book on Analysis or Topology). Since $\alpha_I \chi_U$ is zero on $(U \bigcap V) - supp\chi_U$, we get a C^∞ map β_I from U to $\mathbb{R}$ given by $\beta_I(u) = \alpha_I(u)\chi_U(u)$ if $u \in U \bigcap V$, and $\beta_I(u) = 0$ otherwise. Similarly, we get a C^∞ map γ_I from V to $\mathbb{R}$ given by $\gamma_I(v) = -\alpha_I(v)\chi_V(v)$ if $v \in U \bigcap V$, and $\gamma_I(v) = 0$ otherwise. Evidently,

$$(\Omega(i_1) - \Omega(i_2))\left(\sum_{I \in A_p^n} \beta_I dx^I, \sum_{I \in A_p^n} \gamma_I dx^I\right) = \sum_{I \in A_p^n} \alpha_I dx^I.$$

This shows that $\Omega^p(i_1) - \Omega^p(i_2)$ is surjective. ♯

Corollary 3.4.9 *Under the hypothesis of the above theorem, we have the following exact sequence:*

$$\cdots \overset{\partial^{p-1}}{\to} H_{dR}(W) \overset{(H^p(j_1), H^p(j_2))}{\to} H_{dR}^p(U) \oplus H_{dR}^p(V) \overset{H^p(i_1)-H^p(i_2)}{\to} H_{dR}^p(U \bigcap V) \overset{\partial^p}{\to}$$
$$\cdots.$$

♯

Proof The proof follows from Theorem 1.3.1. ♯

Example 3.4.10 In this example we compute $H_{dR}^p(\mathbb{R}^2 - \{(0,0)\})$. We express $\mathbb{R}^2 - \{(0,0)\}$ as $U \bigcup V$, where $U = \mathbb{R}^2 - \{(x,0) \mid x \geq 0\}$ and $V = \mathbb{R}^2 - \{(x,0) \mid x \leq 0\}$. Clearly, U and V are star-shaped open subsets of $\mathbb{R}^2$. Hence by the Poincare Lemma,

$$H_{dR}^0(U) = \mathbb{R} = H_{dR}^0(V),$$

and

$$H_{dR}^p(U) = 0 = H_{dR}^p(V)$$

for all $p \geq 1$. Further, $U \bigcap V = H^+ \bigcup H^-$, where H^+ is the upper half-plane and H^- is the lower half-plane. Since H^+ and H^- are star-shaped open sets with $H^+ \bigcap H^- = \emptyset$,

$$H_{dR}^0(U \bigcap V) = \mathbb{R} \oplus \mathbb{R},$$

and

$$H_{dR}^p(U \bigcap V) = 0$$

for all $p \geq 1$. In turn, using the above Corollary, for $p > 0$, we get the following exact sequence:

$$0 \longrightarrow H_{dR}^p(U \bigcap V) \overset{\partial^p}{\to} H_{dR}^{p+1}(R^2 - \{(0,0)\}) \longrightarrow 0.$$

It follows that $H_{dR}^p(\mathbb{R}^2 - \{(0,0)\}) = 0$ for all $p \geq 2$. Again, from the above Corollary, we get the exact sequence

$$0 \longrightarrow H^0_{dR}(\mathbb{R}^2 - \{(0,0)\}) \overset{(H^0(j_1), H^0(j_2))}{\longrightarrow} H^0_{dR}(U) \oplus H^0_{dR}(V) \overset{H^0(i_1) - H^0(i_2)}{\longrightarrow}$$

$$H^0_{dR}(U \cap V) \overset{\partial^0}{\rightarrow} H^1_{dR}(\mathbb{R}^2 - \{(0,0)\}) \longrightarrow 0.$$

Since $\mathbb{R}^2 - \{(0,0)\}$, U, V are connected open subsets, $H^0_{dR}(\mathbb{R}^2 - \{(0,0)\}) \approx H^0_{dR}(U) \approx H^0_{dR}(V) \approx \mathbb{R}$. Also $H^0_{dR}(U \cap V) \approx \mathbb{R} \oplus \mathbb{R}$. Thus, we get the following exact sequence of vector spaces:

$$0 \longrightarrow \mathbb{R} \longrightarrow \mathbb{R} \oplus \mathbb{R} \longrightarrow \mathbb{R} \oplus \mathbb{R} \overset{\partial^0}{\rightarrow} H^1_{dR}(\mathbb{R}^2 - \{(0,0)\}) \longrightarrow 0.$$

The dimension considerations show that $H^1_{dR}(\mathbb{R}^2 - \{(0,0)\})$ is of dimension 1. This shows that $H^0_{dR}(\mathbb{R}^2 - \{(0,0)\}) \approx \mathbb{R} \approx H^1_{dR}(\mathbb{R}^2 - \{(0,0)\})$.

Finally, we state (without proof) the theorem of de Rham. The detailed discussion and the proof can be found in the book "Foundations of differentiable manifolds and Lie groups" by Warner. Let U be an open subset (more generally, a smooth manifold) of $\mathbb{R}^n$. Define a map χ_p from $H^p_{dR}(U)$ to the singular co-homology $H^p(M, \mathbb{R}) = Hom_{\mathbb{R}}(H_p(M, \mathbb{R}), \mathbb{R})$ by $\chi_p([\omega])([\sigma]) = \int_{\sigma} \omega$, where ω is a closed p-form on U, σ is a singular chain representing an element of $H_p(M, \mathbb{R})$, and right-hand side represents the integral of the p-form ω around σ (see calculus by Spivak).

Theorem 3.4.11 (de Rham) χ_p *is an isomorphism for all* p.

Remark 3.4.12 The concept of de Rham co-homology can be extended to the category of differentiable manifolds. It is introduced by using a differentiable structure on a manifold. On a manifold, there may be several distinct differential structures. However, Theorem of de Rham asserts that de Rham co-homology depends only on the topology of the manifold, and it is independent of the differential structure on the manifold.

Exercises

3.4.1 Compute $H^p_{dR}(\mathbb{R}^2 - \{(0,0), (1,1)\})$ for all p.

3.4.2 Compute $H^p_{dR}(U)$, where $U = \{(x,y) \in \mathbb{R}^2 \mid 1 < (x^2 + y^2) < 2\}$.

3.4.3 Let $\{U_n \mid n \in \mathbb{N}\}$ and $\{V_n \mid n \in \mathbb{N}\}$ be two monotonic decreasing sequences of open subsets of $\mathbb{R}^2$ such that $\bigcap_{i=1}^{\infty} U_n = S^1 = \bigcap_{i=1}^{\infty} U_n$. Show that $Lim H^p_{dR}(U_n) = Lim H^p_{dR}(V_n)$. Define $H^p_{dR}(S^1) = Lim H^p_{dR}(U_n)$. Compute $H^p_{dR}(S^1)$.

3.4.4 Let U be an open subset which is covered by finitely many convex open sets. Show that $H^p_{dR}(U)$ is finite dimensional.

Chapter 4
Sheaf Co-homology and Its Applications

Sheaf theory and sheaf co-homology are tools which are very efficiently used in topology, number theory, and algebraic geometry. Indeed, these have been successfully used in settling some long-standing conjectures. This chapter is devoted to introducing the language of sheaves, sheaf co-homology, and to have some applications in topology and algebraic geometry.

4.1 Presheaves and Sheaves

In this section, we introduce and describe the category of presheaves, the category of sheaves, and the category of modules over ringed spaces.

Let (X, T) be a topological space. Then (X, T) can be treated as a category whose objects are members of the topology T, and for a pair U, V in T, $Mor(U, V) = \emptyset$ if U is not a subset of V, and if $U \subseteq V$, then $Mor(U, V) = \{i_U^V\}$, where i_U^V is the inclusion map from U to V. Evidently, $Mor(U, U) = \{I_U\}$, and $i_V^W o i_U^V = i_U^W$. This category is called the category of open subsets of X. A **presheaf** of abelian groups on a topological space (X, T) is a contra-variant functor from the category T of open subsets of X to the category AB of abelian groups. Thus, a presheaf F of abelian groups on a topological space X consists of the following: (i) For each open set U of X, there is an abelian group $F(U)$. (ii) For each pair of open sets, U and V with $U \subseteq V$, there is a homomorphism $f_V^U = F(i_U^V)$ from $F(V)$ to $F(U)$ such that

$$f_U^U = I_U, \text{ and } f_W^U = f_V^U o f_W^V, \text{ whenever } U \subseteq V \subset W.$$

Let F be a presheaf of abelian groups on a space X. Suppose that $F(\emptyset) = A$. Then F induces another presheaf $\tilde{F}$ on X defined as $\tilde{F}(U) = Ker f_U^\emptyset$ and $\tilde{f}_V^U = f_V^U | Ker f_V^\emptyset$. Evidently, $\tilde{F}(\emptyset) = 0$. Therefore, without any loss, we may assume

© Springer Nature Singapore Pte Ltd. 2021
R. Lal, *Algebra 3*, Infosys Science Foundation Series,
https://doi.org/10.1007/978-981-33-6326-7_4

that $F(\emptyset) = 0$ for any presheaf F. The members of $F(U)$ are called the sections on U (the terminology will be justified later). The homomorphism f_V^U is called the restriction map from V to U. If $s \in F(V)$, then $f_V^U(s)$ is called the restriction of the section s to U, and it is also denoted by $s|U$.

A presheaf F on X is called a **sheaf** if the following two conditions hold:

(i) If $s, t \in F(U)$ such that for an open cover $\{U_\alpha \mid \alpha \in \Lambda\}$ of U, the restriction s_α of s to U_α is the same as the restriction t_α of t to U_α for each α, then $s = t$, and

(ii) if $\{s_\alpha \in F(U_\alpha) \mid \alpha \in \Lambda\}$ is a family of sections such that the restriction of s_α to $U_\alpha \cap U_\beta$ is the same as the restriction of s_β to $U_\alpha \cap U_\beta$ for each pair α, β in Λ, then there is a section s in $F(\bigcup_{\alpha \in \Lambda} U_\alpha)$ such that the restriction of s to U_α is s_α for each α.

Similarly, we can talk of presheaves and sheaves of rings, R-modules, and more generally presheaves and sheaves of any algebraic structure.

A sheaf of rings is also called a **ringed space**. All the rings are assumed to be commutative rings with identities. Further, given a ringed space ϑ_X on X, a presheaf F_X of abelian groups on X together with left (right) $\vartheta_X(U)$-module structure on $F(U)$ for each U will be termed as presheaf of ϑ_X-modules if the restriction maps respect the corresponding module structures. If F_X is also a sheaf, then we simply term it as ϑ_X-module.

Example 4.1.1 Let X be a topological space, and let A be an abelian group. We have the constant sheaf F_X^A on X defined as follows:

(i) $F_X^A(U) = A$ whenever U is a nonempty open subset of X,
(ii) $F_X^A(\emptyset) = 0$,
(iii) the restriction map $f_U^V = I_A$ whenever $V \neq \emptyset$ and it is a zero map otherwise.Similarly, given a ring R, we have the ringed space ϑ_X^R on X given by $\vartheta_X^R(U) = R$ for all nonempty open sets U and $\vartheta_X^R(\emptyset) = \{0\}$. Further, if M is an R-module, then the constant sheaf F_X^M is ϑ_X^R-module in obvious manner.

Example 4.1.2 Let X be a topological space. For each open subset U of X, take $\vartheta_X(U) = C(U, \mathbb{R})$, where $C(U, \mathbb{R})$ is the additive group of real-valued continuous functions on U. ϑ_X together with obvious restriction maps from $\vartheta_X(U) = C(U, \mathbb{R})$ to $\vartheta_X(V) = C(V, \mathbb{R})$, $V \subseteq U$ is a presheaf of abelian groups. It follows from the patching lemma for continuous maps that ϑ_X is a sheaf. Indeed, since $C(U, \mathbb{R})$ is a ring for all U, ϑ_X is also a ringed space. This sheaf is called the sheaf of germs of continuous functions on X. Further, for each $n \in \mathbb{N}$, put $\vartheta_X^n(U) = C(U, \mathbb{R}^n)$, where $C(U, \mathbb{R}^n)$ denote the group of continuous functions from U to $\mathbb{R}^n$. This together with the obvious restriction maps forms a sheaf of abelian groups. Observe that $\vartheta_X^n(U)$ is a module over $\vartheta_X(U)$ in an obvious manner, and in turn, ϑ_X^n is a ϑ_X-module.

Example 4.1.3 Let M be a C^∞ differentiable manifold M. For each open subset U of M, take $\vartheta_M(U) = C^\infty(U, \mathbb{R})$, where $C^\infty(U, \mathbb{R})$ is the ring of real-valued C^∞ functions on U. This together with the obvious restriction maps from $\vartheta_M(U)$ to

$\vartheta_M(V)$, $V \subseteq U$ forms a ringed space. This sheaf is called the sheaf of germs of C^∞ functions on M. As in the above example, ϑ_M^n is an ϑ_M-module.

Example 4.1.4 Let X be a discrete topological space containing more than one element. Take $F(X) = \mathbb{Z}$ and $F(U) = 0$ for all proper open sets U. Then F defines a presheaf in obvious manner. This presheaf is not a sheaf, for if $\{U_\alpha \mid \alpha \in \Lambda\}$ is an open cover of X consisting of proper open sets, then the restriction of $1 \in \mathbb{Z}$ to U_α is the same as the restriction of 0 to U_α for each α, but $1 \neq 0$.

Example 4.1.5 (*Spec R*) This is an important example which will be referred again and again. Let R be a commutative ring with identity. Consider $Spec R$ with the Zariski topology (see Exercise 1.1.15). Thus, $Spec R = \{P \mid P \text{ is a prime ideal of } R\}$, and an open set U of $Spec R$ is of the form $Spec R - V(A)$, where A is an ideal of R and $V(A) = \{P \in Spec R \mid A \subseteq P\}$. For each prime ideal P of R, let R_P denote the localization at P. More explicitly, R_P is the ring $\{\frac{a}{h} \mid a \in R \text{ and } h \notin P\}$ of fractions. For each open subset U of $Spec R$, let $\vartheta^R(U)$ denote the subring $\{s \in \prod_{P \in U} R_P \mid s \text{ is locally constant}\}$ of the Cartesian product $\prod_{P \in U} R_P$. More explicitly, an element $s \in \prod_{P \in U} R_P$ is a member of $\vartheta^R(U)$ if and only if for each $P \in U$, there is an open subset V of $Spec R$, and elements $a, f \in R$ such that $P \in V \subseteq U$, $f \notin Q$ for each $Q \in V$ and also $s(Q) = \frac{a}{f}$. We have an obvious restriction map j_U^V from $\vartheta^R(U)$ to $\vartheta^R(V)$ whenever $V \subseteq U$. ϑ^R together with these restriction homomorphisms defines a sheaf of rings on $Spec R$. This ringed space is denoted by $(Spec R, \vartheta^R)$, and it is called the **spectrum** of R. ϑ^R is called the structure sheaf of $Spec R$.

Let F be a presheaf of abelian groups on X. Let $x \in X$. Let N_x denote the family of all open subsets of X containing x. Then $(N_x, \leq)$ is a directed set under the relation $\leq$ given by $V \leq U$ if $U \subseteq V$. In turn, we have a directed system $\{(F(V), f_V^U) \mid V \leq U, \ U, V \in N_x\}$ of abelian groups. The direct limit $Lim_\to F(V)$ of this directed system is an abelian group called the **stalk** at x, and it is denoted by F_x. By the construction of the direct limit, it is clear that F_x is the set $\{\bar{a}_U \mid a_U \in F(U), \ U \in N_x\}$ of equivalence classes, where $\bar{a}_U = \bar{b}_V$ if and only if there is a $W \in N_x$, $W \subseteq U \cap V$ such that $f_U^W(a_U) = f_V^W(b_V)$. The operation $+$ in F_x is given by $\bar{a}_U + \bar{b}_V = \bar{c}_W$, where $W = U \cap V$ and $c_W = f_U^W(a_U) + f_V^W(b_V)$. Evidently, $\bar{a}_U$ represents the zero element of F_x if and only if there is a $W \in N_x$, $W \subseteq U$ such that $f_U^W(a_U) = 0$. Similarly, if ϑ is a ringed space on X, the stalk ϑ_x at x is a ring, and if F is a ϑ-module, then the stalk F_x at x is a ϑ_x-module for each $x \in X$.

For all $x \in X$, the stalk of the presheaf F_X^A of Example 4.1.1 at x is the abelian group A, the stalk of the ringed space ϑ_X^R of Example 4.1.1 is the ring R, and the stalk of the ϑ_X^R-module F_M is the R-module M. In Example 4.1.2, the stalk $(\vartheta_X)_x$ is the ring of germs of continuous real-valued functions at x, and the stalk $(\vartheta_X^n)_x$ is the free $(\vartheta_X)_x$-module of rank n. In Example 4.1.3, the stalk $(\vartheta_M)_x$ is the ring of germs of C^∞ real-valued functions at x, and the stalk $(\vartheta_M^n)_x$ is the free $(\vartheta_M)_x$-module of rank n. In case of Example 4.1.4, $F_x = 0$ for all x. In Example 4.1.5, the stalk ϑ_P^R of the ringed space $(Spec R, \vartheta^R)$ at P can be easily seen to be the ring R_P. Evidently, R_P is a local ring in the sense that it has the unique maximal ideal $\{\frac{a}{f} \mid a \in P, f \notin P\}$

consisting of noninvertible elements. A ringed space ϑ on the space X is called a
locally ringed space if all the stalks are local rings. Thus, $(Spec R, \vartheta^R)$ is a locally
ringed space.

Let F and G be presheaves (sheaves) of abelian groups (more generally, of R-
modules) on a topological space X. Then F and G can be thought of as contra-variant
functors from the category of open subsets of X to the category of abelian groups
(R-modules). Similarly, if they are presheaves or sheaves of rings, then they can
be thought of as contra-variant functors from the category of open subsets of X
to the category of rings. A natural transformation η from F to G is also called a
morphism from F to G. Thus, a morphism η from a presheaf (sheaf) F to a presheaf
(sheaf) G is a family $\{\eta_U \in Hom(F(U), G(U)) \mid U \ is \ an \ open \ subset \ of \ X\}$ of
homomorphisms such that whenever $U \subseteq V$, the diagram

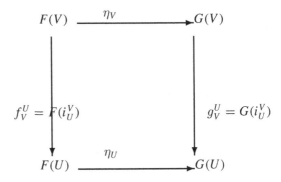

is commutative, where the vertical maps are corresponding restriction maps. This
gives us the category Pr_X (Sh_X) of presheaves (sheaves) on X. Evidently, Sh_X is a
subcategory of Pr_X. The constant sheaf 0 given by $0(U) = 0$ for all U is the zero
object in the category Pr_X (Sh_X).

Let ϑ_X be a ringed space on X. A ϑ_X-morphism from a ϑ_X-module M_X to a
ϑ_X-module N_X is a morphism η from the sheaf M_X of abelian groups to the sheaf
N_X of abelian groups such that η_U is a $\vartheta_X(U)$-module homomorphism for each U.
This gives us a category $\vartheta_X - Mod$ of left ϑ_X-modules. Observe that a ringed space
ϑ_X is a module over itself.

A sub-presheaf K of a presheaf F of abelian groups (R-modules) on X is a family
$\{K(U) \mid K(U) \ is \ a \ subgroup \ (R - submodule) \ of \ F(U) \ for \ all \ U\}$ such that
$f_U^V(K(U)) \subseteq K(V)$ whenever $V \subseteq U$. Similarly, we can talk of sub-presheaf of a
presheaf of rings. Thus, a sub-presheaf is a presheaf at its own right. Observe that a
sub-presheaf of sheaf need not be a sheaf. If a sub-presheaf is also a sheaf, then we
call it a sub-sheaf.

Let ϑ_X be a ringed space on X, and M_X be a left ϑ_X-module. A sub-sheaf N_X
of the sheaf M_X of abelian groups is called a ϑ_X-submodule of M_X if $N_X(U)$ is a
$\vartheta_X(U)$-submodule of $M_X(U)$ for each U. Evidently, N_X is a ϑ_X-module at its own
right.

Let η be a morphism from a presheaf F on X to a presheaf G on X. Let K be a sub-presheaf of F. Then the family $\{\eta_U(K(U)) \mid U \ is \ open \ set\}$ together with the family $\{g_U^V|_{\eta_U(K(U))} \mid V \subseteq U\}$ of restriction homomorphisms defines a sub-presheaf of G. This sub-presheaf of G is called the **image** of η, and it is denoted by $Im(\eta)$ or $\eta(K)$. Observe that the image of a sub-sheaf need not be a sub-sheaf even if F and G are both sheaves. Let L be a sub-presheaf of G. Then the family $\{\eta_U^{-1}(L(U)) \mid U \ is \ open \ set\}$ together with the family $\{f_U^V|_{\eta_U^{-1}(L(U))} \mid V \subseteq U\}$ of restriction homomorphisms defines a sub-presheaf of G. This sub-presheaf of G is called the **inverse image** of η, and it is denoted by $\eta^{-1}(L)$. In particular, the inverse image $\eta^{-1}(0)$ of the constant zero sheaf 0 is a presheaf of F. This presheaf is called the kernel of η, and it is denoted by $Ker \ \eta$. Here again, the inverse image of a sub-sheaf under a presheaf morphism need not be a sheaf. Next, let K be a sub-presheaf of the presheaf F of R-modules on a space X. The family $\{F(U)/K(U) \mid U \ is \ an \ open \ set\}$ of quotient R-modules together with obvious homomorphisms from $F(U)/K(U)$ to $F(V)/K(V)$, $V \subseteq U$ induced by f_U^V forms a presheaf on X. This presheaf is called the quotient presheaf, and it is denoted by F/K. We have the obvious quotient morphism ν from F to F/K whose kernel is the sub-presheaf K.

Let ϑ_X be a ringed space on X. Let η be a ϑ_X-morphism from a ϑ_X-module M_X to a ϑ_X-module N_X. Let K_X be a ϑ_X-submodule of M_X and L_X be a ϑ_X-submodule of N_X. Then the image $\eta(K_X) = \{\eta_U(K_X(U)) \mid U \ is \ open\}$ is a presheaf of ϑ_X-submodules of N_X but, it need not be a ϑ_X-submodule of N_X as it need not be a sub-sheaf of the sheaf N_X. However, $\eta^{-1}(L_X) = \{\eta_U^{-1}(L(U)) \mid U \ is \ open\}$ is a ϑ_X-submodule of M_X, and it is called the inverse image of L_X. In particular, the submodule $\eta^{-1}(0_X)$ is called the kernel of η. If K_X is a ϑ_X-submodule of M_X, then we have the quotient ϑ_X-module M_X/K_X given by $(M_X/K_X)(U) = M_X(U)/K_X(U)$. We have the obvious quotient morphism ν from M_X to M_X/K_X whose kernel is K_X. It is easy to observe that the analogues of the correspondence theorem, fundamental theorem of morphisms, and the isomorphism theorems hold in the category of ϑ_X-modules. Indeed, we shall see that the category ϑ_X-Mod is an abelian category.

A sub-presheaf K of a presheaf F of rings is called an ideal of the presheaf F if $K(U)$ is an ideal of $F(U)$ for all U. Thus, if K is an ideal of a presheaf F on X, then we have a quotient presheaf F/K of rings. The analogues of the fundamental theorem of homomorphism, correspondence theorem, and isomorphism theorems hold for presheaves of rings also. Observe that the category of rings is not an abelian category.

Proposition 4.1.6 (i) *Let η be a morphism from a presheaf F on X to a presheaf G on X. Then $Ker \ \eta$ together with the inclusion morphism i from $Ker \ \eta$ to F represents the kernel of the morphism η in the categorical sense in the category Pr_X of presheaves on a space X.*

(ii) *A morphism η from a presheaf F to a presheaf G is a monomorphism in the category Pr_X if and only if η_U is injective for all U. In particular, η is a monomorphism if and only if the kernel of η is the constant zero sheaf on X.*

(iii) *If F and G are both sheaves, then the inverse image $\eta^{-1}(L)$ of a sub-sheaf L of G is a sub-sheaf of F. In particular, $Ker\ \eta$ is a sub-sheaf of F.*

Proof (i) Evidently, the inclusion morphism i from $Ker\ \eta$ to F is such that $\eta o i$ is the zero morphism. Let L be a presheaf and j be a morphism from L to F such that $\eta o j$ is a zero morphism. Then $\eta_U o j_U$ is a zero homomorphism from $L(U)$ to $G(U)$ for each U. In turn, for each U, we have a unique homomorphism ρ_U from $L(U)$ to $K(U)$ such that $i_U o \rho_U = j_U$ for each U. Clearly, $\rho = \{\rho_U \mid U \text{ is open set}\}$ is a unique morphism from L to K such that $i o \rho = j$. This shows that $Ker\ \eta$ together with i represents the kernel of η in the category Pr_X.

(ii) Let η be a morphism from a presheaf F to a presheaf G. Evidently, if η_U is injective for each U, then η is a monomorphism. Suppose that η_U is not injective for some open set U of X. Then $\eta o i$ and $\eta o 0$ are both zero morphisms from $Ker\ \eta$ to G, but $i \neq 0$.

(iii) Suppose that F and G are sheaves. Let L be a sub-sheaf of G. We need to show that the presheaf $\eta^{-1}(L)$ is a sheaf. Let $\{U_\alpha \mid \alpha \in \Lambda\}$ be an open cover of U, and $\{a_\alpha \in \eta_{U_\alpha}^{-1}(L(U_\alpha)) \mid \alpha \in \Lambda\}$ be a compatible family in the sense that $k_{U_\alpha}^{U_\alpha \cap U_\beta}(a_\alpha) = k_{U_\beta}^{U_\alpha \cap U_\beta}(a_\beta)$ for all $\alpha, \beta \in \Lambda$, where k_U^V denotes the restriction $f_U^V|\eta_U^{-1}(L(U))$. Since F is a sheaf, there is a unique element $a \in F(U)$ such that $f_U^{U_\alpha}(a) = a_\alpha$ for each α. Now $g_U^{U_\alpha}(\eta_U(a)) = \eta_{U_\alpha}(f_U^{U_\alpha}(a)) = \eta_{U_\alpha}(a_\alpha)$ is a member of $L(U_\alpha)$ for each α. Since L is a sheaf, $\eta_U(a) \in L(U)$. Thus, we have a unique $a \in \eta_U^{-1}(L(U))$ such that $k_U^{U_\alpha}(a) = a_\alpha$ for each α. This shows that $\eta^{-1}(L)$ is a sheaf. ♯

Similarly, we have the following proposition.

Proposition 4.1.7 *(i) Let η be a morphism from a presheaf F to a presheaf G of abelian groups (R-modules) on X. Then the quotient presheaf $G/\eta(F)$ together with the quotient morphism ν from G to $G/\eta(F)$ represents the co-kernel of the morphism η in the category Pr_X of presheaves of abelian groups (R-modules) on a space X.*

(ii) A morphism η from a presheaf F to a presheaf G is an epimorphism in the category Pr_X if and only if η_U is surjective for all U. In particular, η is a epimorphism in Pr_X if and only if the co-kernel of η is the constant zero sheaf on X.

(iii) In the category Pr_X of presheaves of abelian groups (R-modules), a monomorphism is a kernel of its co-kernel, and an epimorphism is a co-kernel of its kernel.

Proof The proofs of (i) and (ii) are similar to the proofs of (i) and (ii) in Proposition 4.1.6. We prove the part (iii) of the proposition. Let η be a monomorphism from a presheaf F to a presheaf G in the category Pr_X of abelian groups (R-modules) on X. Then from Proposition 4.1.6 (ii), η_U is an injective homomorphism from $F(U)$

to $G(U)$ for each open set U. The co-kernel of η is the quotient morphism ν from G to $G/\eta(F)$. Already $\nu o\eta$ is a zero morphism. Let ρ be a morphism from a presheaf L to G such that $\nu o\rho$ is a zero morphism. Then $\rho_U(L(U)) \subseteq \eta_U(F(U))$ for each open set U. Since η_U is injective, we have a unique homomorphism μ_U from $L(U)$ to $F(U)$ such that $\eta_U o\mu_U = \rho_U$. The fact that η is a monomorphism ensures that the family $\mu = \{\mu_U \mid U \text{ is open}\}$ is the unique morphism from L to F such that $\eta o\mu = \rho$. This proves that η is a kernel of its co-kernel. Similarly, it can be shown that every epimorphism is a co-kernel of its kernel. ♯

Remark 4.1.8 A monomorphism in the category of presheaves of rings need not be a kernel of a morphism.

Theorem 4.1.9 *The category Pr_X of presheaves of abelian groups (R-modules) on a space X is an abelian category.*

Proof From Propositions 4.1.6 and 4.1.7, it follows that in the category Pr_X, every morphism has a kernel as well as a co-kernel, every monomorphism is a kernel, and every epimorphism is a co-kernel. Thus, it is sufficient to show that the category Pr_X is an exact category. Let η and ρ be two members of $Mor_{Pr_X}(F, G)$. Then $\eta + \rho$ given by $(\eta + \rho)_U = \eta_U + \rho_U$ is again a member of $Mor_{Pr_X}(F, G)$. This defines an addition $+$ in $Mor_{Pr_X}(F, G)$ which makes it an abelian group. It is also easy to observe that the composition law of morphisms is bi-additive. This shows that Pr_X is an abelian category. ♯

Remark 4.1.10 The category Sh_X of sheaves of abelian groups and the category ϑ_X-Mod of ϑ_X-modules are also abelian categories. However, the above proof fails as the image of a sheaf under a morphism need not be a sheaf, and in turn, the image of a ϑ_X-module under ϑ_X-morphism need not be a ϑ_X-module. We need to sheafify the image presheaf of the sheaf. The concept of sheafification of a presheaf will be introduced soon, and then we shall establish the fact that the category Sh_X of sheaves of abelian groups and the category ϑ_X-Mod of ϑ_X-modules are abelian categories.

Proposition 4.1.11 *(i) A morphism η from a presheaf F on X to a presheaf G on X is an isomorphism if and only if for each open subset U of X, η_U is an isomorphism.*

(ii) A ϑ_X-morphism from a ϑ_X-module M_X to a ϑ_X-module N_X is an isomorphism if and only if η_U is a $\vartheta_X(U)$-isomorphism from $M_X(U)$ to $N_X(U)$.

Proof (i) Suppose that η is an isomorphism. Then there is a morphism ρ from G to F such that $\rho o\eta = I_F$ and $\eta o\rho = I_G$. Evidently, $\rho_U o\eta_U = I_{F(U)}$ and $\eta_U o\rho_U = I_{G(U)}$. This shows that η_U is an isomorphism for each open subset U of X. Conversely, suppose that η_U is an isomorphism for each U. Since $\eta_U of_V^U = g_V^U o\eta_V$ for each pair U, V, $\eta_U^{-1} og_V^U = f_V^U o\eta_V^{-1}$ for each pair U, V with $U \subseteq V$. Thus, $\eta^{-1} = \{\eta_U^{-1} \mid U \text{ is open subset of } X\}$ is a morphism from G to F such that $\eta^{-1} o\eta = I_F$ and $\eta o\eta^{-1} = I_G$.

(ii) The proof of this part is similar to that of (i). ♯

Let η be a morphism from a presheaf F on X to a presheaf G on X. Let x be an element of X. Suppose that $\bar{a}_U = \bar{b}_V$ in F_x. Then there is a $W \in N_x$ such that $W \subseteq U \cap V$ and $f_U^W(a_U) = f_V^W(b_V)$. Since η is a morphism, $g_U^W(\eta_U(a_U)) = g_V^W(\eta_V(b_V))$. Thus, η induces a homomorphism η_x from F_x to G_x given by $\eta_x(\bar{a}_U) = \overline{\eta_U(a_U)}$. If ρ is another morphism from the presheaf G to a presheaf H, then it is clear from the definition that $(\rho \circ \eta)_x = \rho_x \circ \eta_x$. Also $(I_F)_x = I_{F_x}$. Thus, for each $x \in X$, we get a functor from the category Pr_X of presheaves on X to the category of abelian groups (R-modules) which associates with each presheaf F, the stalk F_x at x and with each morphism η, the homomorphism η_x. Similarly, given a ringed space ϑ_X on X, for each $x \in X$, we have a natural functor from the category $\vartheta_X\text{-}Mod$ of ϑ_X-modules to the category of $(\vartheta_X)_x$-modules which associates with each ϑ_X-module M_X, the stalk $(M_X)_x$ at x, and with a ϑ_X-morphism η from M_X to N_X, the $(\vartheta_X)_x$-homomorphism from $(M_X)_x$ to $(N_X)_x$. In particular, if η is an isomorphism, then η_x is an isomorphism for each $x \in X$. The following theorem asserts that the converse of this last statement is also true in the categories Sh_X and $\vartheta_X - mod$.

Theorem 4.1.12 *(i) A morphism η from a sheaf F on X to a sheaf G on X is an isomorphism if and only if η_x is an isomorphism from F_x to G_x for each $x \in X$.*
(ii) A ϑ_X-morphism η from M_X to N_X is an isomorphism if and only if η_x is an isomorphism for each $x \in X$.

Proof (i) Let η be a morphism from a sheaf F on X to a sheaf G on X such that η_x is an isomorphism for each $x \in X$. In the light of Proposition 4.1.6, it is sufficient to show that η_U is bijective for each U. Fix an open subset U of X. Suppose that $a \in Ker\ \eta_U$. Then $\eta_U(a) = 0$. Let x be a member of U. Then $\eta_x(\bar{a}) = \overline{\eta_U(a)} = 0$. Since η_x is an isomorphism, $\bar{a} = 0$ in F_x. This means that there is a $W_x \in N_x$, $W_x \subseteq U$ such that $f_U^{W_x}(a) = 0$. Thus, we have an open cover $\{W_x \mid x \in U\}$ of U such that $f_U^{W_x}(a) = 0$ for each x. Since F is a sheaf, $a = 0$. This shows that η_U is injective for each open set U.
Next, we show that η_U is surjective for each U. Fix an open subset U of X. Let b be a member of $G(U)$. For each $x \in U$, let b_x denote the member of G_x determined by b. Since η_x is surjective for each x, there is an element $a_x \in F_x$ such that $\eta_x(a_x) = b_x$. Suppose that $a_x = \overline{a_{V_x}}$, where $V_x \in N_x$, $V_x \subseteq U$, and $a_{V_x} \in F(V_x)$. Then $b_x = \eta_x(a_x) = \overline{\eta_{V_x}(a_{V_x})}$. This means that there is a $W_x \in N_x$, $W_x \subseteq U \cap V_x$ such that

$$g_U^{W_x}(b) = g_{V_x}^{W_x}(\eta_{V_x}(a_{V_x})) = \eta_{W_x}(f_{V_x}^{W_x}(a_{V_x})). \tag{4.1}$$

This gives us an open cover $\{W_x \mid x \in U\}$ of U, the family

$$\{a_{W_x} = f_{V_x}^{W_x}(a_{V_x}) \in F(W_x) \mid x \in U\},$$

and the family

$$\{b_{W_x} = g_U^{W_x}(b) = \eta_{W_x}(f_{V_x}^{W_x}(a_{V_x})) \mid x \in U\}.$$

Now,

$$\eta_{W_x \cap W_y}(f_{W_x}^{W_x \cap W_y}(a_{W_x}))$$
$$= \eta_{W_x \cap W_y}(f_{W_x}^{W_x \cap W_y}(f_{V_x}^{W_x}(a_{V_x})))$$
$$= \eta_{W_x \cap W_y}(f_{V_x}^{W_x \cap W_y}(a_{V_x}))$$
$$= g_{V_x}^{W_x \cap W_y}(\eta_{V_x}(a_{V_x}))$$
$$= g_{W_x}^{W_x \cap W_y}(g_{V_x}^{W_x}(\eta_{V_x}(a_{V_x})))$$
$$= g_{W_x}^{W_x \cap W_y}(g_U^{W_x}(b)) \; by \; (1)$$
$$= g_U^{W_x \cap W_y}(b)$$

for all $x, y \in U$. Similarly,

$$\eta_{W_x \cap W_y}(f_{W_y}^{W_x \cap W_y}(a_{W_y})) \; (by \; 1)$$
$$= g_U^{W_x \cap W_y}(b)$$

for all $x, y \in U$. Since η_U is already seen to be injective for all U,

$$f_{W_x}^{W_x \cap W_y}(a_{W_x}) = f_{W_y}^{W_x \cap W_y}(a_{W_y})$$

for all $x, y \in U$. Since F is a sheaf, there is an element $a \in F(U)$ such that $f_U^{W_x}(a) = a_{W_x}$ for each x. Further, it is clear that

$$g_U^{W_x}(\eta_U(a)) = g_U^{W_x}(b)$$

for all $x \in U$. Again, since G is a sheaf, $\eta_U(a) = b$. This shows that η_U is surjective for all U.

(ii) This follows from (i) if we observe that M_X and N_X are also sheaves of abelian groups and η_x is $(\vartheta_X)_x$-isomorphism if and only if it is a group isomorphism from $(M_X)_x$ to $(N_X)_x$. ♯

Remark 4.1.13 It is clear from the proof of the above theorem that a morphism η in the category Sh_X (ϑ_X-Mod) is a monomorphism if and only if η_x is injective for each $x \in X$.

Our next aim is to show that Sh_X and ϑ_X-Mod are abelian categories.

Proposition 4.1.14 (i) *Let η be a monomorphism from a sheaf F on X to a sheaf G on X in the category Sh_X. Then $\eta(F)$ is a subsheaf of G which is isomorphic to the sheaf F.*

(ii) *Let F be a subsheaf of a sheaf G. Then the quotient presheaf G/F is a sheaf.
Further, the quotient morphism is an epimorphism which is a co-kernel of the
inclusion morphism i from F to G.*

(iii) *The homomorphism η_U need not be a surjective homomorphism even if the morphism η from a sheaf F to a sheaf G is an epimorphism in the category Sh_X.
However, a morphism η in Sh_X is an isomorphism if and only if it is a monomorphism as well as an epimorphism in Sh_X. In particular, if a monomorphism η
in Sh_X is an epimorphism, then η_U is surjective for all U.*

Proof (i) If η is a monomorphism, then $Ker\ \eta_U$ is zero for all U, and in turn, η_U is
an isomorphism from $F(U)$ to $\eta_U(F(U))$ for all U. This means that η induces
an isomorphism from F to $\eta(F)$, and $\eta(F)$ is a subsheaf of G.

(ii) The proof is a straightforward verification.

(iii) Let $\Lambda_\mathbb{C}$ denote the sheaf of germs of analytic function on the complex plane $\mathbb{C}$.
More explicitly, for each open subset U of $\mathbb{C}$, $\Lambda_\mathbb{C}(U)$ is the group of analytic
functions on U, and if $V \subseteq U$, the restriction homomorphism λ_U^V from $\Lambda_\mathbb{C}(U)$
to $\Lambda_\mathbb{C}(V)$ is given by $\lambda_U^V(f) = f|_V$. Evidently, $\Lambda_\mathbb{C}$ is a sheaf of abelian groups
on $\mathbb{C}$. We have a morphism $D = \{D_U \mid U\ is\ open\}$ from $\Lambda_\mathbb{C}$ to itself, where
D_U is the derivation map from $\Lambda_\mathbb{C}(U)$ to itself given by $D_U(f) = f'$. If we
take $U = \mathbb{C}^\star = \mathbb{C} - \{0\}$, then the map ω given by $\omega(z) = \frac{1}{z}$ belongs to
$\Lambda_\mathbb{C}(U)$ but there is no function in $\Lambda_\mathbb{C}(U)$ whose image under D_U is ω. Thus,
D is not an epimorphism in $Pr_\mathbb{C}$. However, we show that D is an epimorphism
in the category $Sh_\mathbb{C}$. Suppose that μ and ν are morphisms from $\Lambda_\mathbb{C}$ to a sheaf L
on $\mathbb{C}$ such that $\mu oD = \nu oD$. Let U be an open subset of $\mathbb{C}$ and f be a member
of $\Lambda_\mathbb{C}(U)$. Then f is analytic on U. For each $z \in U$, there is an open ball U_z
with center z such that $U_z \subseteq U$. Since f is analytic on U_z and U_z is simply
connected, there is a member $g_z \in \Lambda_\mathbb{C}(U_z)$ such that $D_{U_z}(g_z) = f_z = f|_{U_z}$.
In turn, $\mu_{U_z}(f_z) = \nu_{U_z}(f_z)$. Thus, we get an open cover $\{U_z \mid z \in U\}$ of U such
that $\mu_{U_z}(f_z) = \nu_{U_z}(f_z)$ for each $z \in \mathbb{C}$. Since $\Lambda_\mathbb{C}$ is a sheaf, $\mu_U(f) = \nu_U(f)$.
This shows that $\mu = \nu$.

Clearly, an isomorphism is a monomorphism as well as an epimorphism. Conversely, let η be a monomorphism and also an epimorphism from a sheaf F to
a sheaf G in the category Sh_X. Since η is a monomorphism, η_U is injective for
each U, and the quotient presheaf $G/\eta(F)$ is a sheaf. Let ν denote the quotient
morphism from G to $G/\eta(F)$. Then $\nu o\eta = 0 o\eta$ where 0 is the zero morphism
from G to $G/\eta(F)$. Since η is an epimorphism, ν is the zero morphism. This
shows that η_U is surjective for each U. Thus, η_U is an isomorphism for each U.
From Proposition 4.1.11, η is an isomorphism. ♯

Corollary 4.1.15 (i) *Let η be a monomorphism in the category ϑ_X-Mod from a
ϑ_X-module M_X to a ϑ_X-module N_X. Then the image $\eta(M_X)$ is a ϑ_X- submodule
of N_X which is isomorphic to M_X.*

(ii) *If* K_X *is a* ϑ_X-*submodule of* M_X, *then* $M_X/K_X = \{M_X(U)/K_X(U) \mid U$ *is open*$\}$ *is a* ϑ_X-*module. Further the quotient morphism is an epimorphism.*

(iii) *A morphism in the category* ϑ_X-*Mod is an isomorphism if and only if it is a monomorphism as well as an epimorphism.*

Proof (i) From the above proposition, $\eta(M_X)$ is a sub-sheaf of N_X as a sheaf of abelian groups. Further, since η is a ϑ_X-morphism, $\eta_U(M_X(U))$ is a ϑ_X-submodule of $N_X(U)$ for all U. This shows that $\eta(M_X)$ is a ϑ_X-submodule of N_X. Since η_U is injective for all U, M_X and $\eta(M_X)$ are isomorphic.

(ii) Again, from the above proposition, M_X/K_X is a sheaf of abelian groups. Since $K_X(U)$ is a $\vartheta_X(U)$-submodule of $M_X(U)$, $M_X(U)/K_X(U)$ is a $\vartheta_X(U)$-module. This shows that M_X/K_X is a ϑ_X-module. It is evident that the quotient morphism ν is an epimorphism in the category ϑ_X-*Mod*.

(iii) Suppose that η is a monomorphism and also an epimorphism in the category ϑ_X-*Mod*. Then as in the above proposition, η_U is an isomorphism from $M_X(U)$ to $N_X(U)$. It follows that η is an isomorphism. ♯

Sheafification

Definition 4.1.16 A sheafification of a presheaf F on a topological space X is a pair (SF, j), where SF is a sheaf, and j is a morphism from F to SF such that given any pair (G, ρ), where G is a sheaf and ρ is a morphism from F to G, there is a unique morphism μ from SF to G such that $\mu \circ j = \rho$.

Theorem 4.1.17 *The sheafification of a presheaf F on X exists, and it is unique up to natural equivalence in the sense that if (SF, j) and $(S'F, j')$ are sheafifications of F, then there is a unique isomorphism μ from SF to $S'F$ such that $\mu \circ j = j'$.*

Proof Firstly, we prove the uniqueness. Let (SF, j) and $(S'F, j')$ be sheafifications of F. Then there is a unique morphism μ from SF to $S'F$ and a unique morphism μ' from $S'F$ to SF such that $\mu \circ j = j'$ and $\mu' \circ j' = j$. In turn, $(\mu' \circ \mu) \circ j = j = I_{SF} \circ j$. Again, from the universal property of sheafification, $\mu' \circ \mu = I_{SF}$. Similarly, $\mu \circ \mu' = I_{S'F}$.

Now, to prove the existence, we construct the sheafification of a presheaf. Let F be a presheaf on X. For each open subset U of X, let $SF(U)$ denote the subgroup of $\prod_{x \in U} F_x$ consisting of the members $s \in \prod_{x \in U} F_x$ with the property that for each $x \in U$, there is an open set V_x with $x \in V_x$, $V_x \subseteq U$ and an element $a_x \in F(V_x)$ such that a_x represents $s(y)$ in F_y for each $y \in V_x$. If $V \subseteq U$, then we have a homomorphism Sf_U^V from $SF(U)$ to $SF(V)$ given by $Sf_U^V(s) = s|V$, where $s|V$ denotes the restriction of s to V (more explicitly, $s|_V$ is the projection from $\prod_{x \in U} F_x$ to $\prod_{x \in V} F_x$). It follows from the construction that SF together with the family $\{Sf_U^V \mid V \subseteq U\}$ of homomorphisms is a sheaf of abelian groups. Further, for each open subset U of X, we have a homomorphism j_U from $F(U)$ to $SF(U)$ given by $j_U(s)(x) = s_x$, where s_x is the member of F_x determined by $s \in F(U)$. It is also clear from the construction that $j = \{j_U \mid U$ *is open*$\}$ is a morphism from F to SF. Let G be a sheaf, and ρ be a morphism from F to G. For each open subset U of X, we have a homomorphism μ_U

from $SF(U)$ to $G(U)$ defined as follows: Let $s \in SF(U)$. Then for each $x \in U$, we have an open subset V_x together with an element $a_x \in F(V_x)$ such that $x \in V_x \subseteq U$ and a_x represents $s(y)$ for each $y \in V_x$. In turn, we have an open cover $\{V_x \mid x \in U\}$ of U and the family $\{\rho_{V_x}(a_x) \in G(V_x)\}$. Since ρ is a morphism and $s \in SF(U)$, $g_{V_x}^{V_x \cap V_y} \rho_{V_x}(a_x) = g_{V_y}^{V_x \cap V_y} \rho_{V_y}(a_y)$ for each $x, y \in U$. Since G is a sheaf, there is a unique member $\mu_U(s)$ in $G(U)$ such that $g_U^{V_x}(\mu_U(s)) = \rho_{V_x}(a_x)$ for each $x \in U$. This gives us a morphism μ from SF to G such that $\mu o j = \rho$. Further, if μ' is another morphism from SF to G such that $\mu' o j = \rho$, then $g_U^{V_x}(\mu'_U(s)) = \rho_{V_x}(a_x) = g_U^{V_x}(\mu_U(s))$ for each x. Since G is a sheaf, $\mu_U = \mu'_U$ for each U. This completes the construction of a sheafification. ♯

Corollary 4.1.18 *Let F be a presheaf ϑ_X-module, where ϑ_X is a ringed space on X. Let (SF, j) denote the sheafification of F considered as a presheaf of abelian groups. Then we have a unique ϑ_X-module structure on SF such that*

(i) j is a ϑ_X-morphism from F to SF, and
(ii) given any ϑ_X-module G and a ϑ_X-morphism from F to G, there is a unique ϑ_X-morphism ρ from SF to G such that $\rho o j = \eta$.

Proof Let U be an open subset of X and $x \in U$. Suppose that a_U and b_U are members of $F(U)$ such that $\overline{a_U} = \overline{b_U}$ in F_x. Then there is an open set U_x, $x \in U_x \subseteq U$ such that $a_U|_{U_x} = b_U|_{U_x}$. Let α be a member of $\vartheta_X(U)$. Since F is a presheaf ϑ_X-module, $(\alpha a_U)|U_x = (\alpha b_U)|U_x$. This ensures the existence of an external product $\cdot$ from $\vartheta_X(U) \times F_x$ to F_x given by $\alpha \cdot \overline{a_U} = \overline{\alpha a_U}$. It can be easily seen that F_x is a $\vartheta_X(U)$-module with respect to this product. In turn, $\prod_{x \in U} F_x$ is also a $\vartheta_X(U)$-module. From the definition of $SF(U)$, it follows that $SF(U)$ is a $\vartheta_X(U)$-submodule of $\prod_{x \in U} F_x$. This makes $SF(U)$ a ϑ_X-module such that j is a ϑ_X-morphism. The rest is evident.   ♯

Remark 4.1.19 It follows from the construction (also from the universal property of sheafification) that j is a monomorphism, and j_x is an isomorphism from F_x to $(SF)_x$ for each $x \in X$. Evidently, the sheafification of a sheaf is the sheaf itself.

Let η be a morphism from a sheaf F on X to a sheaf G on X. Let $\eta(F)$ denote the image presheaf of η, i the inclusion morphism from $\eta(F)$ to G, and $\hat{\eta}$ the obvious morphism from F to $\eta(F)$ so that $\eta = i o \hat{\eta}$. Let $(S\eta(F), j)$ denote the sheafification of $\eta(F)$. From the universal property of sheafification, there is a unique morphism ρ from $S\eta(F)$ to G such that $\rho o j = i$. Let μ denote the morphism $j o \hat{\eta}$ from F to $S\eta(F)$. Evidently, $\eta = \rho o \mu$. Let ϕ and ψ be morphisms from $S\eta(F)$ to a sheaf L such that $\phi o \mu = \psi o \mu$. This means that $\phi o i o \hat{\eta} = \psi o i o \hat{\eta}$. Since $\hat{\eta}$ is an epimorphism in the category Pr_X, it follows that $\phi o i = \psi o i$. Since L is a sheaf, it follows from the universal property of sheafification that $\phi = \psi$. This shows that μ is an epimorphism. Further, $\rho_x o j_x = i_x$ for each $x \in X$. Since j_x is an isomorphism and i_x is injective

(Remark 4.1.15) for all x, it follows that ρ_x is injective for all $x \in X$. Since $S\eta(F)$ and G are sheaves, ρ is a monomorphism in Sh_X (Remark 4.1.12). We have established the following proposition.

Proposition 4.1.20 *Every morphism η in Sh_X can be expressed as $\eta = \rho \circ \mu$ where ρ is a monomorphism and μ is an epimorphism.* ♯

Similarly, the arguments used to establish the above proposition can be imitated to establish the following proposition.

Proposition 4.1.21 *Every morphism η in ϑ_X-Mod can be expressed as $\eta = \rho \circ \mu$ where ρ is a monomorphism and μ is an epimorphism.* ♯

The morphism $S\eta(F) \xrightarrow{\rho} G$ is called the image of η. By the abuse of language, we also term $S\eta(F)$ as the image of η. The epimorphism $F \xrightarrow{\mu} S\eta(F)$ is called the co-image of η.

Theorem 4.1.22 *The category Sh_X of sheaves of abelian groups and the category ϑ_X-Mod of ϑ_X-modules are abelian categories.*

Proof It follows from Proposition 4.1.6(iii) that every morphism in Sh_X has a kernel. Let η be a morphism from a sheaf F to a Sheaf G on X. Consider the image $S\eta(F) \xrightarrow{\rho} G$ of η. Since ρ is a monomorphism, $\rho(S\eta(F))$ is a subsheaf of G. In turn, the quotient $G/(\rho(S\eta(F)))$ is a sheaf. We show that the quotient morphism $G \xrightarrow{\nu} G/(\rho(S\eta(F)))$ is a co-kernel of η. Evidently, $\nu \circ \eta = \nu \circ \rho \circ \mu$ is a zero morphism, where μ is the co-image as defined above. Let ϕ be a morphism from G to a sheaf L such that $\phi \circ \eta$ is a zero morphism. Then $\phi \circ \rho \circ \mu$ is a zero morphism. Since μ is an epimorphism, $\phi \circ \rho$ is a zero morphism. In turn, there is a unique morphism ψ from $G/(\rho(S\eta(F)))$ to L such that $\psi \circ \nu = \phi$. This shows that every morphism in Sh_X has a co-kernel. Let η be a monomorphism from a sheaf F to a sheaf G. Then $\eta(F)$ is a sheaf, and so $\eta(F) \approx S\eta(F)$. Clearly, η is a kernel of the quotient morphism $G \xrightarrow{\nu} G/\eta(F)$. Thus, every monomorphism is a kernel of its co-kernel.

Next, let η be an epimorphism in the category Sh_X from a sheaf F to a sheaf G. We show that the monomorphism ρ from $S\eta(F)$ to G is an isomorphism. Let ϕ and ψ be morphisms from G to a sheaf L such that $\phi \circ \rho = \psi \circ \rho$. Then $\phi \circ \eta = \phi \circ \rho \circ \mu = \psi \circ \rho \circ \mu = \psi \circ \eta$. Since η is an epimorphism, $\phi = \psi$. This shows that ρ is also an epimorphism. It follows from Proposition 4.1.13(iii) that ρ is an isomorphism. Now, we show that η is the co-kernel of its kernel. Let α be a morphism from a sheaf K to F which represents the kernel of η. Let β be a morphism from F to L such that $\beta \circ \alpha$ is a zero morphism. Then there is a unique morphism γ from the image presheaf $\eta(F)$ to L such that $\gamma \circ \hat{\eta} = \beta$, where $\hat{\eta}$ is the morphism from F to $\eta(F)$ induced by η. Since $(S\eta(F), j)$ is a sheafification of the presheaf $\eta(F)$, we have a unique morphism δ from $S\eta(F)$ to L such that $\delta \circ j = \gamma$. Consequently, we have the unique morphism $\delta \circ \rho^{-1}$ from G to L such that $(\delta \circ \rho^{-1}) \circ \eta = \delta \circ \mu = \delta \circ j \circ \hat{\eta} = \gamma \circ \hat{\eta} = \beta$. This shows that η is a co-kernel of the kernel of η.

Finally, as in Theorem 4.1.9, we observe that Sh_X is an exact category. This completes the proof of the fact that Sh_X is an abelian category.

Similar arguments establish the fact that ϑ_X-Mod is an abelian category. ♯

Let (X, ϑ_X) be a ringed space. Let M_X and N_X be ϑ_X-modules. The set $Hom_{\vartheta_X}(M_X, N_X)$ of all ϑ_X-morphisms from M_X to N_X is a $\vartheta_X(X) = \Gamma_X(\vartheta_X)$-module with respect to the operations $+$ and $\cdot$ given by

$$f + g = \{f_U + g_U \in Hom_{\vartheta_U}(M_X(U), N_X(U)) \mid U \text{ is open subset of } X\},$$

and

$$\alpha \cdot f = \{\alpha|_U \cdot f|_U \mid U \text{ is open subset of } X\},$$

where $f, g \in Hom_{\vartheta_X}(M_X, N_X)$ and $\alpha \in \vartheta_X(X)$. It can be easily observed that $Hom_{\vartheta_X}(-, -)$ is a functor from $(\vartheta_X - Mod)^o \times \vartheta_X - Mod$ to the category $\vartheta_X(X) - Mod$ of $\vartheta_X(X)$-modules. As usual, $Hom_{\vartheta_X}(M_X, -)$ and $Hom_{\vartheta_X}(-, N_X)$ are left exact functors from the category $\vartheta_X - Mod$ to the category of $\vartheta_X(X)$-modules.

We have another functor $SHom_{\vartheta_X}(-, -)$ from the category $(\vartheta_X - Mod)^o \times \vartheta_X - Mod$ to the category $\vartheta_X - Mod$ of ϑ_X-modules which is termed as **sheaf hom** functor, and which is defined as follows: For an open subset U of X put

$$SHom_{\vartheta_X}(M_X, N_X)(U) = Hom_{\vartheta_X|_U}(M_X|_U, N_X|_U),$$

where $\vartheta_X|_U$, $M_X|_U$, and $N_X|_U$ are the restrictions of ϑ_X, M_X, and N_X to U, respectively. Evidently, $SHom_{\vartheta_X}(M_X, N_X)(U)$ is a $\vartheta_X(U)$-module. For $V \subseteq U$, we have a restriction homomorphism from $Hom_{\vartheta_X|_U}(M_X|_U, N_X|_U)$ to $Hom_{\vartheta_X|_V}(M_X|_V, N_X|_V)$ which respects the corresponding module structures. It can be checked that $SHom_{\vartheta_X}(M_X, N_X)$ introduced is a ϑ_X-module. Evidently, $\Gamma_X SHom_{\vartheta_X}(M_X, N_X) \approx Hom_{\vartheta_X}(M_X, N_X)$. It is also easy to observe that $SHom_{\vartheta_X}(-, -)$ is a functor from the category $(\vartheta_X - Mod)^o \times \vartheta_X - Mod$ to the category $\vartheta_X - Mod$ and the functors $SHom_{\vartheta_X}(M_X, -)$ and $SHom_{\vartheta_X}(-, N_X)$ are left exact functors. As in the case of modules over commutative rings, $SHom_{\vartheta_X}(\vartheta_X, M_X)$ is naturally isomorphic to M_X.

Let $\{M_X^\alpha \mid \alpha \in \Lambda\}$ be a family of ϑ_X-modules. Then for each open subset U of X, $\prod_{\alpha \in \Lambda} M_X^\alpha(U)$ and $\oplus \sum_{\alpha \in \Lambda} M_X^\alpha(U)$ are $\vartheta_X(U)$-modules. We have the obvious restriction homomorphisms from $\prod_{\alpha \in \Lambda} M_X^\alpha(U)$ to $\prod_{\alpha \in \Lambda} M_X^\alpha(V)$ and from $\oplus \sum_{\alpha \in \Lambda} M_X^\alpha(U)$ to $\oplus \sum_{\alpha \in \Lambda} M_X^\alpha(V)$ whenever $V \subseteq U$. This defines ϑ_X-modules $\prod_{\alpha \in \Lambda} M_X^\alpha$ and $\oplus \sum_{\alpha \in \Lambda} M_X^\alpha$. The ϑ_X-module $\oplus \sum_{\alpha \in \Lambda} M_X^\alpha$ is called the direct sum of the family $\{M_X^\alpha \mid \alpha \in \Lambda\}$. The direct sum of the family consisting of the ϑ_X-modules M_X and N_X is denoted by $M_X \oplus N_X$. The ϑ_X-module $M_X \oplus N_X$ is the product as well as the co-product of M_X and N_X in the category ϑ_X-Mod of ϑ_X-modules. As in the case of modules over rings, we have

(i) $Hom_{\vartheta_X}(\oplus \sum_{\alpha \in \Lambda} M_X^\alpha, N_X) \approx \oplus \sum_{\alpha \in \Lambda} Hom_{\vartheta_X}(M_X^\alpha, N_X),$
(ii) $Hom_{\vartheta_X}(N_X, \oplus \sum_{\alpha \in \Lambda} M_X^\alpha) \approx \oplus \sum_{\alpha \in \Lambda} Hom_{\vartheta_X}(N_X, M_X^\alpha),$

(iii) $SHom_{\vartheta_X}(\oplus \sum_{\alpha \in \Lambda} M_X^\alpha, N_X) \approx \oplus \sum_{\alpha \in \Lambda} SHom_{\vartheta_X}(M_X^\alpha, N_X)$, and

(iv) $SHom_{\vartheta_X}(N_X, \oplus \sum_{\alpha \in \Lambda} M_X^\alpha) \approx \oplus \sum_{\alpha \in \Lambda} SHom_{\vartheta_X}(N_X, M_X^\alpha)$.

Let S be a set, and (X, ϑ_X) be a ringed space. The direct sum of S copies of the ϑ_X-module ϑ_X is denoted by $\overset{S}{\oplus} \vartheta_X$. Thus, $\overset{S}{\oplus} \vartheta_X(U)$ is the $\vartheta_X(U)$-module consisting of maps from S to $\vartheta_X(U)$ which vanish at all but finitely many points of S. Since $\vartheta_X(U)$ is a commutative ring, $\overset{S}{\oplus} \vartheta_X$ is isomorphic to $\overset{T}{\oplus} \vartheta_X$ if and only if S and T have the same cardinalities. If $S = \{1, 2, \cdots, n\}$, then $\overset{S}{\oplus} \vartheta_X$ is denoted by ϑ_X^n. Thus, ϑ_X^n is isomorphic to ϑ_X^m if and only if $n = m$. A ϑ_X-module M_X is called a **free ϑ_X-module** if M_X is ϑ_X-isomorphic to $\overset{S}{\oplus} \vartheta_X$ for some set S. If M_X is ϑ_X-isomorphic to ϑ_X^n, then we say that it is a free ϑ_X-module of rank n.

Let (X, ϑ_X) be a ringed space. A ϑ_X-module M_X is called a **locally free ϑ_X-module** if for each $x \in X$, there is an open neighborhood U of x such that $M_X|_U$ is a free $\vartheta_X|_U$-module. Let U and V be open subsets such that $M_X|_U$ is a free $\vartheta_X|_U$-module, $M_X|_V$ is a free $\vartheta_X|_V$-module, and $U \cap V \neq \emptyset$. Then the rank of $M_X|_U$ is the same as $M_X|_V$. This enables us to define the rank r of a locally free ϑ_X-module M_X to be the function r from X to a set of cardinal numbers as follows. Let x be a member of X. Let U be an open neighborhood of x such that $M_X|_U$ is a free $\vartheta_X|_U$-module. Define $r(x)$ to be the rank of $M_X|_U$. Evidently, the rank of a locally free ϑ_X-module is a locally constant function. Thus, if X is a connected space, then the rank of a locally free ϑ_X-module is constant. A locally free ϑ_X-module of constant rank 1 is also termed as an **invertible sheaf**.

Let M_X and N_X be ϑ_X-modules. For each open subset U of X, we have $\vartheta_X(U)$-module $M_X(U) \otimes_{\vartheta_X(U)} N_X(U)$. Further, for $V \subseteq U$, we have the restriction homomorphism from $M_X(U) \otimes_{\vartheta_X(U)} N_X(U)$ to $M_X(V) \otimes_{\vartheta_X(V)} N_X(V)$ which respects the $\vartheta_X(U)$ and $\vartheta_X(V)$-module structures. This defines a presheaf which associates with each open subset U the $\vartheta_X(U)$-module $M_X(U) \otimes_{\vartheta_X(U)} N_X(U)$. The sheaf associated with this presheaf is a ϑ_X-module which is denoted by $M_X \otimes_{\vartheta_X} N_X$, and it is called the tensor product of M_X and N_X. As in the case of modules over rings, we can show that tensoring by a fixed ϑ_X-module is a right exact functor from ϑ_X-Mod to itself, and also one can show that $SHom_{\vartheta_X}(-, -)$ and $- \otimes_{\vartheta_X} -$ are adjoint functors.

Let (X, ϑ_X) be a ringed space. Let M_X and N_X be a locally free ϑ_X-modules of rank 1. Let x be a member of X. Then there is a neighborhood U_x and a neighborhood V_x of x such that $M_X|_{U_x} \approx \vartheta_X|_{U_x}$ and $N_X|_{V_x} \approx \vartheta_X|_{V_x}$. Consequently, $M_X|_{U_x \cap V_x} \approx \vartheta_X|_{U_x \cap V_x}$ and $N_X|_{U_x \cap V_x} \approx \vartheta_X|_{U_x \cap V_x}$. Thus, $(M_X \otimes_{\vartheta_X} N_X)|_{U_x \cap V_x} \approx \vartheta_X|_{U_x \cap V_x}$. This shows that $M_X \otimes_{\vartheta_X} N_X$ is also a locally free ϑ_X-module of rank 1. Let $[M_X]$ denote the class of locally free ϑ_X-modules which are ϑ_X-isomorphic to M_X. The class of all locally free ϑ_X-modules of rank 1 do not form a set. However, $\{[M_X] \mid M_X$ is a $locally$ $free$ $\vartheta_X - module$ of $rank$ $1\}$ is a set. We denote this set by $Pic(\vartheta_X)$. We have the operation $\cdot$ on $Pic(\vartheta_X)$ defined by $[M_X] \cdot [N_X] = [M_X \otimes_{\vartheta_X} N_X]$. It is easily observed that $(Pic(\vartheta_X), \cdot)$ is a semigroup with $[\vartheta_X]$ as an identity. If M_X is a locally free ϑ_X-module of rank 1, then the dual ϑ_X-module $M_X^* = SHom_{\vartheta_X}(M_X, \vartheta_X)$ is also a locally free ϑ_X-module of rank 1. Further, as in the case of modules over commutative rings, it can be shown that $M_X^\star \otimes_{\vartheta_X} M_X \approx SHom_{\vartheta_X}(M_X, M_X) \approx \vartheta_X$. Thus, $[M_X^\star] \cdot [M_X]$ is the identity of $Pic(\vartheta_X)$. This shows that $Pic(\vartheta_X)$ is a group. This group is called the **Picard group** of the ringed space ϑ_X. In the next section, we shall describe the Picard group as a sheaf co-homology.

Sheaf Spaces

There is another naturally equivalent formulation of the concept of sheaves in terms of sheaf spaces. More explicitly, we describe another category Sp_X termed as the category of sheaf spaces of abelian groups (rings, modules) on a topological space X which is naturally isomorphic to the category Sh_X of sheaves of abelian groups (rings, modules). Sometimes it is convenient to view a sheaf as a sheaf space. Recall that a continuous map p from a topological space E to a topological space X is called a local homeomorphism if for each $u \in E$, there is an open subset U of E containing u and open subset V of X containing $p(u)$ such that p induces homeomorphism from U to V. A local homeomorphism need not be surjective. However, a local homeomorphism is always an open map.

Definition 4.1.23 A pair (E, p), where p is a surjective continuous local homeomorphism from a topological space E to a topological space X with given abelian group structure on the fiber $p^{-1}(x)$ for each $x \in X$, is called a **sheaf space** of abelian groups over the space X if the subtraction map $(x, y) \mapsto x - y$ from the subspace $E + E = \{(x, y) \in E \times E \mid p(x) = p(y)\}$ of $E \times E$ to E is continuous. Similarly, we can talk of a sheaf space of rings or sheaf space of R-modules or more generally sheaf space of sets.

Evidently, the map $x \mapsto -x$ is a homeomorphism from E to E, and the addition map $(x, y) \mapsto x + y$ from $E + E$ to E is continuous. Since p is a local homeomorphism, the fibers are discrete subspaces of E.

We have a category Sp_X whose objects are sheaf spaces of abelian groups over the space X, and a morphism from a sheaf space (E, p) to a sheaf space (E', p') over X is a continuous map η from E to E' such that (i) $p'o\eta = p$, and (ii) $\eta|p^{-1}(\{x\})$ is a group homomorphism from the group $p^{-1}(\{x\})$ to the group $p'^{-1}(\{x\})$ for each x. This category is termed as the category of sheaf spaces of abelian groups over X.

Let (E, p) be a sheaf space of abelian groups over the space X. Let U be an open subset of X. A continuous map s from U to E is called a section over U if $pos = I_U$. Let $\Gamma(E, p)(U)$ denote the set of all sections of (E, p) over U. If $s, t \in \Gamma(E, p)(U)$, then $s + t$ defined by $(s + t)(x) = s(x) + t(x)$ is also a member of $\Gamma(E, t)(U)$. This defines an addition $+$ on $\Gamma(E, t)(U)$, and it becomes an abelian group with respect to this operation. We have an obvious restriction map from $\Gamma(E, p)(U)$ to $\Gamma(E, p)(V)$ whenever $V \subseteq U$, and the patching lemma for continuous maps ensures that $\Gamma(E, p)$ is a sheaf of abelian groups. A morphism ϕ from a sheaf space (E, p) to a sheaf space (E', p') induces a morphism $\Gamma(\phi) = \{\Gamma(\phi)_U \mid U \text{ is open subset}\}$ from the sheaf $\Gamma(E, p)$ to the sheaf $\Gamma(E', p')$, where $\Gamma(\phi)_U(s) = po\phi o's$. It is easily observed that Γ is a functor from Sp_X to the category Sh_X. We shall see that the functor Γ is an equivalence from the category Sp_X to the category Sh_X.

Let F be a presheaf of abelian groups on a topological space X. Let E_F denote the disjoint union $\coprod_{x \in x} F_x$ of stalks of F. We have the projection map p_F from E_F to X which maps F_x to x for each $x \in X$. For each open subset U of X, let χ_U denote the map from $F(U)$ to the set of sections of p_F on U defined by $\chi_U(s)(x) = s_x$, where s_x is the element of F_x determined by s. We give the largest topology on E_F so that for all open subsets U of X, all the members of $\chi_U(F(U))$ are continuous. This amounts to say that E_F is a topological space with $\{\chi_U(s)(U) \mid s \in F(U) \text{ and } U \text{ is open}\}$ as a basis for the topology on E_F. It follows from the construction that p_F is a surjective local homeomorphism. Let $u, v \in F_x$. Suppose that $\chi_U(s)(U)$, $s \in F(U)$ is a basic open subset of E_F containing $u + v$. Then $\chi_U(s)(x) = u + v$. Let r and t be members of $F(U)$ such that $\chi_U(r)(x) = u$ and $\chi_U(t)(x) = v$. Then $\chi_U(r + t)(x) = u + v = \chi_U(s)(x)$. This means that there is an open subset V containing x such that $V \subseteq U$ and $\chi_V(r + t)(y) = \chi_V(s)(y)$ for each $y \in V$. Thus, $\chi_V(r)(V) + \chi_V(t)(V) \subseteq \chi_U(s)(U)$. It follows that the addition $+$ from $E_F + E_F$ to E_F is continuous. We have established the following proposition.

Proposition 4.1.24 *For any presheaf F of abelian groups, (E_F, p_F) is a sheaf space of abelian groups.* ♯

Let η be a morphism from a presheaf F of abelian groups to a presheaf G of abelian groups on a space X. Then for each $x \in X$, we have a homomorphism η_x from F_x to G_x. This gives us a map $\hat{\eta}$ from E_F to E_G given by $\hat{\eta}(u) = \eta_x(u)$, where $u \in F_x$. Let u be a member of F_x, and $\chi_U(t)(U)$ be a basic neighborhood of $\hat{\eta}(u) = \eta_x(u)$, where $t \in G(U)$. Then $\chi_U(t)(x) = \eta_x(u)$. Now, $u = \chi_U(s)(x)$ for some $s \in F(U)$. In turn, $\chi_U(\eta_U(s))(x) = \hat{\eta}(\chi_U(s)(x)) = \hat{\eta}(u) = \chi_U(t)(x)$. Hence, there is an open set V containing x such that $V \subseteq U$ and $(\eta_U(s)|V)(y) = \chi_V(t|V)(y)$ for all $y \in V$. Evidently, $\eta_U(s)|V = \eta_V(s|V)$. Thus, $\hat{\eta}(\chi_V(s|V)(V)) \subseteq \chi_V(t|V)(V) \subseteq \chi_U(t)(U)$. It follows that the inverse image of a basic open subset of E_G under the map $\hat{\eta}$ is open, and hence, $\hat{\eta}$ is continuous. Consequently, $\hat{\eta}$ is a morphism from (E_F, p_F) to (E_G, p_G). If η is a morphism from F to G and ρ is a morphism from G to H, then $\widehat{\rho o \eta} = \hat{\rho} o \hat{\eta}$ and $\hat{I}_F = I_{E_F}$. Summarizing the above discussion, we get a functor Ω from Pr_X to Sp_X given by $\Omega(F) = (E_F, p_F)$ and $\Omega(\eta) = \hat{\eta}$.

Now, we describe the functor $\Gamma o \Omega$ from Pr_X to Sh_X. Each $s \in F(U)$ determines a section $\hat{s}$ of $\Omega(F)$ on U. In turn, we get a homomorphism ι_U^F from $F(U)$ to $\Gamma(\Omega(F))(U)$ given by $\iota_U^F(s) = \hat{s}$. It can be easily checked that $\iota^F = \{\iota_U^F \mid U \text{ is open}\}$ is a morphism from F to $(\Gamma o \Omega)(F)$.

Proposition 4.1.25 *For any presheaf F, $(\Gamma(\Omega(F)), \iota^F)$ is a sheafification of F.*

Proof Let η be a morphism from the presheaf F to a sheaf G. We need to show that there is a unique morphism ρ from $\Gamma(\Omega(F))$ to G such that $\rho o \iota^F = \eta$. Let U be an open subset of X, and let s be a member of $(\Gamma o \Omega)(F)(U)$. Then s is a continuous section of p_F on U and $s(U)$ is an open subset of E_F. In turn, $s(U)$ is a union of a family of basic open neighborhoods of E_F. Consequently, for each $x \in U$, there is an open neighborhood U_x of x and a member s_x of $F(U_x)$ such that $U_x \subseteq U$ and $\chi_{U_x}(s_x)(U_x) \subseteq s(U)$. This gives us an open cover $\{U_x \mid x \in U\}$ of U. Further, since s and $\chi_{U_x}(s_x)$ are injective, $\chi_{U_x}(s_x)(z) = s(z)$ for each $z \in U_x$. Now, for each $z \in U_x \bigcap U_y$,

$$\chi_{U_x \bigcap U_y}(s_x|U_x \bigcap U_y)(z) = \chi_{U_x}(s_x)(z) = s(z) = \chi_{U_y}(s_y)(z) = \chi_{U_x \bigcap U_y}(s_y|U_x \bigcap U_y)(z).$$

This means that $s_x|U_x \bigcap U_y = s_y|U_x \bigcap U_y$ for all $x, y \in U$. Since η is a morphism from F to G,

$$\eta_{U_x}(s_x)|U_x \bigcap U_y = \eta_{U_y}(s_y)|U_x \bigcap U_y$$

for each $x, y \in U$. Since G is a sheaf, we have a unique member denoted by $\rho_U(s) \in G(U)$ such that $\rho_U(s)|U_x = \eta_{U_x}(s_x)$ for each $x \in U$. This gives us a morphism $\rho = \{\rho_U \mid U \text{ is open}\}$ from $(\Gamma o\Omega)(F)$ to G such that $\rho o \iota^F = \eta$. Let μ be a morphism from $(\Gamma o\Omega)(F)$ to G such that $\mu o \iota^F = \eta$. Then $\mu_U(s)|U_x = \eta_{U_x}(s_x) = \rho_U(s)|U_x$ for each $x \in U$. Since G is a sheaf, $\mu_U = \rho_U$ for each U. This shows the uniqueness of ρ. $\sharp$

Corollary 4.1.26 *The functor Ω when restricted to the category Sh_X of sheaves is an equivalence from Sh_X to Sp_X.*

Proof Let F be a sheaf on X. From the above proposition, we have a unique natural isomorphism ρ^F from $(\Gamma o\Omega)(F)$ to F. Evidently, $\rho = \{\rho^F \mid F \text{ is a sheaf on } X\}$ is a natural isomorphism from $\Gamma o\Omega$ to the identity functor I_{Sh_X} on Sh_X. Similarly, $\Omega o\Gamma$ is naturally isomorphic to the identity functor I_{Sp_X} on Sp_X. $\sharp$

Exercises

4.1.1 Describe the presheaves and sheaves of abelian groups on a discrete space. Describe sheaf space associated with a presheaf on a discrete space. Describe the sheafification of the presheaf of Example 4.1.4.

4.1.2 Describe the ringed spaces $(Spec\mathbb{Z}, \vartheta^{\mathbb{Z}})$, and $(Spec\mathbb{C}[X], \vartheta^{\mathbb{C}[X]})$.

4.1.3 Show that the sheafification functor S from the category Pr_X of presheaves on X to the category Sh_X of sheaves on X is a retraction from Sh_X to Pr_X.

4.1.4 Describe the finite products and co-products in the categories Pr_X and Sh_X.

4.1.5 Let X be a topological space and A be an abelian group. Consider A with the discrete topology. For each open set U of X, let $F(U)$ denote the group of all continuous functions from U to A. Show that $F = \{F(U) \mid U \text{ is open}\}$ together with restriction maps defines a presheaf. Consider the sub sheaf G of F for which $G(U)$ is the group of constant functions from U to A. Is the quotient presheaf F/G a sheaf? If not describe its sheafification.

4.1.6 Let f be a continuous map from a topological space X to a topological space Y. Let F be a presheaf on X. Show that $\{f_\star F(U) = F(f^{-1}(U)) \mid U \text{ is open subset of } Y\}$ together with the obvious restriction map defines a presheaf $f_\star F$ on Y. The presheaf $f_\star F$ is called the direct image of f. Show further that the image $f_\star F$ of a sheaf F is also a sheaf. Show also that the association $F \mapsto f_\star F$ is a functor from Sh_X to Sh_Y.

4.1.7 Let f be a continuous map from a topological space X to a topological space Y. Let G be a presheaf on Y. Consider the sheaf space $\Omega G = (E_G, p_G)$ associated with the presheaf G. Let E denote the subspace $\{(u, v) \in X \times E_G \mid f(u) = p_G(v)\}$ of the product space $X \times E_G$. Let p denote the first projection on X. Show that (E, p) is a sheaf space over X and the fiber $p^{-1}(\{x\})$ is the group $G_{f(x)}$. The sheaf $\Gamma(E, p)$ is called the inverse image of G under f and it is denoted by $f^\star(G)$. Show further that $\Gamma(E, p)(U) = Lim_{f(U)\subseteq V} G(V)$.

4.1.8 Show that $f^\star$ in the above exercise defines a functor from Sh_Y to Sh_X. Show further that $f_\star$ and $f^\star$ are adjoint to each other.

4.1.9 Show that $f_\star$ is left exact whereas $f^\star$ is exact.

4.1.10 Let Y be a subspace of X and F be a sheaf on Y. For each $y \in Y$, show that the stalk $(i_\star Y)_y$ of $i_\star Y$ at y is the same as the stalk F_y of F at y. If x is an exterior point of Y, then show that the stalk $(i_\star Y)_x$ of $i_\star Y$ at x is 0. Deduce that if Y is a closed subset of X, then $(i_\star Y)_y = F_y$ for all $y \in Y$ and $(i_\star F)_x = \{0\}$ for all x outside Y. Such a sheaf is termed as a sheaf on X which is obtained by extending F by zero outside Y. Let $\hat{F}$ be a sheaf on X such that $\hat{F}_x \approx (i_\star F)_x$ for all x. Can we conclude that $\hat{F} \approx i_\star F$? Support.

4.1.11 Let U be an open subset of X and F be a sheaf on U. Define a presheaf F' on X by taking $F'(V) = F(V)$ whenever $V \subseteq U$ and $F'(V) = 0$ otherwise. Let $j_\sharp F$ denote the sheaf on X associated with the presheaf F'. Show that $j_\sharp F$ satisfies the following properties.

(i) $J_\sharp F|_U = F$.
(ii) $(j_\sharp F)_x = F_x$ for all $x \in U$ and $(j_\sharp F)_x = 0$ for all $x \in X - U$.
(iii) $j_\sharp F$ is unique with the properties (i) and (ii) in the sense that if $\hat{F}$ is a sheaf on X such that $\hat{F}|_U = F$, $\hat{F}_x = F_x$ for all $x \in U$, and $\hat{F}_x = 0$ for all $x \in X - U$, then $\hat{F} = j_\sharp F$.

4.1.12 Let F be a sheaf on X, and Y be a closed subset of X. Show that the sequence

$$0 \longrightarrow j_\sharp(F|_{X-Y}) \longrightarrow F \longrightarrow i_\star(F|_Y) \longrightarrow 0$$

of sheaves is exact.

4.1.13 Let X be a topological space and $\{U_\alpha \mid \alpha \in \Lambda\}$ be an open cover of X. For each $\alpha \in \Lambda$, let F_α be a sheaf on U_α. Suppose that for each ordered pair $(\alpha, \beta) \in \Lambda^2$, we have a sheaf isomorphism $\phi_{(\alpha,\beta)}$ from $F_\alpha|_{U_\alpha \cap U_\beta}$ to $F_\beta|_{U_\alpha \cap U_\beta}$ such that

$$(i) \quad \phi_{(\alpha,\alpha)} = I_{F_\alpha}, \text{ and}$$

$$(ii) \quad \phi_{(\alpha,\gamma)}|_{U_\alpha \cap U_\beta \cap U_\gamma} = \phi_{(\beta,\gamma)}|_{U_\alpha \cap U_\beta \cap U_\gamma} o \phi_{(\alpha,\beta)}|_{U_\alpha \cap U_\beta \cap U_\gamma}$$

for all triple $(\alpha, \beta, \gamma) \in \Lambda^3$. Show that there is unique sheaf F_X on X together with a sheaf isomorphism ϕ_α from $F_X|_{U_\alpha}$ to F_α such that

$$\phi_\beta|_{F_X|_{U_\alpha \cap U_\beta}} o (\phi_\alpha|_{F_X|_{U_\alpha \cap U_\beta}})^{-1} = \phi_{(\alpha,\beta)}$$

for all $(\alpha, \beta) \in \Lambda^2$. The sheaf F_X thus obtained is termed as the sheaf obtained by gluing the sheaves F_α, $\alpha \in \Lambda$ through the isomorphisms $\phi_{(\alpha,\beta)}$, $(\alpha, \beta) \in \Lambda^2$.
Hint. Put

$$F_X(U) = \{s \in \prod_{\alpha \in \Lambda} F_\alpha(U \cap U_\alpha) \mid \phi_{(\alpha,\beta)}|_{U \cap U_\alpha \cap U_\beta}(s_\alpha|_{U \cap U_\alpha \cap U_\beta}) = s_\beta|_{U \cap U_\alpha \cap U_\beta}\}.$$

4.1.14 Use the above exercise to show that a locally free ϑ_X-module of rank 1 determines, and it is uniquely (up to isomorphism) determined by an open cover $\{U_\alpha \mid \alpha \in \Lambda\}$ together with the family $\{\phi_{(\alpha,\beta)} \mid (\alpha, \beta) \in \Lambda^2\}$, where $\phi_{(\alpha,\beta)}$ is a $\vartheta_X|_{U_\alpha \cap U_\beta}$-isomorphism from $\vartheta_X|_{U_\alpha \cap U_\beta}$ to itself.

4.1.15 As in the case of modules over commutative rings, introduce the tensor algebra, exterior algebra, and symmetric algebra of modules over ringed spaces. Derive some of its basic properties.

4.1.16 Show that a locally free ϑ_X-module M_X of finite rank is reflexive in the sense that $M_X^{\star\star} \approx M_X$.

4.1.17 Let M_X be a locally free ϑ_X-module of rank 1. Show that $SHom_{\vartheta_X}(M_X, M_X) \approx \vartheta_X$.

4.2 Sheaf Co-homology and Čech Co-homology

Now, we start to develop homological algebra in the abelian category Sh_X of sheaves over a space X, and also in the abelian category ϑ_X-Mod of ϑ_X-modules. We say that a chain

$$\cdots \longrightarrow F_{n+1} \overset{\alpha_{n+1}}{\to} F_n \overset{\alpha_n}{\to} F_{n-1} \longrightarrow \cdots \quad (1)$$

of morphisms in the category Pr_X/Sh_X or ϑ_X-Mod is exact at F_n if the image of α_{n+1} is the kernel of α_n in the category Pr_X/Sh_X or ϑ_X-Mod. Explicitly, this means that

$(\alpha_{n+1})_U(F_{n+1}(U)) = ker\ (\alpha_n)_U$ for each open set U of X in case it is exact in Pr_X, and in case it is exact in Sh_X/ϑ_X-Mod, $(\rho_{n+1})_U(S\alpha_{n+1}(F_{n+1})(U)) = ker\ (\alpha_n)_U$ for each U, where $(S\alpha_{n+1}(F_{n+1}),\ j_{n+1})$ is a sheafification of $\alpha_{n+1}(F_{n+1})$ and ρ_{n+1} is the unique morphism from $S\alpha_{n+1}(F_{n+1})$ to F_n such that $\rho_{n+1}j_{n+1}$ is the inclusion morphism i_{n+1} from $\alpha_{n+1}(F_{n+1})$ to F_n.

The sequence (1) is said to be exact if it is exact at all places. As usual, an exact sequence in $Pr_X/Sh_X/\vartheta_X$-Mod of the form

$$0 \longrightarrow F \xrightarrow{\alpha} G \xrightarrow{\beta} H \longrightarrow 0$$

is called a short exact sequence in $Pr_X/Sh_X/\vartheta_X$-Mod. For each open subset U of X, we have a functor Γ_U from $Pr_X/Sh_X/\vartheta_X$-Mod to the category AB of abelian groups given by $\Gamma_U(F) = F(U)$. Also for each $x \in X$, we have another functor Γ_x from $Pr_X/Sh_X/\vartheta_X$-Mod to the category AB given by $\Gamma_x(F) = F_x$, where F_x is the stalk at x. We wish to describe the effect of these functors on the exact sequence, and in particular on short exact sequences. Recall that a functor Γ from an abelian category Σ to an abelian category Σ' is said to be an exact functor if it takes a short exact sequence in Σ to a short exact sequence in Σ'. It is said to be a left exact functor if given any short exact sequence

$$0 \longrightarrow A \xrightarrow{\alpha} B \xrightarrow{\beta} C \longrightarrow 0$$

in Σ, the sequence

$$0 \longrightarrow \Gamma(A) \xrightarrow{\Gamma(\alpha)} \Gamma(B) \xrightarrow{\Gamma(\beta)} \Gamma(C)$$

is exact in Σ'. Similarly, we can talk of the right exact functor.

Proposition 4.2.1 *Let X be a topological space. Then*

(i) for each open subset U of X, Γ_U is an exact functor from Pr_X to AB,
(ii) for each open subset U of X, Γ_U is a left exact functor from Sh_X/ϑ_s-Mod to AB, and
(iii) a chain

$$F \xrightarrow{\alpha} F' \xrightarrow{\beta} F''$$

of morphisms in the category Sh_X of sheaves of abelian groups is exact if and only if the sequence

$$F_x \xrightarrow{\alpha} F'_x \xrightarrow{\beta} F''_x$$

of homomorphisms on the stalk at x is exact for each $x \in X$.

In particular, for each $x \in X$, the functor Γ_x which associates the stalk F_x with a sheaf F is an exact functor from Sh_X to AB.

Proof (i) Suppose that

$$0 \longrightarrow F \xrightarrow{\alpha} F' \xrightarrow{\beta} F'' \longrightarrow 0$$

is an exact sequence in Pr_X. Then α is a monomorphism, β is an epimorphism which is a co-kernel of α. From Proposition 4.1.6(ii), α_U is injective, and from Proposition 4.1.7, β_U is surjective and $Ker\ \alpha_U = image\ \beta_U$. This shows that Γ_U is an exact functor for each open set U.

(ii) As in the case of presheaves, α_U in injective, and $ker\ \beta_U = image\ \alpha_U$ for all open sets U. However, β_U need not be surjective (Proposition 4.1.14 (iii)).

(iii) Recall that

$$F \xrightarrow{\alpha} F' \equiv F \xrightarrow{\hat{\alpha}} \alpha F \xrightarrow{j} S(\alpha F) \xrightarrow{\rho} F',$$

where αF is the presheaf image of α, $\hat{\alpha}$ is the natural morphism induced by α, j is the sheafification morphism, and ρ is the unique monomorphism from the sheafification $S(\alpha F)$ of αF to F'. Further, $S(\alpha F) \xrightarrow{\rho} F'$ is the image of α. Thus, to say that

$$F \xrightarrow{\alpha} F' \xrightarrow{\beta} F''$$

is exact is to say that

$$0 \longrightarrow S(\alpha F) \xrightarrow{\rho} F' \xrightarrow{\beta} F''$$

is exact. Further, since j_x is a natural isomorphism from F_x to $S(\alpha F)_x$ for each x, it is sufficient to show that

$$0 \longrightarrow F \xrightarrow{\alpha} F' \xrightarrow{\beta} F''$$

is exact if and only if

$$0 \longrightarrow F_x \xrightarrow{\alpha_x} F'_x \xrightarrow{\beta_x} F''_x$$

is exact for each $x \in X$.

Suppose that

$$0 \longrightarrow F \xrightarrow{\alpha} F' \xrightarrow{\beta} F''$$

is exact. Since α is a monomorphism, α_x is injective homomorphism for each $x \in X$. Again, since $\beta o \alpha$ is a zero morphism, $\beta_x o \alpha_x = (\beta o \alpha)_x$ is the zero homomorphism for each $x \in X$. This shows that $image\ \alpha_x \subseteq ker\ \beta_x$ for each $x \in X$. Let a_x be a member of $ker\ \beta_x$. Let a_U be a member of $F'(U)$ which represents a_x. Then $\beta_U(a_U)$ represents the zero element of F''_x. Hence there is a neighborhood V_x of x such that $V_x \subseteq U$ and $\beta_U(a_U)|_{V_x} = 0$. In turn, $\beta_{V_x}(a_U|_{V_x}) = 0$. From (ii), there is an element $b_{V_x} \in F(V_x)$ such that $\alpha_{V_x}(b_{V_x}) = a_U|_{V_x}$. Let b_x be the member of F_x represented by b_{V_x}. Then $\alpha_x(b_x)$ is the element of F'_x which is represented by $a_U|_{V_x} \in F'(V_x)$. Evidently, the element represented by $a_U|_{V_x}$ in F'_x is the element represented by a_U. Hence $\alpha_x(b_x) = a_x$. This shows that

$$0 \longrightarrow F_x \xrightarrow{\alpha_x} F'_x \xrightarrow{\beta_x} F''_x$$

is exact.

Conversely, suppose that the sequence

$$0 \longrightarrow F \xrightarrow{\alpha} F' \xrightarrow{\beta} F''$$

of morphisms is such that

$$0 \longrightarrow F_x \xrightarrow{\alpha_x} F'_x \xrightarrow{\beta_x} F''_x$$

is exact for each $x \in X$. It follows (see Remark 4.1.13) that α is a monomorphism. Let U be an open subset of X. Let s_U be a member of $F(U)$. For each $x \in U$, let s_x denote the element of F_x represented by s_U. Then $(\beta_x \circ \alpha_x)(s_x) = 0$. Evidently, $(\beta_x \circ \alpha_x)(s_x)$ is the element of F''_x represented by $\beta_U(\alpha_U(s_U))$. Hence $\beta_U(\alpha_U(s_U))$ represents the zero element of F''_x for each $x \in U$. Thus, for each $x \in U$, there is an open neighborhood V_x of x such that $V_x \subseteq U$ and $(\beta_{V_x} \circ \alpha_{V_x})(s_U|_{V_x}) = (\beta_U(\alpha_U(s_U)))|_{V_x} = 0$. Since F'' is a sheaf, $(\beta_U \circ \alpha_U)(s_U) = 0$. This shows that $image\ \alpha_U \subseteq ker\ \beta_U$. Next, let t be a member of $ker\ \beta_U$. Then $\beta_U(t)$ is the zero element of $F''(U)$. Hence $\beta_x(t_x) = 0$ for each $x \in U$, where t_x denotes the element of F'_x represented by t. Since

$$0 \longrightarrow F_x \xrightarrow{\alpha_x} F'_x \xrightarrow{\beta_x} F''_x$$

is exact, there is an element s_x in F_x such that $\alpha_x(s_x) = t_x$. Let V_x be an open neighborhood of x and $s_{V_x} \in F(V_x)$ which represents s_x. Evidently, $\alpha_{V_x}(s_{V_x})$ and $t|_{V_x}$ represents the same element t_x in F'_x. Since F' is a sheaf, there is an open neighborhood W_x of x such that $W_x \subseteq V_x$ and also $\alpha_{W_x}(s_{V_x}|_{W_x})$ and $t|_{W_x}$ represent the same member of F_y for each $y \in W_x$. Consider the family $\{s_{W_x} \mid x \in U\}$. Evidently, $\alpha_{W_x \cap W_y}(s_{W_x}|_{W_x \cap W_y})$ and $\alpha_{W_x \cap W_y}(s_{W_y}|_{W_x \cap W_y})$ represent the same member t_z for each $z \in W_x \cap W_y$. Since α_z is injective for all z, it follows that $s_{W_x}|_{W_x \cap W_y} = s_{W_y}|_{W_x \cap W_y}$ for all pairs $x, y \in U$. Since F is a sheaf, there is a unique member $s \in F(U)$ such that $s|_{W_x} = s_{W_x}$ for all $x \in U$. Clearly, $\alpha_U(s) = t$. This proves that

$$0 \longrightarrow F \xrightarrow{\alpha} F' \xrightarrow{\beta} F''$$

is exact. The rest is evident. ♯

Theorem 4.2.2 *Let* (X, ϑ_X) *be a ringed space. Then the category* ϑ_X-*Mod of* ϑ_X-*modules has enough injectives.*

Proof Let (X, M_X) be a ϑ_X-module, where (X, ϑ_X) is a ringed space. For each $x \in X$, the stalk $(M_X)_x$ of (X, M_X) at x is a $(\vartheta_X)_x$-module. Since every module over a ring can be embedded in an injective module (Algebra 2, Theorem 7.2.40), there is an injective $(\vartheta_X)_x$-module I_x and an injective $(\vartheta_X)_x$-homomorphism i_x from $(M_X)_x$ to I_x. Further, for each open neighborhood U of x, I_x is a $\vartheta_X(U)$-module with respect to the scalar product given by $s \cdot a = s_x \cdot x$, where $s \in \vartheta_X(U)$, $a \in I_x$, and s_x is the element of $(\vartheta_X)_x$ represented by s. For each open set U of X, let $I_X(U)$ denote the $\vartheta_X(U)$-module $\prod_{x \in U} I_x$. We have the obvious projection map from $I_X(U)$ to $I_X(V)$ whenever $V \subseteq U$. This gives us a ϑ_X-module (X, I_X). We show that (X, I_X) is an injective ϑ_X-module. Let $j = \{j_U \mid U \text{ is open subset of } X\}$ be a monomorphism from a ϑ_X-module N_X to a ϑ_X-module L_X. This is equivalent (Remark 4.1.13) to say that for each $x \in X$, j_x is an injective $(\vartheta_X)_x$-homomorphism from $(N_X)_x$ to $(L_X)_x$. Let $\rho = \{\rho_U \mid U \text{ is open subset of } X\}$ be a ϑ_X-morphism from N_X to I_X. Since I_x is an injective $(\vartheta_X)_x$-module, ρ_x can be lifted to a $(\vartheta_X)_x$-homomorphism η_x from $(L_X)_x$ to I_x. In turn, we get a homomorphism η_U from $L_X(U)$ to $I_X(U)$ given by $\eta_U(s)(x) = \eta_x(s_x)$, $x \in U$. Evidently, $\eta = \{\eta_U \mid U \text{ is open subset of } X\}$ is a ϑ_X-morphism from L_X to I_X such that $\eta \circ \mathrm{J} = \rho$. This shows that (X, I_X) is an injective object in ϑ_X-*Mod*.

Further, for each open set U, we have a $\vartheta_X(U)$-homomorphism i_U from $M_X(U)$ to $I_X(U)$ defined by $i_U(s)(x) = i_x(s_x)$, where s_x denotes the element of $(M_X)_x$ determined by the element $s \in M_X(U)$. Since i_x is injective for each x and M_X and I_X are sheaves, it follows from Proposition 4.2.1 that $i = \{i_U \mid U \text{ is open subset of } X\}$ is an embedding. ♯

Corollary 4.2.3 *The category* Sh_X *has enough injectives.*

Proof The proof follows from the above theorem as a particular case if we observe that Sh_X is isomorphic to the category $\vartheta_X^{\mathbb{Z}}$-*Mod*, where $\vartheta_X^{\mathbb{Z}}$ is the constant ringed space given by $\vartheta_X^{\mathbb{Z}}(U) = \mathbb{Z}$ for all open sets U. ♯

From Proposition 4.2.1(ii), the global section functor Γ_X from the category Sh_X to the category of abelian groups is a left exact functor. The pth right-derived functors $R^p\Gamma_X$ of the global section functor Γ_X is called the pth **sheaf co-homology** functor on X. We denote this functor by $H^p(X, -)$. If F is a sheaf on X, then $H^p(X, F)$ is called the pth sheaf co-homology of X with coefficient in the sheaf F. Thus, given an injective co-resolution

$$0 \longrightarrow F \xrightarrow{\epsilon} I_0 \xrightarrow{d_0} I_1 \xrightarrow{d_1} \cdots \xrightarrow{d_{n-1}} I_n \xrightarrow{d_n} \cdots$$

of F, $H^p(X, F)$ is the pth co-homology of the co-chain complex

$$0 \longrightarrow \Gamma_X(I_0) \xrightarrow{\Gamma_X(d_0)} \Gamma_X(I_1) \xrightarrow{\Gamma_X(d_1)} \cdots \xrightarrow{\Gamma_X(d_{n-1})} \Gamma_X(I_n) \xrightarrow{\Gamma_X(d_n)} \cdots$$

of abelian groups. Evidently, $H^0(X, F) = \Gamma_X(F) = F(X)$, and for injective sheaf I, $H^p(X, I) = 0$ for all $p \geq 1$. Using the dual of Proposition 2.3.8, in the category Sh_X, we obtain the following.

Proposition 4.2.4 *Let*

$$0 \longrightarrow F \xrightarrow{\alpha} F' \xrightarrow{\beta} F'' \longrightarrow 0$$

be a short exact sequence in Sh_X. Then we have the associated long exact co-homology sequence

$$\cdots \xrightarrow{H^{n-1}(X,\beta)} H^{n-1}(X, F'') \xrightarrow{\partial^{n-1}} H^n(X, F) \xrightarrow{H^n(X,\alpha)} H^n(X, F') \xrightarrow{H^n(X,\beta)} H^n(X, F'') \xrightarrow{\partial^n} \cdots$$

of abelian groups. ♯

Flasque Sheaves
It was Godement who introduced the notion of flasque sheaves in his book on Sheaf Theory to give a convenient description of the sheaf co-homology with the help of the flasque resolution (also termed as the Godement resolution) of sheaves.

Definition 4.2.5 Let (X, ϑ_X) be a ringed space. A ϑ_X-module F_X on a space X is called a **flasque** ϑ_X-module if for each open set U, the restriction homomorphism from $F_X(X)$ to $F_X(U)$ is a surjective homomorphism. In particular, a sheaf F of abelian groups on X is called a **flasque sheaf** if for each open set U of X, the restriction homomorphism f_X^U from $F(X)$ to $F(U)$ is a surjective homomorphism.

Proposition 4.2.6 *Let (X, ϑ_X) be a ringed space. Let (X, F_X) be a flasque ϑ_X-module. Then for each open subset U of X, $F_U = F_X|_U$ is a flasque $\vartheta_U = \vartheta_X|_U$-module. Conversely, suppose that F_X is a ϑ_X-module on X with the property that every point $x \in X$ has an open neighborhood U_x such that $F_{U_x} = F_X|_{U_x}$ is a flasque $\vartheta_{U_x} = \vartheta_X|_{U_x}$-module. Then F_X is a flasque ϑ_X-module.*

Proof Suppose that F_X is a flasque ϑ_X-module on a space X. Let U be an open subset of X. Let V be an open subset contained in U. Let s be a member of $F_U(V) = F_X|_U(V) = F_X(V)$. Since F_X is a flasque ϑ_X-module, there is a member $t \in F_X(X)$ such that $f_X^V(t) = s$. But, then $f_U^V(f_X^U(t)) = f_X^V(t) = s$. This shows that f_U^V is surjective. By definition, it follows that $F_U = F_X|_U$ is a flasque $\vartheta_U = \vartheta_X|_U$-module.

Conversely, let F_X be a ϑ_X-module on X with the property that every point $x \in X$ has an open neighborhood U_x such that F_{U_x} is a flasque ϑ_{U_x}-module and $s \in F_X(U)$, where U is an open subset of X. We have to show the existence of an element $t \in F_X(X)$ such that $t|_U = s$. Let Σ denote the set

$$\{(W, \tau) \mid W \text{ is open, } U \subseteq W, \tau \in F_W(W) = F_X(W), \text{ and, } \tau|_U = s\}.$$

We obtain a partial order $\leq$ on Σ by putting $(W_1, \tau_1) \leq (W_2, \tau_2)$ if $W_1 \subseteq W_2$ and $\tau_2|_{W_1} = \tau_1$. Evidently, $(U, s) \in \Sigma$, and so $(\Sigma, \leq)$ is a nonempty poset. Let

$\{(W_\alpha, s_\alpha) \mid \alpha \in \Lambda\}$ be a chain in Σ. Put $V = \bigcup_{\alpha \in \Lambda} W_\alpha$. Since F_X is a sheaf, there is a unique element $\hat{s} \in F_V(V) = F_X(V)$ such that $\hat{s}|_{W_\alpha} = s_\alpha$ for all $\alpha \in \Lambda$. This means that $(V, \hat{s})$ is an upper bound of the chain. By Zorn's lemma, $(\Sigma, \leq)$ has a maximal element (W_0, s_0) (say). Suppose that $W_0 \neq X$. Let $x \in X - W_0$ and U_x be an open neighborhood of x such that $F_{U_x} = F_X|_{U_x}$ is a flasque $\vartheta_{U_x} = \vartheta_X|_{U_x}$-module. Since F_{U_x} is a flasque ϑ_{U_x}-module, we have an element $t \in F_{U_x}(U_x) = F_X(U_x)$ such that $f_{U_x}^{U_x \cap W_0}(t) = f_{W_0}^{U_x \cap W_0}(s_0)$. Thus, the member $t \in F_X(U_x)$ and the member $s_0 \in F_X(W_0)$ are such that $t|_{U_x \cap W_0} = s_0|_{U_x \cap W_0}$. Since F_X is a ϑ_X-module, there is a member $\hat{t} \in F_X(U_x \bigcup W_0) = F_{U_x \bigcup W_0}(U_x \bigcup W_0)$ such that $\hat{t}|_{W_0} = s_0$ and $\hat{t}|_{U_x} = t$. This means that $(U_x \bigcup W_0, \hat{t}) \in \Sigma$ and $(W_0, s_0) \leq (U_x \bigcup W_0, \hat{t})$. This is a contradiction to the maximality of (W_0, s_0). Hence $W_0 = X$, and it follows that F_X is a flasque ϑ_X-module. ♯

In particular, we have the following corollary.

Corollary 4.2.7 *A sheaf F on a space X is a flasque sheaf if and only if $F_U = F|_U$ is a flasque sheaf on U for all open sets U of X.* ♯

Proposition 4.2.8 *Let (X, ϑ_X) be a ringed space. Then any injective ϑ_X-module (X, M_X) is a flasque ϑ_X-module. In particular, any injective sheaf F on X is a flasque sheaf.*

Proof For each open subset U of X, we have the ringed space $\vartheta_X|_U$ on U. Let $\hat{\vartheta}_U$ denote the sheaf $j_\sharp(\vartheta_X|_U)$ (see Exercise 4.1.11) which is the sheaf $\vartheta_X|_U$ extended to X by 0 outside U. Evidently, $(X, \hat{\vartheta}_U)$ is a ϑ_X-module in an obvious way. Indeed, if V is an open subset of X contained in U, then $\hat{\vartheta}_U(V) = \vartheta_X(V)$ is a $\vartheta_X(V)$-module, and if $V \not\subseteq U$, then again $\hat{\vartheta}_U(V) = 0$ is a $\vartheta_X(V)$-module. These module structures also respect the restriction maps. Further, if (X, N_X) is a ϑ_X-module, then any ϑ_X-homomorphism f from $\hat{\vartheta}_U$ to N_X is uniquely determined by the element $f(1) \in N_X(U)$, where 1 is the identity of the ring $\hat{\vartheta}_U(U) = \vartheta_X(U)$. This means that $Hom_{\vartheta_X}(\hat{\vartheta}_U, N_X)$ is naturally isomorphic to $N_X(U)$ for all open sets U and for all ϑ_X-modules N_X. Evidently, these natural isomorphisms respect the restriction maps. Let V and U be open subsets of X with $V \subseteq U$. Then we have the natural inclusion ϑ_X-homomorphism i_V^U from $\hat{\vartheta}_V$ to $\hat{\vartheta}_U$. Since M_X is an injective ϑ_X-module, i_V^U induces a surjective ϑ_X-homomorphism from $Hom_{\vartheta_X}(\hat{\vartheta}_U, M_X)$ to $Hom_{\vartheta_X}(\hat{\vartheta}_V, M_X)$. In turn, it follows that the restriction map from $M_X(U)$ to $M_X(V)$ is surjective. From Proposition 4.2.6, it follows that M_X is a flasque ϑ_X-module. ♯

Corollary 4.2.9 *For every ϑ_X-module M_X, we have a flasque co-resolution over M_X.* ♯

Our next aim is to show that sheaf co-homology can be computed with the help of flasque co-resolutions.

Proposition 4.2.10 *Let*

$$0 \longrightarrow N_X \overset{\alpha}{\to} M_X \overset{\beta}{\to} L_X \longrightarrow 0$$

be an exact sequence of ϑ_X-modules. Suppose that N_X is a flasque ϑ_X-module. Then

$$0 \longrightarrow N_X(U) \overset{\alpha_U}{\to} M_X(U) \overset{\beta_U}{\to} L_X(U) \longrightarrow 0$$

is exact for all open sets U.

Proof By Proposition 4.2.6, $N_X|_U$ is a flasque $\vartheta_U = \vartheta_X|_U$-module, and also

$$0 \longrightarrow N_X|_U \overset{\alpha|_U}{\to} M_X|_U \overset{\beta|_U}{\to} L_X|_U \longrightarrow 0$$

is an exact sequence of ϑ_U-modules. Thus, it is sufficient to show that

$$0 \longrightarrow N_X(X) \overset{\alpha_X}{\to} M_X(X) \overset{\beta_X}{\to} L_X(X) \longrightarrow 0$$

is exact. In the light of Proposition 4.2.1(ii), we need to show that β_X is surjective. Let s be a member of $L_X(X)$. For each $x \in X$, let s_x denote the element of $(L_X)_x$ determined by s. By Proposition 4.2.1(iii),

$$0 \longrightarrow (N_X)_x \overset{\alpha_x}{\to} (M_X)_x \overset{\beta_x}{\to} (L_X)_x \longrightarrow 0$$

is exact for each x. Hence, there is an open neighborhood V of x and an element $t \in M_X(V)$ such that $\beta_x(t_x) = s_x$. This means that $\beta_V(t) \in L_X(V)$ and $s \in L_X(X)$ determine the same element of $(L_X)_x$. Consequently, there is an open neighborhood U_x of x such that $U_x \subseteq V$ and $\beta_{U_x}(t|_{U_x}) = s|_{U_x}$. Now, consider the set

$$\Sigma = \{(U, u) \mid U \text{ is open}, u \in M_X(U), \text{ and } \beta_U(u) = s|_U\}.$$

It follows from our previous discussion that $\Sigma \neq \emptyset$. We obtain a partial order $\leq$ on Σ by putting $(U, u) \leq (V, v)$ if $U \subseteq V$ and $v|_U = u$. This gives us a nonempty poset $(\Sigma, \leq)$. As in the proof of Proposition 4.2.6, every chain in $(\Sigma, \leq)$ has an upper bound. By Zorn's lemma, it has a maximal element (U_0, u_0) (say). Suppose that $U_0 \neq X$. Let x be a member of $X - U_0$. Using our earlier arguments, we obtain an open neighborhood W of x together with an element $w \in M_X(W)$ such that $\beta_W(w) = s|_W$. Evidently, $s|_{W \cap U_0} = \beta_{W \cap U_0}(w|_{W \cap U_0}) = \beta_{W \cap U_0}(u_0|_{W \cap U_0})$. Hence, $(u_0|_{W \cap U_0} - w|_{W \cap U_0}) \in ker\,\beta_{W \cap U_0}$. Since $ker\,\beta_{W \cap U_0} = image\,\alpha_{W \cap U_0}$, there is an element w' of $N_X(W \cap U_0)$ such that $\alpha_{W \cap U_0}(w') = u_0|_{W \cap U_0} - w|_{W \cap U_0}$. Since N_X is a flasque ϑ_X-module, there is an element w'' of $N_X(X)$ such that $w''|_{W \cap U_0} = w'$. Put $\hat{w} = w + \alpha_W(w''|_W)$. Evidently, $\hat{w}$ is a member of $M_X(W)$ such that $\hat{w}|_{W \cap U_0} = u_0|_{W \cap U_0}$. Since M_X is a sheaf, we have an element $\hat{\hat{w}}$ of $M_X(W \bigcup U_0)$ such that $\hat{\hat{w}}|_W = \hat{w}$ and $\hat{\hat{w}}|_{U_0} = u_0$. Clearly, $(W \bigcup U_0, \hat{\hat{w}}) \in \Sigma$ and $(U_0, u_0) \leq (W \bigcup U_0, \hat{\hat{w}})$. This is a contradiction to the maximality of (U_0, u_0). Hence $U_0 = X$. This completes the proof. ♯

Corollary 4.2.11 *If F is a flasque sheaf on a space X, and*

$$0 \longrightarrow F \overset{\alpha}{\to} G \overset{\beta}{\to} H \longrightarrow o$$

is an exact sequence of sheaves on X, then for each open set U,

$$0 \longrightarrow F(U) \overset{\alpha_U}{\to} G(U) \overset{\beta_U}{\to} H(U) \longrightarrow o$$

is exact. ♯

Corollary 4.2.12 *Let*

$$0 \longrightarrow N_X \overset{\alpha}{\to} M_X \overset{\beta}{\to} L_X \longrightarrow o$$

be an exact sequence of ϑ_X-modules, where N_X and M_X are flasque ϑ_X-modules. Then L_X is also a ϑ_X-module.

Proof Let U be an open subset of X and $s \in L_X(U)$. Since N_X is a flasque ϑ_X-module,

$$0 \longrightarrow N_X(U) \overset{\alpha_U}{\to} M_X(U) \overset{\beta_U}{\to} L_X(U) \longrightarrow o$$

is exact. Hence there is a member t of $M_X(U)$ such that $\beta_U(t) = s$. Again, since M_X is flasque, there is an element $\hat{t}$ of $M_X(X)$ such that $\hat{t}|_U = t$. Thus, the element $\beta_X(\hat{t})$ of $L_X(X)$ is such that $\beta_X(\hat{t})|_U = s$. ♯

Corollary 4.2.13 *Let F be a flasque sheaf of abelian groups on X. Then the sheaf co-homology $H^p(X, F)$ of X with a coefficient in F is zero for all $p \geq 1$.*

Proof We prove the result by induction on p. Observe that $H^0(X, F) = F(X)$. Let F be a flasque sheaf, and

$$0 \longrightarrow F \overset{\epsilon}{\to} I_0 \overset{d_0}{\to} I_1 \overset{d_1}{\to} I_2 \overset{d_2}{\to} \cdots$$

be an injective co-resolution of F. We have a short exact sequence

$$0 \longrightarrow F \overset{\epsilon}{\to} I_0 \overset{\hat{d}_0}{\to} Im\, d_0 \longrightarrow 0, \tag{4.2}$$

where $Im\, d_0$ denote the image of d_0 and $\hat{d}_0$ is the obvious morphism induced by d_0. We have also an exact sequence

$$0 \longrightarrow Im\, d_0 \overset{i}{\to} I_1 \overset{d_1}{\to} I_2 \overset{d_2}{\to} I_3 \overset{d_3}{\to} . \tag{4.3}$$

Since F is a flasque sheaf, by Corollary 4.2.11, we have the short exact sequence

$$0 \longrightarrow \Gamma_X(F) \overset{\Gamma_X(\epsilon)}{\rightarrow} \Gamma_X(I_0) \overset{\Gamma_X(\hat{d_0})}{\rightarrow} \Gamma_X(Im\ d_0) \longrightarrow 0. \qquad (4.4)$$

Again, since Γ_X is a left exact functor, we have also the exact sequence

$$0 \longrightarrow \Gamma_X(Im\ d_0) \overset{\Gamma_X(i)}{\rightarrow} \Gamma_X(I_1) \overset{\Gamma_X(d_1)}{\rightarrow} \Gamma_X(I_2). \qquad (4.5)$$

It follows that $ker\ \Gamma_X(d_1) = image\ \Gamma_X(d_0)$. This shows that $H^1(X, F) = \frac{ker\ \Gamma_X(d_1)}{image\ \Gamma_X(d_0)} = \frac{\Gamma_X(Im\ d_0)}{image\ \Gamma_X(d_0)} = \frac{\Gamma_X(image\ d_0)}{\Gamma_X(image\ d_0)} = 0$. This proves the result for $p = 1$. Assume that the result is true for $m \geq 1$. Since F and I_0 (being injective) are flasque sheaves, it follows from Corollary 4.2.12 that $Im\ d_0$ is a flasque sheaf. Further, we have the long exact co-homology sequence

$$\overset{\partial^{p-1}}{\rightarrow} H^p(X, F) \overset{H^p(X, \epsilon)}{\rightarrow} H^p(X, I_0) \overset{H^p(X, \hat{d_0})}{\rightarrow} H^p(X, Im\ d_0) \overset{\partial^p}{\rightarrow}$$
$$\times H^{p+1}(X, F) \overset{H^p(X, \epsilon)}{\rightarrow} H^{p+1}(X, I_0) \overset{H^{p+1}(X, \hat{d_0})}{\rightarrow} \cdots$$

associated with the short exact sequence (1). Since I_0 is injective, $H^q(X, I_0) = 0$ for all $q \geq 1$. It follows that $H^{p+1}(X, F) \approx H^p(X, Im\ d_0)$ for all $p \geq 1$. Since $Im\ d_0$ is flasque, by the induction hypothesis $H^m(X, Im\ d_0) = 0$. Hence $H^{m+1}(X, F) = 0$. ♯

The following corollary is immediate.

Corollary 4.2.14 *Flasque sheaves are acyclic for the global section functor and the sheaf co-homology can be computed by using flasque resolutions.* ♯

Now, consider the global section functor Γ_X from the category ϑ_X-Mod to the category of abelian groups (forget the ϑ_X-module structure). A ϑ_X-module M_X can be treated as a sheaf of abelian groups. Indeed, we have the forgetful functor Ω from the category ϑ_X-Mod to the category Sh_X. Under this forgetful functor, injective objects correspond and also a flasque ϑ_X-module corresponds to a flasque sheaf. Consequently, we have the following proposition.

Proposition 4.2.15 *The pth right-derived functor $R^p\Gamma_X$ of the global section functor Γ_X from ϑ_X-Mod to AB is the same as the composite of the forgetful functor Ω with the sheaf co-homology functor $H^p(X, -)$. More explicitly, $R^p\Gamma_X(M_X)$ is naturally the same as $H^p(X, M_X)$.* ♯

Čech Co-homology

Čech co-homology is another important tool which has an enormous amount of applications in algebraic topology, differential geometry, and also in algebraic geometry. It has its connection with sheaf co-homology. Indeed, it agrees and gives us a convenient and practical method to compute sheaf co-homology for some important and special types of sheaves, viz., quasi-coherent sheaves. We describe it as follows.

First, we associate a nonnegative co-chain complex $\check{C}^\star(X, F, U)$ of R-modules and the pth co-homology module $\check{H}^p(X, F, U)$ with a triple (X, F, U), where X is a topological space, F is a presheaf/sheaf on X, and U is an open covering of X. $\check{C}^\star(X, F, U)$ is termed as the Čech co-chain complex and $\check{H}^p(X, F, U)$ is termed as p^{th} Čech co-homology associated with the triple (X, F, U). In turn, we look at a directed system $(D, \leq)$ of open covers of X, where each open cover is equivalent to an open cover in D, and then introduce the pth Čech co-homology $\check{H}^p(X, F)$ of X with values in a presheaf/sheaf F as $Lim_{\to U \in D}\, \check{H}^p(X, F, U)$.

Recall that a pair (Λ, A), where Λ is a set and A is a map from Λ to the power set $\wp(X)$ of X is called a family of subsets of X. The image $A(\alpha)$ of $\alpha \in \Lambda$ under the map A is denoted by A_α, and the family (Λ, A) is also denoted by $\{A_\alpha \subseteq X \mid \alpha \in \Lambda\}$. Sometimes we say that the map A itself is a family of subsets of X. If X is a topological space and A_α is an open subset of X for each $\alpha \in \Lambda$, then it is said to be a family of open subsets of X. If in addition to this $\bigcup_{\alpha \in \Lambda} A_\alpha = X$, then we say that it is an open cover of X. Thus, given a subset Σ of the power set $\wp(X)$ of X, the pair $(\Sigma, i_\Sigma^{\wp(X)})$ is a family of subsets of X, where $i_\Sigma^{\wp(X)}$ is the inclusion map from Σ to $\wp(X)$. This family may also be denoted by $\{A \mid A \in \Sigma\}$. Incidentally, the class Ω of all families of subsets of X is not a set. We overcome this logical difficulty as follows.

We say that a family (Λ, A) of subsets of X is equivalent to the family (Γ, B) of subsets of X if there is a map η from Λ to Γ and a map ρ from Γ to Λ such that $A_\alpha = B_{\eta(\alpha)}$ and $B_\gamma = A_{\rho(\gamma)}$ for all $\alpha \in \Lambda$ and $\gamma \in \Gamma$. This defines an equivalence relation on Ω. Let $\wp(\hat{X})$ denote the set $\{(\Sigma, i_\Sigma^{\wp(X)}) \mid \Sigma \subseteq \wp(X)\}$ of families of subsets of X. Then any family (Λ, A) of subsets of X is equivalent to a unique family $(A(\Lambda), i_{A(\Lambda)}^{\wp(X)})$ of subsets of X in the set $\wp(\hat{X})$. Thus, without any ambiguity and without any loss of generality, we can talk of a set of all families (up to equivalence) of subsets of X. We can also talk of a set of all open covers (up to equivalence) of a topological space X.

Consider a fixed triple (X, F, U), where X is a topological space, F is a presheaf/sheaf of R-modules on X, and $U = \{U_\alpha \mid \alpha \in \Lambda\}$ is an open cover of X. For each $p \geq 0$, let I^p denote the set $\Lambda^{p+1} = \{\overline{\alpha} = (\alpha_0, \alpha_1, \cdots, \alpha_p) \mid \alpha_i \in \Lambda\}$ of ordered $p+1$-tuples of members of the indexing set Λ. For each $\overline{\alpha} \in I^p$, let $U_{\overline{\alpha}}$ denote the open set $U_{\alpha_0} \cap U_{\alpha_1} \cap \cdots \cap U_{\alpha_p}$. Again, for each $i, 0 \leq i \leq p$ and $\overline{\alpha} \in I^p$, let $\sigma_i(\overline{\alpha})$ denote the element of I^{p-1} obtained by removing the ith entry from $\overline{\alpha}$. Thus, $\sigma_i(\alpha_0, \alpha_1, \cdots, \alpha_p) = (\beta_0, \beta_1, \cdots, \beta_{p-1})$, where $\beta_j = \alpha_j$ for all $j < i$ and $\beta_j = \alpha_{j+1}$ for all $j \geq i$. Symbolically, we denote $\sigma_i(\overline{\alpha})$ by $(\alpha_0, \alpha_1, \cdots, \alpha_{i-1}, \hat{\alpha}_i, \alpha_{i+1}, \cdots, \alpha_p)$. This gives us a map σ_i from I^p to I^{p-1} for each $i, 0 \leq i \leq p, p \geq 1$.

Let $\check{C}^p(X, F, U)$ denote the R-module $\prod_{\overline{\alpha} \in I^p} F(U_{\overline{\alpha}})$. Thus, $\check{C}^p(X, F, U)$ is the module of all maps f from I^p to $\bigcup_{\overline{\alpha} \in I^p} F(U_{\overline{\alpha}})$ such that $f(\overline{\alpha}) \in F(U_{\overline{\alpha}})$ for all $\overline{\alpha} \in I^p$. The R-module $\check{C}^p(X, F, U)$ will be termed as the R-module of Čech p-co-chains on X associated with the open covering U of X and with a coefficient in the presheaf/sheaf F on X. Thus, $\check{C}^0(X, F, U) = \prod_{\alpha \in \Lambda} U_\alpha$ and $\check{C}^1(X, F, U) = \prod_{(\alpha,\beta) \in \Lambda^2} F(U_\alpha \cap U_\beta)$ and so on. For each $p \geq 1$, we have a unique homomorphism $\check{d}^{p-1}$ from $\check{C}^{p-1}(X, F, U)$ to $\check{C}^p(X, F, U)$ given by

$$\check{d}^{p-1}(f)(\overline{\alpha}) = \sum_{i=0}^{p}(-1)^i f(\sigma_i(\overline{\alpha}))|_{U_{\overline{\alpha}}},$$

where $\overline{\alpha} \in I^p$. Thus,

$$\check{d}^0(f)((\alpha_0, \alpha_1)) = f(\alpha_1)|_{U_{\alpha_0} \cap U_{\alpha_1}} - f(\alpha_0)|_{U_{\alpha_0} \cap U_{\alpha_1}}$$

and

$$\check{d}^1(f)((\alpha_0, \alpha_1, \alpha_2)) = f((\alpha_1, \alpha_2))|_{U_{\alpha_0} \cap U_{\alpha_1} \cap U_{\alpha_2}} - f((\alpha_0, \alpha_2))|_{U_{\alpha_0} \cap U_{\alpha_1} \cap U_{\alpha_2}} + f((\alpha_0, \alpha_1))|_{U_{\alpha_0} \cap U_{\alpha_1} \cap U_{\alpha_2}}.$$

In turn,

$$\begin{aligned}
&(\check{d}^1 o \check{d}^0)(f)((\alpha_0, \alpha_1, \alpha_2)) \\
&= \check{d}^0(f)((\alpha_1, \alpha_2))|_{U_{\alpha_0} \cap U_{\alpha_1} \cap U_{\alpha_2}} - \check{d}^0(f)((\alpha_0, \alpha_2))|_{U_{\alpha_0} \cap U_{\alpha_1} \cap U_{\alpha_2}} \\
&\quad + \check{d}^0(f)((\alpha_0, \alpha_1))|_{U_{\alpha_0} \cap U_{\alpha_1} \cap U_{\alpha_2}} \\
&= (f(\alpha_2)|_{U_{\alpha_1} \cap U_{\alpha_2}} - f(\alpha_1)|_{U_{\alpha_1} \cap U_{\alpha_2}})|_{U_{\alpha_0} \cap U_{\alpha_1} \cap U_{\alpha_2}} \\
&\quad - (f(\alpha_2)|_{U_{\alpha_0} \cap U_{\alpha_2}} - f(\alpha_0)|_{U_{\alpha_0} \cap U_{\alpha_2}})|_{U_{\alpha_0} \cap U_{\alpha_1} \cap U_{\alpha_2}} \\
&\quad + (f(\alpha_1)|_{U_{\alpha_0} \cap U_{\alpha_1}} - f(\alpha_0)|_{U_{\alpha_0} \cap U_{\alpha_1}})|_{U_{\alpha_0} \cap U_{\alpha_1} \cap U_{\alpha_2}} \\
&= f(\alpha_2)|_{U_{\alpha_0} \cap U_{\alpha_1} \cap U_{\alpha_2}} - f(\alpha_1)|_{U_{\alpha_0} \cap U_{\alpha_1} \cap U_{\alpha_2}} - f(\alpha_2)|_{U_{\alpha_0} \cap U_{\alpha_1} \cap U_{\alpha_2}} + \\
&\quad f(\alpha_0)|_{U_{\alpha_0} \cap U_{\alpha_1} \cap U_{\alpha_2}} + f(\alpha_1)|_{U_{\alpha_0} \cap U_{\alpha_1} \cap U_{\alpha_2}} - f(\alpha_0)|_{U_{\alpha_0} \cap U_{\alpha_1} \cap U_{\alpha_2}} \\
&= 0
\end{aligned}$$

for all $(\alpha_0, \alpha_1, \alpha_2) \in I^2$. This means that $\check{d}^1 o \check{d}^0 = 0$. More generally, it can be easily checked that $\check{d}^p o \check{d}^{p-1} = 0$ for all $p \geq 1$. This gives us a co-chain complex

$$\check{C}^*(X, F, U) \equiv 0 \longrightarrow \check{C}^0(X, F, U) \xrightarrow{\check{d}^0} \check{C}^1(X, F, U) \xrightarrow{\check{d}^1} \cdots \xrightarrow{\check{d}^{n-1}} \check{C}^n(X, F, U) \xrightarrow{\check{d}^n} \cdots$$

of R-modules. The co-chain complex $\check{C}^*(X, F(U))$ is termed as **Čech co-chain complex** associated with the open covering U and with a coefficient in the presheaf/sheaf F. The p_{th} co-homology of $\check{C}^*(X, F, U)$ denoted by $\check{H}^p(X, F, U)$ is called the p_{th} **Čech co-homology** of X associated with the open covering U and with a coefficient in the presheaf/sheaf F.

Proposition 4.2.16 *If F is sheaf, then $\check{H}^0(X, F, U) \approx \Gamma_X(F) \approx H^0(X, F)$ for all open covers $U = \{U_\alpha \mid \alpha \in \Lambda\}$ of X. More generally, for any presheaf F, $\check{H}^0(X, F, U) \approx \Gamma_X(SF) \approx H^0(X, SF)$, where SF is the sheafification of F.*

Proof A member s of $\Gamma_X(F)$ determines a unique member $\hat{s}$ of $\check{C}^0(X, F, U)$ given by $\hat{s}(\alpha) = s|_{U_\alpha}$. Evidently, $\check{d}^0(\hat{s})((\alpha, \beta)) = \hat{s}(\beta)|_{U_\alpha \cap U_\beta} - \hat{s}(\alpha)|_{U_\alpha \cap U_\beta} = s|_{U_\beta \cap U_\alpha} - s|_{U_\alpha \cap U_\beta} = 0$. Thus, we have a homomorphism ϕ from $\Gamma_X(F)$ to $\check{H}^0(X, F, U)$ defined by $\phi(s) = \hat{s}$. Since F is a sheaf, ϕ is injective. Let t be a member of $\check{H}^0(X, F, U)$. Then $t \in Ker\ \check{d}^0$. This means that the family $\{t(\alpha) \in F(U_\alpha) \mid \alpha \in \Lambda\}$ is such that $t(\alpha)|_{U_\alpha \cap U_\beta} = t(\beta)|_{U_\alpha \cap U_\beta}$ for all pairs α, β in Λ. Since F is a sheaf, there is a unique s in $\Gamma_X(F)$ such that $\hat{s} = t$. This shows that ϕ is an isomorphism. Similarly, the rest of the assertion follows. ♯

The Čech co-chain complex $\check{C}^*(X, F, U)$ associated with the open cover U and the presheaf/sheaf F is a huge co-chain complex. We have a convenient, natural, and smaller co-chain subcomplex $\check{C}^*_{alt}(X, F, U)$ of $\check{C}^*(X, F, U)$ consisting of alternating Čech co-chains such that the inclusion is a chain equivalence from

$\check{C}_{alt}^*(X, F, U)$ to $\check{C}^*(X, F, U)$. A p-Čech co-chain f is called an alternating p-Čech co-chain if (i) $f(\alpha_0, \alpha_1, \cdots, \alpha_p) = 0$ whenever $\alpha_i = \alpha_j$ for some $i \neq j$, and (ii) $f(\alpha_{\tau(0)}, \alpha_{\tau(1)}, \cdots, \alpha_{\tau(p)}) = -f(\alpha_0, \alpha_1, \cdots, \alpha_p)$ for all transpositions τ. Let $\check{C}_{alt}^p(X, F, U)$ denote the submodule of $\check{C}^p(X, F, U)$ consisting of alternating Čech p-co-chains. Then it can be easily seen that

$$\check{C}_{alt}^*(X, F, U) \equiv 0 \longrightarrow \check{C}_{alt}^0(X, F, U) \xrightarrow{\check{d}^0} \check{C}_{alt}^1(X, F, U) \xrightarrow{\check{d}^1} \cdots \xrightarrow{\check{d}^{n-1}} \check{C}_{alt}^n(X, F, U) \xrightarrow{\check{d}^n} \cdots$$

is a co-chain subcomplex of $\check{C}^*(X, F, U)$. Indeed, this co-chain subcomplex is chain equivalent to $\check{C}^*(X, F, U)$. In turn, the p_{th} co-homology of $\check{C}_{alt}^*(X, F, U)$ is the same as $\check{H}^p(X, F, U)$.

Fixing an open cover U of X, we get functors $\check{H}^p(X, -, U)$ from the category of presheaves/sheaves to the category of abelian groups. However, it need not describe a genuine co-homology theory with coefficients in presheaves/sheaves. Indeed, for a short exact sequence

$$0 \longrightarrow F \xrightarrow{\alpha} F' \xrightarrow{\beta} F'' \longrightarrow 0$$

of sheaves on X, we may not have a connecting homomorphism ∂^p from $\check{H}^p(X, F, U)$ to $\check{H}^{p+1}(X, F, U)$ so that the sequence

$$\cdots \xrightarrow{\partial^{p-1}} \check{H}^p(X, F, U) \xrightarrow{\check{H}^p(X, \alpha, U)} \check{H}^p(X, F', U) \xrightarrow{\check{H}^p(X, \beta, U)} \check{H}^p(X, F'', U) \xrightarrow{\partial^p} \check{H}^{p+1}(X, F, U) \xrightarrow{\check{H}^p(X, \alpha, U)} \cdots$$

is exact. For example, if $U = \{X\}$ and F is a sheaf, then the above sequence need not be exact as Γ_X is not an exact functor.

Next, we make our discussions independent of a particular open covering by passing through a limit.

Definition 4.2.17 Let $U = \{U_\alpha \mid \alpha \in \Lambda\}$ and $V = \{V_\mu \mid \mu \in \Gamma\}$ be open coverings of a topological space X. We say that V is a **refinement** of U if there is a map χ from Γ to Λ such that $V_\mu \subseteq U_{\chi(\mu)}$ for all $\mu \in \Gamma$. The map χ is called a refinement map.

Proposition 4.2.18 *Let* $V = \{V_\mu \mid \mu \in \Gamma\}$ *be an open coverings of a topological space* X *which is a refinement of an open covering* $U = \{U_\alpha \mid \alpha \in \Lambda\}$ *of* X *with a refinement map* χ. *Let* F *be a presheaf/sheaf on* X. *Then for each* $p \geq 0$, χ *induces a homomorphism* $\overline{\chi_p}$ *from* $\check{H}^p(X, F, U)$ *to* $\check{H}^p(X, F, V)$. *Further, if* χ' *is another refinement map, then* $\overline{\chi_p} = \overline{\chi'_p}$ *for each* p.

Proof For each $p \geq 0$, χ induces a map χ_p from $\check{C}^p(X, F, U)$ to $\check{C}^p(X, F, V)$ which is given by

$$\chi_p(f)((\mu_0, \mu_1, \cdots, \mu_p)) = f(\chi(\mu_0), \chi(\mu_1), \cdots, \chi(\mu_p))|_{V_{\mu_0} \cap V_{\mu_1} \cap \cdots \cap V_{\mu_p}}.$$

Evidently, the family $\{\chi_p \mid p \geq 0\}$ defines a chain map from $\check{C}^*(X, F, U)$ to $\check{C}^*(X, F, V)$. This, in turn, induces a homomorphism $\overline{\chi_p}$ from $\check{H}^p(X, F, U)$ to

$\check{H}^p(X, F, V)$ for each p. Further, if χ' is another refinement map, the fact that F is a presheaf implies that $\chi_p = \chi'_p$ for each p. In turn, $\overline{\chi_p} = \overline{\chi'_p}$ for each p. ♯

Let (X, T) be a topological space. Let Ω denote the set $\{(\Sigma, i^T_\Sigma) \mid \Sigma \subseteq T, \bigcup_{A \in \Sigma} A = X$, and i^T_Σ is the inclusion map$\}$ of open covers of X. Note that any open cover of X is equivalent to a unique member of Ω. For simplicity, we shall denote the open cover (Σ, i^T_Σ) of X by Σ itself. We have a partial order $\leq$ on Ω given by $\Sigma \leq \Sigma'$ if Σ' is a refinement of Σ. This is, indeed, equivalent to say that there is a map λ from Σ' to Σ such that $A \subseteq \lambda(A)$ for all $A \in \Sigma'$. Let Σ and Σ' be members of Ω. Consider the subset $\Sigma \bigwedge \Sigma' = \{A \bigcap B \mid A \in \Sigma \ and \ B \in \Sigma'\}$ of T. Clearly, $\Sigma \bigwedge \Sigma'$ is a member of Ω such that $\Sigma \leq \Sigma \bigwedge \Sigma'$ and $\Sigma' \leq \Sigma \bigwedge \Sigma'$. This shows that $(\Omega, \leq)$ is a directed system of open coverings of X. Consequently, given a presheaf F on X, we get a directed system $\{\check{C}^*(X, F, \Sigma) \mid \Sigma \in \Omega\}$ of co-chain complexes of R-modules and the directed system $\{\check{H}^p(X, F, \Sigma) \mid \Sigma \in \Omega\}$ of co-homology modules for each $p \geq 0$. The $Lim_{\Sigma \in \Omega} \check{H}^p(X, F, \Sigma)$ is called the p_{th} Čech co-homology of X with a coefficient in the presheaf F, and it is denoted by $\check{H}^p(X, F)$.

The following proposition is immediate from Proposition 4.2.16.

Proposition 4.2.19 $\check{H}^0(X, F) \approx \Gamma_X(SF)$. ♯

Proposition 4.2.20 *Let Σ be a member of Ω. Then $\check{C}^*(X, -, \Sigma)$ defines an exact functor from the category Pr_X to the category of co-chain complexes of abelian groups.*

Proof Let

$$0 \longrightarrow F \xrightarrow{\alpha} F' \xrightarrow{\beta} F'' \longrightarrow 0$$

be a short exact sequence in Pr_X. It follows from Proposition 4.2.1(i) that

$$0 \longrightarrow F(U) \xrightarrow{\alpha_U} F'(U) \xrightarrow{\beta_U} F''_U \longrightarrow 0$$

is exact for each open set U. Thus, for each $p + 1$-tuple $(U_0, U_1, \cdots, U_p)$ in Σ^{p+1}, we have a short exact sequence

$$0 \longrightarrow F(U_0 \bigcap U_1 \bigcap \cdots \bigcap U_p) \xrightarrow{\alpha_{U_0 \bigcap U_1 \bigcap \cdots \bigcap U_p}} F'(U_0 \bigcap U_1 \bigcap \cdots \bigcap U_p) \xrightarrow{\beta_{U_0 \bigcap U_1 \bigcap \cdots \bigcap U_p}}$$
$$\times F''(U_0 \bigcap U_1 \bigcap \cdots \bigcap U_p) \longrightarrow 0.$$

In turn, for each $p \geq 0$, we get a homomorphism α^p from $\check{C}^p(X, F, \Sigma)$ to $\check{C}^p (X, F', \Sigma)$, and a homomorphism β^p from $\check{C}^p(X, F', \Sigma)$ to $\check{C}^p(X, F'', \Sigma)$ such that

$$0 \longrightarrow \check{C}^p(X, F, \Sigma) \xrightarrow{\alpha^p} \check{C}^p(X, F', \Sigma) \xrightarrow{\beta^p} \check{C}^p(X, F'', \Sigma) \longrightarrow 0$$

is exact. Evidently, $\{\alpha^p \mid p \geq o\}$ is a chain transformation from $\check{C}^*(X, F, \Sigma)$ to $\check{C}^*(X, F', \Sigma)$ and $\{\beta^p \mid p \geq o\}$ is a chain transformation from $\check{C}^*(X, F', \Sigma)$ to $\check{C}^*(X, F'', \Sigma)$. Hence

$$0 \longrightarrow \check{C}^*(X, F, \Sigma) \overset{\alpha^*}{\to} \check{C}^*(X, F', \Sigma) \overset{\beta^*}{\to} \check{C}^*(X, F, \Sigma) \longrightarrow 0$$

is an exact sequence of co-chain complexes. ♯

Corollary 4.2.21 *Given a short exact sequence*

$$0 \longrightarrow F \overset{\alpha}{\to} F' \overset{\beta}{\to} F'' \longrightarrow 0$$

in Pr_X, for each $p \geq 0$, we get a natural connecting homomorphism ∂^p from $\check{H}^p(X, F'')$ to $\check{H}^{p+1}(X, F)$ such that the sequence

$$\cdots \overset{\partial^{p-1}}{\to} \check{H}^p(X, F) \overset{\check{H}^p(X, \alpha)}{\to} \check{H}^p(X, F') \overset{\check{H}^p(X, \beta)}{\to} \check{H}^p(X, F'') \overset{\partial^p}{\to} \cdots$$

is exact.

Proof Applying the functor $Lim_{\Sigma \in \Omega}(-)$ to the short exact sequence

$$0 \longrightarrow \check{C}^*(X, F, \Sigma) \overset{\alpha^*}{\to} \check{C}^*(X, F', \Sigma) \overset{\beta^*}{\to} \check{C}^*(X, F, \Sigma) \longrightarrow 0$$

of co-chain complexes, we obtain the short exact sequence

$$0 \longrightarrow \check{C}^*(X, F) \overset{\alpha^*}{\to} \check{C}^*(X, F') \overset{\beta^*}{\to} \check{C}^*(X, F) \longrightarrow 0$$

of the co-chain complex of abelian groups. Applying Theorem 1.3.1, we get the desired exact sequence. ♯

Our next aim is to relate *Čech* co-homology and sheaf co-homology. Let (X, T) be a topological space, and let Σ be a member of Ω. Let F be a sheaf on X. For each open subset U of X, let $\Sigma|_U$ denote the open cover $\{U \bigcap A \mid A \in \Sigma\}$ of U and $\check{C}^p(U, F|_U, \Sigma|_U)$ denote the group of *Čech* p-co-chains of U associated with the open covering $\Sigma|_U$ and the sheaf $F|_U$. For each $p \geq 0$, we have a sheaf $\check{C}^p(F, \Sigma)$ on X given by $\check{C}^p(F, \Sigma)(U) = \check{C}^p(U, F|_U, \Sigma|_U)$ together with the obvious restriction homomorphisms. The co-boundary map $\check{d}^p$ induces a sheaf morphism $\check{d}^p(F, \Sigma)$ from $\check{C}^p(F, \Sigma)$ to $\check{C}^{p+1}(F, \Sigma)$. Further, we have a sheaf morphism $\epsilon = \{\epsilon_U \mid U \in T\}$ from F to $\check{C}^0(F, \Sigma)$, where ϵ_U is given by $\epsilon_U(s)(U \bigcap A) = s|_{U \bigcap A}$.

Proposition 4.2.22 *Given a sheaf F on X, for every open covering Σ of X in Ω, we have a co-resolution of F given by*

$$0 \longrightarrow F \overset{\epsilon}{\to} \check{C}^0(F, \Sigma) \overset{\check{d}^0(F, \Sigma)}{\to} \check{C}^1(F, \Sigma) \overset{\check{d}^1(F, \Sigma)}{\to} \cdots \overset{\check{d}^{p-1}(F, \Sigma)}{\to} \check{C}^p(F, \Sigma) \overset{\check{d}^p(F, \Sigma)}{\to} \cdots .$$

Proof The fact that F is a sheaf implies that ϵ is a monomorphism and the image of ϵ is $ker\ \check{d}^0(F, \Sigma)$. In the light of Proposition 4.2.1(iii), it is sufficient to show that

$$\check{C}^+(F, \Sigma) \equiv \check{C}^0(F, \Sigma)_x \overset{\check{d}^0(F, \Sigma)_x}{\to} \check{C}^1(F, \Sigma)_x \overset{\check{d}^1(F, \Sigma)_x}{\to} \cdots \overset{\check{d}^{p-1}(F, \Sigma)_x}{\to} \check{C}^p(F, \Sigma)_x \overset{\check{d}^p(F, \Sigma)_x}{\to} \cdots$$

is exact for all $x \in X$. For that, we define a chain homotopy $\chi = \{\chi_p \mid p \geq 1\}$ from $I_{\check{C}^+(F,\Sigma)}$ to $0_{\check{C}^+(F,\Sigma)}$. We define a homomorphism χ^p from $\check{C}^p(F,\Sigma)_x$ to $\check{C}^{p-1}(F,\Sigma)_x$ as follows: Let a_x be a member of $\check{C}^p(F,\Sigma)_x$. Suppose that $s_V \in \check{C}^p(F,\Sigma)(V) = \check{C}^p(V, F|_V, \Sigma|_V)$ represents a_x, where V is an open neighborhood of x. Define a homomorphism χ_V^p from $\check{C}^p(F,\Sigma)(V)$ to $\check{C}^{p-1}(F,\Sigma)(V)$ by putting

$$\chi_V^p(s_V)(U_0, U_1, \cdots, U_{p-1}) = s_V|_{V \cap U_0 \cap U_1 \cap \cdots \cap U_{p-1}}.$$

If $s_W \in \check{C}^p(F,\Sigma)(W) = \check{C}^p(W, F|_W, \Sigma|_W)$ also represents a_x, then it is clear that $\chi_V^p(s_V)$ and $\chi_W^p(s_W)$ represent the same element in $\check{C}^{p-1}(F,\Sigma)_x$. Thus, we have a homomorphism χ_x^p from $\check{C}^p(F,\Sigma)_x$ to $\check{C}^{p-1}(F,\Sigma)_x$ by putting $\chi_x^p(a_x)$ to be the element of $\check{C}^{p-1}(F,\Sigma)_x$ which is represented by $\chi_V^p(s_V)$, where s_V represents a_x. From the construction, it is clear that

$$\check{d}_x^{p+1}\chi_x^p + \chi_x^{p+1}\check{d}_x^p = I_{\check{C}^p(F,\Sigma)_x} = I_{\check{C}^p(F,\Sigma)_x} - 0_{\check{C}^p(F,\Sigma)_x}$$

for each p and each $x \in X$ the result follows. ♯

Corollary 4.2.23 *Let F be a flasque sheaf on X. Then $\check{H}^p(X, F, \Sigma) = 0$ for all open covering Σ of X and for all $p \geq 1$.*

Proof Since F is a flasque sheaf, by Corollary 4.2.13, $H^p(X, F) = 0$. Since the restrictions of flasque sheaves are flasque sheaves and since the direct products of flasque sheaves are flasque sheaves, $\check{C}^p(F,\Sigma)$ are flasque sheaves for all $p \geq 0$. By Proposition 4.2.14, the sheaf co-homology $H^p(X, F)$ of X with values in a sheaf F can be computed with the help of the co-resolution of F given in the previous proposition. Thus, $H^p(X, F)$ is the pth co-homology of the co-chain complex

$$\Gamma_X\check{C}^\star(F,\Sigma) \approx 0 \longrightarrow \Gamma_X\check{C}^0(F,\Sigma) \overset{\check{d}^0(F,\Sigma)}{\to} \Gamma_X\check{C}^1(F,\Sigma) \overset{\check{d}^1(F,\Sigma)}{\to} \cdots \overset{\check{d}^{p-1}(F,\Sigma)}{\to} \Gamma_X\check{C}^p(F,\Sigma) \overset{\check{d}^p(F,\Sigma)}{\to} \cdots.$$

Evidently, $\Gamma_X\check{C}^p(F,\Sigma) = \check{C}^p(X, F, \Sigma)$. Hence $\check{H}^p(X, F, \Sigma) = \check{H}^p(X, F) = 0$ for all $p \geq 0$. ♯

Let F be a sheaf on X. Let Σ be an open covering in Ω. Let

$$0 \longrightarrow F \overset{\eta}{\to} I_0 \overset{d_0}{\to} I_1 \overset{d1}{\to} \cdots \overset{d^{p-1}}{\to} I_p \overset{d^p}{\to} \cdots$$

be an injective co-resolution of F. By Proposition 4.2.22, we have a co-resolution

$$0 \longrightarrow F \overset{\epsilon}{\to} \check{C}^0(F,\Sigma) \overset{\check{d}^0(F,\Sigma)}{\to} \check{C}^1(F,\Sigma) \overset{\check{d}^1(F,\Sigma)}{\to} \cdots \overset{\check{d}^{p-1}(F,\Sigma)}{\to} \check{C}^p(F,\Sigma) \overset{\check{d}^p(F,\Sigma)}{\to} \cdots$$

of F. From Proposition 2.1.3, there is a co-chain transformation $f(F,\Sigma) = \{f^p(F,\Sigma) \mid p \geq 0\}$ from $\check{C}^+(F,\Sigma)$ to I such that $f^0(F,\Sigma) \circ \epsilon = \eta$, where

$$\check{C}^+(F,\Sigma) \equiv \check{C}^0(F,\Sigma) \overset{\check{d}^0(F,\Sigma)}{\to} \check{C}^1(F,\Sigma) \overset{\check{d}^1(F,\Sigma)}{\to} \cdots \overset{\check{d}^{p-1}(F,\Sigma)}{\to} \check{C}^p(F,\Sigma) \overset{\check{d}^p(F,\Sigma)}{\to} \cdots$$

and

$$I \equiv I_0 \xrightarrow{d_0} I_1 \xrightarrow{d_1} \cdots \xrightarrow{d^{p-1}} I_p \xrightarrow{d^p} \cdots.$$

Applying the section functor Γ_X, and then taking co-homologies, we obtain natural homomorphisms $H^p(f(F, \Sigma))$ from $\check{H}^p(X, F, \Sigma)$ to $H^p(X, \Sigma)$. Further, if Σ' is a refinement of Σ and χ is a refinement map, then it can be easily observed that $H^p(f(F, \Sigma))o\overline{\chi_p} = H^p(f(F, \Sigma))$. In turn, taking the direct limit, we obtain a natural homomorphism H^p_F from $\check{H}^p(X, F)$ to $H^p(X, F)$ which is functorial in F and also in X.

Proposition 4.2.24 H^0_F and H^1_F are isomorphisms.

Proof From Proposition 4.2.16, $\check{H}^0(X, F, \Sigma)$ is naturally isomorphic to $H^0(X, F)$. It follows that $\check{H}^0(X, F) = Lim_{\Sigma \in \Omega} \check{H}^0(X, F, \Sigma)$ is isomorphic to $H^0(X, F)$.

Next, we show that H^1_F is an isomorphism. Since Sh_X has enough injectives, F can be embedded in an injective sheaf F'. In turn, we get a sequence

$$0 \longrightarrow F \xrightarrow{\alpha} F' \xrightarrow{\beta} F'' \longrightarrow 0$$

which is exact in Pr_X. From Corollary 4.2.21, we have a long exact sequence

$$0 \longrightarrow \check{H}^0(X, F) \xrightarrow{\check{H}^0(X,\alpha)} \check{H}^0(X, F') \xrightarrow{\check{H}^0(X,\beta)} \check{H}^0(X, F'') \xrightarrow{\partial^0} \check{H}^1(X, F) \xrightarrow{\check{H}^1(X,\alpha)}$$

$$\times \check{H}^1(X, F') \xrightarrow{\check{H}^1(X,\beta)} \check{H}^1(X, F'') \xrightarrow{\partial^1} \cdots.$$

Further, we have the exact sequence

$$0 \longrightarrow F \xrightarrow{\alpha} F' \xrightarrow{\hat{\beta}} SF'' \longrightarrow 0$$

in Sh_X, where SF'' denote the sheafification of F'' and $\hat{\beta}$ is the morphism induced by β. In turn, from Proposition 4.2.4, we have a long exact sequence

$$0 \longrightarrow H^0(X, F) \xrightarrow{H^0(X,\alpha)} H^0(X, F') \xrightarrow{H^0(X,\hat{\beta})} H^0(X, SF'') \xrightarrow{\partial^0} H^1(X, F) \xrightarrow{H^1(X,\alpha)}$$

$$\times H^1(X, F') \xrightarrow{H^1(X,\hat{\beta})} H^1(X, SF'') \xrightarrow{\partial^0} \cdots.$$

The homomorphisms H^p_F, $H^p_{F'}$, and $H^p_{F''}$, $p \geq 0$, make the diagram involving the relevant long exact sequences commutative. Further, from Proposition 4.2.16, H^0_F, $H^0_{F'}$, and $H^0_{F''}$ are isomorphisms. Since F' is flasque (being injective), it follows from Proposition 4.2.23 that $\check{H}^1(X, F) = 0$. Also from Corollary 4.2.13, it follows that $H^1(X, F') = 0$. Chasing the said commutative diagram, we obtain that H^1_F is an isomorphism. ♯

However, for $p \geq 2$, H^p_F need not be injective, and it may not be surjective also. We state (without proof) a sufficient condition under which H^p_F is an isomorphism for all $p \geq 0$. Recall that an open covering Σ is termed as a **locally finite covering** if every point $x \in X$ has an open neighborhood U which intersects only a finitely many members of Σ. A Hausdorff topological space X is called a **Paracompact Space** if every open cover has a locally finite refinement. For example, a metric space is a paracompact space.

Theorem 4.2.25 *If X is a paracompact space, then H_F^p is an isomorphism for each $p \geq 0$.* ♯

Further, the following theorem expresses a singular co-homology group $H^p(X, A)$ of a space X with a coefficient in a module A as a sheaf co-homology and also as a Čech co-homology.

Theorem 4.2.26 *Let X be the second countable topological manifold, and A be a R-module. Then $H^p(X, A) \approx \check{H}^p(X, F_A) \approx H^p(X, F_A)$, where F_A is the constant sheaf with $F_A(U) = A$ for all $U \neq \emptyset$ and $F_A(\emptyset) = \{0\}$.* ♯

The proofs of the above two theorems can be found in "Algebraic Topology" by Spanier.

Let (X, ϑ_X) be a ringed space. We have sheaf $\vartheta_X^\star$ of abelian groups defined by $\vartheta_X^\star(U) = \vartheta_X(U)^\star$, where $\vartheta_X(U)^\star$ is the group of units of $\vartheta_X(U)$ and the restriction homomorphisms are those induced by the restriction homomorphisms in ϑ_X. The following proposition describes the Picard group of a ringed spaces as sheaf co-homology.

Proposition 4.2.27 $Pic((X, \vartheta_X)) \approx H^1(X, \vartheta_X^\star)$.

Proof Let M_X be a locally free ϑ_X-module of rank 1. Let Σ be an open cover of X together with $\vartheta_X|_U$-isomorphism ϕ_U from $\vartheta_X|_U$ to $M_X|_U$ for each $U \in \Sigma$. For each pair $(U, V) \in \Sigma^2$, we have a $\vartheta_X|_{U \cap V}$-isomorphism $\phi_U^{-1} o \phi_V$ from $\vartheta_X|_{U \cap V}$ to itself. In turn, it determines a unique element $\hat{\phi}_{(U,V)} \in \vartheta_X|_{U \cap V}^\star$ given by $(\phi_U^{-1} o \phi_V)(a) = \hat{\phi}_{(U,V)}a$, $a \in \vartheta_X|_{U \cap V}$. Consequently, we get an element $\hat{\phi} = \{\hat{\phi}_{(U,V)} \mid (U, V) \in \Sigma^2\}$ of $\check{C}^1(X, \vartheta_X^\star, \Sigma)$. Further, since

$$(\phi_V^{-1} o \phi_W)|_{U \cap V \cap W} o (\phi_W^{-1} o \phi_U)|_{U \cap V \cap W} o (\phi_U^{-1} o \phi_V)|_{U \cap V \cap W} = I_{\vartheta_X|_{U \cap V \cap W}},$$

we have

$$\check{d}^1(\hat{\phi})(U, V, W) = \hat{\phi}(V, W)|_{U \cap V \cap W}(\hat{\phi}(U, W))^{-1}|_{U \cap V \cap W}\hat{\phi}(U, V)|_{U \cap V \cap W} = 1.$$

(Note that we are using multiplicative notation for $\vartheta_X^\star$.) This shows that $\hat{\phi} \in ker \, \hat{d}^1$. In turn, it determines an element $\rho(\hat{\phi})$ of $\hat{H}^1(X, \vartheta_X^\star, \Sigma)$. For each $U \in \Sigma$, let ψ_U be another choice of $\vartheta_X|U$-isomorphism from $\vartheta_X|_U$ to $M_X|_U$. Then for each $U \in \Sigma$, there is an element $\alpha_U \in \vartheta_X(U)$ such that $\phi_U(a) = \alpha_U \psi_U(a)$ for all $a \in \vartheta_X(U)$. Clearly, $\alpha = \{\alpha_U \mid U \in \Sigma\}$ is a member of $\check{C}^0(X, \vartheta_X^\star, \Sigma)$ such that $\hat{d}^0(\alpha) = \hat{\psi}^{-1}\hat{\phi}$. This shows that $\rho(\hat{\phi}) = \rho(\hat{\psi})$. If $\eta = \{\eta_U \mid U \in T\}$ is an isomorphism from M_X

to N_X, and ϕ_U is a $\vartheta_X|_U$-isomorphism from $\vartheta_X|_U$ to $M_X|_U$, then $\psi_U = \eta|_U \circ \phi_U$ is a $\vartheta_X|_U$-isomorphism from $\vartheta_X|_U$ to $N_X|_U$. As before $\rho(\hat{\phi}) = \rho(\hat{\psi})$. Consequently, ρ induces a map $\overline{\rho}$ from $Pic(X, \vartheta_X)$ to $\check{H}^1(X, \vartheta_X^\star, \Sigma)$ given by $\overline{\rho}([M_X]) = \rho(\hat{\phi})$, where ϕ is described as above. It can be easily checked that $\overline{\rho}$ is a homomorphism. It further induces a homomorphism $\hat{\rho}$ from $Pic(X, \vartheta_X)$ to $\check{H}^1(X, \vartheta_X^\star)$ which can be seen to be an isomorphism (see Exercises 4.1.13 and 4.1.14). The result follows from Proposition 4.2.24. ♯

Exercises

4.2.1 Let $\mathbb{Z}_{S^1}$ denote the constant sheaf on S^1 given by $\mathbb{Z}_{S^1}(U) = \mathbb{Z}$ for all nonempty open subsets of S^1. Show that $H^1(S^1, \mathbb{Z}_{S^1}) \approx \mathbb{Z}$.

4.2.2 Compute $H^1(S^1, F_{S^1})$, where F_{S^1} is the sheaf of germs of real-valued continuous functions on S^1.

4.2.3 Let X be a topological space, and F_X be a sheaf of abelian groups on X. Let Y be a closed subset of X. Let $F_{X|_Y}(U) = \{s \in F_X(U) \mid s \text{ represents } 0 \text{ in } (F_X)_x \text{ for each } x \notin Y \bigcap U\}$. Show that the association $U \mapsto F_{X|_Y}(U)$ defines a subsheaf $F_{X|_Y}$ of F_X. The subsheaf $F_{X|_Y}$ is called the sub-sheaf of F_X with support in Y. Show further that the functor $\Gamma_X|_Y$ from Sh_X to AB given by $\Gamma_X|_Y(F) = \Gamma_X(F_{X|_Y})$ is a left exact functor. The right-derived functors $\{R^p \Gamma_X|_Y, p \geq 0\}$ of $\Gamma_X|_Y$ are called the **sheaf co-homology functors on X with support in Y**, and they are denoted by $\{H_Y^p(X, -)\}$. We term $H_Y^p(X, F_X)$ as the sheaf co-homology of X with coefficient in F_X and with support in Y.

4.2.4 Let Y be a closed subset of X which is contained in an open subset U of X. Show that $H_Y^p(X, F_X)$ is naturally isomorphic to $H_Y^p(U, F_X|_U)$.

4.2.5 Let Y and Z be closed subsets of X. Show that

$$0 \longrightarrow F_{X|_{Y \bigcap Z}} \overset{(i_1, i_2)}{\to} F_{X|_Y} \oplus F_{X|_Z} \overset{j_1 + j_2}{\to} F_{X|_{Y \bigcup Z}} \longrightarrow 0$$

is a short exact sequence of sheaves on X, where i_1, i_2, j_1, and j_2 are natural inclusions. Deduce that the Mayer–Vietoris sequence

$$\longrightarrow H_{Y \bigcap Z}^p(X, F) \overset{(H^p(i_1), H^p(i_2))}{\to} H_Y^p(X, F) \oplus H_Z^p(X, F) \overset{H^p(j_1) + H^p(j_2)}{\to} H_{Y \bigcup Z}^p(X, F)$$
$$\overset{\partial^p}{\to} H_{Y \bigcap Z}^{p+1}(X, F) \longrightarrow \cdots$$

is exact.

4.2.6 Show that the natural homomorphism from $\check{H}^2(X, F_X)$ to $H^2(X, F)$ is injective.

4.2.7 Discover a nonzero element of $H^1(\mathbb{C}^\star, \mathbb{C}_{\mathbb{C}^\star})$, where $\mathbb{C}_{\mathbb{C}^\star}$ is the constant sheaf on $\mathbb{C}^\star$ given by $\mathbb{C}_{\mathbb{C}^\star}(U) = \mathbb{C}$ for all nonempty open subsets U of $\mathbb{C}^\star$.

4.3 Algebraic Varieties

Our aim is to have some applications of sheaf co-homology in algebraic geometry. The present section and the following section are devoted to introducing the basics in algebraic geometry,viz., the theory of algebraic varieties, the theory of schemes, and their relationships.

We recall some basic definitions and properties of affine and projective varieties. Let K be an algebraically closed field. The set K^n is called the affine n-set over K. The affine n-set K^n is also denoted by A_K^n. In vector notation, an element $(\alpha_1, \alpha_2, \cdots, \alpha_n) \in K^n$ is denoted by $\overline{\alpha}$. A polynomial $f(X_1, X_2, \cdots, X_n)$ in $K[X_1, X_2, \cdots, X_n]$ will be denoted by $f(\overline{X})$, and $f(\alpha_1, \alpha_2, \cdots, \alpha_n)$ will be denoted by $f(\overline{\alpha})$. Let $Map(K^n, K)$ denote the set of all maps from K^n to K. Evidently, $Map(K^n, K)$ is a commutative algebra over K. We have the evaluation map $\hat{ev}$ from the polynomial ring $K[X_1, X_2, \cdots, X_n]$ to $Map(K^n, K)$ given by $\hat{ev}(f(\overline{X}))(\overline{\alpha}) = f(\overline{\alpha})$. Evidently, $\hat{ev}$ is an algebra homomorphism which is injective provided that K is infinite (Corollary 7.7.3, Algebra 1). In particular, if K is algebraically closed, then $\hat{ev}$ is injective. The members of the image of $\hat{ev}$ are called the **polynomial functions**. They are also called the **regular functions** on A_K^n.

We have a map V from the power set $\wp(K[X_1, X_2, \cdots, X_n])$ of $K[X_1, X_2, \cdots, X_n]$ to the power set $\wp(K^n)$ of K^n defined by $V(A) = \{\overline{\alpha} \in K^n \mid f(\overline{\alpha}) = 0 \ for \ all \ f(\overline{X}) \in A\}$, where A is a subset of $K[X_1, X_2, \cdots, X_n]$. Thus, $V(A)$ is the set of all common zeros of A. Evidently, $V(A) = V(<A>)$, where $<A>$ is the ideal generated by A. Recall that for an ideal B of a commutative ring R with identity, the subset $\{a \in R \mid a^n \in B \ for \ some \ n \in \mathbb{N}\}$ of R is an ideal of R. This ideal is denoted by $\sqrt{B}$ and it is called the **radical** of B. Clearly, $V(A) = V(\sqrt{<A>})$. Note that $V(\emptyset) = A_K^n$, $V(\{1\}) = \emptyset$, $\bigcap_{\alpha \in \Lambda} V(A_\alpha) = V(\bigcup_{\alpha \in \Lambda} A_\alpha)$, and $V(A_1) \bigcup V(A_2) = V(A_1 A_2)$. Thus, the family $\{V(A) \mid A \subseteq K[X_1, X_2, \cdots, X_n]\}$ of zero sets of polynomials forms a family of closed sets for a unique topology on K^n (Exercise 7.7.1, Algebra 1). This topology is called the **Zariski topology** on the affine n-set A_K^n, and A_K^n with the Zariski topology is called the affine n-space. A set of the form $V(A)$ is called an **affine algebraic subset**. An ideal A of a ring is called a **radical ideal** if $\sqrt{A} = A$. Since $\sqrt{\sqrt{<A>}} = \sqrt{<A>}$, V defines a surjective map from the set of radical ideals of $K[X_1, X_2, \cdots, X_n]$ to the set of all closed subsets of A_K^n. We have the map I from the set of all subsets of A_K^n to the set of radical ideals of $K[X_1, X_2, \cdots, X_n]$ defined by $I(Y) = \{f(\overline{X}) \mid f(\overline{\alpha}) = 0 \ for \ all \ \overline{\alpha} \in Y\}$. Note that $I(Y) = I(\overline{Y})$. If K is an algebraically closed field, the Hilbert Nullstellensatz asserts that $IV(A) = \sqrt{<A>}$. Thus, if K is an algebraically closed field, then V defines an inclusion reversing bijective map from the set of all radical ideals of $K[X_1, X_2, \cdots, X_n]$ to the set of all affine algebraic subsets of A_K^n. Also, $V(I(Y)) = \overline{Y}$, where $\overline{Y}$ is the closure of Y in the Zariski topology.

A topological space X is said to be **irreducible** if it cannot be expressed as the union of two proper closed subsets. A subset Y of a topological space X is said to be irreducible if it is irreducible as a subspace of X. It can be easily verified that the closure of an irreducible set is irreducible, while an open subspace of an irreducible space is always irreducible and also dense. An irreducible closed subset of A_K^n is called an **affine subvariety**. An open subset of an affine variety is called a **quasi-affine variety**. Thus, a quasi-affine variety is also irreducible. If Z is an irreducible closed subset of a space Y, then an element $z \in Z$ is called a **generic** point of Z if $\overline{\{z\}} = Z$.

Proposition 4.3.1 *An affine algebraic subset Y of A_K^n is an affine subvariety of A_K^n if and only if $I(Y)$ is a prime ideal of $K[X_1, X_2, \cdots, X_n]$. In particular, $A_K^n = V(\{0\})$ is an affine variety.*

Proof Suppose that Y is an affine subvariety of A_K^n. Suppose that $f(\overline{X})g(\overline{X}) \in I(Y)$. Then $Y \subseteq V(\{f(\overline{X})g(\overline{X})\}) = V(\{f(\overline{X})\}) \bigcup V(\{g(\overline{X})\})$. In turn, $Y = (Y \bigcap V(\{f(\overline{X})\})) \bigcup (Y \bigcap V(\{g(\overline{X})\}))$. Since Y is irreducible, $Y \bigcap V(\{f(\overline{X})\}) = Y$ or $Y \bigcap V(\{g(\overline{X})\}) = Y$. This means that $f(\overline{X}) \in I(Y)$ or $g(\overline{X}) \in I(Y)$. It follows that $I(Y)$ is a prime ideal. Conversely, suppose that $P = I(Y)$ is a prime ideal, where Y is a closed subset of A_K^n. Suppose that $Y = V(P) = Y' \bigcup Z'$, where Y' and Z' are closed sets. Then $P = I(Y) = I(Y') \bigcap I(Z')$. Since P is a prime ideal, $P = I(Y')$ or $P = I(Z')$. Since Y' and Z' are closed, $Y = V(I(Y')) = Y'$ or $Y = V(I(Z')) = Z'$. This shows that Y is irreducible. ♯

Let Y be an affine algebraic set in A_K^n. The difference ring $K[X_1, X_2, \cdots, X_n]/I(Y)$ is called the **coordinate ring** of Y and it is denoted by $\Gamma(Y)$. Since $I(Y)$ is a radical ideal, the coordinate ring $\Gamma(Y)$ is a reduced ring in the sense that it has no nonzero nilpotent elements. Indeed, the coordinate ring $\Gamma(Y)$ is finitely generated reduced commutative algebra over K. Conversely, let $R = K[\alpha_1, \alpha_2, \cdots, \alpha_n]$ be a finitely generated commutative algebra generated by $\{\alpha_1, \alpha_2, \cdots, \alpha_n\}$ which is reduced as a ring. Then we have surjective algebra homomorphism η from $K[X_1, X_2, \cdots, X_n]$ to R given by $\eta(f(\overline{X})) = f(\overline{\alpha})$ whose kernel A is a radical ideal. Thus, R is isomorphic to the coordinate ring $\Gamma(Y)$ of Y, where $Y = V(A)$. Further, it follows from Proposition 4.3.1 that Y is an affine variety if and only if its coordinate ring is an integral domain. A function ϕ from Y to K is called a **regular function** on Y if there is a polynomial $f(\overline{X}) \in K[X_1, X_2, \cdots, X_n]$ such that $\phi(\overline{\alpha}) = f(\overline{\alpha})$ for all $\overline{\alpha} \in Y$.

Proposition 4.3.2 *Every regular function from an affine algebraic subset Y of A_K^n to $K = A_K^1$ is continuous.*

Proof Let ϕ be a regular function from Y to K. By definition, there is a polynomial $f(\overline{X}) \in K[X_1, X_2, \cdots, X_n]$ such that $\phi(\overline{\alpha}) = f(\overline{\alpha})$ for all $\overline{\alpha} \in Y$. All proper closed subsets of A_K^1 are finite subsets of $K - \{0\}$. Thus, it is sufficient to observe that $\phi^{-1}(\{a\}) = \{\overline{\alpha} \in Y \mid f(\overline{\alpha}) = a\} = Y \bigcap V(\{f(\overline{X}) - a\})$ is a closed subset of Y for all nonzero elements a of K. ♯

The set of all regular functions on Y is denoted by $\vartheta(Y)$. Evidently, $\vartheta(Y)$ is a ring under obvious addition and multiplication. The evaluation map $\overline{ev}$ induces a surjective homomorphism from $K[X_1, X_2, \cdots, X_n]$ to $\vartheta(Y)$ whose kernel is $I(Y)$.

Thus, the coordinate ring $\Gamma(Y)$ of Y is also naturally isomorphic to the ring $\vartheta(Y)$ of regular functions on Y.

A map η from an affine algebraic subset Y to an affine algebraic subset Z over K is called an **affine morphism** (also called a polynomial map) if $\phi o \eta \in \vartheta(Y)$ for all $\phi \in \vartheta(Z)$. This is equivalent to say that η induces a homomorphism η^* from $\vartheta(Z)$ to $\vartheta(Y)$ given by $\eta^*(\phi) = \phi o \eta$. This gives us the category of affine algebraic sets over K. It can be easily seen that ϑ defines a contra-variant equivalence from this category to the category of finitely generated reduced commutative algebras over K. In turn, ϑ also induces a contra-equivalence from the category AV of affine varieties over K to the category of finitely generated commutative algebras over K which are integral domains.

Recall that a ring R is called a Noetherian ring if it satisfies the ascending chain condition for ideals. On the same lines, a topological space X is called a **Noetherian space** if it satisfies ascending chain condition for open sets or equivalently it satisfies the descending chain condition for closed sets.

Proposition 4.3.3 *Let Y be an affine algebraic set. Then $\Gamma(Y)$ is a Noetherian ring and Y is a Noetherian space. In particular, A_K^n is a Noetherian space.*

Proof By the Hilbert basis theorem, $K[X_1, X_2, \cdots, X_n]$ is Noetherian. Since the quotient of a Noetherian ring is a Noetherian ring, $\Gamma(Y)$ is Noetherian. Let

$$Y_1 \supseteq Y_2 \supseteq \cdots \supseteq Y_r \supseteq Y_{r+1} \supseteq \cdots$$

be a descending chain of closed subsets of Y. Then we have an ascending chain

$$I(Y_1)/I(Y) \subseteq I(Y_2)/I(Y) \subseteq \cdots \subseteq I(Y_r)/I(Y) \subseteq I(Y_{r+1})/I(Y) \subseteq \cdots$$

of ideals of $\Gamma(Y)$. Since $\Gamma(Y)$ is Noetherian, there is a natural number m such that $I(Y_r)/I(Y) = I(Y_{r+1})/I(Y)$ for all $r \geq m$. It follows that $I(Y_r) = I(Y_{r+1})$ for all $r \geq m$. Hence $Y_r = V(I(Y_r)) = V(I(Y_{r+1})) = Y_{r+1}$ for all $r \geq m$. ♯

Proposition 4.3.4 *Let X be a Noetherian space. Then every nonempty closed set Y of X can be expressed as*

$$Y = Y_1 \bigcup Y_2 \bigcup \cdots \bigcup Y_n,$$

where each Y_i is irreducible and $Y_i \not\subseteq Y_j$ for all $i \neq j$. Further, this representation is unique up to rearrangements.

Proof We first show that for every nonempty nonirreducible closed subset Y of X, there is an irreducible closed subset Y_1 and a proper closed subset Z_1 of Y such that $Y = Y_1 \bigcup Z_1$. Let Y be a nonempty nonirreducible closed subset of X. Then there are proper closed subset Y_1 and Z_1 of Y such that $Y = Y_1 \bigcup Z_1$. If Y_1 is irreducible, we are done. If not, there are proper closed subsets Y_2 and Z_2 of Y_1 such that $Y_1 = Y_2 \bigcup Z_2$. If Y_2 is irreducible, then $Y = Y_2 \bigcup (Z_2 \bigcup Z_1)$ and we are done.

If not proceed. Since X satisfies the descending chain condition for closed sets, this process terminates after finitely many steps giving an irreducible closed subset Y_r of Y together with closed subsets $Z_r, Z_{r-1}, \cdots, Z_1$ of Y such that

$$Y = Y_r \bigcup (Z_r \bigcup Z_{r-1} \bigcup \cdots \bigcup Z_1),$$

and we are done.

Next, we show that every nonempty closed subset Y of X is expressible as the union of finitely many irreducible closed sets. Let Y be a nonempty closed subset of X. If Y is irreducible, then there is nothing to do. Suppose that Y is not irreducible. From our earlier observation, there is an irreducible proper closed subset Y_1 and a nonempty proper closed set Z_1 of Y such that $Y = Y_1 \bigcup Z_1$. If Z_1 is irreducible, then there is nothing to do. Suppose that Z_1 is not irreducible. Again, from our previous observation, there is an irreducible closed subset Y_2 of Z_1 (and so of Y) and a closed subset Z_2 such that $Z_1 = Y_2 \bigcup Z_2$. In turn, $Y = Y_1 \bigcup Y_2 \bigcup Z_2$, where Y_1 and Y_2 are irreducible. If Z_2 is irreducible, then again we are done. If Z_2 is not irreducible, proceed as above. This process terminates after finitely many steps expressing Y as the union of finitely many irreducible closed sets, because of the descending chain condition for closed sets in X.

Now, suppose that

$$Y = Y_1 \bigcup Y_2 \bigcup \cdots \bigcup Y_n,$$

where each Y_i is irreducible. Removing Y_i from the above representation whenever $Y_i \subseteq Y_j$, and proceeding inductively we arrive at a representation of Y of the desired type. Now to establish the uniqueness of the representation, suppose that

$$Y = Y_1 \bigcup Y_2 \bigcup \cdots \bigcup Y_n = Y_1' \bigcup Y_2' \bigcup \cdots \bigcup Y_m',$$

where Y_i and Y_k' are irreducible, $Y_i \not\subseteq Y_j$, and $Y_k' \not\subseteq Y_l'$ for all $i \neq j$ and $k \neq l$. We use induction on $max(m, n)$ to show that $m = n$, and after some rearrangement $Y_i = Y_i'$ for all i. If $max(m, n) = 1$, then $m = 1 = n$ and $Y = Y_1 = Y_1'$. Assume the induction hypothesis. Clearly,

$$Y_1' = (Y_1 \bigcap Y_1') \bigcup (Y_2 \bigcap Y_1') \bigcup \cdots \bigcup (Y_n \bigcap Y_1').$$

Since Y_1' is irreducible, $Y_i \bigcap Y_1' = Y_1'$ for some i. After rearranging, we may assume that $Y_1 \bigcap Y_1' = Y_1'$. This means that $Y_1' \subseteq Y_1$. Again, using the same argument for Y_1 we see that $Y_1 \subseteq Y_k'$ for some k. But then $Y_1' \subseteq Y_k'$. Hence $k = 1$ and $Y_1' = Y_1$. Clearly,

$$\overline{Y - Y_1} = Y_2 \bigcup Y_3 \bigcup \cdots \bigcup Y_n = Y_2' \bigcup Y_3' \bigcup \cdots \bigcup Y_m'.$$

By the induction hypothesis, $m = n$ and after some rearrangement $Y_i = Y_i'$ for all i. ♯

Corollary 4.3.5 *Every affine algebraic set has a unique representation as a union of finitely many affine algebraic varieties as described in the above proposition.* ♮

Definition 4.3.6 Let X be a topological space. Then $sup\{n \in \mathbb{N} \bigcup \{0\} \mid there\ is\ a$ $chain\ Y_0 \subset Y_1 \subset Y_2 \subset \cdots \subset Y_n\ of\ distinct\ irreducible\ closed\ sets\}$, if exists, is called the **dimension** of X. The dimension of X is denoted by $dim X$. If the supremum does not exist, then we say that X is of infinite dimension.

Since the singletons are the only irreducible closed subsets of a discrete space, a discrete space is 0 dimensional. An infinite co-finite space is of dimension 1. In particular the dimension of A_K^1 is 1 whenever K is infinite.

Definition 4.3.7 Let P be a prime ideal of a commutative ring R. The **height** $ht(P)$ of the prime ideal P is defined to be $sup\{n \in \mathbb{N} \bigcup \{0\} \mid there\ is\ a\ chain\ P_0 \subset P_1 \subset$ $P_2 \subset \cdots \subset P_n = P\ of\ distinct\ prime\ ideals\}$. The **Krull dimension** of R is defined to be the supremum of heights of prime ideals of R.

Thus, the height of a nonzero prime ideal of a PID is 1, and hence the Krull dimension of a PID is 1. Evidently, the Krull dimension of a field is 0. The following theorem is useful in computing the Krull dimension of a ring, and its proof can be found in "Introduction to Commutative Algebra" by Atiyah and Mcdonald.

Theorem 4.3.8 *Let R be an integral domain which is finitely generated algebra over a field K. Then we have the following:*

(i) *The Krull dimension of R is the transcendence degree of the field extension $K(R)$ over K, where $K(R)$ is the field of fractions of R.*

(ii) *Given any prime ideal P of R, the Krull dimension of R is $ht(P) + Dim R/P$.* ♮

Corollary 4.3.9 *The Krull dimension of $K[X_1, X_2, \cdots, X_n]$ is n.*

Proof Since the transcendence degree of $K(X_1, X_2, \cdots, X_n)$ over K is n, the result follows. ♮

Proposition 4.3.10 *The dimension of the affine algebraic set Y over K is the same as the Krull dimension of the coordinate ring $\Gamma(Y)$ of Y.*

Proof Observe that the correspondence $Y \mapsto I(Y)$ is a bijective inclusion reversing map from the set of all irreducible closed subsets of A_K^n to the set of all prime ideals of $K[X_1, X_2, \cdots, X_n]$. Consequently, we have a bijective inclusion reversing map $Z \mapsto I(Z)/I(Y)$ from the set of irreducible closed subsets of an affine algebraic set Y to the set of prime ideals of $\Gamma(Y)$. The result follows. ♮

Corollary 4.3.11 *The dimension of A_K^n is n.*

Proof The proof follows from the above proposition and the corollary. ♮

The proof of the following theorems can also be found in "Introduction to Commutative Algebra" by Atiyah and Mcdonald.

Theorem 4.3.12 *Let f be an element of a Noetherian ring which is neither a zero divisor nor a unit. Let P be a minimal prime ideal containing f. Then $ht(P) = 1$.* ♯

Theorem 4.3.13 *Let R be a Noetherian integral domain. Then R is a UFD if and only if every prime ideal of R of height 1 is a principal ideal.* ♯

Corollary 4.3.14 *An affine subvariety Y of A_K^n is of dimension $n - 1$ if and only if Y is a hyper-surface $V(\{f(\overline{X})\})$, where $f(\overline{X})$ is an irreducible polynomial in $K[X_1, X_2, \cdots, X_n]$.*

Proof Suppose that $Y = V(\{f(\overline{X})\})$, where $f(\overline{X})$ is an irreducible polynomial in $K[X_1, X_2, \cdots, X_n]$. Then $I(Y)$ is the prime ideal generated by $f(\overline{X})$. It follows from Theorem 4.3.12 that $ht(I(Y)) = 1$. By Proposition 4.3.10, $dim Y$ is the Krull dimension of $\Gamma(Y)$. Finally by Theorem 4.3.8, the Krull dimension of $\Gamma(Y)$ is $n - 1$.

Conversely, Let Y be an affine subvariety of A_K^n of dimension $n - 1$. Then the Krull dimension of $\Gamma(Y) = K[X_1, X_2, \cdots, X_n]/I(Y)$ is $n - 1$. This means that the height of the prime ideal $I(Y)$ in $K[X_1, X_2, \cdots, X_n]$ is 1. Since $K[X_1, X_2, \cdots, X_n]$ is UFD, it follows from Theorem 4.3.13 that $I(Y)$ is a principal ideal $< f(\overline{X}) >$ generated by an irreducible polynomial $f(\overline{X})$. Evidently, $Y = V(I(Y)) = V(\{f(\overline{X})\})$. ♯

Projective Varieties

Let K be an algebraically closed field. Define a relation $\approx$ on $A_K^{n+1} - \{\overline{0}\}$ by putting $\overline{\alpha} \approx \overline{\beta}$ if there is a $\lambda \in K - \{0\}$ such that $\lambda\overline{\alpha} = \overline{\beta}$. Clearly, $\approx$ is an equivalence relation. The equivalence class determined by $\overline{\alpha} = (\alpha_0, \alpha_1, \cdots, \alpha_n)$ will be denoted by $\hat{\alpha}$. We shall term $\overline{\alpha}$ as a **homogeneous coordinate** of $\hat{\alpha}$. The quotient set thus obtained is denoted by P_K^n, and it is called the **projective n-set** over K. The points in P_K^n can be viewed as affine lines in A_K^{n+1} passing through the origin. For each i, $0 \leq i \leq n$, we have an injective map η_i from A_K^n to P_K^n given by $\eta_i(\alpha_1, \alpha_2, \cdots, \alpha_n) = \hat{\beta^i}$, where $\overline{\beta^i} = (\alpha_1, \alpha_2, \cdots \alpha_{i-1}, 1, \alpha_i, \alpha_{i+1}, \cdots, \alpha_n)$. We shall denote the image of η_i in P_K^n by U_i. It can be easily observed that $\bigcup_{i=0}^n U_i = P_K^n$.

Recall that a ring R together with the internal direct sum decomposition $\oplus \sum_{n \in \mathbb{N} \bigcup \{0\}} R_n$ of the abelian group $(R, +)$ is called a graded ring if $R_n R_m \subseteq R_{n+m}$. The members of R_n are called the homogeneous elements of R of degree n. Evidently, every element $a \in R - \{0\}$ can be uniquely expressed as $a = a_{m_1} + a_{m_2} + \cdots + a_{m_r}$, where $a_{m_i} \in R_{m_i} - \{0\}, m_1 < m_2 < \cdots < m_r$. An ideal A of R is called a **homogeneous ideal** if whenever $a = a_{m_1} + a_{m_2} + \cdots + a_{m_r}, m_1 < m_2 < \cdots < m_r$ is a member of A, each $a_{m_i} \in A$. Equivalently, an ideal A of R is a homogeneous ideal if $(A, +)$ is the direct sum $\oplus \sum_{n \in \mathbb{N} \bigcup \{0\}} R_n \bigcap A$. It can be easily seen that the sum, intersection, and product of homogeneous ideals are homogeneous. The radical of a homogeneous ideal is also homogeneous. A homogeneous ideal P is a prime ideal if and only if whenever the product of two homogeneous elements belongs to P, at least one of them is in P.

The most important example of a graded ring in which we shall be interested is the polynomial ring $K[X_0, X_1, \cdots, X_n]$. Let M_d denote the vector subspace of $K[X_0, X_1, \cdots, X_n]$ generated by the the monomials of degree d. Then

$K[X_0, X_1, \cdots, X_n] = \oplus \sum_{d \geq 0} M_d$. Let $f(\overline{X})$ be a homogeneous polynomial of degree d. Then $f(\lambda \overline{\alpha}) = \lambda^d f(\overline{\alpha})$ for all $\lambda \in K$. Thus, given a homogeneous polynomial $f(\overline{X})$ and $\overline{\alpha} \approx \overline{\beta}$, $f(\overline{\alpha}) = 0$ if and only if $f(\overline{\beta}) = 0$. Let S be a set of homogeneous polynomials. The set $V(S) = \{\hat{\alpha} \mid f(\overline{\alpha}) = 0 \ for \ all \ f \in S\}$ is called a **projective algebraic set**. If A is a homogeneous ideal of $K[X_0, X_1, \cdots, X_n]$, then $V(A)$ is defined to be $V(S)$, where S is the set of all homogeneous elements in A. Evidently, $V(S) = V(< S >)$ for all sets of homogeneous elements. Since $K[X_0, X_1, \cdots, X_n]$ is Noetherian, a homogeneous ideal A is generated by finitely many polynomials. Taking the homogeneous components of the generating set of polynomials, we can find a finite set $S = \{f_1(\overline{X}), f_2(\overline{X}), \cdots, f_r(\overline{X})\}$ of homogeneous polynomials in A such that $V(A) = V(S)$. As in the case of affine n-set A_K^n, the family of projective algebraic subsets of P_K^n forms a family of closed sets for a topology on P_K^n. This topology is called the Zariski topology, and P_K^n with this topology is called the **Projective n-space**. Irreducible closed subsets of P_K^n are called **projective varieties**. Open subsets of projective varieties are termed as **quasi-projective varieties**.

Proposition 4.3.15 *For each $i, 0 \leq i \leq n$, the map η_i from A_K^n to P_K^n is a homeomorphism on to the image $U_i = P_K^n - V(\{X_i\})$.*

Proof We have already seen that η_i is an injective map. Evidently, the image of η_i is contained in $P_K^n - V(\{X_i\})$. Further, let $\hat{\alpha}$ be an element of $P_K^n - V(\{X_i\})$, where $\alpha_i \neq 0$. Then $\hat{\alpha} = \hat{\beta}$, where $\beta_j = \frac{\alpha_j}{\alpha_i}$. Clearly, $\hat{\beta}$ is in the image of η_i. This shows that η_i is a bijective map from A_K^n to U_i. Evidently, each U_i is an open subset of P_K^n. Further, it is a straightforward verification to show that η_i and its inverse are closed maps. ♯

Corollary 4.3.16 *Any projective (quasi-projective) subvariety of P_K^n can be covered by open subsets $U_0, U_1, \cdots, U_n$, each being homeomorphic to affine (quasi-affine) subvarieties of A_K^n.* ♯

Let Y be a projective algebraic subset of P_K^n. The **homogeneous ideal** $I(Y)$ is defined to be the ideal of $K[X_0, X_1, \cdots, X_n]$ generated by the set $\{f(\overline{X}) \mid f(\overline{X}) \ is \ homogeneous \ and \ f(\overline{\alpha}) = 0 \ for \ all \ \hat{\alpha} \in Y\}$. The difference ring $K[X_0, X_1, \cdots, X_n]/I(Y)$ is called the **homogeneous coordinate ring** of Y, and it is denoted by $\Gamma_h(Y)$.

Unlike ideals, the projective algebraic set $V(A)$ determined by a proper homogeneous ideal may be an empty set. For example, $V(\{X_0^r, X_1^r, \cdots, X_{n+1}^r\}) = \emptyset$ for $r > 0$. Thus, if A is a proper homogeneous ideal which contains $M_r, r > 0$ (for example, $A = M_+ = \oplus \sum_{r>0} M_r$), then $V(A) = \emptyset$. Observe that M_+ is a maximal ideal of $K[X_0, X_1, \cdots, X_n]$ but unlike the affine case, it corresponds to no points in P_K^n. Of course, it corresponds to $\overline{0}$ in A_K^{n+1}. This homogeneous ideal is termed as an irrelevant maximal ideal for the projective n-space. The following proposition is an analogue of Nullstellensatz for homogeneous polynomials.

Proposition 4.3.17 *Let A be a proper homogeneous ideal of $K[X_0, X_1, \cdots, X_n]$, and $f(\overline{X})$ be a homogeneous polynomial of positive degree such that $f(\overline{X}) \in I(V(A))$. Then $f(\overline{X}) \in \sqrt{A}$.*

Proof As already observed, there is a finite set $\{f_1(\overline{X}), f_2(\overline{X}), \cdots, f_r(\overline{X})\}$ of homogeneous elements of A such that $A = <\{f_1(\overline{X}), f_2(\overline{X}), \cdots, f_r(\overline{X})\}>$ and $V(A) = V(\{f_1(\overline{X}), f_2(\overline{X}), \cdots, f_r(\overline{X})\})$. Since $f(\overline{X}) \in I(V(A))$, $f(\overline{X}) \in I(\{\overline{\alpha} \in A_K^{n+1} \mid f_i(\overline{\alpha}) = 0 \ for \ all \ i\}$. From the Hilbert Nullstellensatz, $f(\overline{X}) \in \sqrt{<\{f_1(\overline{X}), f_2(\overline{X}), \cdots, f_r(\overline{X})\}>} = \sqrt{A}$. ♯

Proposition 4.3.18 *Let A be a homogeneous ideal of $K[X_0, X_1, \cdots, X_n]$. Then the following conditions are equivalent:*

1. $V(A) = \emptyset$.
2. $\sqrt{A} = K[X_0, X_1, \cdots, X_n]$ or $\sqrt{A} = M_+$.
3. $M_r \subseteq A$ for some $r > 0$.

Proof Assume 1. Since $V(A) = \emptyset$, any homogeneous polynomial $f(\overline{X})$ belongs to $I(V(A))$. In particular, $X_i \in I(V(A))$ for each i. From the above proposition, $X_i \in \sqrt{A}$ for each i. Since $\sqrt{A}$ is also homogeneous, $M_+ \subseteq \sqrt{A}$. Since M_+ is a maximal ideal, $\sqrt{A} = K[X_0, X_1, \cdots, X_n]$ or $\sqrt{A} = M_+$.

2. For each i, there is a natural number r_i such that $X_i^{r_i} \in A$. Take $r = max\{r_i, 0 \leq i \leq n\}$. Then $X_i^r \in A$ for all i. Since A is homogeneous, $M_r \subseteq A$.

3. Then $\{X_i^r \mid 0 \leq i \leq n\} \subseteq A$. Hence $V(A) \subseteq V(\{X_i^r \mid 0 \leq i \leq n\})$. Evidently, $V(\{X_i^r \mid 0 \leq i \leq n\}) = \emptyset$. ♯

Corollary 4.3.19 *Let A be a homogeneous ideal in $K[X_0, X_1, \cdots, X_n]$ such that the projective algebraic set $V(A)$ is a nonempty set. Then $I(V(A)) = \sqrt{A}$.*

Proof By definition, $I(V(A))$ is generated by the set of homogeneous polynomials $f(\overline{X})$ such that $f(\overline{\alpha}) = 0$ for all $\hat{\alpha} \in V(A)$. Since $V(A) \neq \emptyset$, $f(\overline{X})$ is of positive degree. From Proposition 4.3.17, $f(\overline{X}) \in \sqrt{A}$. This shows that $I(V(A)) \subseteq \sqrt{A}$. Since A is homogeneous, $\sqrt{A}$ is homogeneous. This means that $\sqrt{A}$ is generated by the set of homogeneous polynomials in $\sqrt{A}$. Again, since $V(A) \neq \emptyset$, $\sqrt{A}$ is properly contained in M_+. Let $f(\overline{X})$ be a homogeneous polynomial in $\sqrt{A}$. Then $(f(\overline{X}))^r \in A$ for some $r > 0$. Evidently, $(f(\overline{X}))^r$ is also homogeneous. This means that $(f(\overline{\alpha}))^r = 0$ for all $\hat{\alpha} \in V(A)$. In turn, $f(\overline{\alpha}) = 0$ for all $\hat{\alpha} \in V(A)$. By the Hilbert Nullstellensatz, $f(\overline{X}) \in I(V(A))$. ♯

Corollary 4.3.20 *V defines a bijective correspondence from the set of all homogeneous radical ideals different from M_+ to the set of all projective algebraic sets. Further, under this correspondence, prime homogeneous ideals and projective algebraic varieties correspond. In particular, P_K^n is a projective variety.* ♯

Though the concept of variety is more general and abstract, for the time being, a variety will mean a quasi-affine or a quasi-projective variety. Our next aim is to introduce the concept of morphisms between varieties, and in turn, the category of varieties.

Definition 4.3.21 Let Y be an quasi-affine subvariety of A_K^n. A map ϕ from Y to A_K^1 is called a **regular function** at a point $\overline{p}$ in Y if there is an open subset U of Y containing $\overline{p}$ together with a pair of polynomials $f(\overline{X}), g(\overline{X})$ in $K[X_1, X_2, \cdots, X_n]$ such that $g(\overline{q}) \neq 0$ for all $\overline{q} \in U$ and $\phi(\overline{q}) = \frac{f(\overline{q})}{g(\overline{q})}$ for all $\overline{q} \in U$. We say that it is a regular function on Y if it is regular at each point of Y.

We shall see soon that this definition of regular function agrees with our earlier definition of regular functions from affine varieties.

Proposition 4.3.22 *Every regular function from a quasi-affine sub variety Y of A_K^n to A_K^1 is continuous.*

Proof Let Y be a regular function from a quasi-affine variety Y to A_K^1. Let $\overline{p}$ be a member of Y. We show that ϕ is continuous at $\overline{p}$. Suppose that $\phi(\overline{p})$ belongs to $A_K^1 - \{a\}$. Since ϕ is regular at $\overline{p}$, there is an open subset U of Y containing $\overline{p}$ and polynomials $f(\overline{X})$ and $g(\overline{X})$ in $K[X_1, X_2, \cdots, X_n]$ such that $g(\overline{q}) \neq 0$ for all $\overline{q} \in U$ and $\phi(\overline{q}) = \frac{f(\overline{q})}{g(\overline{q})}$. In particular, $\frac{f(\overline{p})}{g(\overline{p})} \neq a$. Evidently, $\phi((Y - V(f(\overline{X}) - ag(\overline{X}))) \bigcap U) \subseteq A_K^1 - \{a\}$. This means that $\phi^{-1}(A_K^1 - \{a\})$ is open for all $a \in K$. Since the family $\{A_K^1 - \{a\} \mid a \in K\}$ forms a subbasis for the topology of A_K^1, it follows that ϕ is continuous at $\overline{p}$. ♯

Now, we introduce the notion of regular functions on a quasi-projective variety. Let $f(\overline{X})$ and $g(\overline{X})$ be homogeneous polynomials in $K[X_0, X_1, \cdots, X_n]$ of the same degrees. Then $\frac{f(\overline{X})}{g(\overline{X})}$ defines a map ϕ from $P_K^n - V(g(\overline{X}))$ to A_K^1 given by $\phi(\hat{\alpha}) = \frac{f(\overline{\alpha})}{g(\overline{\alpha})}$.

Definition 4.3.23 A map ϕ from a quasi-projective sub variety Y of P_K^n to A_K^1 is said to be a regular function at $\hat{p} \in Y$ if there is an open subset U of Y containing $\hat{p}$ together with a pair of homogeneous polynomials $f(\overline{X})$ and $g(\overline{X})$ in $K[X_0, X_1, \cdots, X_n]$ of the same degrees such that $g(\hat{q}) \neq 0$ for all $\hat{q} \in U$ and $\phi(\hat{q}) = \frac{f(\overline{q})}{g(\overline{q})}$. We say that ϕ is regular on Y if it is regular at each point of Y.

As in Proposition 4.3.22, we can easily show that a regular function on a quasi-projective variety is continuous.

Proposition 4.3.24 *Let ϕ and ψ be two regular functions on a variety Y which agree with a nonempty open subset U of Y. Then $\phi = \psi$.*

Proof Evidently, ϕ and ψ agree on the closed set $(\phi - \psi)^{-1}\{0\}$ which contains U. Thus, ϕ and ψ agree on the closure of U. Since an open subset of an irreducible closed set is dense, $\phi = \psi$ on Y. ♯

Proposition 4.3.25 *Let Y be a variety. Then the sum and the product of regular functions on Y are regular functions, and the set $\vartheta(Y)$ of regular functions on Y is a commutative integral domain with respect to the addition and multiplication of regular functions. It is also a commutative algebra over K.*

Proof Let Y be a quasi-affine variety. Let ϕ and ψ be regular functions on Y. Then for each point $\overline{p}$ of Y, there is an open subset U of Y containing $\overline{p}$ together with polynomials $f(\overline{X}), g(\overline{X}), h(\overline{X})$, and $k(\overline{X})$ such that for each $\overline{q} \in U$, $g(\overline{q}) \neq 0 \neq k(\overline{q})$, $\phi(\overline{q}) = \frac{f(\overline{q})}{g(\overline{q})}$, and $\psi(\overline{q}) = \frac{h(\overline{q})}{k(\overline{q})}$. In turn, we have polynomials $u(\overline{X}) = f(\overline{X})k(\overline{X}) + g(\overline{X})h(\overline{X})$, $v(\overline{X}) = g(\overline{X})k(\overline{X})$, and $w(\overline{X}) = f(\overline{X})h(\overline{X})$ such that $(\phi + \psi)(\overline{q}) = \frac{u(\overline{q})}{v(\overline{q})}$ and $(\phi\psi)(\overline{q}) = \frac{w(\overline{q})}{v(\overline{q})}$. This shows that $\phi + \psi$ and $\phi\psi$ are regular functions. The result for the projective variety also follows for, if $f(\overline{X}), g(\overline{X})$ are homogeneous of the same degrees and $h(\overline{X}), k(\overline{X})$ are homogeneous of the same degrees, then $u(\overline{X}), v(\overline{X})$, and $w(\overline{X})$ are also homogeneous of the same degrees. Next, suppose that $\phi \cdot \psi = 0$, where ϕ and ψ are regular functions. Suppose that $\phi(P) \neq 0$ at some point P of Y. Since ϕ is continuous, there is an open neighborhood W of P such that $\phi(Q) \neq 0$ for all $Q \in W$. But then $\psi(Q) = 0$ for all $Q \in W$. Thus, the zero regular function and the regular function ψ agree on a nonempty open set W. From Proposition 4.3.24, $\psi = 0$. This shows that $\vartheta(Y)$ is an integral domain. The scalar multiplication on $\vartheta(Y)$ is defined in an obvious way. $\sharp$

Let Y be a variety over an algebraically closed field K. For each open subset U of Y, we have a finitely generated K algebra $\vartheta_Y(U)$ of regular functions defined on U. This together with natural restriction homomorphisms defines a natural sheaf ϑ_Y of finitely generated K algebras over Y.

Definition 4.3.26 A continuous map η from a variety Y over K to a variety Y' over K is called a **morphism** from Y to Y' if for every open subset U' of Y' and for every regular function ϕ from U' to A_K^1, the map $(\phi o \eta)|\eta^{-1}(U')$ from $\eta^{-1}(U')$ to A_K^1 is a regular function on $\eta^{-1}(U')$. Evidently, the composition of two morphisms is a morphism. This gives us a category VAR of varieties over K. Two varieties Y and Y' are said to be isomorphic if there is an isomorphism from Y to Y' in this category. Thus, a morphism η from a variety Y over K to a variety Y' over K induces naturally a morphism $(\eta, \eta^{\sharp})$ from the ringed space (Y, ϑ_Y) to $(Y', \vartheta_{Y'})$, where $\eta_{U'}^{\sharp}$ is the homomorphism from $\vartheta_{Y'}(U')$ to $(\eta_* \vartheta_Y)(U') = \vartheta_Y(\eta^{-1}(U'))$ given by $\eta_U^{\sharp}(\phi) = \phi o \eta | \eta^{-1}(U)$.

One of the guiding problems in algebraic geometry is to classify varieties up to isomorphism. As usual, one is always in search of invariants of varieties, and uses them to roam around this problem.

Proposition 4.3.27 ϑ *defines a contra-variant functor from the category VAR to the category of commutative algebras over K. In particular, $\vartheta(Y)$ is an invariant of Y.*

Proof Given any morphism η from Y to Y', we have a map $\vartheta(\eta)$ from $\vartheta(Y')$ to $\vartheta(Y)$ given by $\vartheta(\eta)(\phi) = \phi o \eta$. It can be easily seen that $\vartheta(\eta)$ is an algebra homomorphism. Evidently, $\vartheta(\rho o \eta) = \vartheta(\eta) o \vartheta(\rho)$ and $\vartheta(I_Y) = I_{\vartheta(Y)}$. $\sharp$

Local Ring of a Variety at a Point

Let Y be a variety, and let P be a point in Y. Let us denote the set of germs of the regular functions on Y at P by $\vartheta_{Y,P}$. More explicitly, $\vartheta_{Y,P}$ is the set

$$\{\overline{(U, \phi)} \mid U \text{ is open set containing } P \text{ and } \phi \text{ is regular function on } U\}$$

of equivalence classes, where $\overline{(U, \phi)} = \overline{(V, \psi)}$ if and only if there is an open set W such that $P \in W \subseteq U \bigcap V$ and $\phi|W = \psi|W$. It can be seen easily that $\overline{(U, \phi)} + \overline{(V, \psi)} = \overline{(U \bigcap V, \phi|(U \bigcap V) + \psi|(U \bigcap V))}$, $\overline{(U, \phi)} \cdot \overline{(V, \psi)} = \overline{(U \bigcap V, \phi|(U \bigcap V) \cdot \psi|(U \bigcap V))}$, and $\alpha(\overline{(U, \phi)}) = \overline{(U, \alpha\phi)}$ define operations in $\vartheta_{Y,P}$ with respect to which it is a commutative algebra over K. The ring $\vartheta_{Y,P}$ is called the **local ring of Y at P**. We have a map χ from $\vartheta_{Y,P}$ to K given by $\chi(\overline{(U, \phi)}) = \phi(P)$. Evidently, χ is a surjective homomorphism whose kernel $M_{Y,P}$ is a maximal ideal of $\vartheta_{Y,P}$. Suppose that $\overline{(U, \phi)} \in \vartheta_{Y,P} - M_{Y,P}$. Then $\phi(P) \neq 0$. Since ϕ is continuous at P, there is an open neighborhood W of P contained in U such that $\phi(Q) \neq 0$ for all $Q \in W$. Evidently, the map ψ from W to K given by $\psi(Q) = (\phi(Q))^{-1}$ is a regular function on W. Clearly, $\overline{(W, \psi)} \cdot \overline{(U, \phi)}$ is the identity $\overline{(Y, 1)}$ of $\vartheta_{Y,P}$, where 1 is the constant function. This shows that $\vartheta_{Y,P}$ is a local ring in the sense that it has unique maximal ideal consisting of noninvertible elements. We have the natural embedding ι from $\vartheta(Y)$ to $\vartheta_{Y,P}$ given by $\iota(\phi) = \overline{(Y, \phi)}$. We can treat $\vartheta(Y)$ as a subring of $\vartheta_{Y,P}$ through this embedding. Indeed, $\vartheta_{Y,P}$ can be naturally identified with the localization of $\vartheta(Y)$ at the prime ideal consisting of those regular functions on Y which vanish at P.

Function Field of a Variety

Let Y be a variety. Let $K(Y)$ denote the set

$$\{\overline{(U, \phi)} \mid U \text{ is nonempty open subset of } Y \text{ and } \phi \text{ is regular function on } U\}$$

of equivalence classes, where $\overline{(U, \phi)} = \overline{(V, \psi)}$ if there is a nonempty open subset W contained in $U \bigcap V$ such that $\phi|W = \psi|W$. The members of $K(Y)$ are called the rational functions on Y. Let $\overline{(U, \phi)}$ and $\overline{(V, \psi)}$ be members of $K(Y)$. Then U and V are nonempty open subsets of Y. Since Y is irreducible, $U \bigcap V \neq \emptyset$. As in the case of $\vartheta_{Y,P}$, we have addition $+$ and the multiplication $\cdot$ in $K(Y)$ given by $\overline{(U, \phi)} + \overline{(V, \psi)} = \overline{(U \bigcap V, \phi|(U \bigcap V) + \psi|(U \bigcap V))}$, and $\overline{(U, \phi)} \cdot \overline{(V, \psi)} = \overline{(U \bigcap V, \phi|(U \bigcap V) \cdot \psi|(U \bigcap V))}$, with respect to which it is a commutative ring. Let $\overline{(U, \phi)}$ be a nonzero element of $K(Y)$. Then ϕ is nonzero on U. Since ϕ is continuous, there is a nonempty open subset W contained in U such that $\phi(Q) \neq 0$ for all $Q \in W$. Evidently, the function ψ from W to K given by $\psi(Q) = \phi(Q)^{-1}$ is a regular function on W. It is easy to observe that $\overline{(W, \psi)} \cdot \overline{(U, \phi)}$ is the identity $\overline{(Y, 1)}$ of $K(Y)$. This shows that $K(Y)$ is a field. This field is called the **function field** of Y.

Evidently, $\vartheta_{Y,P}$ is a subring of $K(Y)$. In turn, ι is an embedding of $\vartheta(Y)$ to $K(Y)$. We show that $K(Y)$ is naturally isomorphic to the field $F(\vartheta(Y))$ of fractions of $\vartheta(Y)$. The map ι is an embedding of the integral domain $\vartheta(Y)$ into the field, and hence it induces an injective homomorphism $\bar{\iota}$ from $F(\vartheta(Y))$ to $K(Y)$ which is given by $\bar{\iota}(\frac{\phi}{\psi}) = \overline{(Y, \phi)}\overline{(Y, \psi)}^{-1}$. First let us suppose that Y is a quasi-affine variety. Let $\overline{(U, \phi)}$ be a nonzero member of $K(Y)$. Then there is an open set W contained in U together with polynomials $f(\overline{X})$ and $g(\overline{X})$ such that $\phi(\overline{\alpha}) = \frac{f(\overline{\alpha})}{g(\overline{\alpha})}$ for all $\overline{\alpha} \in W$. We have regular functions μ and ν on Y given by $\mu(\overline{\alpha}) = f(\overline{\alpha})$ and $\nu(\overline{\alpha}) = g(\overline{\alpha})$. Clearly, $\overline{(U, \phi)} = \overline{\iota}(\frac{\mu}{\nu})$. This shows that ι is an isomorphism. If Y is a quasi-projective variety, then we can take polynomials $f(\overline{X})$ and $g(\overline{X})$ to be homogeneous polynomials of the same degrees, and the proof goes as above. Observe that the function field $K(Y)$ is also an invariant of the variety.

Proposition 4.3.28 *Let $Y \subseteq A_K^n$ be an affine variety, where K is an algebraically closed field. Then (i) there is a natural isomorphism η from the coordinate ring $\Gamma(Y)$ to the ring $\vartheta(Y)$ of regular functions on Y, and (ii) the function field $K(Y)$ of Y is a field extension of K in a natural manner with the transcendence degree equal to the dimension of the variety Y.*

Proof A member $f(\overline{X}) + I(Y)$ of $\Gamma(Y)$ gives a map $\hat{f}$ from Y to K defined by $(\hat{f})(\overline{\alpha}) = f(\overline{\alpha})$. Evidently, $(\hat{f}) \in \vartheta(Y)$. The map η from $\Gamma(Y)$ to $\vartheta(Y)$ defined by $\eta(f(\overline{X}) + I(Y)) = (\hat{f})$ is easily seen to be an injective homomorphism from the ring $\Gamma(Y)$ to the ring $\vartheta(Y)$. Let ϕ be a member of $\vartheta(Y)$. Then for each point $P \in Y$, we have an open subset U_P of Y together with a pair $f_P(\overline{X})$ and $g_P(\overline{X})$ of polynomials in $K[\overline{X}]$ such that $g_P(\overline{\alpha}) \neq 0$ and $\phi(\overline{\alpha}) = \frac{f_P(\overline{\alpha})}{g_P(\overline{\alpha})}$ for all $\overline{\alpha} \in U_P$. Evidently, $U_P \subseteq Y - V(\{g_P(\overline{X})\})$ for each $P \in Y$. Since Y is compact, we have finitely many points $P_1, P_2, \cdots, P_r$ in Y such that $Y = \bigcup_{i=1}^r U_{P_i} = \bigcup_{i=1}^r (Y - V(\{g_{P_i}(\overline{X})\})) = Y - \bigcap_{i=1}^r V(\{g_{P_i}(\overline{X})\}) = Y - V(\{g_{P_1}(\overline{X}), g_{P_2}(\overline{X}), \cdots, g_{P_r}(\overline{X})\})$. This means that $Y \bigcap V(\{g_{P_1}(\overline{X}), g_{P_2}(\overline{X}), \cdots, g_{P_r}(\overline{X})\}) = \emptyset$. By the Hilbert Nullstellensatz, $I(Y) + <\{g_{P_1}(\overline{X}), g_{P_2}(\overline{X}), \cdots, g_{P_r}(\overline{X})\}> = K[\overline{X}]$. Hence there is a polynomial $h(\overline{X}) \in I(Y)$ and polynomials $h_1(\overline{X}), h_2(\overline{X}), \cdots, h_r(\overline{X})$ in $K[\overline{X}]$ such that

$$1 = h(\overline{X}) + h_1(\overline{X})g_{P_1}(\overline{X}) + h_2(\overline{X})g_{P_2}(\overline{X}) + \cdots + h_r(\overline{X})g_{P_r}(\overline{X}).$$

Thus, $h_1(\overline{X})g_{P_1}(\overline{X}) + h_2(\overline{X})g_{P_2}(\overline{X}) + \cdots + h_r(\overline{X})g_{P_r}(\overline{X})$ restricted to Y is the constant function 1 on Y. Multiplying with ϕ, we obtain that $\phi(\overline{\alpha}) = p(\overline{\alpha})$ for all $\overline{\alpha} \in Y$, where $p(\overline{X})$ is the polynomial $h_1(\overline{X})f_{P_1}(\overline{X}) + h_2(\overline{X})f_{P_2}(\overline{X}) + \cdots + h_r(\overline{X})f_{P_r}(\overline{X})$. This shows that $\eta(p(\overline{X}) + I(Y)) = \phi$. The rest follows from Theorem 4.3.8, Proposition 4.3.10, and the fact that η is an isomorphism. ♯

Tangent Space and Singularities

Let $Y \subseteq A_K^n$ be an affine variety, where K is an algebraically closed field. Let $\overline{a} \in Y$. We say that a vector $\overline{v} \in K^n$ is a **tangent vector** to Y at the point $\overline{a}$ if $\frac{df(\overline{a}+t\overline{v})}{dt}|_{t=0} = 0$ for all $f \in I(Y)$. This is equivalent to say that every polynomial f in $I(Y)$ is of the form

$$f = w_1(X_1 - a_1) + w_2(X_2 - a_2) + \cdots + w_n(X_n - a_n) + g,$$

where $w_1 v_1 + w_2 v_2 + \cdots + w_n v_n = 0$ and $g \in (M_{\overline{a}})^2$, $M_{\overline{a}}$ being the maximal ideal determined by the point $\overline{a}$. The set $T_{\overline{a}}(Y)$ of tangent vectors to Y at the point $\overline{a}$ forms a vector space over K, which, of course, depends on a particular embedding of Y in an affine space.

Ler R be a commutative K-algebra, and M be a K-space which is also an R-module. A K-linear map d from R to M is called a **derivation** if $d(ab) = d(a)b + ad(b)$ for all $a, b \in R$. Each $\bar{v} \in A_K^n$ determines a map $D_{\bar{v}}$ from $K[X_1, X_2, \cdots, X_n]$ $= \Gamma(A_K^n)$ to itself which is given by

$$D_{\bar{v}}(f) = v_1 \frac{\partial f}{\partial X_1} + v_2 \frac{\partial f}{\partial X_2} + \cdots + v_n \frac{\partial f}{\partial X_n}.$$

Evidently, $D_{\bar{v}}$ is a derivation on the K-algebra $K[X_1, X_2, \cdots, X_n]$. To say that $\bar{v}$ is a tangent vector to Y at a point $\bar{a} \in Y$ is to say that $D_{\bar{v}}(f)(\bar{a}) = 0$ for all $f \in I(Y)$. We have a $\Gamma(Y)$-module structure on $(K, +)$ given by $(f + I(Y))\alpha = f(\bar{a})\alpha$. This module is denoted by $K_{\bar{a}}$. Since $D_{\bar{v}}(f)(\bar{a}) = 0$ for all $f \in I(Y)$, it induces a K-linear map $\hat{D}_{\bar{v}}$ from $\Gamma(Y)$ to $K_{\bar{a}}$ given by $\hat{D}_{\bar{v}}(f + I(Y)) = D_{\bar{v}}(f)(\bar{a})$. It is easy to observe that $\hat{D}_{\bar{v}}$ is a derivation. The map ρ from $T_{\bar{a}}(Y)$ to $Der(\Gamma(Y), K_{\bar{a}})$ given by $\rho(\bar{v}) = \hat{D}_{\bar{v}}$ is an K-linear map. Let d be a member of $Der(\Gamma(Y), K_{\bar{a}})$. Suppose that $d((X_i - a_i) + I(Y)) = v_i$, $1 \le i \le n$. We show that $d = \hat{D}_{\bar{v}} = \rho(\bar{v})$, where $\bar{v} = (v_1, v_2, \cdots, v_n)$. Since $d(1) = d(1 \cdot 1) = d(1)1 + 1d(1)$, it follows that $d(1) = 0$. In turn, $d(\alpha) = 0$ for all constant polynomials α. Further,

$$d(((x_i - a_i) + I(Y))(((X_j - a_j) + I(Y)))) = d(((X_i - a_i) + I(Y)))(X_j - a_j)|_{\bar{a}}$$
$$+ (X_i - a_i)|_{\bar{a}}d(((X_j - a_j) + I(Y))) = 0$$

for all i, j. Using induction, we can show that $d((f + I(Y))) = 0$, whenever f is a polynomial in $\{(X_1 - a_1), X_2 - a_2), \cdots, (X_n - a_n)\}$ of degree at least 2. Let $f + I(Y)$ be a member of $\Gamma(Y)$. The Taylor representation of f at $\bar{a}$ is given by

$$f = f(\bar{a}) + \sum_{i=1}^{n}(X_i - a_i)\frac{\partial f}{\partial X_i}|_{\bar{a}} + g,$$

where g is a polynomial in $\{(X_1 - a_1), X_2 - a_2), \cdots, (X_n - a_n)\}$ of degree at least 2. It follows that $d(f + I(Y)) = \hat{D}_{\bar{v}}(f + I(Y))$. This shows that ρ is a surjective linear transformation.

The above discussion prompts us to have the following definition.

Definition 4.3.29 Let Y be an affine variety over an algebraically closed field K, and $\bar{a} \in Y$. Then $Der(\Gamma(Y), K_{\bar{a}})$ is called the **Tangent space** to Y at the point $\bar{a}$.

Note that the definition of the tangent space is independent of the embedding. We also view and interpret the tangent space from a different angle as follows.

Proposition 4.3.30 *There is a natural K-isomorphism from $Der(\Gamma(Y), K_{\bar{a}})$ to $Hom_K(\frac{M_{\bar{a}}}{(M_{\bar{a}})^2}, K)$, where $M_{\bar{a}} = \{f + I(Y) \mid f(\bar{a}) = 0\}$ is the maximal ideal of $\Gamma(Y)$ determined by the point $\bar{a}$.*

Proof Let $d \in Der(\Gamma(Y), K_{\overline{a}})$. Evidently, $d|_{M_{\overline{a}}}$ is a linear transformation from $M_{\overline{a}}$ to $K_{\overline{a}}$ such that $(M_{\overline{a}})^2 \subseteq Ker\, d|_{M_{\overline{a}}}$. Thus, d induces a linear transformation $\hat{d}$ from $\frac{M_{\overline{a}}}{(M_{\overline{a}})^2}$ to K (note that $K_{\overline{a}}$ is the same as K considered as a vector space over K). This gives us a natural vector space homomorphism μ from $Der(\Gamma(Y), K_{\overline{a}})$ to $Hom_K(\frac{M_{\overline{a}}}{(M_{\overline{a}})^2}, K)$ given by $\mu(d) = \hat{d}$. We show that μ is an isomorphism. Suppose that $\hat{d} = 0$. Then $d|_{M_{\overline{a}}}$ is a zero transformation from $M_{\overline{a}}$ to $K_{\overline{a}}$. Using the Taylor expansion about $\overline{a}$, we observe that any $f \in K[X_1, X_2, \cdots, X_n]$ can be uniquely expressed as $f = f(\overline{a}) + f^+$, where $f^+ \in M_{\overline{a}}$. Clearly, $d(f + I(Y)) = d(f^+ + I(Y)) = 0$. This shows that μ is injective. Next, let ϕ be a homomorphism from $\frac{M_{\overline{a}}}{(M_{\overline{a}})^2}$ to K. Then ϕ is induced by a homomorphism ψ from $M_{\overline{a}}$ to K such that $\psi((M_{\overline{a}})^2) = \{0\}$. Define a map d from $\Gamma(Y)$ to K by putting $d(f + I(Y)) = \psi(f^+)$. It can be seen that $d \in Der(\Gamma(Y), K_{\overline{a}})$. Evidently $\mu(d) = \phi$. This shows that μ is an isomorphism. ♯

As such, without any loss, we can also term the dual space $Hom_K(\frac{M_{\overline{a}}}{(M_{\overline{a}})^2}, K)$ of $\frac{M_{\overline{a}}}{(M_{\overline{a}})^2}$ as the tangent space of Y at the point $\overline{a}$.

In general, $Dim\, T_{\overline{a}}(Y) \geq Dim\, Y$. We say that $\overline{a}$ is a nonsingular point of Y if $Dim\, T_{\overline{a}}(Y) = Dim\, Y$. A point which is not nonsingular is called a singular point. A variety Y is said to be a **nonsingular variety** if all points of Y are nonsingular. It can be checked that the set $Sing(Y)$ of singular points of Y forms a proper closed subset or equivalently, the set of nonsingular points of an affine variety forms a nonempty open set.

Let Y and Y' be affine varieties over an algebraically closed field K. Let ϕ be a morphism from Y to Y'. Then ϕ induces a K-algebra homomorphism $\Gamma(\phi)$ from $\Gamma(Y')$ to $\Gamma(Y)$ which is given by $\Gamma(\phi)(f') = f'o\phi$. Let $y \in Y$ and $y' = \phi(y)$. Then, we have a linear map $d\phi_y$ from $T_y(Y) = Der(\Gamma(Y), K_y)$ to $T_{y'}(Y') = Der(\Gamma(Y'), K_{y'})$ which is given by $d\phi_y(d)(f') = d(\Gamma(\phi)(f')) = d(f'o\phi)$. The map $d\phi_y$ is called the **differential** of ϕ at y. Indeed, we have a functor $\Im$ from the category VAR^* of pointed varieties over an algebraically closed field K to the category $VECT_K$ of vector spaces over K which associates with each pointed variety (Y, y) the tangent space $T_y(Y)$, and with each morphism ϕ from (Y, y) to (Y', y'), the linear map $\Im(\phi) = d\phi_y$ described above.

A Noetherian local ring R with the maximal ideal M is said to be a **regular local ring** if the dimension of M/M^2 considered as a vector space over R/M is the dimension of the ring R. It may be mentioned here that every regular local ring is a UFD. Let Y be a variety. It is clear from the above discussion that a point y of Y is a nonsingular point if and only if the local ring $\vartheta_{Y,y}$ of Y at the point y is a regular local ring. Thus, a point y of Y is a nonsingular point of Y if $Dim_{\vartheta_{Y,y}/M_{Y,y}} M_{Y,y}/M_{Y,y}^2 = Dim\vartheta_{Y,y} = Dim\vartheta(Y) = DimY$.

Example 4.3.31 Let $\overline{\alpha}$ be a point in A_K^n, where K is an algebraically closed field. Then the local ring $\vartheta_{A_K^n, \overline{\alpha}}$ of A_K^n at $\overline{\alpha}$ is the subring $\{\frac{f(\overline{X})}{g(\overline{X})} \mid g(\overline{\alpha}) \neq 0\}$ of the function field $K(\overline{X})$ with the unique maximal ideal $M_{A_K^n, \overline{\alpha}} = \{\frac{f(\overline{X})}{g(\overline{X})} \mid f(\overline{\alpha}) = 0 \text{ and } g(\overline{\alpha}) \neq 0\}$. Clearly, $\{(X_i - \alpha_i) + (M_{A_K^n, \overline{\alpha}})^2 \mid 1 \leq i \leq n\}$ is a basis of the vector space $M_{A_K^n, \overline{\alpha}}/(M_{A_K^n, \overline{\alpha}})^2$ over the field $\vartheta_{A_K^n, \overline{\alpha}}/M_{A_K^n, \overline{\alpha}}$. Thus, the dimension of the vector space $M_{A_K^n, \bar{\alpha}}/(M_{A_K^n, \bar{\alpha}})^2$ over the field $\vartheta_{A_K^n, \bar{\alpha}}/M_{A_K^n, \bar{\alpha}}$ is the same as the dimension of A_K^n. This means that the affine variety A_K^n is a nonsingular variety.

Further, consider the projective variety $P^n(K)$. Clearly, $P^n(K) = \bigcup_{i=0}^{n} A_i$, where $A_i = P_K^n - V(\{X_i\})$ is an open subset of P_K^n. Note that A_i is isomorphic to the affine variety A_K^n for each i. Since $dim P_K^n = n$, it follows that all points of P_K^n are nonsingular points. This ensures that P_K^n is also a nonsingular projective variety.

Exercises

The field K considered in the following exercises are algebraically closed fields.

4.3.1 Show that the coordinate ring $\Gamma(V(\{Y^2 - X\}))$ of $V(\{Y^2 - X\})$ is isomorphic to the polynomial ring in one variable over K. Describe the function field of $V(\{Y^2 - X\})$, and also the local ring at the point $(1, 1)$.

4.3.2 Describe the coordinate ring and the function field of $V(XY - 1)$, where $XY - 1 \in K[X, Y]$. Also describe the local ring at $(1, 1)$.

4.3.3 Let Y be a subspace of a topological space X. Show that $Dim Y \le Dim X$. Show further that if Y is a closed subspace of a finite-dimensional subspace X such that $Dim Y = Dim X$, then $Y = X$.

4.3.4 Let $V(A)$ be an affine variety in $A_n(K)$ of dimension r, and $H = V(\{f(\overline{X})\})$ be a hyper-surface in $A_n(K)$ which does not contain $V(A)$. Show that each irreducible component of $H \cap V(A)$ is of dimension $r - 1$.

4.3.5 Let A be an ideal of $K[X_1, X_2, \cdots, X_n]$ which is generated by a set containing r elements. Show that every irreducible component of $V(A)$ has dimension $\ge n - r$.

4.3.6 Give an example of an irreducible polynomial in $f(X, Y)$ in $\mathbb{R}[X, Y]$ such that $V(\{f(X, Y)\})$ is not irreducible.

4.3.7 Show that a product exists in the category of varieties by actually constructing it.

4.3.8 An algebraic variety G together with a group structure on G is termed as an **algebraic group**, also termed as **Group variety**, if the map $(a, b) \mapsto ab^{-1}$ is a morphism from $G \times G$ to G. If in addition G is an affine variety, then we term it as an affine algebraic group. Similarly, we have the concept of the Ring variety. Let G be an algebraic group and X be a variety. Show that $Mor(X, G)$ has a natural group structure. If R is a ring variety, show that $Mor(X, R)$ has a natural ring structure.

4.3.9 Show that A_K^1 is an algebraic group with respect to the usual addition. This algebraic group is denoted by G_a. Similarly, $A_K^1 - \{0\}$ is also an algebraic group with respect to the usual multiplication, and it is denoted by G_m. Also observe that A_K^1 is a ring variety with respect to the usual addition and multiplication. Show further that the ring $Mor(X, A_K^1)$ is isomorphic to the ring $\vartheta(X)$ of regular functions on X.

4.3.10 Let K be a field. Treat the general linear group $GL(n, K)$ as an affine alge-braic subvariety $V(\{X Det [X_{ij}] - 1\})$ of $A_K^{n^2+1}$, where $X Det [X_{ij}] - 1$ is the polyno-mial in $n^2 + 1$ variables $\{X_{ij} \mid 1 \leq i \leq n, 1 \leq j \leq n\} \bigcup \{X\}$. Show that $GL(n, K)$ is an affine algebraic group. A closed subgroup of $GL(n, K)$, is called a *linear algebraic group*.

4.3.11 Describe $SL(n, K)$ as an algebraic group over K. Describe $U(n)$ and $SU(n)$ as linear algebraic groups over $\mathbb{C}$.

4.3.12 Develop the basic theory of algebraic groups such as isomorphism theorems, and also the basic language of solvable and nilpotent algebraic groups.

4.3.13 Find the ring $\vartheta(Y)$ of regular functions of Y and also the local rings $\vartheta_{Y,P}$ at different points of Y, where $Y = V(X_2 - X_1^2) \subseteq A_K^2$. Do the same for $Y = V(\{X_2^2 - X_1^3 + X_1\}) \subseteq A_K^2$. Show that both are nonsingular curves.

4.3.14 Show that $(0, 0)$ is a singular point of $Y = V(\{X_2^2 - X_1^3\}) \subseteq A_K^2$, whereas all other points are nonsingular points.

4.3.15 Show that an irreducible curve $V(\{f(X_1, X_2)\}) \subseteq A_K^2$ has only finitely many singular points.

4.3.16 Show that the singular points of a variety form a proper closed set. Deduce that every algebraic group is a nonsingular variety.

4.4   Schemes

The ringed spaces $(Spec R, \vartheta^R)$ are the building blocks of schemes. The ringed space $(Spec R, \vartheta^R)$ is also called the spectrum of R. We first discuss the topology of $Spec R$. If R is an integral domain, then $\{0\}$ is a prime ideal, and so it is a member of $Spec R$. Since every prime ideal contains $\{0\}$, the only closed set of $Spec R$ containing $\{0\}$ is the whole space $Spec R$. Thus, $Spec R$ need not be T_1- space. Clearly, $Spec R$ is a T_1- space if and only if all prime ideals are maximal ideals. Indeed, the closed points in $Spec R$ are precisely maximal ideals of R. Thus, for an integral domain R, $Spec R$ is T_1-space if and only if R is a field, and in that case $Spec R$ is a singleton space. Usually, $Spec R$ is non-Hausdorff (characterize rings R for which $Spec R$ is Hausdorff).

Let f be a non-nilpotent element of R. The open subset $D(f) = Spec R - V(\{f\}) = \{P \in Spec R \mid f \notin P\}$ is called a principal open subset of $Spec R$.

Proposition 4.4.1 *The set $\{D(f) \mid f$ is a non $-$ nilpotent elemnt of $R\}$ of prin-cipal open sets form a basis for the topology of $Spec R$.*

Proof Let $Spec R - V(A)$ be an open subset of $Spec R$ containing a prime ideal P. Then A is not a subset of P. Let $f \in A - P$. Since P is a prime ideal, f is non-nilpotent and $V(A) \subseteq V(\{f\})$. Thus, $P \in D(f) \subseteq Spec R - V(A)$. ♯

Proposition 4.4.2 *Spec R is compact.*

Proof Since the set of principal open subsets of $Spec\, R$ forms a basis for the topology of $Spec\, R$, it is sufficient to show that any open cover of $Spec\, R$ consisting of principal open sets has a finite subcover. Suppose that $\{D(f_\alpha) \mid \alpha \in \Lambda\}$ is an open cover of $Spec\, R$. Then $V(A) = \emptyset$, where A is the ideal generated by the set $\{f_\alpha \mid \alpha \in \Lambda\}$. Since every proper ideal is contained in a maximal ideal (Theorem 7.5.31, Algebra 1), it follows that $A = R$. Thus, there exist $a_1, a_2, \cdots, a_r$ in R and $\alpha_1, \alpha_2, \cdots, \alpha_r$ in Λ such that

$$1 = a_1 f_{\alpha_1} + a_2 f_{\alpha_2} + \cdots a_r f_{\alpha_r}.$$

But, then

$$V(f_{\alpha_1}) \bigcap V(f_{\alpha_2}) \bigcap \cdots \bigcap V(f_{\alpha_r}) = \emptyset.$$

Consequently, $D(f_{\alpha_1}) \bigcup D(f_{\alpha_2}) \bigcup \cdots \bigcup D(f_{\alpha_r}) = Spec\, R$. ♯

Let f be a non-nilpotent element of R. Then $S = \{f^n \mid n \in \mathbb{N} \bigcup \{0\}\}$ is a multiplicative closed subset of R. The localization $S^{-1} R$ is denoted by R_f. Thus, R_f is the set $\{\frac{a}{f^n} \mid a \in R \text{ and } n \in \mathbb{N} \bigcup \{0\}\}$ of equivalence classes, where $\frac{a}{f^n} = \frac{b}{f^m}, n, m \geq 0$ if and only if there is a natural number r such that $f^r(af^m - bf^n) = 0$. We have the obvious addition and multiplication in R_f with respect to which R_f is a ring.

Proposition 4.4.3 *Let f be a non-nilpotent element of R. Then the subspace $D(f)$ of Spec R is homeomorphic to Spec R_f.*

Proof Let P be a member of $D(f)$. Then P is a prime ideal of R such that $f \notin P$. Let $\hat{P}$ denote the subset $\{\frac{a}{f^n} \mid a \in P, n \geq 0\}$ of R_f. Clearly, $\hat{P}$ is an ideal of R_f. Suppose that $\frac{a}{f^n} \frac{b}{f^m} \in \hat{P}; a, b \in R;$ and $n, m \geq 0$. Then $\frac{ab}{f^{n+m}} = \frac{c}{f^r}$ for some $c \in P, r \geq 0$. This means that $f^t(abf^r - cf^{n+m}) = 0$ for some $t \geq 0$. Hence $abf^{t+r} \in P$. Since $f^{t+r} \notin P$ and P is a prime ideal, $a \in P$ or $b \in P$. This shows that $\frac{a}{f^n} \in \hat{P}$ or $\frac{b}{f^m} \in \hat{P}$. Hence $\hat{P}$ is a prime ideal of R_f. Define a map ϕ from $D(f)$ to $Spec\, R_f$ by $\phi(P) = \hat{P}$. Suppose that $\hat{P} = \hat{Q}$, where P, Q are prime ideals of R such that $f \notin P \bigcup Q$. Let $a \in P$. Then $\frac{a}{1} \in \hat{Q}$. Hence there is an element $b \in Q$ and $n \geq 0$ such that $\frac{a}{1} = \frac{b}{f^n}$. In turn, $f^m(af^n - b) = 0$ for some $m \geq 0$. This means that $f^{n+m} a \in Q$. Since Q is a prime ideal and $f^{n+m} \notin Q, a \in Q$. This shows that $P \subseteq Q$. Similarly, $Q \subseteq P$. Thus, ϕ is an injective map. Let $\wp$ be a prime ideal of R_f. Consider $P = \{a \in R \mid \frac{a}{1} \in \wp\} = \iota^{-1}(\wp)$, where ι is the homomorphism from R to R_f given by $\iota(a) = \frac{a}{1}$. Since $\wp$ is a prime ideal of R_f and $\frac{f}{1}$ is an invertible element of R_f, $\frac{f}{1} \notin \wp$. This means that $f \notin P$. Since $\wp$ is a prime ideal of R_f, P is a prime ideal of R. Hence $P \in D(f)$. If $a \in P$, then $\frac{a}{1} \in \wp$. Since $\wp$ is an ideal of R_f, $\frac{a}{f^n} = \frac{a}{1} \frac{1}{f^n}$ belongs to $\wp$. Thus $\phi(P) = \hat{P} \subseteq \wp$. Evidently, $\wp \subseteq \hat{P}$. Thus, $\phi(P) = \wp$. This shows that ϕ is a bijective map. Finally, we need to prove the continuity of ϕ and its inverse. A closed subset of $D(f)$ is of the form $D(f) \bigcap V(A)$, where $A \subseteq R$. Clearly, $\phi(D(f) \bigcap V(A)) = V(\iota(A))$ is a closed subset of $Spec\, R_f$. Also for any subset B of $Spec R_f$, $\phi^{-1}(V(B)) = D(f) \cap V(\phi^{-1}(B))$ is closed in $D(f)$. This shows that ϕ is an homeomorphism. ♯

The following corollary follows from Proposition 4.4.2 and Proposition 4.4.3.

Corollary 4.4.4 $D(f)$ *is an open compact subspace of* $Spec R$. ♯

Proposition 4.4.5 *Consider the spectrum* $(Spec R, \vartheta^R)$ *of the ring* R. *Then the stalk* ϑ_P^R *at* P *is isomorphic to* R_P *for all* $P \in Spec R$.

Proof Recall Example 4.1.5. Fix $P \in Spec R$. By definition,

$$\vartheta_P^R = Lim_{\rightarrow} \{(\vartheta^R(U), j_U^V) \mid V \subseteq U, \ P \in V\}.$$

Thus, ϑ_P^R is the group $\{\overline{s_U} \mid s_U \in \vartheta^R(U), \ P \in U\}$ of equivalence classes, where $\overline{s_U} = \overline{t_V}$ if and only if there is an open subset W of $Spec R$ such that $P \in W \subseteq (U \cap V)$ and $s_U|W = t_V|W$. Let U be an open subset of $Spec R$ containing P and $s_U \in \vartheta^R(U)$. Then there is an open neighborhood V of P together with elements $a, f \in R$ such that $V \subseteq U$, for each $Q \in V$, $f \notin Q$ and $s_U(Q) = \frac{a}{f} \in R_Q$. Define a map χ_U from $\vartheta^R(U)$ to R_P by putting $\chi_U(s_U) = s_U(P) = \frac{a}{f} \in R_P$. It is easily observed that χ_U is a ring homomorphism for each open set U containing P, and $s_U(P) = t_V(P)$ whenever $\overline{s_U} = \overline{t_V}$. This defines a homomorphism χ from ϑ_P^R to R_P by putting $\chi(\overline{s_U}) = s_U(P)$. Let $\frac{a}{f}$ be a member of R_P. Then $P \in D(f)$. The map $s_{D(f)}$ from $D(f)$ to $\prod_{Q \in D(f)} R_Q$ given by $s_{D(f)}(Q) = \frac{a}{f} \in R_Q$ is a member of $\vartheta^R(D(f))$ such that $\chi(\overline{s_{D(f)}}) = \frac{a}{f}$. This shows that χ is surjective homomorphism.

Next, we show that χ is injective. Suppose that $\chi(\overline{s_U}) = \chi(\overline{t_V})$, where $s(U) \in \vartheta^R(U)$ and $t(V) \in \vartheta^R(V)$. It follows from the definition of ϑ^R that there exists an open set W together with $a, b \in R$, $f, g \notin P \bigcup Q$ such that $P \in W \subseteq U \cap V$ and for each $Q \in W$, $s_U(Q) = \frac{a}{f}$ in R_P and $t_V(Q) = \frac{b}{g}$ in R_Q. Since $\chi(\overline{s_U}) = \chi(\overline{t_V})$, $s_U(P) = t_V(P)$. This means that $\frac{a}{f} = \frac{b}{g}$ in R_P. Hence, there is an element $c \notin P$ such that $c(ag - bf) = 0$. But then $\frac{a}{f} = \frac{b}{g}$ for all $Q \in W \cap D(c)$. Consequently, $s_U|(W \cap D(c)) = t_V|(W \cap D(c))$. By definition $\overline{s_U} = \overline{t_V}$. ♯

Proposition 4.4.6 *Let* R *be a ring. Then* $\vartheta^R(D(f))$ *is isomorphic to* R_f *for all non-nilpotent elements* f *of the ring* R.

Proof Define a map η from R_f to $\vartheta^R(D(f))$ by taking $\eta(\frac{a}{f^n})(Q)$ to be the element of R_Q which is represented by $\frac{a}{f^n}$, $Q \in D(f)$. Clearly, η is a homomorphism. Suppose that $\eta(\frac{a}{f^n}) = 0$. Then for each $Q \in D(f)$, $\frac{a}{f^n}$ represents 0 elements of R_Q. Hence for each $Q \in D(f)$, there is an element $h_Q \notin Q$ such that $h_Q a = 0$. Let A denote the annihilator of a. Then $h_Q \in A - Q$. This means that $Q \notin V(A)$ for each $Q \in D(f)$. Hence $D(f) \cap V(A) = \emptyset$. Thus, f belongs to each prime ideal containing A. In turn, $f \in \sqrt{A}$ (Proposition 7.5.39, Algebra 1), and so $f^n \in A$ for some $n \in \mathbb{N}$. Consequently, $f^n a = 0$, and $\frac{a}{f^n}$ represents zero in R_f. This shows that η is injective.

Next, we show that η is surjective. Let s be a member of $\vartheta^R(D(f))$. We need to show the existence of an element $a \in R$ and an element $n \in \mathbb{N} \bigcup \{0\}$ such that the element $s(Q)$ in R_Q is represented by $\frac{a}{f^n}$ for each $Q \in D(f)$. By definition, s is an element of $\prod_{P \in D(f)} R_P$ such that for each $P \in D(f)$, there is an open subset U_P of $D(f)$ containing P together with elements a_P, b_P in R such that for each $Q \in U_P$, $b_P \notin Q$ and $\frac{a_P}{b_P}$ represents the element $s(Q)$ of R_Q. It further means that $U_P \subseteq D(b_P)$. Since $\{D(g) \mid g \ is \ non-nilpotent \ element \ of \ R\}$ is a basis for the topology of $Spec R$, we may assume that $U_P = D(g_P) \subseteq D(b_P)$. In turn, $V(<b_P>) = V(b_P) \subseteq V(g_P) = V(<g_P>)$. Hence $\sqrt{<g_P>} \subseteq \sqrt{<b_P>}$. Consequently, $(g_P)^n = ab_P$ for some $n \in \mathbb{N}$ and $a \in R$. Thus, $\frac{a_P}{b_P} = \frac{aa_P}{(g_P)^n}$. Put $h_P = (g_P)^n$ and $c_P = ag_P$. Then $D(g_P) = D(h_P)$ and $s(Q)$ is represented by $\frac{c_P}{h_P}$ in R_Q for each $Q \in D(h_P)$. More explicitly, we have an open cover $\{D(h_P) \mid P \in D(f)\}$ of $D(f)$ together with a family $\{c_P \mid P \in D(f)\}$ of members of R such that s is represented by $\frac{c_P}{h_P}$ on $D(h_P)$ for each $P \in D(f)$. Since $D(f)$ is compact (Corollary 4.4.4), we have a finite subset $\{P_1, P_2, \cdots, P_r\}$ of $D(f)$ such that $D(f) = D(h_{P_1}) \bigcup D(h_{P_2}) \bigcup \cdots \bigcup D(h_{P_r})$ and s on $D(h_{P_i})$ is represented by $\frac{c_{P_i}}{h_{P_i}}$ for each i, $1 \le i \le r$. Further, $D(h_{P_i} h_{P_j}) = D(h_{P_i}) \bigcap D(h_{P_j})$ for each pair i, j. Thus, the element s restricted to $\vartheta^R(D(h_{P_i} h_{P_j}))$ is represented by $\frac{c_{P_i}}{h_{P_i}}$ and also by $\frac{c_{P_j}}{h_{P_j}}$. Earlier, we have seen that the map η from $R_{h_{P_i} h_{P_j}}$ to $\vartheta^R(D(h_{P_i} h_{P_j}))$ given by $\eta(\frac{a}{(h_{P_i} h_{P_j})^n})(Q) = \frac{a}{(h_{P_i} h_{P_j})^n}$ in R_Q is an injective map. As already observed, $\eta(\frac{c_{P_i}}{h_{P_i}}) = \eta(\frac{c_{P_j}}{h_{P_j}})$. Hence $\frac{c_{P_i}}{h_{P_i}} = \frac{c_{P_j}}{h_{P_j}}$ in $R_{h_{P_i} h_{P_j}}$ for each pair i, j. In turn, there is a $n_{ij} \in \mathbb{N}$ such that $(h_{P_i} h_{P_j})^{n_{ij}} (h_{P_j} c_{P_i} - h_{P_i} c_{P_j}) = 0$. Take $N = max\{n_{ij}\}$. Then $(h_{P_j})^{N+1}((h_{P_i})^N c_{P_i}) - (h_{P_i})^{N+1}((h_{P_j})^N c_{P_j}) = 0$ for all i, j. Put $(h_{P_i})^N c_{P_i} = d_{P_i}$ and $(h_{P_i})^{N+1} = k_{P_i}$. Then $D(k_{P_i}) = D(h_{P_i}) = D(g_{P_i})$ and $k_{P_j} d_{P_i} = k_{P_i} d_{P_j}$ for all i, j. This shows the existence of members $d_{P_1}, d_{P_2}, \cdots, d_{P_r}$ of R together with non-nilpotent elements $k_{P_1}, k_{P_2}, \cdots, k_{P_r}$ such that
(i) $D(f) = D(k_{P_1}) \bigcup D(k_{P_2}) \bigcup \cdots \bigcup D(k_{P_r})$,
(ii) on each $D(k_{P_i})$, s is represented by $\frac{d_{P_i}}{k_{P_i}}$, and
(iii) $k_{P_j} d_{P_i} = k_{P_i} d_{P_j}$ for all i, j.
From (i), it follows that $V(<\{k_{P_i} \mid 1 \le i \le r\}>) \subseteq V(f)$. Consequently, $f \in \sqrt{<\{k_{P_i} \mid 1 \le i \le r\}>}$, and so there is a natural number n together with elements $\alpha_1, \alpha_2, \cdots, \alpha_r$ of R such that

$$f^n = \alpha_1 k_{P_1} + \alpha_2 k_{P_2} + \cdots + \alpha_r k_{P_r}.$$

Take

$$a = \alpha_1 d_{P_1} + \alpha_2 d_{P_2} + \cdots + \alpha_r d_{P_r}.$$

Evidently, $ak_{P_j} = f^n d_{P_j}$ for each j. This means that $\frac{a}{f^n}$ and $\frac{d_{P_j}}{k_{P_j}}$ represent the same element of R_Q for each $Q \in D(k_{P_j})$. Consequently, s is represented by $\frac{a}{f^n}$ on $D(f)$. This shows that $\eta(\frac{a}{f^n}) = s$. $\sharp$

Corollary 4.4.7 *Let R be a ring. Then $\vartheta^R(Spec R) \approx R$.*

Proof The proof follows from the above proposition if we observe that $Spec R = D(1)$ and the fact that $R_1 = R$. ♯

Example 4.4.8 We describe the spectrum $(Spec\mathbb{Z}, \vartheta^{\mathbb{Z}})$ of the ring $\mathbb{Z}$ of integers. Every nonzero prime ideal is uniquely expressible as $p\mathbb{Z}$, where p is a positive prime. Thus, the set $Spec\mathbb{Z}$ can be identified with the set $P \bigcup \{0\}$ in a natural manner, where P denotes the set of positive primes. Again, since every prime ideal contains the prime ideal $\{0\}$, it follows that the singleton $\{0\}$ is dense in $Spec\mathbb{Z} \approx P \bigcup \{0\}$. Since every nonzero prime ideal of $\mathbb{Z}$ is a maximal ideal, all singleton subsets, and so all finite subsets of P, are closed. Every nonzero ideal of $\mathbb{Z}$ is of the form $m\mathbb{Z}$, $m \neq 0$, and a prime ideal $p\mathbb{Z}$ contains $m\mathbb{Z}$ if and only if p divides m. Thus, the proper closed subsets of $Spec\mathbb{Z}$ are precisely finite subsets of P. It is also clear that any open subset of $Spec\mathbb{Z}$ is of the form $D(m)$, where $m0$ is a product of distinct primes. From Proposition 4.4.6, $\vartheta^{\mathbb{Z}}(D(m)) = \mathbb{Z}_{(p_1 p_2 \cdots p_r)} = \{\frac{a}{b} \mid (p_i, b) = 1 \text{ } for \text{ } each \text{ } i\}$, where $p_1, p_2, \cdots, p_r$ are distinct primes dividing m. In the same way, we can describe $(Spec R, \vartheta^R)$ for a principal ideal domain R. In particular, we can describe $(Spec K[x], \vartheta^{K[x]})$ which is termed as an **affine line** over K, and it is also denoted by A^1_K. If K is an algebraically closed field, then A^1_K can be identified with K, where open sets are precisely compliments of finite subsets of $K^\star$.

Example 4.4.9 In this example, we describe the **affine plane** $A^2_K = (Spec K[x, y], \vartheta^{K[x,y]})$ over an algebraically closed field K. Here also the singleton subset of A^2_K containing the prime ideal $\{0\}$ is dense in A^2_K. The point $\{0\}$ is called the **generic point** of A^2_K. Further, given a point (α, β) in K^2, the ideal $M_{(\alpha,\beta)}$ generated by the pair of elements $x - \alpha$ and $y - \beta$ is a maximal ideal of $K[x, y]$. Since K is an algebraically closed field, the zero set of every proper ideal is a nonempty set. Thus, every maximal ideal of $K[x, y]$ is of the form $M_{(\alpha,\beta)}$, and so the set of closed points of A^2_K can be faithfully identified by K^2. It is also clear that K^2 treated as a subspace of A^2_K is the affine variety whose closed sets are precisely the algebraic subsets. Apart from these, given a prime ideal P generated by an irreducible polynomial $f(x, y)$, the closure $\overline{\{P\}}$ of the singleton $\{P\}$ is the set of all prime ideals containing P, and it is precisely $\{P\} \bigcup \{M_{(\alpha,\beta)} \mid f(\alpha, \beta) = 0\}$. The point P is termed as the generic point of $\overline{\{P\}}$.

Ringed spaces which appear locally as a spectrum of rings are termed as schemes. We make it more precise as follows.

Definition 4.4.10 A morphism from a ringed space (X, ϑ_X) to a ringed space (Y, ϑ_Y) is a pair $(f, f^\sharp)$, where f is a continuous map from X to Y and $f^\sharp$ is a morphism from the sheaf ϑ_Y to the sheaf $f_\star \vartheta_X$. More explicitly, $f^\sharp$ is the family $\{f^\sharp_U \in Hom(\vartheta_Y(U), \vartheta_X(f^{-1}(U))) \mid U \text{ } is \text{ } open \text{ } subset \text{ } of \text{ } Y\}$ of ring homomorphisms which respect the restriction maps. This defines the category RS of ringed spaces.

Let $(f, f^\sharp)$ be a morphism from a ringed space (X, ϑ_X) to a ringed space (Y, ϑ_Y). For each $x \in X$, the directed family $\{U \mid U \text{ is an open set containing } x\}$ is cofinal in the directed family $\{f^{-1}(V) \mid V \text{ is an open set containing } f(x)\}$. In turn, the restriction map induces a homomorphism j_x from the stalk $(f_*\vartheta_X)_{f(x)} = Lim_\to \vartheta_X(f^{-1}(V))$ to the stalk $(\vartheta_X)_x = Lim_\to \vartheta_X(U)$. Also the morphism $f^\sharp$ from the sheaf ϑ_Y to the sheaf $f_*\vartheta_X$ induces a homomorphism $f^\sharp_{f(x)}$ from the stalk $(\vartheta_Y)_{f(x)}$ to the stalk $(f_*\vartheta_X)_{f(x)}$. We have the induced composite homomorphism $f^\sharp_x = j_x o f^\sharp_{f(x)}$ from $(\vartheta_Y)_{f(x)}$ to $(\vartheta_X)_x$.

Recall that a ringed space (X, ϑ_X) is called a locally ringed space if all the stalks are local rings. A morphism from $(f, f^\sharp)$ from a locally ringed space (X, ϑ_X) to a locally ringed space (Y, ϑ_Y) is called a **local morphism** if the induced homomorphism $f^\sharp_x$ from the stalk $(\vartheta_Y)_{f(x)}$ to the stalk $(\vartheta_X)_x$ is a local homomorphism (recall that a homomorphism from a local ring R to a local ring S is called a local homomorphism if the inverse image of the maximal ideal of S is that of R). This gives us a subcategory LRS of locally ringed spaces whose objects are locally ringed spaces and morphisms are local morphisms. An isomorphism in LRS is called a local isomorphism.

An **affine scheme** is a locally ringed space (X, ϑ_X) which is locally isomorphic to $(SpecR, \vartheta^R)$ for some ring R. We have the category $ASCH$ of affine schemes which is a subcategory of LRS. Further, a **scheme** is a ringed space (X, ϑ_X) which is locally an affine scheme. More explicitly, a ringed space (X, ϑ_X) is termed as a scheme if for all $x \in X$, we have an open subset U containing x such that the induced ringed space $(U, \vartheta^U = \vartheta_X|U)$ is an affine scheme. The category of schemes is denoted by SCH. Thus, $ASCH \subseteq SCH \subseteq LRS \subseteq RS$.

Proposition 4.4.11 *Spec defines a contra-variant equivalence from the category $RING$ of rings to the category $ASCH$ of affine schemes.*

Proof Let f be a homomorphism from a ring R to a ring S. Then f defines a map $\hat{f}$ from $SpecS$ to $SpecR$ by putting $\hat{f}(Q) = f^{-1}(Q)$, where Q is a prime ideal of S. Evidently, $\hat{f}^{-1}(P) = \{Q \in SpecS \mid f^{-1}(Q) = P\}$. Thus, $\hat{f}^{-1}(V(A)) = \{Q \in SpecS \mid f^{-1}(Q) \in V(A)\} = \{Q \in SpecS \mid f^{-1}(Q) \supset A\} = \{Q \in SpecS \mid Q \supset f(A)\} = V(f(A))$. This shows that $\hat{f}$ is a continuous map. Next, we introduce a morphism $\hat{f}^\sharp$ from the sheaf ϑ^R to the sheaf $\hat{f}_*\vartheta^S$ so that $(\hat{f}, \hat{f}^\sharp)$ becomes a morphism from the affine scheme $(SpecS, \vartheta^S)$ to the affine scheme $(SpecR, \vartheta^R)$. Let U be an open subset of $SpecR$. Then $\hat{f}^{-1}(U) = \{Q \in SpecS \mid \hat{f}(Q) = f^{-1}(Q) \in U\}$. For any $Q \in SpecS$, f induces a homomorphism f_Q from $R_{f^{-1}(Q)}$ to S_Q. Let s be a member of $\vartheta^R(U)$. Then the map $\hat{f}^\sharp_U(s)$ from $\hat{f}^{-1}(U)$ to $\prod_{Q \in \hat{f}^{-1}(U)} S_Q$ defined by $\hat{f}^\sharp(s)(Q) = f_Q(s(f^{-1}(Q)))$ can be easily seen to be a member of $\vartheta^S(\hat{f}^{-1}(U)) = (\hat{f}_*\vartheta^S)(U)$. This gives us a homomorphism $\hat{f}^\sharp_U$ from $\vartheta^R(U)$ to $(\hat{f}_*\vartheta^S)(U)$. It can be easily seen that the family $\hat{f}^\sharp = \{\hat{f}^\sharp_U \mid U \text{ is open in } SpecR\}$ respects the corresponding restriction maps, and hence it is a morphism from ϑ^R to $\hat{f}_*\vartheta^S$. Evidently, the homomorphism induced by $\hat{f}^\sharp$ on the stalk at Q is the localization map which associates $\frac{a}{x}$ in $R_{f^{-1}(Q)}$ with the element $\frac{f(a)}{f(x)}$ in S_Q. Clearly, this is a local homomorphism. This gives us a contra-variant functor $Spec$ from the category of rings to the category $ASCH$ which associates with each ring R the affine scheme $(Spec R, \vartheta^R)$ and with each homomorphism f from R to S, $Spec(f) = (\hat{f}, \hat{f}^\sharp)$.

Using the axiom of choice, we observe that the category $ASCH$ is equivalent to its full subcategory whose objects are of the form $(Spec R, \vartheta^R)$. Thus, we need to show that any local morphism from $(\phi, \phi^\sharp)$ and $(Spec S, \vartheta^S)$ to $(Spec R, \vartheta^R)$ is of the form $(\hat{f}, \hat{f}^\sharp)$ for some homomorphism f from R to S. Now, $(\phi_\star \vartheta^S)(Spec R) = \vartheta^S(\phi^{-1}(Spec R)) = \vartheta^S(Spec S)$. By Corollary 4.4.7, $\vartheta^R(Spec R)$ is naturally isomorphic to the ring R and $\vartheta^S(Spec S)$ is naturally isomorphic to S. Hence $\phi^\sharp_{Spec R}$ gives us a homomorphism f from R to S in a natural manner. From the relevant definitions, it follows that $\phi(Q) = \{a \in R \mid \phi^\sharp_{Spec R}(a) \in Q\} = f^{-1}(Q) = \hat{f}(Q)$. This shows that $\phi = \hat{f}$. In turn, it follows that $\hat{f}^\sharp = \phi^\sharp$. ♯

Example 4.4.12 (i) For every ring R, we have a unique morphism from $(Spec R, \vartheta^R)$ to $(Spec \mathbb{Z}, \vartheta^\mathbb{Z})$ which is induced by the unique ring homomorphism from $\mathbb{Z}$ to R. Indeed, for every scheme (Y, ϑ_Y), we have a unique morphism from the scheme (Y, ϑ_Y) to $(Spec \mathbb{Z}, \vartheta^\mathbb{Z})$. In other words, $(Spec \mathbb{Z}, \vartheta^\mathbb{Z})$ is the terminal object in the category SCH.

(ii) Let K be a field. Then $Spec K$ is the singleton space $\{x_0\}$, where x_0 represents the prime ideal $\{0\}$ of K. Evidently, $\vartheta^K(Spec K) = \vartheta^K(\{x_0\}) \approx K$. A morphism $(\phi, \phi^\sharp)$ from $(Spec K, \vartheta^K)$ to a scheme (X, ϑ_X) determines a unique point $x = \phi(x_0)$ of X. Further, for an open subset U of X, $(\phi_\star \vartheta^K)(U) = \vartheta^K(\phi^{-1}(U))$. Thus, $(\phi_\star \vartheta^K)(U) = K$ if $x \in U$, and it is an trivial ring $\{0\}$ otherwise. This means that $\phi^\sharp$ is uniquely determined by the nonzero homomorphism from the stalk $(\vartheta_X)_x$ to K. Since $(\vartheta_X)_x$ is a local ring with the maximal ideal $(M_X)_x$, it is determined uniquely by the unique injective homomorphism from the field $(\vartheta_X)_x/(M_X)_x$ to the field K. The field $(\vartheta_X)_x/(M_X)_x$ is called the **residue field** of X at x, and it is denoted by $k_X(x)$. Thus, a morphism from $(Spec K, \vartheta^K)$ to (X, ϑ_X) is uniquely determined by a point $x \in X$ and an injective homomorphism from the residue field $k_X(x)$ to K.

More generally, we have the following.

Proposition 4.4.13 *Given a scheme (Y, ϑ_Y) and a ring R, we have a natural map $\phi_{Y,R}$ from $Mor_{Sch}((Y, \vartheta_Y), (Spec R, \vartheta^R))$ to $Hom_{RING}(R, \vartheta_Y(Y))$.*

Proof Given a morphism $(f, f^\sharp)$ from (Y, ϑ_Y) to $(Spec R, \vartheta^R)$, we have the homomorphism $f^\sharp_{Spec R}$ from $R \approx \vartheta^R(Spec R)$ to $\vartheta_Y(Y)$. Put $\phi_{Y,R}((f, f^\sharp)) = f^\sharp_{Spec R}$. The naturality of $\phi_{Y,R}$ is easy to observe. ♯

Definition 4.4.14 A scheme (X, ϑ_X) is said to be

(i) a **connected scheme** if the space X is connected,
(ii) an **irreducible scheme** if X is an irreducible space,
(iii) a **reduced scheme** if $\vartheta_X(U)$ is a reduced ring for each open subset U of X,
(iv) an **integral scheme** if $\vartheta_X(U)$ is an integral domain for each open subset U of X,

(v) a **Noetherian scheme** if X is compact, and it can be covered by finitely many open sets $U_1, U_2, \cdots, U_r$ such that $(U_i, \vartheta_X|U_i)$ is isomorphic to $(Spec R_i, \vartheta^{R_i})$ for each i, where each R_i is a Noetherian ring.

Example 4.4.15 Let R be a commutative ring. Suppose that Spec R is disconnected. Then we have a pair A, B of proper ideals such that $V(A) \cap V(B) = \emptyset$ and $V(A) \cup V(B) = Spec R$. Consequently, $A \cap B = \{0\}$ and $A + B = R$. Using the Chinese remainder theorem, we see that $R \approx \frac{R}{A} \times \frac{R}{B}$, where R/A and R/B are nonzero rings. Conversely, if $R \approx R_1 \times R_2$, where R_1 and R_2 are nonzero rings, then $Spec R$ is disconnected. Indeed, $V(R_1 \times \{0\})$ is a proper clopen subset of $Spec R$. Thus, the affine scheme $(Spec R, \vartheta^R)$ is connected if and only if R is indecomposable.

Example 4.4.16 Suppose that the radical $P = \sqrt{R}$ of R is a prime ideal. Then every closed subset of $Spec R$ containing P is $Spec R$. This means that $Spec R$ cannot be expressed as a union of two proper closed sets. This means that $(Spec R, \vartheta^R)$ is an irreducible affine scheme. The converse is also easy to observe.

Proposition 4.4.17 *An affine scheme $(Spec R, \vartheta^R)$ is a reduced scheme if and only if R is a reduced ring.*

Proof Suppose that $(Spec R, \vartheta^R)$ is a reduced scheme. Then $R \approx \vartheta^R(Spec R)$ is a reduced ring. Conversely, suppose that R is a reduced ring. Let $s \in \vartheta^R(U)$, where U is an open subset of $Spec R$. Then by the definition $s \in \prod_{P \in U} R_P$ such that for all $P \in U$, there is an open neighborhood V of P together with elements $a, f \in R$ such that $f \notin Q$ for all $Q \in V$ and $s(Q)$ is the member of R_Q represented by $\frac{a}{f}$. Suppose that $s^n = 0$. Then $s(Q)^n = (\frac{a}{f})^n = \frac{a^n}{f^n}$ represents 0 in R_Q for all $Q \in V$. This means that there is an element $h \notin Q$ such that $ha^n = 0$ in R. Evidently, $h^n \notin Q$ and $h^n a^n = 0$ in R. Since R is reduced, $ha = 0$. This means that $s(Q)$ represented by $\frac{a}{f}$ is the zero element of R_Q. Thus, $\vartheta^R(U)$ is a reduced ring for all open subset U of R. ♯

Imitating the proof of the above proposition, we can easily establish the following proposition.

Proposition 4.4.18 *An affine scheme $(Spec R, \vartheta^R)$ is an integral scheme if and only if R is an integral domain.* ♯

If ϕ is a homomorphism from a ring A to a ring B, then B is an A-algebra with respect to the scalar product $\cdot$ given by $a \cdot b = \phi(a)b$ (note that all rings considered are commutative rings with identities). If in addition B is a finitely generated A-algebra, then ϕ is termed as a homomorphism of finite type. A morphism $(\phi, \phi^\sharp)$ from an affine scheme $(Spec B, \vartheta^B)$ to $(Spec A, \vartheta^A)$ is said to be a morphism of finite type if B is a finitely generated algebra over A with respect to the ring homomorphism from A to B induced by $(\phi, \phi^\sharp)$. Observe that a finitely generated A-algebra need not be finitely generated as an A-module. For example, the polynomial algebra $K[X_1, X_2, \cdots, X_n]$ is a finitely generated K-algebra but it is an infinite-dimensional vector space over K. A homomorphism ϕ from A to B is termed as a finite homomorphism if B turns out to be a finitely generated module over A. Thus, a morphism $(\phi, \phi^\sharp)$ from the affine scheme $(Spec B, \vartheta^B)$ to $(Spec A, \vartheta^A)$ is termed as a finite morphism if the induced homomorphism from A to B is a finite homomorphism. More generally, we have the following.

Definition 4.4.19 A morphism $(\phi, \phi^\sharp)$ from a scheme (X, ϑ_X) to a scheme (Y, ϑ_Y) is called a morphism of **locally finite type** if for each $y \in Y$, there is an open neighborhood U of y together with a ring A and a local isomorphism $(\rho, \rho^\sharp)$ from $(U, \vartheta_U = \vartheta_Y|U)$ to $(Spec A, \vartheta^A)$ such that $\phi^{-1}(U)$ has an open cover $\{V_\alpha \mid \alpha \in \Lambda\}$ together with rings $\{B_\alpha \mid \alpha \in \Lambda\}$ and local isomorphisms $(\phi_\alpha, \phi_\alpha^\sharp)$ from $(V_\alpha, \vartheta_{V_\alpha} = \vartheta_X|_{V_\alpha})$ to $(Spec B_\alpha, \vartheta^{B_\alpha})$ for each α such that B_α is a finitely generated A-algebra with respect to the homomorphism from A to B_α induced by the local morphism $\rho \circ \phi|_{V_\alpha} \circ \phi_\alpha^{-1}$ from $(Spec B_\alpha, \vartheta^{B_\alpha})$ to $(Spec A, \vartheta^A)$. If in addition to that the family $\{B_\alpha \mid \alpha \in \Lambda\}$ is finite, then $(\phi, \phi^\sharp)$ is termed as a morphism of **finite type**. A morphism $(\phi, \phi^\sharp)$ from a scheme (X, ϑ_X) to a scheme (Y, ϑ_Y) is called a **finite morphism** if for each $y \in Y$, there is an open neighborhood U of y together with rings A and B and local isomorphisms $(\rho, \rho^\sharp)$ from $(U, \vartheta_U = \vartheta_Y|_U)$ to $(Spec A, \vartheta^A)$ and $(\eta, \eta^\sharp)$ from $(\phi^{-1}(U), \vartheta_X|_{\phi^{-1}(U)})$ to $(Spec B, \vartheta^B)$ such that the homomorphism from A to B induced by $\rho \circ \phi|_{\phi^{-1}(U)} \circ \eta^{-1}$ is a finite morphism.

Thus, the morphism from $(Spec K[x_1, X_2, \cdots, X_n], \vartheta^{K[X_1, X_2, \cdots, X_n]})$ to $(Spec K, \vartheta^K)$ which is induced by the embedding of K into $K[X_1, X_2, \cdots, X_n]$ is a morphism of finite type. The morphism from $(Spec \mathbb{Z}[\sqrt{2}], \vartheta^{\mathbb{Z}[\sqrt{2}]})$ to $(Spec \mathbb{Z}, \vartheta^{\mathbb{Z}})$ induced by the inclusion homomorphism from $\mathbb{Z}$ to $\mathbb{Z}[\sqrt{2}]$ is a finite morphism (as $\mathbb{Z}[\sqrt{2}]$ is a finitely generated $\mathbb{Z}$-module).

Definition 4.4.20 A morphism $(\phi, \phi^\sharp)$ from a scheme (X, ϑ_X) to a scheme (Y, ϑ_Y) is termed as **open immersion** if $\phi(X) = U$ is an open subset of Y and $(\phi, \phi^\sharp)$ induces isomorphism from (X, ϑ_X) to $(U, \vartheta_U = \vartheta_Y|_U)$. A scheme (Y, ϑ_Y) is called a **closed subscheme** of (X, ϑ_X) if Y is a closed subset of X, and $(i, i^\sharp)$ is an epimorphism from (Y, ϑ_Y) to (X, ϑ_X). A morphism $(\phi, \phi^\sharp)$ from (Y, ϑ_Y) to (X, ϑ_X) is called a **closed immersion** if $(\phi, \phi^\sharp)$ induces an isomorphism from (Y, ϑ_Y) to a closed subscheme of (X, ϑ_X).

Thus, given a ring R and a non-nilpotent element f of R, the morphism from $(Spec R_f, \vartheta^{R_f})$ to $(Spec R, \vartheta^R)$ which is induced by the inclusion ring homomorphism i from R to R_f is an open immersion. If R is a ring and A is an ideal of R, then the morphism from $(Spec R/A, \vartheta^{R/A})$ to $(Spec R, \vartheta^R)$ which is induced by the quotient map ν from R to R/A is easily seen to be a closed immersion.

The Pullback (see Definition 1.1.40) exists in the category SCH of schemes. Although the proof and the constructions are straightforward, it is lengthy and painstaking. We shall outline the proof and leave the details to be filled by the reader as an exercise.

Let ϕ be a homomorphism from a ring R to a ring A and ψ be a homomorphism from R to a ring B. Then A and B both can be treated as R-algebras with respect to the scalar products given by $a \cdot x = \phi(a)x, a \in R, x \in A$ and $a \cdot y = \psi(a)y, a \in R, y \in B$. Consider the tensor product $A \otimes_R B$ which is an R-module. Clearly, the map χ from $A \times B \times A \times B$ to $A \otimes_R B$ given by $\chi(a, b, c, d) = (ac) \otimes (bd)$ is R-linear in each coordinate. In turn, this induces an R-homomorphism $\overline{\chi}$ from $(A \otimes B) \otimes_R (A \otimes_R B)$ to $A \otimes_R B$. Consequently, we have an R-bilinear map η from $(A \otimes_R B) \times (A \otimes_R B)$ to $A \otimes_R B$ given by $\eta(a \otimes b, c \otimes d) = (ac) \otimes (bd)$. Thus, we have a product $\cdot$ in $A \otimes_R B$ given by $(\sum_{i=1}^{r}(a_i \otimes b_i)) \cdot (\sum_{j=1}^{s}(c_j \otimes d_j)) = \sum_{i=1}^{r}(\sum_{j=1}^{s}((a_i c_j) \otimes (b_i d_j)))$ with respect to which it is a commutative ring with identity $1 \otimes 1$. We have also the scalar product given by $\alpha(\sum_{i=1}^{r}(a_i \otimes b_i)) = \sum_{i=1}^{r}((\alpha a_i) \otimes b_i) = \sum_{i=1}^{r}(a_i) \otimes (\alpha b_i))$ with respect to which $A \otimes_R B$ is an R-algebra. Further, we have an R-homomorphism i_1 from A to $A \otimes_R B$ given by $i_1(a) = a \otimes 1$ and also an R-homomorphism i_2 from B to $A \otimes_R B$ given by $i_2(b) = 1 \otimes b$ such that the diagram

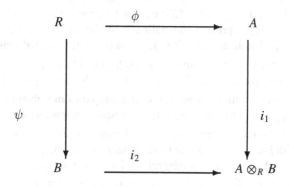

is commutative. If we have another commutative diagram

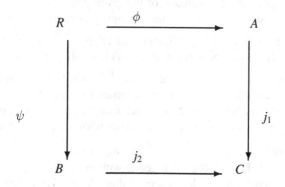

of R-homomorphisms, then we have the unique R-homomorphism μ from $A \otimes_R B$ to C given by $\mu(a \otimes b) = j_1(a)j_2(b)$ such that $\mu o i_1 = j_1$ and $\mu o i_2 = j_2$. This means that Pushout exists in the category of commutative rings with identities. Since a contra-variant equivalence takes a Pushout diagram to a Pullback diagram, it follows from Proposition 4.4.11 that Pullback exists in the category $ASCH$ of affine schemes. Using Proposition 4.4.13, it further follows that a Pullback diagram in $ASCH$ is also a Pullback diagram in SCH. Thus, we have the following proposition.

Proposition 4.4.21 *Let* $(X, \vartheta_X) \overset{(\phi,\phi^\sharp)}{\to} (S, \vartheta_S)$, *and* $(Y, \vartheta_Y) \overset{(\psi,\psi^\sharp)}{\to} (S, \vartheta_S)$ *be morphisms, where* (X, ϑ_X), (Y, ϑ_Y) *and* (S, ϑ_S) *are affine schemes. Then the pullback of the pair* $((\phi, \phi^\sharp), (\psi, \psi^\sharp))$ *exists in the category SCH of schemes.* ♯

Next, let $(X, \vartheta_X) \overset{(\phi,\phi^\sharp)}{\to} (S, \vartheta_S)$, and $(Y, \vartheta_Y) \overset{(\psi,\psi^\sharp)}{\to} (S, \vartheta_S)$ be morphisms, where (Y, ϑ_Y), and (S, ϑ_S) are affine schemes. By definition, we have the family $\{(X_i, \vartheta_{X_i}) \mid i \in I\}$ of open affine subschemes of (X, ϑ_X) such that $\{X_i \mid i \in I\}$ is an open cover of X. From the above proposition, the pullback of the pair $((\phi|_{X_i}, \phi^\sharp|_{X_i}), (\psi, \psi^\sharp))$ of morphisms exists for each $i \in I$. Let $X_i \times_S Y$ together with the morphism $(\eta_i, \eta_i^\sharp)$ from $(X_i \times_S Y, \vartheta_{X_i \times_S Y})$ to (X_i, ϑ_{X_i}) and the morphism $(\rho_i, \rho_i^\sharp)$ from $(X_i \times_S Y, \vartheta_{X_i \times_S Y})$ to (Y, ϑ_Y) represent the pullback of the pair $((\phi|_{X_i}, \phi^\sharp|_{X_i}), (\psi, \psi^\sharp))$. For $i \neq j$, let U_{ij} denote $\eta_i^{-1}(X_i \cap X_j)$. It can be easily verified that $(U_{ij}, \vartheta_{U_{ij}} = \vartheta_{X_i \times_S Y}|_{U_{ij}})$ together with morphisms $(\eta_i|_{U_{ij}}, \eta_i^\sharp|_{U_{ij}})$ and $(\rho_i|_{U_{ij}}, \rho_i^\sharp|_{U_{ij}})$ represents the pullback of the pair $((\phi|_{X_i \cap X_j}, \phi^\sharp|_{X_i \cap X_j}), (\psi, \psi^\sharp))$. Since the pullback, if exists, is unique up to natural isomorphisms, we have a unique isomorphism $(\phi_{ij}, \phi_{ij}^\sharp)$ from $(U_{ij}, \vartheta_{U_{ij}})$ to $(U_{ji}, \vartheta_{U_{ji}})$ which respects the corresponding projection morphisms. We take $(\phi_{ji}, \phi_{ji}^\sharp)$ to be the inverse of $(\phi_{ij}, \phi_{ij}^\sharp)$. It can be further checked that for each triple i, j, k with $i \neq j \neq k \neq i$, the isomorphisms $\{\phi_{pq} \mid p, q \in \{i, j, k\}\}$ are compatible in a natural sense. Gluing (Exercise 4.1.13) the family $\{(X_i \times_S Y, \vartheta_{X_i \times_S Y}) \mid i \in I\}$ with the help of the family $\{(\phi_{ij}, \phi_{ij}^\sharp) \mid i, j \in I, i \neq j\}$ of isomorphisms, we obtain a scheme $(X \times_S Y, \vartheta_{X \times_S Y})$ together with a morphism $(p_1, p_1^\sharp)$ from $(X \times_S Y, \vartheta_{X \times_S Y})$ to (X, ϑ_X), and a morphism $(p_2, p_2^\sharp)$ from $(X \times_S Y), \vartheta_{X \times_S Y})$ to (Y, ϑ_Y) which represents the pullback of the given pair $((\phi, \phi^\sharp), (\psi, \psi^\sharp))$ of morphisms. Using the same argument after interchanging the role of X and Y, we observe that the pullback of $(X, \vartheta_X) \overset{(\phi,\phi^\sharp)}{\to} (S, \vartheta_S)$, and $(Y, \vartheta_Y) \overset{(\psi,\psi^\sharp)}{\to} (S, \vartheta_S)$ for arbitrary (X, ϑ_X) and (Y, ϑ_Y), exists provided that (S, ϑ_S) is an affine scheme.

Finally, let $(X, \vartheta_X) \overset{(\phi,\phi^\sharp)}{\to} (S, \vartheta_S)$ and $(Y, \vartheta_Y) \overset{(\psi,\psi^\sharp)}{\to} (S, \vartheta_S)$ be morphisms, where (X, ϑ_X), (Y, ϑ_Y), and (S, ϑ_S) are arbitrary schemes. Let $\{(S_i, \vartheta_{S_i}) \mid i \in I\}$ be a family of open affine subschemes of (S, ϑ_S) such that $\{S_i \mid i \in I\}$ is an open cover of S. Let $\phi^{-1}(S_i) = X_i$ and $\psi^{-1}(S_i) = Y_i$. From what we proved, the pullback of the pair of morphisms $(X_i, \vartheta_{X_i}) \overset{(\phi|_{X_i},\phi^\sharp|_{X_i})}{\to} (S_i, \vartheta_{S_i})$ and $(Y_i, \vartheta_{Y_i}) \overset{(\psi|_{Y_i},\psi^\sharp|_{Y_i})}{\to} (S_i, \vartheta_{S_i})$ exists for each $i \in I$. Let $(X_i \otimes_{S_i} Y_i)$ together with morphisms $(\eta_i, \eta_i^\sharp)$ and $(\rho_i, \rho_i^\sharp)$ represent the pullback. It can be observed that $(X_i \otimes_{S_i} Y_i)$ together with morphisms $(\eta_i, \eta_i^\sharp)$ and $(\iota \circ \rho_i, (\iota \circ \rho_i)^\sharp)$ represents the pullback of the pair of morphisms $(X_i, \vartheta_{X_i}) \overset{(\phi|_{X_i},\phi^\sharp|_{X_i})}{\to} (S, \vartheta_S)$ and $(Y_i, \vartheta_{Y_i}) \overset{(\psi|_{Y_i},\psi^\sharp|_{Y_i})}{\to} (S, \vartheta_S)$, where ι is the inclusion morphism from (Y_i, ϑ_{Y_i}) to (Y, ϑ_Y). Again using the earlier argument, we find that the pullback of the pair of morphisms $(X, \vartheta_X) \overset{(\phi,\phi^\sharp)}{\to} (S, \vartheta_S)$ and $(Y, \vartheta_Y) \overset{(\psi,\psi^\sharp)}{\to} (S, \vartheta_S)$ exists. This establishes the following theorem.

Proposition 4.4.22 *Pullback exists in the category SCH of schemes. More precisely (see Definition 1.1.40), given schemes (X, ϑ_X), (Y, ϑ_Y), and (S, ϑ_S) together with morphisms $(\phi, \phi^\sharp)$ from (X, ϑ_X) to (S, ϑ_S) and $(\psi, \psi^\sharp)$ from (Y, ϑ_Y) to (S, ϑ_S), there is a scheme $(X \times_S Y, \vartheta_{X \times_S Y})$ together with a morphism $(p_1, p_1^\sharp)$ from $(X \times_S Y, \vartheta_{X \times_S Y})$ to (X, ϑ_X) and a morphism $(p_2, p_2^\sharp)$ from $(X \times_S Y, \vartheta_{X \times_S Y})$ to (Y, ϑ_Y) giving the commutative diagram*

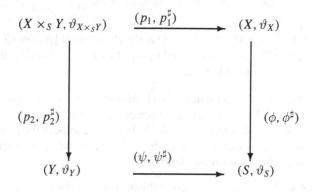

such that given any commutative diagram

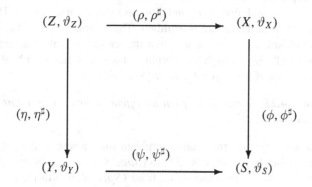

in the category SCH, there is a unique morphism $(\chi, \chi^\sharp)$ from (Z, ϑ_Z) to $(X \otimes_S Y, \vartheta_{X \times_S Y})$ such that $(p_1, p_1^\sharp)(\chi, \chi^\sharp) = (\rho, \rho^\sharp)$ and $(p_2, p_2^\sharp)(\chi, \chi^\sharp) = (\eta, \eta^\sharp)$. ♯

Example 4.4.23 (i) Let $(\phi, \phi^\sharp)$ be a morphism from a scheme (X, ϑ_X) to a scheme (Y, ϑ_Y). Let y be a member of Y which is in the image of ϕ. The identity map from $k_Y(y)$ to $k_Y(y)$ determines a unique natural morphism $(\eta_y^Y, (\eta_y^Y)^\sharp)$ from $(Spec\ k_Y(y), \vartheta^{gk_Y(y)})$ to (Y, ϑ_Y) (see Example 4.4.12 (ii)). Consider the pullback

$X \times_Y Spec\ k_Y(y)$ of the pair $((\phi, \phi^\sharp), (\eta_y^Y, (\eta_y^Y)^\sharp))$ of morphisms. Let ι denote the inclusion map from $\phi^{-1}(\{y\})$ to X, and j denote the constant map from $\phi^{-1}(\{y\})$ to $Spec\ k_Y(y)$. Evidently, $\phi o \iota = \eta_y^Y o j$. From the universal property of pullback, we get a unique morphism $(\rho, \rho^\sharp)$ from $(\phi^{-1}(\{y\}), \vartheta_X|_{\phi^{-1}(\{y\})})$ to $X \times_Y Spec\ k_Y(y)$ which respects the projection morphisms. Using again the universal property of pullback, it can be seen that $(\rho, \rho^\sharp)$ is an isomorphism. This prompts us to term $X \times_Y Spec\ k_Y(y)$ as a fiber of ϕ at y.

(ii) Consider $(Spec\ \mathbb{Z}, \vartheta^\mathbb{Z})$. Evidently, the stalk at the generic point $\{o\}$ is the field $\mathbb{Q}$ of rational numbers. Let $(\phi, \phi^\sharp)$ be a surjective morphism from a scheme (X, ϑ_X) to $(Spec\ \mathbb{Z}, \vartheta^\mathbb{Z})$. Thus, the fiber $X \otimes_{Spec\ \mathbb{Z}} k_{Spec\ \mathbb{Z}}^{\{0\}}$ of $(\phi, \phi^\sharp)$ at the generic point of $(Spec\ \mathbb{Z}, \vartheta^\mathbb{Z})$ is a scheme over $\mathbb{Q}$, and it is denoted by $X_\mathbb{Q}$. Also the fiber $X \otimes_{Spec\ \mathbb{Z}} k_{Spec\ \mathbb{Z}}^{\{p\mathbb{Z}\}}$ of $(\phi, \phi^\sharp)$ at the closed point $\{p\mathbb{Z}\}$ of $(Spec\ \mathbb{Z}, \vartheta^\mathbb{Z})$ is a scheme over $\mathbb{Z}_p$, and it is denoted by X_p. This fiber X_p is called the **reduction mod p** of the scheme (X, ϑ_X).

Observe that a topological space X is a Hausdorff space if and only if the diagonal $\Delta = \{(x, x) \mid x \in X\}$ is a closed subspace of the product space $X \times X$. However, for a scheme (X, ϑ_X), X is rarely Hausdorff. To capture some of the properties which are analogous to the properties of Hausdorff spaces, we introduce a special type of schemes as given in the following definition .

Definition 4.4.24 Let $(\phi, \phi^\sharp)$ be a morphism from a scheme (X, ϑ_X) to a scheme (Y, ϑ_Y). Let $(X \times_Y X, \vartheta_{X \times_Y X})$ together with projection morphisms $(p_1, p_1^\sharp)$ and $(p_2, p_2^\sharp)$ represent the pullback of the pair $((\phi, \phi^\sharp), (\phi, \phi^\sharp))$ of morphisms. From the universal property of pullback, we get a unique morphism $(\Delta, \Delta^\sharp)$ from (X, ϑ_X) to $(X \times_Y X, \vartheta_{X \times_Y X})$ such that $\Delta o p_1$ and $\Delta o p_2$ are identity morphisms. The morphism $(\Delta, \Delta^\sharp)$ is called the **Diagonal morphism**. The morphism $(\phi, \phi^\sharp)$ is called a **separated morphism** if $(\Delta, \Delta^\sharp)$ is a closed immersion. We also say that (X, ϑ_X) is separated over (Y, ϑ_Y) through the morphism $(\phi, \phi^\sharp)$. A scheme (X, ϑ_X) is called a **separated scheme** if it is separated over $(Spec\ \mathbb{Z}, \vartheta_\mathbb{Z})$.

Proposition 4.4.25 *A morphism from an affine scheme to an affine scheme is a separated morphism.*

Proof If ρ is a surjective ring homomorphism from a ring S to a ring R, then the induced morphism from $(Spec\ R, \vartheta^R)$ to $(Spec\ S, \vartheta^S)$ is easily seen to be a closed immersion. Let $(\phi, \phi^\sharp)$ be a morphism from $(Spec\ R, \vartheta^R)$ to $(Spec\ S, \vartheta^S)$ which is induced by a homomorphism η from S to R. Then R is an S-algebra through the homomorphism η, and $(Spec\ (R \otimes_S R), \vartheta^{R \otimes_S R})$ together with obvious projection morphisms induced by the homomorphisms i_1 and i_2 from R to $R \otimes_S R$ represents the pullback of the pair $((\phi, \phi^\sharp), (\phi, \phi^\sharp))$ of morphisms. The diagonal morphism $(\Delta, \Delta^\sharp)$ from $(Spec\ R, \vartheta_R)$ to $(Spec\ (R \otimes_S R), \vartheta^{R \otimes_S R})$ is induced by the co-diagonal homomorphism ∇ from $R \otimes_S R$ to R given by $\nabla(a \otimes b) = ab$. The result follows, since ∇ is a surjective ring homomorphism. ♯

Corollary 4.4.26 *An affine scheme is a separated scheme.* ♯

Proposition 4.4.27 *A morphism* $(\phi, \phi^\sharp)$ *from* (X, ϑ_X) *to* (Y, ϑ_Y) *is a separated morphism if and only if the image of the associated diagonal map* Δ *is a closed subset of* $X \otimes_Y X$.

Proof Suppose that $(\phi, \phi^\sharp)$ is a separated morphism. Then the induced diagonal morphism $(\Delta, \Delta^\sharp)$ is a closed immersion. In particular the image of Δ is closed. Conversely, suppose that the image of Δ is closed. We need to show that $(\Delta, \Delta^\sharp)$ is a closed immersion. Since $\Delta(X)$ is closed, we need to show that Δ is a homeomorphism from X to the image $\Delta(X)$, and $\Delta^\sharp$ is a surjective morphism from the sheaf $\vartheta_{X \times_Y X}$ to $\Delta_* \vartheta_X$. If $(p_1, p_1^\sharp)$ is the first projection morphism from $X \times_Y X$ to X, then $p_1 \circ \Delta = I_X$. This shows that Δ induces homeomorphism from X to $\Delta(X)$. For each point x of X, there is an affine neighborhood U of x such that $\phi(U)$ is a subset of an affine open subset V of Y. In turn, $U \times_V U$ is an affine open neighborhood of $\Delta(x)$. From the above proposition $(\Delta, \Delta^\sharp)$ restricted to U is a closed immersion of U into $U \times_V U$. This shows that $\Delta^\sharp$ is surjective in a neighborhood of x. ♯

We have the concept of proper maps and that of universally closed maps in the category of topological spaces. We describe their counterparts in the category SCH of schemes.

Let (S, ϑ_S) be a scheme. We have a category SCH_S whose objects are morphisms $(X, \vartheta_X) \overset{(\phi, \phi^\sharp)}{\to} (S, \vartheta_S)$ from schemes to the scheme S, and a morphism from $(X, \vartheta_X) \overset{(\phi, \phi^\sharp)}{\to} (S, \vartheta_S)$ to $(Y, \vartheta_Y) \overset{(\psi, \psi^\sharp)}{\to} (S, \vartheta_S)$ is a morphism $(\eta, \eta^\sharp)$ from (X, ϑ_X) to (Y, ϑ_Y) such that $(\eta, \eta^\sharp)(\psi, \psi^\sharp) = (\phi, \phi^\sharp)$. The category SCH_S is termed as the category of schemes with the base scheme S. If $(S', \vartheta_{S'}) \overset{(\rho, \rho^\sharp)}{\to} (S, \vartheta_S)$ is a fixed object in SCH_S, then for any object $(X, \vartheta_X) \overset{(\phi, \phi^\sharp)}{\to} (S, \vartheta_S)$ of SCH_S, we have an object $(X \times_S S', \vartheta_{X \times_S S'}) \overset{(p_2, p_2^\sharp)}{\to} (S', \vartheta_{S'})$ in $SCH_{S'}$. Indeed, this defines a functor from the category SCH_S to $SCH_{S'}$. This is called the **base extension** functor. For a ring R, we shall denote $SCH_{Spec R}$ by SCH_R.

Definition 4.4.28 A morphism $(\phi, \phi^\sharp)$ from (X, ϑ_X) to (S, ϑ_S) is called a **closed morphism** if the map ϕ is a closed map. It is said to be **universally closed** morphism if for any morphism $(\psi, \psi^\sharp)$ from $(S', \vartheta_{S'})$ to (S, ϑ_S), the projection $(p_2, p_2^\sharp)$ from $(X \times_S S', \vartheta_{X \times_S S'})$ to $(S', \vartheta_{S'})$ is a closed morphism. A universally closed separated morphism of finite type is called a **proper morphism**.

Example 4.4.29 Let K be a field. The obvious morphism from $(Spec\ K[X], \vartheta^{K[X]})$ to $(Spec\ K, \vartheta^K)$ is a separated morphism (Proposition 4.4.25) of finite type. However, it is not proper. Clearly, $(Spec\ K[X] \times_{Spec\ K} K[X], \vartheta_{Spec\ K[X] \times_{Spec\ K} K[X]})$ is the affine plane $(Spec\ K[X, Y], \vartheta^{K[X,Y]})$. The projection morphism is the surjective morphism from the affine plane to the affine line $(Spec\ K[T], \vartheta^{K[T]})$ which is not a closed morphism. Indeed, $V(< XY - 1 >)$ is the closed subset of $Spec\ K[X, Y]$ consisting of the set $\{< \{X - \alpha, Y - \beta\} > | \alpha\beta = 1\} \bigcup \{< XY - 1 >\}$ of prime ideals containing $< XY - 1 >$. The image of this closed set under the projection morphism is $Spec\ K[T] - \{0\}$ which is not a closed set.

Recall that a functor F from a category Σ to a category Γ is called a **fully faithful functor** if the induced natural map from $Mor_\Sigma(A, B)$ to $Mor_\Gamma(F(A), F(B))$ is bijective. Our next aim is to describe a natural fully faithful functor $\hat{\Omega}$ from the category VAR_K of varieties over an algebraically closed field K to the category SCH_K of schemes over K. This we do by adding generic points to varieties.

Theorem 4.4.30 *There is a natural fully faithful functor from the category VAR_K of varieties over an algebraically closed field K to the category SCH_K of schemes over K.*

Proof Let Y be a variety over an algebraically closed field K. For any closed subset Z of Y, let $\Omega(Z)$ denote the set of all nonempty irreducible closed subsets of Z. Let Y_1 and Y_2 be closed subsets of Y. Evidently, $(\Omega(Y_1) \bigcup \Omega(Y_2)) \subseteq \Omega(Y_1 \bigcup Y_2)$. Further, let Z be a nonempty irreducible closed subset contained $Y_1 \bigcup Y_2$. Then $Z = (Z \bigcap Y_1) \bigcup (Z \bigcap Y_2)$. Since Z is irreducible, $Z \bigcap Y_1 = Z$ or $Z \bigcap Y_2 = Z$. This means that $Z \subseteq Y_1$ or $Z \subseteq Y_2$. Thus, $(\Omega(Y_1) \bigcup \Omega(Y_2)) = \Omega(Y_1 \bigcup Y_2)$. It is also evident that $\bigcap_{\alpha \in \Lambda} \Omega(Y_\alpha) = \Omega(\bigcap_{\alpha \in \Lambda} Y_\alpha)$, where $\{Y_\alpha \mid \alpha \in \Lambda\}$ is a family of closed subsets of Y. This shows that the family $\{\Omega(Z) \mid Z \text{ is a closed subset of } Y\}$ forms a family of closed sets for a unique topology on $\Omega(Y)$. Define a map η from Y to $\Omega(Y)$ by $\eta(y) = \overline{\{y\}}$. Clearly,

$$\eta^{-1}(\Omega(Z)) = \{u \in Y \mid \overline{\{u\}} \subseteq Z\} = \{u \in Z \mid \overline{\{u\}} \subseteq Z\} = Z.$$

This shows that η is continuous. Further, the correspondence $(Y - Z) \mapsto (\Omega(Y) - \Omega(Z))$ defines a bijective correspondence between the set of open subsets of Y to the set of open subsets of $\Omega(Y)$. Put $\hat{\Omega}(Y) = (\Omega(Y), \eta_*\vartheta_Y)$. We first show that $\hat{\Omega}(Y)$ is a scheme. Since every variety is a finite union of open affine subvarieties, it is sufficient to assume that Y is an affine variety. Let R denote the ring $\vartheta(Y)$ of regular functions on Y. We have a map ρ from Y to $Spec R$ defined by $\rho(P) = M_{Y,P}$, where $M_{Y,P}$ is the maximal ideal of $\vartheta(Y)$ consisting of the regular functions on Y vanishing at P. Clearly, ρ is a homeomorphism from Y to the subspace of $spec R$ consisting of closed points. Now, for each point $P \in Y$, the stalk of ϑ^R at $\rho(P)$ is the localization of R at the maximal ideal $M_{Y,\rho(P)}$. Thus, for each $P \in Y$, we have a natural homomorphism χ_P from the stalk $R_{M_{Y,\rho(P)}}$ at $\rho(P)$ to K. For each open subset U of $spec R$ and an element $s \in \vartheta^R(U)$, we have a map μ_s from $\rho^{-1}(U)$ to K given by $\mu_s(P) = \chi_P \bar{s}$, where $\bar{s}$ represents the element of $R_{M_{Y,\rho(P)}}$ determined by s. It can be easily seen that μ_s is a regular function on $\rho^{-1}(U)$. In turn, we get a homomorphism $\rho_U^\sharp$ from $\vartheta^R(U)$ to $\vartheta_Y(\rho^{-1}(U))$ given by $\rho_U^\sharp(s) = \mu_s$. Further, the irreducible closed subsets of Y are in bijective correspondence with the points in $spec R$, and it follows that $(Spec R, \vartheta^R)$ is isomorphic to $\hat{\Omega}(Y) = (\Omega(Y), \eta_*\vartheta_Y)$. This shows that $\hat{\Omega}(Y)$ is a scheme. Treating the elements of K as constant functions on Y, we see that $\hat{\Omega}(Y)$ is a scheme over K. The rest of the statements can be easily verified. ♯

It can be further checked that the scheme $\hat{\Omega}(Y)$ is an integral separated scheme of finite type over K for any variety Y over an algebraically closed field K. This prompts have the following definition.

Definition 4.4.31 A integral separated scheme of finite type over an algebraically closed field K is called an **abstract variety** over K. If in addition it is proper over K, then we term it as a **complete abstract variety**.

Thus, an affine or a projective variety Y over an algebraically closed K can be viewed as an abstract variety given by $\hat{\Omega}(Y)$.

Exercises

4.4.1 Give a proof of Proposition 4.4.18.

4.4.2 Discuss the categorical properties of the category of schemes.

4.4.3 Show that $(Spec\mathbb{Z}, \vartheta^{\mathbb{Z}})$ is a terminal object in the category of schemes. Does it have an initial object?

4.4.4 Describe $Spec(\frac{\mathbb{Z}[X,Y]}{(Y^2-X^3-5)})$.

4.5 Weil Conjectures and l-adic Co-homology

M. Artin and Grothendieck introduced étale co-homology, and, in particular, the l-adic co-homology theory needed to prove the Weil conjectures. Indeed, Grothendieck proved it partially, and it was finally settled by Deligne. The l-adic co-homology was further used to construct Deligne–Lusztig virtual characters of finite groups of Lie types in terms of maximal tori T and a character θ of T, and thereby settling the conjectures of Macdonald. The purpose of this section is to introduce the l-adic co homology theory and demonstrate as to how Deligne and Lusztig used it to describe the characters of finite groups of Lie types.

The Weil Conjectures

Let X be a nonsingular projective variety defined over a finite field F_q containing q elements. For $r \geq 1$, let $X(F_{q^r})$ denote the set of rational points over F_{q^r}. The sequence $\{\mid X(F_{q^r}) \mid r \geq 1\}$ reflects some important arithmetical properties of the scheme associated with X. Weil associated a power series $Z(X, t)$ with X which is defined as

$$Z(X,t) = exp(\sum_{r=1}^{\infty} \mid X(F_{q^r}) \mid \frac{t^r}{r}).$$

The function $Z(X, t)$ is called the **zeta function** of X. Evidently, $Z(X, t) \in \mathbb{Q}[[t]]$, where $\mathbb{Q}[[t]]$ denotes the power series ring over $\mathbb{Q}$. The power series $Z(X, t)$ can also be expressed by

$$\frac{d}{dt}(logZ(X,t)) = \sum_{r=1}^{\infty} \mid X(F_{q^r}) \mid t^{r-1}$$

with $Z(X, o) = 1$.

Example 4.5.1 Let K denote the algebraic closure of the prime field $\mathbb{Z}_p$. We try to describe the zeta function $Z(P_K^n, t)$ of the projective space P_K^n. Evidently, P_K^n is defined over any finite field F_q, where q is a power of p. By definition, the rational points of P_K^n over F_{q^r} constitute the projective space $P_{F_{q^r}}^n = \frac{(F_{q^r})^{n+1} - \overline{0}}{\approx}$, where $\overline{\alpha} \approx \overline{\beta}$ if and only if there is a nonzero member $\lambda \in F_{q^r} - 0$ such that $\lambda \overline{\alpha} = \overline{\beta}$. Thus,

$$| P_{F_{q^r}}^n | = \frac{(q^r)^{n+1} - 1}{q^r - 1} = 1 + q^r + (q^r)^2 + \cdots + (q^r)^n.$$

Hence

$$\frac{d}{dt}(\log Z(P_K^n, t)) = \sum_{r=1}^{\infty} | P_{(F_{q^r})}^n | \, t^{r-1} = \frac{1}{1-t} + \frac{q}{1-qt} + \frac{q^2}{1-q^2t} + \cdots + \frac{q^n}{1-q^nt}.$$

Integrating and putting $Z(P_K^n, 0) = 1$, we obtain

$$Z(P_K^n, t) = \frac{1}{(1-t)(1-qt)(1-q^2t)\cdots(1-q^nt)}.$$

Looking at the the zeta functions of a few more nonsingular projective varieties (e.g., elliptic curves), A. Weil made the following conjectures.

Let X denote the nonsingular projective variety of dimension n defined over a finite field F_q containing $q = p^m$ elements. Then the following hold:

1. The zeta function $Z(X, t)$ is a rational function over $\mathbb{Z}$.

2. Let Y be a variety defined over a ring R of algebraic integers such that X is obtained by reducing Y modulo a prime ideal $\wp$ of R containing p. Let $Y_{\mathbb{C}}$ denote the complex manifold obtained by treating the variety Y over $\mathbb{C}$. Let B_i denote the i_{th} Betti number of $Y_{\mathbb{C}}$ (with the usual topology on the complex space). Then

$$Z(X, t) = \frac{P_1(t) P_3(t) \cdots P_{2n-1}(t)}{P_0(t) P_2(t) \cdots P_{2n}(t)},$$

where $P_i(t)$ is a polynomial in $\mathbb{Z}[t]$ of degree B_i, and it is expressible as

$$P_i(t) = \prod_{j=1}^{B_i} (1 - \alpha_{ij} t),$$

where α_{ij} are algebraic integers with $| \alpha_{ij} | = q^{\frac{i}{2}}$ for all i.

3. $Z(X, q^{-n}t^{-1}) = (-q^{\frac{n}{2}}t)^{\chi(Y_{\mathbb{C}})} Z(X, t)$, where $\chi(Y_{\mathbb{C}})$ is the Euler characteristic $\sum_{i=0}^{2n} (-1)^i B_i$ of $Y_{\mathbb{C}}$.

Example 4.5.1 verifies the Weil conjectures for the projective n space.

Our next aim is to demonstrate as to how the Weil conjectures follow from the existence of certain co-homology theory on varieties.

Theorem 4.5.2 *Suppose that we have co-homology functors* $\{H^r \mid r \geq 0\}$ *from the category of nonsingular projective varieties defined over finite fields of characteristic p to the category of finite-dimensional vector spaces over a field of characteristic 0 and which satisfy the following properties:*

(i) *Dim* $X = n$ *implies that* $H^r(X) = 0$ *for all r unless* $0 \leq r \leq 2n$.

(ii) *Dim* $H^{2n}(X) = 1$, *and we have a perfect pairing* $\sqcup$ *from* $H^r(X) \times H^{2n-r}(X)$ *to* $H^{2n}(X)$ *in the sense that* $\sqcup$ *is bilinear and the map* χ *from* $H^r(X)$ *to* $Hom(H^{2n-r}(X), H^{2n}(X))$ *given by* $\chi(a)(b) = \sqcup(a, b)$ *is an isomorphism.*

(iii) *If X is obtained from a variety Y over a ring of algebraic integers in the manner described in the statement of the Weil conjectures, then* $Dim \, H^r(X) = Dim \, H^r(Y_{\mathbb{C}}, \mathbb{C})$, *where* $H^r(Y_{\mathbb{C}}, \mathbb{C})$ *is the singular co-homology of* $Y_{\mathbb{C}}$ *in the dimension r with a coefficient in the field* $\mathbb{C}$ *of complex numbers.*

(iv) *Let F denote the* **Frobenius morphism** *from X to itself given by* $F_q(a_0, a_1, \cdots, a_n) = (a_0^q, a_1^q, \cdots, a_n^q)$, *where* $q = p^t$. *Then F is compatible with* $\sqcup$, *and*

$$| X(F_{q^m}) | = | X^{F^m} | = \sum_{r=0}^{2n} (-1)^r Tr(H^r(F^m)),$$

where F^m *is the* m^{th} *power of F (note that* F^m *is the Frobenius morphism associated with* q^m), X^{F^m} *is the fixed-point set of* F^m, *and Tr denotes the trace function.*

(v) *The eigenvalue of* $H^{2n}(F)$ *is* q^n *(note that* $H^{2n}(F)$ *is multiplication by a scalar) and the eigenvalues of* $H^r(F)$ *are algebraic integers* α_{ij}, $1 \leq j \leq B_r$ *with* $| \alpha_{ij} | = q^{\frac{r}{2}}$, $0 \leq r \leq 2n$, *where* B_r *denotes the dimension of* $H^r(X)$.

Then the Weil conjectures hold good.

Proof From (iv) and (v) of the hypothesis of the theorem,

$\frac{d}{dt} log Z(X, t)$

$= \sum_{m=0}^{\infty} | X(F_{q^m}) |$

$= \sum_{m=0}^{\infty} (\sum_{r=0}^{2n} (-1)^r Tr(H^r(F^m))) t^{m-1}$

$= \sum_{m=0}^{\infty} (\sum_{r=0}^{2n} (-1)^r \sum_{j=1}^{B_r} (\alpha_{rj})^m) t^{m-1}$

$= \sum_{r=0}^{2n} (-1)^r \sum_{j=1}^{B_r} \frac{\alpha_{rj}}{1-\alpha_{rj}t}$.

Integrating and using the fact that $Z(X, 0) = 1$, we obtain that

$$Z(X, t) = \frac{P_1(t) P_3(t) \cdots P_{2n-1}(t)}{P_0(t) P_2(t) \cdots P_{2n}(t)},$$

where $P_r(t)$ is a polynomial in $\mathbb{Z}[t]$ of degree B_r, and it is expressible as

$$P_r(t) = \prod_{j=1}^{B_r} (1 - \alpha_{rj} t).$$

This proves 1 and 2 of the Weil conjectures.

Now, we use the the perfect pairing $\sqcup$ to establish part 3 of the Weil conjectures. Since $Dim \, H^{2n}(X) = 1$, $B_r = Dim \, H^r(X) = Dim \, H^{2n-r}(X) = B_{2n-r}$ and

since F is compatible with the cup product $\sqcup$, $\alpha_{ij}\alpha_{(2n-i)j} = q^n$ for all i and $j \leq B_i$.
Now,

$P_r(\frac{1}{q^n t}) = \prod_{j=1}^{B_r}(1 - \frac{\alpha_{rj}}{q^n t})$

$= \frac{1}{(q^n t)^{B_r}}(-1)^{B_r} \prod_{j=1}^{B_r} \alpha_{ij} \prod_{j=1}^{B_r}(1 - \frac{q^n t}{\alpha_{rj}})$

$= (-1)^{B_r} \frac{1}{(q^n t)^{B_r}} det\, H^r(F) \prod_{j=1}^{B_r}(1 - \alpha_{(2n-r)j})$

$= (-1)^{B_r} \frac{1}{(q^n t)^{B_r}} det\, H^r(F) P_{2n-r}(t).$

Substituting the values of $P_r(\frac{1}{q^n t})$ in the expression for $Z(X, \frac{1}{q^n t})$, simplifying, and observing that $det\, H^r(F) \cdot det\, H^{2n-r}(F) = (q^n)^{B_r}$, we obtain that

$$Z(X, \frac{1}{q^n t}) = (-1)^{\chi(Y_\mathbb{C})}(q^{\frac{n}{2}}t)^{\chi(Y_\mathbb{C})} Z(X, t).$$

This establishes the Weil conjectures. $\sharp$

As already mentioned, M. Artin and Grothendieck introduced a co-homology theory satisfying the hypothesis of the theorem. Now, we shall introduce it.

Étale Sheaf Theory and Co-homolgy

The concept of étale sheaf, and, in turn, that of étale co-homology was introduced by Grothendieck in order to introduce a suitable co-homology theory needed to establish the Weil conjectures.

Definition 4.5.3 Let (X, ϑ_X) be a separated scheme of finite type over an algebraically closed field K. A morphism $(f, f^\sharp)$ from a scheme (U, ϑ_U) to (X, ϑ_X) is called an *étale* if (i) $f^{-1}(\{x\})$ is finite for each closed point x of X, and (ii) for each closed point x of X and $y \in f^{-1}(\{x\})$, the local homomorphism $f_x^\sharp$ from the stalk $\vartheta_{X,x}$ to the the stalk $\vartheta_{Y,y}$ induces isomorphism $\hat{f}_x^\sharp$ from the $M_{X,x}$-adic completion $\hat{\vartheta}_{X,x}$ of $\vartheta_{X,x}$ to the $M_{U,y}$-adic completion $\hat{\vartheta}_{U,y}$ of $\vartheta_{U,y}$ (see Exercise 1.1.39 for the definition of A-adic completion associated with an ideal A).

Example 4.5.4 If U is an open subset of X, where (X, ϑ_X) is a separated scheme of finite type over an algebraically closed field K, then $(i, i^\sharp)$ is an étale morphism from (U, ϑ_U) to (X, ϑ_X). However, there may be an étale morphism from a subset (not necessarily open) of X (give an example).

We develop the theory of étale sheaves in the manner in which the sheaf theory was developed by replacing the inclusion morphisms between open subsets of X by the étale morphisms into X. Let (X, ϑ_X) be a separated scheme of finite type over an algebraically closed field K. We have a category $X_{\acute{e}t}$ of étale morphisms in to X whose objects are étale morphisms $U \xrightarrow{\rho} X$ and a morphism from $U \xrightarrow{\rho} X$ to $V \xrightarrow{\eta} X$ is a morphism $U \xrightarrow{h} V$ of schemes such that $\eta \circ h = \rho$. An *étale presheaf* F of abelian groups (rings) is a contra-variant functor from $X_{\acute{e}t}$ to the category of abelian groups (rings). More explicitly, for each étale morphism $U \xrightarrow{\rho} X$, we have an abelian group (ring) $F(U \xrightarrow{\rho} X)$ and to each morphism $U \xrightarrow{h_\rho^\eta} V$ from $U \xrightarrow{\rho} X$ to $V \xrightarrow{\eta} X$, there is a homomorphism $F(h_\rho^\eta)$ from $F(V \xrightarrow{\eta} X)$ to $F(U \xrightarrow{\rho} X)$ such that $F(h_\eta^\mu 0 h_\rho^\eta) = F(h_\rho^\eta) o F(h_\eta^\mu)$ and $F(I_\rho^\rho) = I_{F(U \overset{\rho}{\to} X)}$. A morphism from an étale presheaf F to an étale presheaf G is a natural transformation χ from F to G. More explicitly, χ is the family $\{F(U \overset{\rho}{\to} X) \overset{\chi_\rho}{\to} G(U \overset{\rho}{\to} X) \mid U \overset{\rho}{\to} X \in Obj X_{\acute{e}t}\}$ of homomorphisms such that $G(h_\eta^\rho) o \chi_\rho = \chi_\eta o F(h_\eta^\rho)$.

Now, we introduce the notion of étale sheaf in an analogous manner as we introduced the notion of sheaf.

Definition 4.5.5 An étale presheaf F on X is called an **étale sheaf** if given any étale morphism $U \overset{\rho}{\to} X$ together with a finite family $\{U_\alpha \overset{\phi_\alpha}{\to} U \mid \alpha \in \Lambda\}$ of étale morphisms such that $\bigcup_{\alpha \in \Lambda} \phi_\alpha(U_\alpha) = U$, and also the family $\{s_\alpha \in F(U_\alpha \overset{\rho o \phi_\alpha}{\to} X) \mid \alpha \in \Lambda\}$ of elements satisfying the condition

$$F(U_\alpha \times_U U_\beta \overset{\rho o \phi_\alpha o p_1}{\to} X)(s_\alpha) = F(U_\alpha \times_U U_\beta \overset{\rho o \phi_\beta o p_2}{\to} X)(s_\beta)$$

for all $\alpha, \beta \in \Lambda$, where $U_\alpha \times_U U_\beta$ denotes the fiber product and p_1, p_2 the natural projections, then there exists a unique $s \in F(U \overset{\rho}{\to} X)$ such that $F(U_\alpha \overset{\phi_\alpha}{\to} U)(s) = s_\alpha$ for all $\alpha \in \Lambda$.

As in the case of sheaf theory, with a little care, we can construct the étale sheafification functor $\acute{E}S$ from the category $\acute{e}t Pr_X$ of étale presheaves of abelian groups on X to the category $\acute{e}t Sh_X$ of étale sheaves of abelian groups. More explicitly, we have a functor $\acute{E}S$ from $\acute{e}t Pr_X$ to $\acute{e}t Sh_X$ which is adjoint to the forgetful functor from $\acute{e}t Sh_X$ to $\acute{e}t Pr_X$. Even more, we can observe that $\acute{e}t Pr_X$ and $\acute{e}t Sh_X$ are abelian categories with enough injectives. The global section functor Γ from $\acute{e}t Sh_X$ to AB given by $\Gamma(F) = F(X \overset{I_X}{\to} X)$ is a left exact functor. The rth derived functor of Γ is denoted by $H^r(X_{\acute{e}t}, -)$ and it is called the r_{th} étale co-homology functor on X. $H^r(X_{\acute{e}t}, F)$ is termed as the r_{th} étale co-homology of X with a coefficient in the étale sheaf F. More explicitly, given an étale sheaf F and the injective resolution

$$0 \longrightarrow F \overset{\epsilon}{\to} I_0 \overset{d^0}{\to} I_1 \overset{d^1}{\to} \cdots \overset{d^{n-1}}{\to} I_n \overset{d^n}{\to} \cdots$$

of F, $H^r(X_{\acute{e}t}, F)$ is the r_{th} co-homology of the co-chain complex

$$0 \longrightarrow \Gamma(I_0) \overset{\Gamma(d^0)}{\to} \Gamma(I_1) \overset{\Gamma(d^1)}{\to} \cdots \overset{\Gamma(d^{n-1})}{\to} \Gamma(I_n) \overset{\Gamma(d^n)}{\to} \cdots.$$

Recall that a **geometric point** of a scheme X is a morphism $\hat{\sigma}$ from $Spec\ \hat{k}$ to X, where $\hat{k}$ is an algebraically closed field. Observe that $Spec\ \hat{k}$ is a single point. The image $x_{\hat{\sigma}}$ of $\hat{\sigma}$ is called the center of the geometric point $\hat{\sigma}$. Let $Spec\ \hat{k} \overset{\hat{\sigma}}{\to} X$ be a geometric point of X. A pair $(U \overset{\rho}{\to} X,\ Spec\ \hat{k} \overset{\eta}{\to} U)$ is called an **étale neighborhood** of $\hat{\sigma}$ if $\rho o \eta = \hat{\sigma}$. Define a relation $\leq$ on the family $N_{\hat{\sigma}}$ of all étale neighborhoods of $\hat{\sigma}$ by putting

$$(U \overset{\rho}{\to} X, Spec\ \hat{k} \overset{\hat{\eta}}{\to} U) \le (V \overset{\lambda}{\to} X, Spec\ \hat{k} \overset{\hat{\mu}}{\to} U)$$

if and only if there is a morphism ν from V to U such that $\rho o \nu = \lambda$. In turn, for each étale sheaf F of abelian groups on X, we have the directed family

$$\{F(U \overset{\rho}{\to} X) \mid (U \overset{\rho}{\to} X, Spec\ \hat{k} \overset{\hat{\eta}}{\to} U)\ is\ étale\ neighborhood\ \hat{\sigma}\ in\ X\}$$

of abelian groups. The direct limit of this directed system of abelian groups is called **stalk** of F at the geometric point $\hat{\sigma}$, and it is denoted by $F_{\hat{\sigma}}$.

Étale Co-homology with Compact Support

Recall (Definition 4.4.28) that a scheme X over a field K is said to be universally closed if for any scheme Y over K, the projection p_2 from $X \times_K Y$ to Y is a closed morphism. It may be observed that a scheme X over $\mathbb{C}$ is universally closed if and only if the associated complex space is compact with usual topology. Before introducing the étale co-homology with compact support, we state (without proof) the following theorem due to Nagata.

Theorem 4.5.6 (Nagata's Compactification Theorem) *Let X be a separated scheme of finite type over an algebraically closed field K. Then there exists a universally closed separated scheme $\tilde{X}$ of finite type over K together with a monomorphism i from X to $\tilde{X}$ such that $i(X)$ is open in $\tilde{X}$ and i is an isomorphism from X to $i(X)$.* ♯

Let X be a separated scheme of finite type over an algebraically closed field K. Let $\tilde{X}$ and i be as described in the above Nagata compactification theorem. Let F be an étale sheaf of torsion abelian groups on X. Consider the étale sheaf $i_\sharp F$ on $\tilde{X}$ which is the sheafification of the étale presheaf $\hat{i}F$ on $\tilde{X}$ defined as follows: Put $\hat{i}F(U \overset{\rho}{\to} \tilde{X}) = F(U \overset{\eta}{\to})X$ provided that $\eta o i = \rho$ and 0 otherwise. We have the obvious restriction morphisms. Evidently, $i_\sharp F$ is the étale sheaf of torsion abelian groups on $\tilde{X}$. Let $\hat{\sigma}$ be a geometric point of $\tilde{X}$. Then $i_\sharp F_{\hat{\sigma}} \approx F_{\hat{\tau}}$ if there is a geometric point $\hat{\tau}$ in X such that $i o \hat{\tau} = \hat{\sigma}$, and it is 0 if there is no such $\hat{\tau}$. It can be shown that if $\hat{X}$ together with j is another compactification of X, then $H^r(\tilde{X}_{ét}, i_\sharp F)$ is naturally isomorphic to $H^r(\hat{X}_{ét}, j_\sharp F)$. The r_{th}, étale co-homology $H^r_c(X_{ét}, F)$ of X with compact support and with a coefficient in the torsion étale sheaf F is defined as

$$H^r_c(X_{ét}, F) = H^r(\tilde{X}_{ét}, i_\sharp F),$$

where $(\tilde{X}, i)$ is a Nagata compactification of X. In particular, if X is a universally closed separated scheme of finite type over an algebraically closed field K, then $H^r_c(X_{ét}, F) = H^r(X_{ét}, F)$.

l-adic Co-homology

Let l be a positive prime integer. Recall (Exercise 1.1.34) the ring $\mathbb{Z}_{(l)}$ of l-adic integers which is the inverse limit of the inverse system $\{\mathbb{Z}_{l^r} \overset{\nu_r^s}{\to} \mathbb{Z}_{l^s} \mid r \ge s\}$, where ν_r^s is the obvious ring homomorphism. As mentioned in Exercise 1.1.34, $\mathbb{Z}_{(l)}$ is an integral domain. The field $\mathbb{Q}_{(l)}$ is called the field of l-adic numbers. Evidently, $\mathbb{Q}_{(l)}$ is a field extension of $\mathbb{Q}$. $\overline{\mathbb{Q}_{(l)}}$ will denote the algebraic closure of $\mathbb{Q}_{(l)}$.

By definition of inverse limit, for each r, we have a unique ring homomorphism ϕ^r from $\mathbb{Z}_{(l)}$ to $\mathbb{Z}_{l^r}$ such that $r \geq s$ implies $\nu_r^s o \phi^r = \phi^s$. In turn, each $\mathbb{Z}_{l^r}$ is a $\mathbb{Z}_{(l)}$-module. Now, for each r, let $\hat{\mathbb{Z}}_{l^r}$ denote the constant étale sheaf on X of $\mathbb{Z}_{(l)}$-module given by $\hat{\mathbb{Z}}_{l^r}(U \xrightarrow{\rho} X) = \mathbb{Z}_{l^r}$. This gives us an inverse system

$$\cdots \xrightarrow{\hat{\nu}_{r+2}^{r+1}} \hat{\mathbb{Z}}_{l^{r+1}} \xrightarrow{\hat{\nu}_{r+1}^{r}} \hat{\mathbb{Z}}_{l^r} \xrightarrow{\hat{\nu}_r^{r-1}} \cdots \xrightarrow{\hat{\nu}_2^1} \hat{\mathbb{Z}}_l \longrightarrow 0$$

of torsion étale sheaves on X. For each m, we have two inverse systems

$$\cdots \xrightarrow{H^m(\hat{\nu}_{r+2}^{r+1})} H^m(X_{\acute{e}t}, \hat{\mathbb{Z}}_{l^{r+1}}) \xrightarrow{H^m(\hat{\nu}_{r+1}^{r})} H^m(X_{\acute{e}t}, \hat{\mathbb{Z}}_{l^r}) \xrightarrow{H^m(\hat{\nu}_r^{r-1})} \cdots \xrightarrow{H^m(\hat{\nu}_2^1)} H^m(X_{\acute{e}t}, \hat{\mathbb{Z}}_l) \longrightarrow 0$$

and

$$\cdots \xrightarrow{H_c^m(\hat{\nu}_{r+2}^{r+1})} H_c^m(X_{\acute{e}t}, \hat{\mathbb{Z}}_{l^{r+1}}) \xrightarrow{H_c^m(\hat{\nu}_{r+1}^{r})} H_c^m(X_{\acute{e}t}, \hat{\mathbb{Z}}_{l^r}) \xrightarrow{H_c^m(\hat{\nu}_r^{r-1})} \cdots \xrightarrow{H_c^m(\hat{\nu}_2^1)} H_c^m(X_{\acute{e}t}, \hat{\mathbb{Z}}_l) \longrightarrow 0$$

of $\mathbb{Z}_{(l)}$-modules. The inverse limit $Lim_{\leftarrow} H^m(X_{\acute{e}t}, \hat{\mathbb{Z}}_{l^r})$ is denoted by $H^m(X, \mathbb{Z}_{(l)})$, and $Lim_{\leftarrow} H_c^m(X_{\acute{e}t}, \hat{\mathbb{Z}}_{l^r})$ is denoted by $H_c^m(X, \mathbb{Z}_{(l)})$. Evidently, $H^m(X, \mathbb{Z}_{(l)})$ and $H_c^m(X, \mathbb{Z}_{(l)})$ are $\mathbb{Z}_{(l)}$-modules. The $\mathbb{Q}_{(l)}$-vector space $H^m(X, \mathbb{Z}_{(l)}) \otimes_{\mathbb{Z}_{(l)}} \mathbb{Q}_{(l)}$ is called m_{th} l-adic co-homology of X, and it is denoted by $H^m(X, \mathbb{Q}_{(l)})$. Further, the $\mathbb{Q}_{(l)}$-vector space $H_c^m(X, \mathbb{Z}_{(l)}) \otimes_{\mathbb{Z}_{(l)}} \mathbb{Q}_{(l)}$ is called m_{th} l-adic co-homology of X with compact support, and it is denoted by $H_c^m(X, \mathbb{Q}_{(l)})$.

For the purpose of applications in the representation theory, we need to consider vector spaces over algebraically closed fields. As such, we shall be interested in the co-homologies $H^m(X, \overline{\mathbb{Q}_{(l)}}) = H^m(X, \mathbb{Q}_{(l)}) \otimes_{\mathbb{Q}_{(l)}} \overline{\mathbb{Q}_{(l)}}$ and $H_c^m(X, \overline{\mathbb{Q}_{(l)}}) = H_c^m(X, \mathbb{Q}_{(l)}) \otimes_{\mathbb{Q}_{(l)}} \overline{\mathbb{Q}_{(l)}}$.

Some Properties of l-adic Co-homology with Compact Support

We state the following theorem (without proof) which describes some important properties of l-adic co-homology with compact support. Some of them follow from the properties of co-homology functors and the properties of Lefschetz numbers. However, the proof can be found in S.G.A 4, S.G.A $4\frac{1}{2}$, and S.G.A(5). The reader may also refer to Lusztig's "Representations of Chevalley Groups", CBMS Regional Conference Series in Mathematics, (AMS)39(1970). These properties will be used in Chap. 5 of Algebra 4 which deals with the representations of finite groups of Lie types.

Theorem 4.5.7 *1. $H_c^m(-, \overline{\mathbb{Q}_{(l)}})$ defines a contra-variant functor from the category whose objects are schemes and morphisms are finite morphisms. In particular, we have a representation ρ of $Aut(X)$ on the space $H_c^m(X, \overline{\mathbb{Q}_{(l)}})$ given by $\rho(\sigma) =$*

$(H_c^m(\sigma))^t$(note that an automorphism of a scheme is a finite morphism). Further, $H_c^m(X, \overline{\mathbb{Q}_{(l)}})$ may be nonzero only when $0 \leq m \leq 2DimX$.

2. If X is an affine space A_K^n, then $H_c^{2n}(X, \overline{\mathbb{Q}_{(l)}}) \approx \overline{\mathbb{Q}_{(l)}}$ and $H_c^m(X, \overline{\mathbb{Q}_{(l)}}) = \{0\}$ for $m \neq 2n$.

3. Let p be a prime and $q = p^r$. Let A be a set of polynomials in $F_q[X_1, X_2, \cdots, X_n]$ defined over the Galois field F_q. Let K be a field extension of F_q, and $Y(K) = V(A)$ denote the variety determined by A over K. The map F from $Y(K)$ to $Y(K)$ given by $F(\alpha_1, \alpha_2, \cdots, \alpha_n) = (\alpha_1^q, \alpha_2^q, \cdots, \alpha_n^q)$ is called the **Standard Frobenius map**. Evidently, $Y(K)^F = Y(F_q)$. More generally, an F_q structure on a scheme (X, ϑ_X) is a scheme (X_0, ϑ_{X_0}) over F_q together with an isomorphism from (X, ϑ_X) to the fiber product $X_0 \times_{Spec\ F_q} Spec\ K$ of the schemes (X_0, ϑ_{X_0}) and $(Spec\ K, \vartheta^K)$ over $(Spec\ F_q, \vartheta^{F_q})$. Let F_0 denote the morphism $(I_{X_0}, F_0^\sharp)$ from (X_0, ϑ_{X_0}) to itself given by $(F_0^\sharp)_U(a) = a^q$ for all $a \in \vartheta_{X_0}(U)$ and for all open subsets U of X_0. The morphism $F_0 \times 1$ from $X_0 \times_{Spec\ F_q} Spec\ K$ to itself induces a morphism F from (X, ϑ_X) to itself. This morphism is called the F_q-Frobenius map on (X, ϑ_X).

Grothendieck Trace Formula. Let l be a prime different from p. Then

$$\mid X^F \mid = \sum\nolimits_{r=0}^{2n}(-1)^r trace(F, H_c^r(X, \overline{\mathbb{Q}_{(l)}})),$$

where $DimX = n$ and F is the F_q-Frobenius map described above. Evidently, F^m is the F_{q^m}-Frobenius map. Observe that this was an essential requirement (hypothesis (iv) of Theorem 4.5.2.) to establish a part of the Weil conjectures.

4. If g is an automorphism of X of finite order, then the Lefschetz number $L(g, X)$ given by

$$L(g, X) = \sum\nolimits_{m=0}^{2DimX}(-1)^m Trace(g, H_c^m(X, \overline{\mathbb{Q}_{(l)}}))$$

is an integer which is independent of l.

5. Let X and Y be algebraic varieties and f be a morphism from X to Y such that each fiber $f^{-1}(\{y\})$ is an affine variety isomorphic to A_K^n. Let g and g' be automorphisms of X and Y, respectively, which are of finite orders. Suppose that $f \circ g = g' \circ f$. Then $L(g, X) = L(g', X)$.

6. Let $\{X_i \mid 1 \leq i \leq n\}$ be a pairwise disjoint finite family of locally closed subsets of X such that $X = \bigcup_{i=1}^n X_i$. Let g be an automorphism of X of finite order such that $g(X_i) = X_i$ for each i. Then

$$L(g, X) = \sum\nolimits_{i=1}^n L(g, X_i).$$

Further, suppose that $\bigcup_{i=1}^m X_i$ is closed for each $m \leq n$. Let G be a finite group of automorphisms of X such that each X_i is invariant under G. Let $H_c^r(X_i, \overline{\mathbb{Q}_{(l)}})_\Theta$ be a subspace of $H_c^r(X_i, \overline{\mathbb{Q}_{(l)}})$ affording an irreducible character Θ of G. Suppose that $H_c^r(X_i, \overline{\mathbb{Q}_{(l)}})_\Theta = 0$ for all r and i. Then $H_c^r(X, \overline{\mathbb{Q}_{(l)}})_\Theta = 0$ for all r.

7. *Let $\{X_i \mid 1 \le i \le n\}$ be a pairwise disjoint finite family of closed subsets of X such that $X = \bigcup_{i=1}^{n} X_i$. Let g be an automorphism of X of finite order such that for each pair i, j, there is a $g \in G$ such that $g(X_i) = X_j$. Let $H = \{g \in G \mid g(X_1) = X_1\}$ be the subgroup of G. Then the generalized character $g \mapsto L(g, X)$ of G is induced by the generalized character $h \mapsto L(h, X_1)$ of H in the sense that*

$$L(g, X) = \frac{1}{\mid H \mid} \Sigma_{x \in G, xgx^{-1} \in H} L(xgx^{-1}, X_1)$$

8. *Let X be an affine variety and G be a finite group of automorphisms of X. Then*

$$H_c^m(X/G, \overline{\mathbb{Q}_{(l)}}) \approx H_c^m(X, \overline{\mathbb{Q}_{(l)}})^G,$$

where $H_c^m(X, \overline{\mathbb{Q}_{(l)}})^G$ is the subspace of the fixed points of the action of G on $H_c^m(X, \overline{\mathbb{Q}_{(l)}})$. Further, if f is a G-equivariant automorphism of X of finite order and g is an automorphism of X/G such that $\nu o f = f o g$, then

$$L(g, X/G) = \frac{1}{\mid G \mid} \sum_{x \in G} L(f o g, X).$$

9. *If X and Y are algebraic varieties, then we have the Kunneth formula*

$$H_c^m(X \times Y, \overline{\mathbb{Q}_{(l)}}) = \oplus \sum_{r+s=m} (H_c^r(X, \overline{\mathbb{Q}_{(l)}}) \otimes H_c^r(X, \overline{\mathbb{Q}_{(l)}})).$$

    Further,

$$L(g \times g', X \times Y) = L(g, X) L(g', Y),$$

    where g and g' are automorphisms of X and Y, respectively, which are of finite orders.

10. *Let g be an automorphism of X of finite order. Suppose that $g = su = us$, where the order of s is a co-prime to p and order of u is a power of p, p being a prime. Let X^s denote the fixed-point set of s. Then*

$$L(g, X) = L(u, X^s).$$

    In particular, if $X^s = \emptyset$, then $L(g, X) = 0$.

11. *If X is finite and g is an automorphism of X, then $L(g, X) = \mid X^g \mid$.*

12. *If G is a connected algebraic group which acts as a group of automorphisms of X, then the induced action of G on $H_c^m(X, \overline{\mathbb{Q}_{(l)}})$ is the trivial action.* ♯

Hypotheses (ii) and (iii) of Theorem 4.5.2 are also satisfied by the $l - adic$ co-homology.

The fact that $l - adic$ co-homology satisfies hypothesis (v) of Theorem 4.5.2 was established by Deligne settling the conjectures of Weil.

Application to the Representations of Finite Groups of Lie Types

The Lefschetz number of automorphisms of schemes/varieties plays a very important role in the Deligne–Lusztig representation theory of finite groups of Lie type and we introduce it. In the discussions to follow, we shall leave the proofs of certain assertions in the discussions to follow. The reader may refer to "Linear Algebraic groups" by Armand Borel for the details and the proofs.

Recall that an algebraic variety G together with a group structure on G is called an algebraic group if the map $(a, b) \to ab^{-1}$ from $G \times G$ to G is a morphism of variety. Thus, for example, $GL(n, K)$ is an algebraic group over K (see Exercises 4.3.8–4.3.10). A sub-algebraic group of a general linear group is called a linear algebraic group. Let G be a connected Linear algebraic group over an algebraically closed field K. A maximal closed connected solvable subgroup of G is called a **Borel subgroup** of G. Clearly, a connected linear algebraic group G has a Borel subgroup and any closed connected solvable subgroup is contained in a Borel subgroup. For example, the subgroup $B(n, K)$ of upper triangular matrices in $GL(n, K)$ forms a Borel subgroup of $GL(n, K)$. Let B be a Borel subgroup of G. Then the quotient variety G/B is a projective variety. Every projective variety X is a complete variety in the sense that for every variety Y, the second projection p_2 from $X \times Y$ to Y is a closed map. Thus, G/B is a complete variety. We state (without proof) the following fixed-point theorem due to Borel.

Theorem 4.5.8 (Borel fixed-point theorem) *Suppose that G is a connected solvable algebraic group acting on a complete variety X. Then there is a fixed point of the action.* ♯

Corollary 4.5.9 *Let G be a connected linear algebraic group. Then any two Borel subgroups of G are conjugate to each other. The set Ω of all Borel subgroups of G has a structure of projective variety isomorphic to G/B.*

Proof Let B_1 and B_2 be Borel subgroups of G. Then G/B_2 is a complete variety on which the connected solvable subgroup B_1 acts through left multiplication. From the Borel fixed-point theorem, it has a fixed point xB_2 (say). Then $bxB_2 = xB_2$ for all $b \in B_1$. This means that $x^{-1}B_1x \subseteq B_2$. Since $x^{-1}B_1x$ is also a Borel subgroup, $x^{-1}B_1x = B_2$. Thus, any two Borel subgroups of G are conjugate. In turn, if B is a Borel subgroup of G, then the map $gB \mapsto gBg^{-1}$ is a natural bijective map from G/B to Ω. The last assertion also follows. ♯

Since the product of closed connected solvable normal algebraic subgroups of G is a closed connected solvable normal algebraic subgroup, G has the largest closed connected solvable normal algebraic subgroup. This subgroup is called the **radical** of G and it is denoted by $R(G)$. A connected linear algebraic group G is called a **semi-simple algebraic group** if $R(G)$ is trivial. An element a of a linear algebraic group G is called a **unipotent element** of G if all its eigenvalues are 1. A subgroup consisting of unipotent elements is called a **unipotent subgroup**. Thus, the subgroup $U(n, K)$ of uni-upper triangular matrices in $GL(n, K)$ is a unipotent subgroup. Observe that all unipotent subgroups are nilpotent. It can also be shown that every connected linear algebraic group G has the largest unipotent closed normal subgroup. This subgroup is called the **unipotent radical** of G and it is denoted by $R_u(G)$. Evidently, $R_u(G) \subseteq R(G)$. A connected linear algebraic group G is called a **reductive** group if $R_u(G)$ is trivial. Thus, a semi-simple group is reductive. However, a reductive group need not be semi-simple. For example, the diagonal subgroup T of $GL(n, \mathbb{C})$ is a reductive group which is not semi-simple. $GL(n, \mathbb{C})$ is also reductive but it is not semi-simple.

Now, let us recall the Jordan–Chevalley decomposition (see Theorem 6.3.17, Corollary 6.3.19, and Corollary 6.3.21 in Algebra 2): If T is a linear transformation on a finite-dimensional vector space V over an algebraically closed field K (or at least all characteristic roots of T are in K), then it can be expressed uniquely as $T = T_s + T_n$, where T_s is semi-simple (i.e., V has a basis consisting of eigenvectors of T_s), T_n is nilpotent, and T_s and T_n commute. Further, there are polynomials $g(X)$ and $h(X)$ without constant terms such that $g(T) = T_s$ and $h(T) = T_n$. In particular, a square matrix A with entries in K can be expressed uniquely as $A = A_s + A_n$, where A_s is diagonalizable, A_n is nilpotent, and A_s and A_n commute. Further, there exist polynomials $g(X)$ and $h(X)$ without constant terms such that $A_s = g(A)$, and $A_n = h(A)$. If in addition A is nonsingular, then A_s is also nonsingular, and as such $A = A_s(I + A_s^{-1}A_n)$. Since $A_s^{-1}A_n$ is nilpotent, $A_u = I + A_s^{-1}A_n$ is unipotent in the sense that all its eigenvalues are 1. Thus, every nonsingular matrix A can be uniquely expressed as $A = A_s A_u$, where A_s is semi-simple, A_u is unipotent, and A_s and A_u commute. This decomposition is called the multiplicative Jordan–Chevalley decomposition.

Let G be a linear algebraic group over K and η be an embedding of G into $GL(n, K)$. Let x be an element of G such that $\eta(x)$ is semi-simple in $GL(n, K)$. Then for any other embedding μ of G, $\mu(x)$ is again a semi-simple element. This allows us to call an element x of G to be a semi-simple element of G if $\eta(x)$ is a semi-simple element for some embedding η. Similarly, we have the concept of unipotent elements in a linear algebraic group. We have the Jordan decomposition for an arbitrary linear algebraic G. More explicitly, every element $x \in G$ is uniquely expressible as $x = x_s x_u$, where x_s is semi-simple, x_u is unipotent, and x_s and x_u commute. x_s is called the semi-simple part and x_u is called the unipotent part of x. Further, under an algebraic homomorphism of groups, semi-simple and nilpotent parts are preserved.

An algebraic group of the form $\underbrace{K^* \times K^* \times \cdots \times K^*}_{n}$ is called a Torus of rank n.

Thus, the subgroup $D(n, K)$ of diagonal matrices is a torus in $GL(n, K)$. Clearly, a connected linear algebraic group has a maximal torus, and a maximal Torus subgroup is contained in a Borel subgroup. If B is a Borel subgroup containing a maximal torus T and U is a unipotent radical of B, then $B = U \rtimes T$ and all maximal torus contained in B are conjugate in B. Since any two Borel subgroups are conjugate, it follows that all maximal tori are conjugate.

Let G be a closed connected reductive algebraic subgroup of $GL(n, K)$, where K is an algebraically closed field of characteristic p. Fix $q = p^m$. Evidently, $[a_{ij}] \in G$ implies that $[b_{ij}] \in G$, where $b_{ij} = a_{ij}^q$. The morphism F from G to G defined by

$F([a_{ij}]) = [b_{ij}]$, where $b_{ij} = a_{ij}^q$, is called the F_q- **Standard Frobenius morphism** on G. F is said to be a Frobenius morphism if some power of F is a standard Frobenius morphism. The morphism L from G to G given by $L(g) = g^{-1}F(g)$ is called the **Lang map**. A theorem of Lang asserts that L is a surjective morphism. Evidently, $G^F = L^{-1}(\{1\})$ is finite, and it is called a **finite group of Lie type**. More generally, a theorem of Lang–Steinberg asserts that if F is a surjective homomorphism from G to G such that G^F is finite, then L is surjective. The group G^F, also termed as finite Chevalley group, has another alternative description (Chap. 4, Algebra 4).

For convenience and simplicity in arguments, we restrict our attention to the connected reductive algebraic group $GL(n, K)$, where K is an algebraically closed field of characteristic p. However, everything can be done on an arbitrary connected linear reductive algebraic group G over an algebraically closed field K. Evidently, $G^F = GL(n, F_q)$, where F is the F_q-Frobenius map on $GL(n, K)$. A group isomorphic to $\underbrace{K^\star \otimes K^\star \times \cdots \times K^\star}_{r}$ for some r is called a **torus** of rank r. A subgroup of $GL(n, K)$ which is isomorphic to a torus is called a **toral subgroup**. Thus, the diagonal subgroup T of $GL(n, K)$ is a toral subgroup of $GL(n, K)$ of rank n, and indeed, it is a maximal toral subgroup of $GL(n, K)$. All conjugates of T are maximal toral subgroups of $GL(n, K)$, and conversely, any maximal toral subgroup of $GL(n, K)$ is conjugate to T. Let P_n denote the group of permutation matrices. Then P_n is a subgroup of $GL(n, K)$ such that $P_n T = T P_n$, where T is the maximal torus $D(n, K)$. Indeed, $T P_n = N_{GL(n,K)}(T) = T \succ P_n$ (prove it). Thus, $N_{GL(n,K)}(T)/T \approx S_n \approx N_{GL(n,K)}(T^g)/T^g$ for all $g \in GL(n, K)$. Consequently, for all maximal toral subgroup $\hat{T}$ of $GL(n, K)$, $N_{GL(n,K)}(\hat{T})/\hat{T}$ is isomorphic to S_n (for arbitrary connected linear reductive algebraical group G, $N_G(T)/T$ is isomorphic to a subgroup of S_n). The group $N_{GL(n,K)}(T)/T$ is called the **Weyl group** of $GL(n, K)$. Further the subgroup $B_+(n, K)$ of upper triangular matrices in $GL(n, K)$ is a maximal solvable subgroup of $GL(n, K)$. Obviously all conjugates, in particular, the subgroup $B_-(n, K)$ is also a maximal solvable subgroup of $GL(n, K)$. Conversely, all maximal solvable subgroups of $GL(n, K)$ are conjugate to each other. A maximal solvable subgroup of $GL(n, K)$ is a **Borel subgroup**. Thus, $B_+(n, K)$ is a Borel subgroup of $GL(n, K)$. The subgroup $U_+(n, K)$ of uni-upper triangular matrices in $GL(n, K)$ is a unipotent subgroup which is a maximal unipotent subgroup of $GL(n, K)$. Observe that $U_+(n, K)$ is unipotent radical of $B_+(n, K)$, and $B_+(n, K) = T U_+(n, K)$. A subgroup H of $GL(n, K)$ is said to be an F-stable subgroup if $F(H) = H$. The standard maximal torus T is F-stable. The Borel subgroup $B_+(n, K)$ is F-stable. In general, a maximal torus T^g need not be F-stable. To say that T^g is F-stable is to say that $F(gTg^{-1}) = gTg^{-1}$. This is equivalent to say that $g^{-1}Fg \in N_{GL(n,K)}(T)$.

Let $\hat{T}$ be an F-stable maximal torus of $GL(n, K)$. Since $\hat{T}$ is a solvable subgroup of $GL(n, K)$, it is contained in a Borel subgroup $\hat{B}$ of $GL(n, K)$. Note that $\hat{B}$ need not be F-stable. If $\hat{T}$ lies in an F-stable Borel subgroup, then we call it a **maximally split torus**. Let $\hat{U}$ denote the unipotent radical of $\hat{B}$. Consider the Lang map L from $GL(n, K)$ to $GL(n, K)$. Evidently, $GL(n, F_q) = GL(n, K)^F = L^{-1}(\{1\})$.

Since F is a morphism and $\hat{U}$ is an algebraic subset of $GL(n, K)$, $L^{-1}(\hat{U})$ is an affine variety over K. Let $\hat{X}$ be the scheme associated with $L^{-1}(\hat{U})$. Let $g \in GL(n, F_q)$ and $x \in L^{-1}(\hat{U})$. Then $L(gx) = (gx)^{-1}F(gx) = x^{-1}g^{-1}F(g)F(x) = x^{-1}F(x) \in \hat{U}$. This shows that $gx \in L^{-1}(\hat{U})$. Hence $GL(n, F_q)$ acts on $\hat{X}$ from left. Next, if $t \in \hat{T}^F$ and $x \in L^{-1}(\hat{U})$, then $L(xt) = (xt)^{-1}F(xt) = t^{-1}x^{-1}F(x)F(t) \in t^{-1}\hat{U}t = \hat{U}$, since $\hat{U}$ is normal in $\hat{B}$. Hence $xt \in L^{-1}(\hat{U})$. This shows that $\hat{T}^F$ acts on $\hat{X}$ from right. Since $(gx)t = g(xt)$, it induces a $(GL(n, F_q), \hat{T}^F)$ bi-module structure on the $l - adic$ co-homology $H_c^r(\hat{X}, \overline{Q_{(l)}})$.

Let θ be an irreducible representation of $\hat{T}^F$ (see Chap. 9 of Algebra 2 for the basic language of representations) over the field $\mathbb{C}$ of complex numbers. Then θ is a homomorphism from $\hat{T}^F$ to $\mathbb{C}^*$ (θ is also an irreducible character). Evidently, the image of the members of $\hat{T}^F$ are algebraic integers in $\mathbb{C}$. Since $\overline{Q_{(l)}}$ is an algebraically closed field containing $\mathbb{Q}$, θ can be realized as a homomorphism from $\hat{T}^F$ to $\overline{Q_{(l)}}^*$. Let $H_c^r(\hat{X}, \overline{Q_{(l)}})_\theta$ denote the $\overline{Q_{(l)}}$-subspace $\{v \in H_c^r(\hat{X}, \overline{Q_{(l)}}) \mid vt = v\theta(t) \forall t \in \hat{T}^F\}$ of $H_c^r(\hat{X}, \overline{Q_{(l)}})$. Clearly, $H_c^r(\hat{X}, \overline{Q_{(l)}})_\theta$ is a $GL(n, F_q)$-submodule of $H_c^r(\hat{X}, \overline{Q_{(l)}})$. Let ρ_r denote the corresponding representation of $GL(n, F_q)$ and χ_r the character associated with ρ_r. Thus, $\chi_r(g) = Trace\ \rho_r(g)$, $g \in GL(n, F_q)$. We have a generalized character $R_{\hat{T},\theta}$ of $GL(n, F_q)$ given by

$$R_{\hat{T},\theta}(g) = \sum_{r=0}^{2Dim\ \hat{X}} (-1)^r \chi_r(g).$$

There may be several Borel subgroups containing an F-stable maximal torus. For example, $B_+(n, K)$ and $B_-(n, K)$ are Borel subgroups containing T. It appears that the generalized character $R_{\hat{T},\theta}$ depends on the choice $\hat{B}$ of a Borel subgroup containing $\hat{T}$. Indeed, it is independent of the choice $\hat{B}$ of a Borel subgroup containing $\hat{T}$. The generalized character $R_{\hat{T},\theta}$ of $GL(n, F_q)$ associated with a maximal F-stable torus $\hat{T}$ and a character θ of $\hat{T}^F$ is called the **Deligne–Lusztig Character** of $GL(n, F_q)$ associated with the pair $(\hat{T}, \theta)$.

The following are a few among several properties and facts about Deligne–Lusztig characters which make it extremely useful, interesting, and applicable.

1. If an F-stable maximal torus $\hat{T}$ is a split torus in the sense that there is an F-stable Borel subgroup $\hat{B}$ containing $\hat{T}$, then $R_{\hat{T},\theta} = Ind_{B^F}^{GL(n,F_q)}\hat{\theta}$, where $\hat{\theta}$ is the obvious character induced by θ on $\hat{T}^F \approx \hat{B}^F/\hat{U}^F$.

2. Every irreducible character of $GL(n, F_q)$ is a constituent of some Deligne–Lusztig character. More explicitly, given any irreducible character χ of $GL(n, F_q)$, there is an F-stable maximal torus $\hat{T}$ together with a character θ of $\hat{T}^F$ such that $< \chi, R_{\hat{T},\theta} > \neq 0$.

3. The constituents of $R_{\hat{T}}^1$ are called **unipotent characters** and they are important objects in the representation theory.

Deligne-Lusztig characters will be studied further in detail in Chap. 5 of Algebra 4.

Exercises

4.5.1 This exercise is meant to determine the conjugacy classes of elements of $GL(2, F_q)$.
(i) Show that $\{aI \mid a \in F_q^\star\}$ is the set of all distinct singleton conjugacy classes of elements $GL(2, F_q)$, where I denotes the identity matrix.
(ii) For $a \in F_q^\star$, put

$$A_a = \begin{bmatrix} a & 1 \\ 0 & a \end{bmatrix}.$$

Show that A_a is conjugate to A_b if and only if $a = b$. Show further that the centralizer $C_{GL(2,F_q)}(A_a)$ of A_a consists of the matrices of the type

$$\begin{bmatrix} u & v \\ 0 & u \end{bmatrix},$$

$u \in F_q^\star$ and $v \in F_q$. Deduce that the conjugacy class $\overline{A_a}$ determined by A_a contains $q^2 - 1$ elements.
(iii) For $a, b \in F_q^\star$, $a \neq b$, put

$$A_{a,b} = \begin{bmatrix} a & 0 \\ 0 & b \end{bmatrix}.$$

Show that $A_{a,b}$ is conjugate to $A_{c,d}$ if and only if $\{a, b\} = \{c, d\}$. Show further that the centralizer $C_{GL(2,F_q)}(A_{a,b}$ of $A_{a,b}$ is precisely the diagonal subgroup of $GL(2, F_q)$. Deduce that the conjugacy class $\overline{A_{a,b}}$ contains $q(q + 1)$ elements.
(iv) Assume that q is odd. Let ξ be a generator of the cyclic group $F_q^\star$. Show that for any $a \in F_q$ and $b \in F_q^\star$, the matrix

$$B_{a,b} = \begin{bmatrix} a & b\xi \\ b & a \end{bmatrix}$$

is a member of $GL(2, F_q)$. Show that $B_{a,b}$ is conjugate to $B_{c,d}$ if and only if $a = c$ and $b = \pm d$. Show that $X^2 - \xi$ is an irreducible polynomial in $F_q[X]$. Let ς be a root of $X^2 - \xi$ in the quadratic extension F_{q^2} of F_q. Show that the map η from $F_{q^2}^\star$ to $GL(2, F_q)$ given by $\eta(a + b\varsigma) = B_{a,b}$ is an injective homomorphism of groups. Deduce that $H = \{B_{a,b} \mid (a, b) \in F_q \times F_q^\star\}$ is a cyclic subgroup of $GL(2, F_q)$ of order $q^2 - 1$. Show further that $H = C_{GL(2,F_q)}(B_{a,b})$ for all $(a, b) \in F_q \times F_q^\star$. Deduce that the number of conjugates to $B_{a,b}$ is $q^2 - q$.

Counting and summing up the number of elements in the conjugacy classes described above and comparing them with the order of $GL(2, F_q)$ gives the complete list of conjugacy classes of $GL(2, F_q)$. There are four types of conjugacy classes described above, and in total there are $q^2 - 1$ conjugacy classes of $GL(2, F_q)$.

4.5.2 Modify the arguments in the above exercise to find the complete list of conjugacy classes of $GL(2, F_q)$, where q is even.

4.5.3 Consider $GL(2, F_q)$, where q is odd. Let $\overline{F}$ denote the algebraic closure of F_q, and F denote the F_q-Frobenius. Let T denote the standard maximal torus, viz., the diagonal subgroup of $GL(2, \overline{F})$ consisting of the diagonal matrices. Evidently, T is F-stable. Describe the variety $L^{-1}(U)$ and the respective $\hat{X}$, where L is the Lang map, U is the group of uni-upper triangular matrices in $GL(2, \overline{F})$. Determine R_T^1 by giving its values on each of the conjugacy classes of $GL(2, F_q)$ described in Exercise 4.5.1, where 1 is the trivial character of T^F. Is it a genuine character? If so check it for irreducibility. Let ξ denote a generator of the cyclic group F_q^*, and ρ the homomorphism from F_q^* to $\overline{\mathbb{Q}_{(l)}}^*$ given by $\rho(\xi) = \zeta$, where ζ is a primitive $q - 1$ root of unity in $\overline{\mathbb{Q}_{(l)}}$. Determine R_T^χ, where χ is the character of T^F given by $\chi(Diag(a, b)) = \rho(a)\rho(b)$, $a, b \in F_q^*$.

4.5.4 Consider $SL(2, F_q)$, where q is odd. Show the following:

(i) There are exactly two distinct singleton conjugacy classes, viz., $\{I\}$ and $\{-I\}$.
(ii) Show that $E_{12}^1, E_{12}^\xi, -E_{12}^{-1}$, and $-E_{12}^{-\xi}$ all determine distinct conjugacy classes containing $\frac{q^2-1}{2}$ elements each.
(iii) Show that there are $\frac{q-3}{2}$ distinct conjugacy classes determined by the elements of the type

$$\begin{bmatrix} a & 0 \\ 0 & a^{-1} \end{bmatrix},$$

$a \neq \pm 1$ each containing $q^2 + q$ elements.
(iv) Show that there are $\frac{q-1}{2}$ distinct conjugacy classes determined by the elements of the type

$$\begin{bmatrix} a & b \\ \xi b & a \end{bmatrix},$$

$a \neq \pm 1$ each containing $q^2 - q$ elements.
   Finally, counting the number of elements, show that these are the only conjugacy classes of $SL(2, F_q)$, and so there are $q + 4$ distinct conjugacy classes of elements of $SL(2, F_q)$.

4.5.5 As in Exercise 4.5.3, determine virtual character R_T^1, where T is the standard F-stable maximal torus, and 1 is the trivial character of T^F. Is it a genuine character?

4.5.6 Describe the Deligne–Lusztig generalized characters of $GL(2, q)$ by computing their values on each conjugacy class.

Bibliography

1. Artin et al., *SGA 4 Lecture Notes in Mathematics* (Springer, 1972–73), pp. 269, 270, 305
2. Atiyah-Macdonald, *Introduction to Commutative Algebra* (Addition-Wesley, 1969)
3. A. Borel, *Linear Algebraic Groups* (Benjamin, New York, 1969)
4. G.E. Bredon, *Sheaf Theory* (Springer, 1967)
5. R.W. Carter, *Simple Groups of Lie Type* (Wiley, London, 1972)
6. R.W. Carter, *Finite Groups of Lie Type*, Conjugacy classes and Complex characters (Wiley, New York, 1985)
7. C. Chevalley, Sur certains groupes simples. Tohoku Math. J. **7**, 14–66 (1955)
8. P. Deligne, La conjecture Weil I. Publ. Math. IHES **43**, 203–307 (1974)
9. P. Deligne, La conjecture Weil II. Publ. Math. IHES **52**, 137–252 (1980)
10. P. Deligne, *SGA $\frac{1}{4}$, Cohomology étale*. Springer, Lecture Notes in Mathematics (1977), p. 569
11. P. Deligne, G. Lusztig, Representations of reductive groups over finite fields. Ann. Math. **103**, (1976)
12. P. Freyd, *Abelian Categories* (Harper and Row, 1964)
13. R. Godement, *Theorie Des Faisceaux* (Herman Paris, 1964)
14. R. Hartshorne, *Algebraic Geometry* (Springer, GTM, 1977)
15. Hilton, Stambach, *A course in Homological Algebra* (Springer, 1997)
16. S. Mac Lane, *Homology GTM* (Springer, 1963)
17. S. Mac Lane, *Categories for the Working Mathematician* (Springer, GTM, 1969)
18. B. Mitchell, *Theory of Categories* (Academic, 1965)
19. M. Suzuki, *Group Theory I and II* (Springer, 1980)

© Springer Nature Singapore Pte Ltd. 2021
R. Lal, *Algebra 3*, Infosys Science Foundation Series,
https://doi.org/10.1007/978-981-33-6326-7

Index

© Springer Nature Singapore Pte Ltd. 2021
R. Lal, *Algebra 3*, Infosys Science Foundation Series,
https://doi.org/10.1007/978-981-33-6326-7

Printed in the United States
by Baker & Taylor Publisher Services

Printed in the United States
by Baker & Taylor Publisher Services